Universitext

Universitext

Universitext is a series of textbooks that presents material from a wide variety of mathematical disciplines at master's level and beyond. The books, often well class-tested by their author, may have an informal, personal, even experimental approach to their subject matter. Some of the most successful and established books in the series have evolved through several editions, always following the evolution of teaching curricula, into very polished texts.

Thus as research topics trickle down into graduate-level teaching, first textbooks written for new, cutting-edge courses may make their way into *Universitext*.

For further volumes:
http://www.springer.com/series/223

Ralf Schindler

Set Theory

Exploring Independence and Truth

Springer

Ralf Schindler
Institut für Mathematische Logik und
Grundlagenforschung
Universität Münster
Münster
Germany

ISSN 0172-5939 ISSN 2191-6675 (electronic)
ISBN 978-3-319-06724-7 ISBN 978-3-319-06725-4 (eBook)
DOI 10.1007/978-3-319-06725-4
Springer Cham Heidelberg New York Dordrecht London

Library of Congress Control Number: 2014938475

Mathematics Subject Classification: 03-01, 03E10, 03E15, 03E35, 03E45, 03E55, 03E60

Printed on acid-free paper

Springer is part of Springer Science+Business Media (www.springer.com)

To Julia, Gregor, and Joana, with love

Preface

Set theory aims at proving interesting true statements about the mathematical universe. Different people interpret "interesting" in different ways. It is well known that set theory comes from real analysis. This led to descriptive set theory, the study of properties of definable sets of reals, and it certainly is an important area of set theory. We now know that the theory of large cardinals is a twin of descriptive set theory. I find the interplay of large cardinals, inner models, and properties of definable sets of reals very interesting.

We give a complete account of the Solovay-Shelah Theorem according to which having all sets of reals to be Lebesgue measurable and having an inaccessible cardinal are equiconsistent. We give a modern account of the theory of $0^{\#}$, produce Jensen's Covering Lemma, and prove the Martin-Harrington Theorem according to which the existence of $0^{\#}$ is equivalent with Σ^1_1 determinacy. We also produce the Martin-Steel Theorem according to which Projective Determinacy follows from the existence of infinitely many Woodin cardinals.

I started learning logic by reading a script of my Master's thesis' advisor, Ulrich Blau, on a nude beach by the Ammersee near Munich back in 1989. It was a very enjoyable way of learning a fascinating and exciting subject, and I then decided to become a logician (In the meantime, Blau's script appeared as [6]). We shall assume in what follows that the reader has some familiarity with mathematical logic, to the extent of e.g. [11]. We are not going to explain the key concepts of first order logic.

I thank David Asperó, Fabiana Castiblanco, William Chan, Gabriel Fernandes, Daisuke Ikegami, Marios Koulakis, Paul Larson, Stefan Miedzianowski, Haimanti Sarbadhikari, Shashi Srivastava, Sandra Uhlenbrock, Yong Cheng, and the anonymous referees for their many helpful comments on earlier versions of this book.

I thank my father and my mother. I thank my academic teachers, Ulrich Blau, Ronald Jensen, Peter Koepke, and John Steel. I thank all my colleagues, especially Martin Zeman. And I thank my wife, Marga López Arpí, for all her support over the last years.

Berkeley, Girona, and Münster, February 2014 Ralf Schindler

Contents

Chapter 1
Naive Set Theory

GEORG CANTOR (1845–1918) discovered set theory. Prior to CANTOR, people often took it to be paradoxical that there are sets which can be put into a bijective correspondence with a proper subset of themselves. For instance, there is a bijection from $\mathbb{N}$ onto the set of all prime numbers. Hence, it seemed, on the one hand the set of all primes is "smaller than" $\mathbb{N}$, whereas on the other hand it is "as big as" $\mathbb{N}$.

CANTOR's solution to this "paradox" was as follows. Let X and Y be arbitrary sets. Define "X is smaller than or of the same size as Y" (or, "Y is not bigger than X") as: there is an injection $f : X \to Y$. Write this as $X \leq Y$. Define "X is of the same size as Y" as: there is a bijection $f : X \to Y$. Write this as $X \sim Y$. Obviously, $X \sim Y$ implies $X \leq Y$. The theorem of CANTOR–SCHRÖDER–BERNSTEIN (cf. Theorem 1.4) will say that $X \sim Y$ follows from $X \leq Y$ and $Y \leq X$. We write $X < Y$ if $X \leq Y$ but not $Y \leq X$.

Notice that if $X \leq Y$, i.e., if there is an injection $f : X \to Y$, then there is a surjection $g : Y \to X$. This is clear if f is already bijective. If not, then pick $a_0 \in X$ (we may assume X to be non–empty). Define $g : Y \to X$ by $g(b) = f^{-1}(b)$, if b is in the range of f, and $g(b) = a_0$ otherwise.

Conversely, if $f : X \to Y$ is surjective then there is an injection $g : Y \to X$, i.e., $Y \leq X$. This is shown by *choosing* for each $b \in Y$ some $a \in X$ with $f(a) = b$ and setting $g(b) = a$. This argument is in need of the Axiom of Choice, AC, which we shall present in the next chapter and discuss in detail later on.

To a certain extent, set theory is the study of the cardinality of arbitrary sets, i.e., of the relations $\leq$ and $\sim$ as defined above. The proof of the following theorem may be regarded as the birth of set theory.

Theorem 1.1 (Cantor)

$$\mathbb{N} < \mathbb{R}.$$

Proof $\mathbb{N} \leq \mathbb{R}$ is trivial. We show that $\mathbb{R} \leq \mathbb{N}$ does not hold.

Assume that there is an injection from $\mathbb{R}$ to $\mathbb{N}$, so that there is then also a surjection $f : \mathbb{N} \to \mathbb{R}$. Write x_n for $f(n)$. In particular, $\mathbb{R} = \{x_n : n \in \mathbb{N}\}$.

R. Schindler, *Set Theory*, Universitext, DOI: 10.1007/978-3-319-06725-4_1,

Let us now recursively define a sequence of closed intervals $[a_n, b_n] = \{x : a_n \leq x \leq b_n\}$ as follows. Put $[a_0, b_0] = [0, 1]$. Suppose $[a_n, b_n]$ has already been defined. Pick $[a_{n+1}, b_{n+1}]$ so that $a_n \leq a_{n+1} < b_{n+1} \leq b_n$, $b_{n+1} - a_{n+1} \leq \frac{1}{n+1}$ and $x_n \notin [a_{n+1}, b_{n+1}]$.

Now $\bigcap_{n \in \mathbb{N}} [a_n, b_n] = \{x\}$ for some $x \in \mathbb{R}$ by the Nested Interval Principle. Obviously, $x \neq x_n$ for every n, as $x_n \notin [a_{n+1}, b_{n+1}]$ and $x \in [a_{n+1}, b_{n+1}]$. Hence $x \notin \{x_n : n \in \mathbb{N}\} = \mathbb{R}$. Contradiction! □

It is not hard to verify that the sets of all integers, of all rationals, and of all algebraic numbers are each of the same size as $\mathbb{N}$ (cf. Problem 1.1). In particular, Theorem 1.1 immediately gives the following.

Corollary 1.2 *There are transcendental numbers.*

For arbitrary sets X and Y, we write $Y \subset X$ for: Y is a (not necessarily proper) subset of X, i.e., every element of Y is also an element of X, and we let $\mathscr{P}(X) = \{Y : Y \subset X\}$ denote the *power set of* X, i.e., the set of all subsets of X. Problem 1.2 shows that $\mathscr{P}(\mathbb{N}) \sim \mathbb{R}$. The following is thus a generalization of Theorem 1.1.

Theorem 1.3 *For every X, $X < \mathscr{P}(X)$.*

Proof We have $X \leq \mathscr{P}(X)$, because $f : X \to \mathscr{P}(X)$ is injective where $f(x) = \{x\}$ for $x \in X$.

We have to see that $\mathscr{P}(X) \leq X$ does not hold true. Given an arbitrary $f : X \to \mathscr{P}(X)$, consider $Y = \{x \in X : x \notin f(x)\} \subset X$. If Y were in the range of f, say $Y = f(x_0)$, then we would have that $x_0 \in Y \Longleftrightarrow x_0 \notin f(x_0) = Y$. Contradiction! In particular, f cannot be surjective, which shows that $\mathscr{P}(X) \leq X$ is false. □

Theorem 1.4 (Cantor–Schröder–Bernstein) *Let X and Y be arbitrary. If $X \leq Y$ and $Y \leq X$, then $X \sim Y$.*

Proof Let both $f : X \to Y$ and $g : Y \to X$ be injective. We are looking for a bijection $h : X \to Y$. Let $x \in X$. An X–orbit of x is a finite or infinite sequence of the form

$$g^{-1}(x), f^{-1}(g^{-1}(x)), g^{-1}(f^{-1}(g^{-1}(x))), \ldots$$

For each $n \in \mathbb{N} \cup \{\infty\}$ there is obviously at most one X–orbit of x of length n. Let $n(x)$ be the maximal $n \in \mathbb{N} \cup \{\infty\}$ so that there is an X–orbit of x of length n. We put $x \in X_0$ iff $n(x) = \infty$, $x \in X_1$ iff $n(x) \in \mathbb{N}$ is even, and $x \in X_2$ iff $n(x) \in \mathbb{N}$ is odd.

For $y \in Y$ we define the concept of a Y–orbit in an analoguous way, i.e., as a finite or infinite sequence of the form

$$f^{-1}(y), g^{-1}(f^{-1}(y)), f^{-1}(g^{-1}(f^{-1}(y))), \ldots$$

We write $n(y)$ for the maximal $n \in \mathbb{N} \cup \{\infty\}$ so that there is a Y–orbit of y of length n. We set $y \in Y_0$ iff $n(y) = \infty$, $y \in Y_1$ iff $n(y) \in \mathbb{N}$ is odd, and $y \in Y_2$ iff $n(y) \in \mathbb{N}$ is even.

Let us now define $h : X \to Y$ by

$$h(x) = \begin{cases} f(x) & \text{if } x \in X_0 \cup X_1, \text{ and} \\ g^{-1}(x) & \text{if } x \in X_2. \end{cases}$$

The function h is well-defined, as X is the disjoint union of X_0, X_1, and X_2, and because for every $x \in X_2$ there is an X–orbit of x of length 1, i.e., $g^{-1}(x)$ is defined.

The function h is injective: Let $x_1 \neq x_2$ with $h(x_1) = h(x_2)$. Say $x_1 \in X_0 \cup X_1$ and $x_2 \in X_2$. Then obviously $h(x_1) = f(x_1) \in Y_0 \cup Y_1$ and $h(x_2) = g^{-1}(x_2) \in Y_2$. But Y is the disjoint union of Y_0, Y_1, and Y_2. Contradiction!

The function h is surjective: Let $y \in Y_0 \cup Y_1$. Then $y = f(x)$ for some $x \in X_0 \cup X_1$; but then $y = h(x)$. Let $y \in Y_2$. Then $g(y) \in X_2$, so $y = g^{-1}(g(y)) = h(g(y))$. □

CANTOR's *Continuum Problem* is the question if there is a set A of real numbers such that

$$\mathbb{N} < A < \mathbb{R}.$$

This problem has certainly always been one of the key driving forces of set theory. A set A is called *at most countable* if $A \leq \mathbb{N}$. A is called *countable* if $A \sim \mathbb{N}$, and A is called *finite* iff $A < \mathbb{N}$. A is called *uncountable* iff $\mathbb{N} < A$.

CANTOR's *Continuum Hypothesis* says that the Continuum Problem has a negative answer, i.e., that for every uncountable set A of real numbers, $A \sim \mathbb{R}$.

CANTOR initiated the project of proving the Continuum Hypothesis by an induction on the "complexity" of the sets A in question. There is indeed a hierarchy of sets of reals which we shall study in Chap. 7. The open and closed sets sit at the very bottom of this hierarchy.

Let $A \subset \mathbb{R}$. A is called *open* iff for every $a \in A$ there are $c < a$ and $b > a$ with $(c, b) = \{x : c < x < b\} \subset A$. A is called *closed* iff $\mathbb{R} \setminus A$ is open.

It is easy to see that if $A \subset \mathbb{R}$ is any non–empty open set, then $\mathbb{R} \leq A$. As $A \leq \mathbb{R}$ is trivial for every $A \subset \mathbb{R}$, we immediately get that $A \sim \mathbb{R}$ for every non–empty open $A \subset \mathbb{R}$ with the help of the Theorem 1.4 of CANTOR–SCHRÖDER–BERNSTEIN. Theorem 1.9 of CANTOR–BENDIXSON will say that $A \sim \mathbb{R}$ for every uncountable closed set $A \subset \mathbb{R}$, which may be construed as a first step towards a realization of CANTOR's project. We shall later prove much more general results (cf. Theorem 12.11 and Corollary 13.8) which have a direct impact on CANTOR's project.

Lemma 1.5 *Let $A \subset \mathbb{R}$. The following are equivalent:*

(1) *A is closed.*
(2) *For all $x \in \mathbb{R}$, if $a < x < b$ always implies $(a, b) \cap A \neq \emptyset$, then $x \in A$.*

Proof (1) $\Longrightarrow$ (2): Let $x \notin A$. Let $a < x < b$ be such that $(a, b) \subset \mathbb{R} \setminus A$. Then $(a, b) \cap A = \emptyset$.

(2) $\Longrightarrow$ (1): We prove that $\mathbb{R} \setminus A$ is open. Let $x \in \mathbb{R} \setminus A$. Then there are $a < x < b$ so that $(a, b) \cap A = \emptyset$, i.e., $(a, b) \subset \mathbb{R} \setminus A$. □

Let $A \subset \mathbb{R}$. x is called an *accumulation point* of A iff for all $a < x < b$, $(a, b) \cap (A \setminus \{x\}) \neq \emptyset$ (here, x itself need not be an element of A). The set of all accumulation points of A is called the *(first) derivative* of A and is abbreviated by A'. Lemma 1.5 readily gives:

Lemma 1.6 *Let $A \subset \mathbb{R}$. The following are equivalent:*

(1) *A is closed.*
(2) $A' \subset A$.

Let $A \subset \mathbb{R}$. A set $B \subset A$ is called *dense in* A iff for all $a, b \in \mathbb{R}$ with $a < b$ and $[a, b] \cap A \neq \emptyset$, $[a, b] \cap B \neq \emptyset$. $B \subset \mathbb{R}$ is called *dense* iff B is dense in $\mathbb{R}$. It is well–known that $\mathbb{Q}$ is dense.

Definition 1.7 A set $A \subset \mathbb{R}$ is called *perfect* iff $A \neq \emptyset$ and $A' = A$.

Theorem 1.8 *Let $A \subset \mathbb{R}$ be perfect. Then $A \sim \mathbb{R}$.*

Proof $A \leq \mathbb{R}$ is trivial. It thus remains to be shown that $\mathbb{R} \leq A$. We shall make use of the fact that $\mathbb{R} \sim {}^{\mathbb{N}}\{0, 1\}$, where ${}^{\mathbb{N}}\{0, 1\}$ is the set of all infinite sequences of 0's and 1's. (Cf. Problem 1.2.) We aim to see that ${}^{\mathbb{N}}\{0, 1\} \leq A$.

Let ${}^*\{0, 1\}$ be the set of all non-empty finite sequences of 0's and 1's, i.e., of all $s : \{0, \ldots, n\} \to \{0, 1\}$ for some $n \in \mathbb{N}$. Let us define a function Φ from ${}^*\{0, 1\}$ to closed intervals as follows.

Let $s_0 : \{0\} \to \{0\}$ and $s_1 : \{0\} \to \{1\}$. As $A \neq \emptyset$ and $A \subset A'$ we easily find

$$a_{s_0} < b_{s_0} < a_{s_1} < b_{s_1}$$

so that

$$(a_{s_0}, b_{s_0}) \cap A \neq \emptyset \text{ and } (a_{s_1}, b_{s_1}) \cap A \neq \emptyset.$$

Set $\Phi(s_0) = [a_{s_0}, b_{s_0}]$ and $\Phi(s_1) = [a_{s_1}, b_{s_1}]$.

Now let $s \in {}^*\{0, 1\}$ and suppose that $\Phi(s)$ is already defined, where $\Phi(s) = [a_s, b_s]$ with $a_s < b_s$ and $(a_s, b_s) \cap A \neq \emptyset$.

Let $s : \{0, \ldots, n\} \to \{0, 1\}$. For $h = 0, 1$ write $s^\frown h$ for the unique $t : \{0, \ldots, n+1\} \to \{0, 1\}$ with $t(i) = s(i)$ for $i \leq n$ and $t(n + 1) = h$. Because $A \subset A'$, we easily find

$$a_s < a_{s^\frown 0} < b_{s^\frown 0} < a_{s^\frown 1} < b_{s^\frown 1} < b_s,$$

so that

$$(a_{s^\frown 0}, b_{s^\frown 0}) \cap A \neq \emptyset, (a_{s^\frown 1}, b_{s^\frown 1}) \cap A \neq \emptyset,$$

$$b_{s^\frown 0} - a_{s^\frown 0} \leq \frac{1}{n+1}, \text{ and } b_{s^\frown 1} - a_{s^\frown 1} \leq \frac{1}{n+1}.$$

Set $\Phi(s^\frown h) = [a_{s^\frown h}, b_{s^\frown h}]$ for $h = 0, 1$.

We may now define an injection $F : {}^{\mathbb{N}}\{0,1\} \to A$. Let $f \in {}^{\mathbb{N}}\{0,1\}$. Then

$$\bigcap_{n\in\mathbb{N}} [a_{f\upharpoonright\{0,\ldots,n\}}, b_{f\upharpoonright\{0,\ldots,n\}}] = \{x\}$$

for some $x \in \mathbb{R}$ by the Nested Interval Principle. Set $F(f) = x$. Obviously, $F(f) \in A$, as $F(f)$ is an accumulation point of A and $A' \subset A$. Also, F is certainly injective. □

Theorem 1.9 (Cantor–Bendixson) *Let $A \subset \mathbb{R}$ be closed. Then there are sets $A_0 \subset \mathbb{R}$ and $P \subset \mathbb{R}$ so that:*

(1) *A is the disjoint union of A_0* and *P,*
(2) *A_0 is at most countable, and*
(3) *P is perfect, unless $P = \emptyset$.*

Corollary 1.10 *Let $A \subset \mathbb{R}$ be closed. Then $A \leq \mathbb{N}$ or $A \sim \mathbb{R}$.*

Proof of Theorem 1.8. An $x \in \mathbb{R}$ is called a *condensation point* of A iff $(a,b) \cap A$ is uncountable for all $a < x < b$.

Let P be the set of all condensation points of A, and let $A_0 = A \setminus P$. As A is closed, $P \subset A' \subset A$. It remains to be shown that (2) and (3) both hold true. We shall make use of the fact that $\mathbb{Q} \sim \mathbb{N}$ (cf. Problem 1.1) and that $\mathbb{Q}$ is dense, so that that for all $x, y \in \mathbb{R}$ with $x < y$ there is some $z \in \mathbb{Q}$ with $x < z < y$.

Let $x \in A_0$. Then there are $a_x < x < b_x$ with $a_x, b_x \in \mathbb{Q}$ and such that $(a_x, b_x) \cap A$ is at most countable. Therefore,

$$A_0 \subset \bigcup_{x\in A_0} (a_x, b_x) \cap A.$$

As $\mathbb{Q} \sim \mathbb{N}$, there are at most countably many sets of the form $(a_x, b_x) \cap A$, and each of them is at most countable. Hence A_0 is at most countable (cf. Problem 1.4).

Suppose that $P \neq \emptyset$. We first show that $P \subset P'$. Let $x \in P$. Let $a < x < b$. We have that $(a,b) \cap A$ is uncountable. Suppose that $(a,b) \cap (P \setminus \{x\}) = \emptyset$. For each $y \in ((a,b)\setminus\{x\}) \cap A$ there are then $a_y < y < b_y$ with $a_y, b_y \in \mathbb{Q}$ so that $(a_y, b_y) \cap A$ is at most countable. But then we have that

$$(a,b) \cap A \subset \{x\} \cup \bigcup_{y\in(a,b)\setminus\{x\}} (a_y, b_y) \cap A$$

is at most countable. Contradiction!

Let us finally show that $P' \subset P$. Let $x \in P'$. Then $(a,b) \cap (P\setminus\{x\}) \neq \emptyset$ for all $a < x < b$. Let $y \in (a,b) \cap (P\setminus\{x\})$, where $a < x < b$. Then $(a,b) \cap A$ is uncountable. Hence $x \in P$. □

There is a different proof of the Theorem of CANTOR–BENDIXSON which brings the concept of an "ordinal number" into play. Let $A \subset \mathbb{R}$ be closed. Define A^1 as A',

A^2 as A'', etc., i.e., A^{n+1} as $(A^n)'$ for $n \in \mathbb{N}$. It is easy to see that each A^n is closed, and

$$\ldots \subset A^{n+1} \subset A^n \subset \ldots \subset A^1 \subset A.$$

If there is some n with $A^{n+1} = A^n$ then $P = A^n$ and $A_0 = A \setminus P$ are as in the statement of Theorem 1.9. Otherwise we have to continue this process into the transfinite. Let

$$A^\infty = \bigcap_{n \in \mathbb{N}} A^n,\ A^{\infty+1} = (A^\infty)', \ldots, A^{\infty+n+1} = (A^{\infty+n})',$$

$$A^{\infty+\infty} = \bigcap_{n \in \mathbb{N}} A^{\infty+n}, \ldots \text{etc.}$$

It can be shown that there is a "number" α so that $A^{\alpha+1} = A^\alpha$. For such an α, $A \setminus A^\alpha$ is at most countable, and if $A^\alpha \neq \emptyset$, then A^α is perfect.

Such "numbers" are called *ordinal numbers* (cf. Definition 3.3). We need an axiomatization of set theory (to be presented in Chap. 2), though, in order to be able to introduce them rigorously. With their help we shall be able to prove much stronger forms of the Theorem of CANTOR–BENDIXSON (cf. Theorems 7.15 and 12.11).

Definition 1.11 A set $A \subset \mathbb{R}$ is called *nowhere dense* iff $\mathbb{R} \setminus A$ has an open subset which is dense in $\mathbb{R}$. A set $A \subset \mathbb{R}$ is called *meager* (or of *first category*) iff there are $A_n \subset \mathbb{R}$, $n \in \mathbb{N}$, such that $A = \bigcup_{n \in \mathbb{N}} A_n$ and each A_n is nowhere dense. If $A \subset \mathbb{R}$ is not meager, then it is of *second category*.

It is not hard to see that A is nowhere dense iff for all $a, b \in \mathbb{R}$ with $a < b$ there are $a', b' \in \mathbb{R}$ with $a \leq a' < b' \leq b$ and $[a', b'] \cap A = \emptyset$ (cf. Problem 1.8(a)). Of course, every countable set of reals is meager, and in fact the countable union of meager sets is meager, but there are nowhere dense sets which have the same size as $\mathbb{R}$ (cf. Problem 1.8 (c)).

Theorem 1.12 (Baire Category Theorem) *If each $A_n \subset \mathbb{R}$ is open and dense, $n \in \mathbb{N}$, then $\bigcap_{n \in \mathbb{N}} A_n$ is dense.*

Proof Let $a < b, a, b \in \mathbb{R}$ be arbitrary. We need to see that $[a, b] \cap \bigcap_{n \in \mathbb{N}} A_n \neq \emptyset$. Let us define $[a_n, b_n]$, $n \in \mathbb{N}$, recursively as follows. We set $[a_0, b_0] = [a, b]$. Suppose $[a_n, b_n]$ is already chosen. As A_n is dense, $(a_n, b_n) \cap A_n \neq \emptyset$, say $x \in (a_n, b_n) \cap A_n$. As A_n is open, we may pick c, d with $a_n < c < x < d < b_n$ and $(c, d) \subset A_n$. Let a_{n+1}, b_{n+1} be such that $c < a_{n+1} < b_{n+1} < d$, so that $[a_{n+1}, b_{n+1}] \subset A_n \cap [a_n, b_n]$. But now $\emptyset \neq \bigcap_{n \in \mathbb{N}} [a_n, b_n] \subset [a, b] \cap \bigcap_{n \in N} A_n$, as desired. □

The BAIRE Category Theorem implies that $\mathbb{R}$ is of second category, in fact that the complement of a meager set is dense in $\mathbb{R}$ (cf. Problem 1.8 (b)).

If $a, b \in \mathbb{R}$, $a < b$, then we call $b - a$ the *length of* the closed interval $[a, b]$. As $\mathbb{Q}$ is dense in $\mathbb{R}$, any union of closed intervals may be written as a union of closed intervals with *rational* endpoints (cf. the proof of Theorem 1.9) and thus as a union of countably many closed intervals which in addition may be picked to be pairwise disjoint. If $A \subset [0, 1]$, $A = \bigcup_{n \in \mathbb{N}}[a_n, b_n]$, where $a_n < b_n$ for each $n \in \mathbb{N}$ and the $[a_n, b_n]$ are pairwise disjoint, then we write

$$\mu(A) = \sum_{n \in \mathbb{N}} b_n - a_n$$

and call it the *measure* of A. One can show that $\mu(A)$ is independent from the choice of the pairwise disjoint intervals $[a_n, b_n]$ with $A = \bigcup_{n \in \mathbb{N}}[a_n, b_n]$ (cf. Problem 1.7).

Definition 1.13 Let $A \subset [0, 1]$. Then A is called a *null set* iff for all $\epsilon > 0$ there is a countable union $A = \bigcup_{n \in \mathbb{N}}[a_n, b_n]$ of closed intervals $[a_n, b_n] \subset [0, 1]$ such that $\mu(A) \leq \varepsilon$.

Of course, every countable subset of [0, 1] is null, and in fact the countable union of null sets is null, but there are null sets which have the same size as $\mathbb{R}$ (cf. Problem 1.8(b)).

1.1 Problems

1.1. Show that the sets of all finite sets of natural numbers, of all integers, of all rationals, and of all algebraic numbers are each countable, i.e., of the same size as $\mathbb{N}$.

1.2. Show that $\mathbb{R} \sim {}^{\mathbb{N}}\{0, 1\}$, where ${}^{\mathbb{N}}\{0, 1\}$ is the set of all infinite sequences of 0's and 1's.

1.3. If A, B are sets of natural numbers, then A and B are called *almost disjoint* iff $A \cap B$ is finite. A collection D of sets of natural numbers is called *almost disjoint* iff any two distinct elements of D are almost disjoint. Show that there is an almost disjoint collection D of sets of natural numbers such that $D \sim \mathbb{R}$. [Hint: Use a bijection between the set of finite 0–1–sequences and $\mathbb{N}$.]

1.4. Let, for each $n \in \mathbb{N}$, A_n be a countable set. Show that $\bigcup_{n \in \mathbb{N}} A_n$ is countable. (This uses AC, the Axiom of Choice, cf. Theorem 6.69.)

1.5. Let $n \in \mathbb{N}$. Construct a set $A \subset \mathbb{R}$ such that $A^n \neq \emptyset$, but $A^{n+1} = \emptyset$. Also construct a set $A \subset \mathbb{R}$ such that $A^{\infty+n} \neq \emptyset$, but $A^{\infty+n+1} = \emptyset$.

1.6. Let $A \subset \mathbb{R}$ be closed. Show that the pair (A_0, P) as in the statement of Theorem 1.9 of CANTOR–BENDIXSON is unique.

1.7. Show that if $A \subset [0, 1]$, $A = \bigcup_{n \in \mathbb{N}}[a_n, b_n]$, where the $[a_n, b_n]$ are pairwise disjoint, then $\mu(A)$ as defined above is independent from the choice of the pairwise disjoint intervals $[a_n, b_n]$ with $A = \bigcup_{n \in \mathbb{N}}[a_n, b_n]$.

1.8. (a) Show that $A \subset \mathbb{R}$ is nowhere dense iff for all $a, b \in \mathbb{R}$ with $a < b$ there are $a', b' \in \mathbb{R}$ with $a \leq a' < b' \leq b$ and $[a', b'] \cap A = \emptyset$.

(b) Show that $\mathbb{R}$ is not meager. In fact, the complement of a meager set $A \subset \mathbb{R}$ is dense in $\mathbb{R}$.
(c) For $a, b \in \mathbb{R}$ with $a < b$ let

$$[a, b]^{\frac{2}{3}} = [a, \frac{2}{3}a + \frac{1}{3}b] \cup [\frac{1}{3}a + \frac{2}{3}b, b],$$

and for $a_0, b_0, \ldots, a_k, b_k \in \mathbb{R}$ with $a_i < b_i$ for all $i \leq k$ let

$$([a_0, b_0] \cup \ldots \cup [a_k, b_k])^{\frac{2}{3}} = [a_0, b_0]^{\frac{2}{3}} \cup \ldots \cup [a_k, b_k]^{\frac{2}{3}}.$$

Finally, let, for $a, b \in \mathbb{R}$ with $a < b$, $[a, b]_0 = [a, b]$, $[a, b]_{n+1} = ([a, b]_n)^{\frac{2}{3}}$, and

$$[a, b]_\infty = \bigcap_{n \in \mathbb{N}} [a, b]_n.$$

$[a, b]_\infty$ is called CANTOR's *Discontinuum*. Show that for all $a, b \in \mathbb{R}$ with $a < b$, $[a, b]_\infty$ is dense in $[a, b]$, and $[a, b]_\infty$ is perfect, nowhere dense, and a null set.

Chapter 2
Axiomatic Set Theory

ERNST ZERMELO (1871–1953) was the first to find an axiomatization of set theory, and it was later expanded by ABRAHAM FRAENKEL (1891–1965).

2.1 Zermelo–Fraenkel Set Theory

The language of set theory, which we denote by $\mathscr{L}_\in$, is the usual language of first order logic (with one type of variables) equipped with just one binary relation symbol, $\in$. The intended domain of set theoretical discourse (i.e., the range of the variables) is the universe of all sets, and the intended interpretation of $\in$ is "is an element of." We shall use $x, y, z, \ldots, a, b, \ldots$, etc. as variables to range over sets.

The standard axiomatization of set theory, ZFC (ZERMELO–FRAENKEL set theory with choice), has infinitely many axioms. The first one, the *axiom of extensionality*, says that two sets are equal iff they contain the same elements.

$$\forall x \forall y (x = y \leftrightarrow \forall z (z \in x \leftrightarrow z \in y)). \tag{Ext}$$

A set x is a *subset* of y, abbreviated by $x \subset y$, if $\forall z(z \in x \rightarrow z \in y)$. (Ext) is then logically equivalent to $\forall x \forall y (x \subset y \wedge y \subset x \rightarrow x = y)$. We also write $y \supset x$ for $x \subset y$. x is a *proper subset* of y, written $x \subsetneq y$, iff $x \subset y$ and $x \neq y$.

The next axiom, the *axiom of foundation*, says that each nonempty set has an $\in$-minimal member.

$$\forall x (\exists y\, y \in x \rightarrow \exists y (y \in x \wedge \neg \exists z (z \in y \wedge z \in x))). \tag{Fund}$$

This is easier to grasp if we use the following abbreviations: We write $x = \emptyset$ for $\neg \exists y\, y \in x$ (and $x \neq \emptyset$ for $\exists y\, y \in x$), and $x \cap y = \emptyset$ for $\neg \exists z (z \in x \wedge z \in y)$. (Fund) then says that

$$\forall x (x \neq \emptyset \rightarrow \exists y (y \in x \wedge y \cap x = \emptyset)).$$

R. Schindler, *Set Theory*, Universitext, DOI: 10.1007/978-3-319-06725-4_2,

(Fund) plays an important technical role in the development of set theory.

Let us write $x = \{y, z\}$ instead of

$$y \in x \wedge z \in x \wedge \forall u(u \in x \rightarrow (u = y \vee u = z)).$$

The *axiom of pairing* runs as follows.

$$\forall x \forall y \exists z \; z = \{x, y\}. \tag{Pair}$$

We also write $\{x\}$ instead of $\{x, x\}$.

In the presence of (Pair), (Fund) implies that there cannot be a set x with $x \in x$: if $x \in x$, then x is the only element of $\{x\}$, but $x \cap \{x\} \neq \emptyset$, as $x \in x \cap \{x\}$. A similar argument shows that there cannot be sets $x_1, x_2, \ldots, x_k$ such that $x_1 \in x_2 \in \cdots \in x_k \in x_1$ (cf. Problem 2.1).

Let us write $x = \bigcup y$ for

$$\forall z(z \in x \leftrightarrow \exists u(u \in y \wedge z \in u)).$$

The *axiom of union* is the following one.

$$\forall x \exists y \; y = \bigcup x. \tag{Union}$$

Writing $z = x \cup y$ for

$$\forall u(u \in z \leftrightarrow u \in x \vee u \in y),$$

(Pair) and (Union) prove that $\forall x \forall y \exists z(z = x \cup y)$, as $x \cup y = \bigcup\{x, y\}$.

The *power set axiom*, (Pow), says that for every set x, the set of all subsets of x exists. We write $x = \mathscr{P}(y)$ for

$$\forall z(z \in x \leftrightarrow z \subset y)$$

and formulate

$$\forall x \exists y \; y = \mathscr{P}(x). \tag{Pow}$$

The *axiom of infinity*, (Inf), tells us that there is a set which contains all of the following sets as members:

$$\emptyset, \{\emptyset\}, \{\emptyset, \{\emptyset\}\}, \{\emptyset, \{\emptyset\}, \{\emptyset, \{\emptyset\}\}\}, \ldots.$$

To make this precise, we call a set x *inductive* iff

$$\emptyset \in x \wedge \forall y(y \in x \rightarrow y \cup \{y\} \in x).$$

We then say:

$$\exists x(x \text{ is inductive}). \tag{Inf}$$

We now need to formulate the separation and replacement schemas.

A *schema* is an infinite set of axioms which is generated in a simple (recursive) way.

Let φ be a formula of $\mathcal{L}_\in$ in which exactly the variables $x, v_1, \ldots, v_p$ (which all differ from b) occur freely. The *axiom of separation*, or "*Aus*sonderung," corresponding to φ runs as follows.

$$\forall v_1 \ldots \forall v_p \forall a \exists b \forall x\, (x \in b \leftrightarrow x \in a \wedge \varphi). \tag{Aus$_\varphi$}$$

Let us write $b = \{x \in a : \varphi\}$ for $\forall x(x \in b \leftrightarrow x \in a \wedge \varphi)$. If we suppress $v_1, \ldots, v_p$, (Aus$_\varphi$) then says that

$$\forall a \exists b\, b = \{x \in a : \varphi\}.$$

Writing $z = x \cap y$ for

$$\forall u(u \in z \leftrightarrow u \in x \wedge u \in y),$$

(Aus$_{x \in c}$) proves that $\forall a \forall c \exists b\, b = a \cap c$. Writing $z = x \backslash y$ for

$$\forall u(u \in z \leftrightarrow u \in x \wedge \neg u \in y),$$

(Aus$_{\neg x \in c}$) proves that $\forall a \forall c \exists b\, b = a \backslash c$. Also, if we write $x = \bigcap y$ for

$$\forall z(z \in x \leftrightarrow \forall u(u \in y \rightarrow z \in u)),$$

then (Aus$_{\forall u(u \in y \rightarrow z \in u)}$), applied to any member of y proves that

$$\forall y(y \neq \emptyset \rightarrow \exists x\, x = \bigcap y).$$

The *separation schema* (Aus) is the set of all (Aus$_\varphi$). It says that we may separate elements from a given set according to some well-defined device to obtain a new set.

Now let φ be a formula of $\mathcal{L}_\in$ in which exactly the variables $x, y, v_1, \ldots, v_p$ (all different from b) occur freely. The *replacement axiom* corresponding to φ runs as follows.

$$\forall v_1 \ldots \forall v_p\, (\forall x \exists y' \forall y(y = y' \leftrightarrow \varphi) \rightarrow \forall a \exists b \forall y(y \in b \leftrightarrow \exists x(x \in a \wedge \varphi))). \tag{Rep$_\varphi$}$$

The *replacement schema* (Rep) is the set of all (Rep$_\varphi$). It says that we may replace elements from a given set according to some well-defined device by other sets to obtain a new set.

We could not have crossed out "$x \in a$" in (Aus$_\varphi$). If we did cross it out in (Aus$_\varphi$) and let φ be $\neg x \in x$, then we would get

$$\exists b \forall x (x \in b \leftrightarrow \neg x \in x),$$

which is a *false* statement, because it gives $b \in b \leftrightarrow \neg b \in b$. This observation sometimes runs under the title of "RUSSELL's *Antinomy*."

In what follows we shall write $x \notin y$ instead of $\neg x \in y$, and we shall write $x \neq y$ instead of $\neg x = y$.

A trivial application of the separation schema is the existence of the empty set $\emptyset$ which may be obtained from any set a by separating using the formula $x \neq x$ as φ, in other words,

$$\exists b \forall x (x \in b \leftrightarrow x \neq x).$$

With the help of (Pair) and (Union) we can then prove the existence of each of the following sets:

$$\emptyset, \{\emptyset\}, \{\{\emptyset\}\}, \{\emptyset, \{\emptyset\}\}, \ldots.$$

In particular, we will be able to prove the existence of each member of the intersection of all inductive sets. This will be discussed in the next chapter.

The *axiom of choice* finally says that for each family of pairwise disjoint non-empty sets there is a "choice set," i.e.

$$\forall x (\forall y (y \in x \rightarrow y \neq \emptyset) \wedge \forall y \forall y' (y \in x \wedge y' \in x \wedge y \neq y' \rightarrow y \cap y' = \emptyset)$$
$$\rightarrow \exists z \forall y (y \in x \rightarrow \exists u \forall u' (u' = u \leftrightarrow u' \in z \cap y))). \tag{AC}$$

In what follows we shall always abbreviate $\forall y (y \in x \rightarrow \varphi)$ by $\forall y \in x\ \varphi$ and $\exists y (y \in x \wedge \varphi)$ by $\exists y \in x\ \varphi$. We may then also formulate (AC) as

$$\forall x (\forall y \in x\ y \neq \emptyset \wedge \forall y \in x \forall y' \in x (y \neq y' \rightarrow y \cap y' = \emptyset)$$
$$\rightarrow \exists z \forall y \in x \exists u\ z \cap y = \{u\}),$$

i.e., for each member of x, z contains exactly one "representative."

One may also formulate (AC) in terms of the existence of choice functions (cf. Problem 2.6).

The theory which is given by the axioms (Ext), (Fund), (Pair), (Union), (Pow), (Inf) and (Aus_φ) for all φ is called ZERMELO's *set theory*, abbreviated by Z. The theory which is given by the axioms of Z together with (Rep_φ) for all φ is called ZERMELO–FRAENKEL *set theory*, abbreviated by ZF. The theory which is given by the axioms of ZF together with (AC) is called ZERMELO–FRAENKEL *set theory with choice*, abbreviated by ZFC. This system, ZFC, is the standard axiomatization of set theory. Most questions of mathematics can be decided in ZFC, but many questions of set theory and other branches of mathematics are independent from ZFC. The theory which is given by the axioms of Z together with (AC) is called ZERMELO *set theory with choice* and is often abbreviated by ZC. We also use ZFC^- to denote ZFC without (Pow), and we use $\text{ZFC}^{-\infty}$ to denote ZFC without (Inf).

Modulo ZF, (AC) has many equivalent formulations. In order to formulate some of them, we first have to introduce basic notations of axiomatic set theory, though.

For sets x, y we write (x, y) for $\{\{x\}, \{x, y\}\}$. It is easy to verify that for all x, y, x', y', if $(x, y) = (x', y')$, then $x = x'$ and $y = y'$. The set (x, y) can be shown to exist for every x, y by applying the pairing axiom three times; (x, y) is called the *ordered pair* of x and y.

We also write $\{x_1, \dots, x_{n+1}\}$ for $\{x_1, \dots, x_n\} \cup \{x_{n+1}\}$ and $(x_1, \dots, x_{n+1})$ for $((x_1, \dots, x_n), x_{n+1})$. If $(x_1, \dots, x_{n+1}) = (x'_1, \dots, x'_{n+1})$, then $x_1 = x'_1$, ..., and $x_{n+1} = x'_{n+1}$.

The *Cartesian product* of two sets a, b is defined to be

$$a \times b = \{(x, y) : x \in a \wedge y \in b\}.$$

Lemma 2.1 *For all a, b, $a \times b$ exists, i.e., $\forall a \forall b \exists c \; c = a \times b$.*

Proof $a \times b$ may be separated from $\mathscr{P}(\mathscr{P}(a \cup b))$. □

We also define $a_1 \times \cdots \times a_{n+1}$ to be $(a_1 \times \cdots \times a_n) \times a_{n+1}$ and

$$a^n = \underbrace{a \times \cdots \times a}_{n\text{-times}}.$$

An *n-ary relation* r is a subset of $a_1 \times \cdots \times a_n$ for some sets $a_1, \dots, a_n$. The n-ary relation r is *on* a iff $r \subset a^n$. If r is a binary (i.e., 2-ary) relation, then we often write $x \; r \; y$ instead of $(x, y) \in r$, and we define the *domain of* r as

$$\mathrm{dom}(r) = \{x : \exists y \; x \; r \; y\}$$

and the *range of* r as

$$\mathrm{ran}(r) = \{y : \exists x \; x \; r \; y\}.$$

A relation $r \subset a \times b$ is a *function* iff

$$\forall x \in \mathrm{dom}(r) \exists y \forall y' (y' = y \leftrightarrow x \; r \; y').$$

If $f \subset a \times b$ is a function, and if $x \in \mathrm{dom}(f)$, then we write $f(x)$ for the unique $y \in \mathrm{ran}(f)$ with $(x, y) \in f$.

A function f is a *function from* d to b iff $d = \mathrm{dom}(f)$ and $\mathrm{ran}(f) \subset b$ (sic!), which we also express by writing

$$f : d \to b.$$

The set of all functions from d to b is denoted by ${}^d b$.

Lemma 2.2 *For all d, b, ${}^d b$ exists.*

Proof ${}^d b$ may be separated from $\mathscr{P}(d \times b)$. □

If $f: b \to d$ and $g: d \to e$, then we write $g \circ f$ for the function from b to e which sends $x \in b$ to $g(f(x)) \in e$.

If $f: d \to b$, then f is *surjective* iff $b = \text{ran}(f)$, and f is *injective* iff

$$\forall x \in d \forall x' \in d(f(x) = f(x') \to x = x').$$

f is *bijective* iff f is surjective and injective.

If $f: d \to b$ and $a \subset d$, then $f \upharpoonright a$, the *restriction of f to a*, is that function $g: a \to b$ such that $g(x) = f(x)$ for every $x \in a$. We write $f''a$ for the *image of a under f*, i.e., for the set $\{y \in \text{ran}(f): \exists x \in a\ y = f(x)\}$. Of course, $f''a = \text{ran}(f \upharpoonright a)$.

If $f: d \to b$ is injective, and if $y \in \text{ran}(f)$, then we write $f^{-1}(y)$ for the unique $x \in \text{dom}(f)$ with $f(x) = y$. If $c \subset b$, then we write $f^{-1}{}''c$ for the set $\{x \in \text{dom}(f): f(x) \in c\}$.

A binary relation $\leq$ on a set a is called a *partial order on a* iff $\leq$ is *reflexive* (i.e., $x \leq x$ for all $x \in a$), $\leq$ is *symmetric* (i.e., if $x, y \in a$, then $x \leq y \wedge y \leq x \to x = y$), and $\leq$ is *transitive* (i.e., if $x, y, z \in a$ and $x \leq y \wedge y \leq z$, then $x \leq z$). In this case we call $(a, \leq)$ (or just a) a *partially ordered set*. If $\leq$ is a partial order on a, then $\leq$ is called *linear* (or *total*) iff for all $x \in a$ and $y \in a$, $x \leq y$ or $y \leq x$.

If $(a, \leq)$ is a partially ordered set, then we also write $x < y$ iff $x \leq y \wedge x \neq y$. Notice that $x \leq y$ iff $x < y \vee x = y$. We shall also call $<$ a partial order.

Let $(a, \leq)$ be a partially ordered set, and let $b \subset a$. We say that x is a *maximal element of b* iff $x \in b \wedge \neg\exists y \in b\ x < y$. We say that x is *the maximum of b*, $x = \max(b)$, iff $x \in b \wedge \forall y \in b\ y \leq x$. We say that x is a *minimal element* of b iff $x \in b \wedge \neg\exists y \in b\ y < x$, and we say that x is *the minimum of b*, $x = \min(b)$, iff $x \in b \wedge \forall y \in b\ x \leq y$. Of course, if $x = \max(b)$, then x is a maximal element of b, and if $x = \min(b)$, then x is a minimal element of b. We say that x is an *upper bound of b* iff $y \leq x$ for each $y \in b$, and we say that x is a *strict upper bound of b* iff $y < x$ for each $y \in b$; x is the *supremum of b*, $x = \sup(b)$, iff x is the minimum of the set of all upper bounds of b, i.e., if x is an upper bound and

$$\forall y \in a(\forall y' \in b\ y' \leq y \to x \leq y).$$

If $x = \max(b)$, then $x = \sup(b)$. We say that x is a *lower bound of b* iff $x \leq y$ for each $y \in b$, and we say that x is a *strict lower bound of b* iff $x < y$ for all $y \in b$; x is the *infimum of b*, $x = \inf(b)$, iff x is the maximum of the set of all lower bounds of b, i.e., if x is a lower bound and

$$\forall y \in a(\forall y' \in b\ y \leq y' \to y \leq x).$$

If $x = \min(b)$, then $x = \inf(b)$. If $\leq$ is not clear from the context, then we also say "$\leq$-maximal element," "$\leq$-supremum," "$\leq$-upper bound," etc.

Let $(a, \leq_a)$, $(b, \leq_b)$ be partially ordered sets. A function $f: a \to b$ is called *order-preserving* iff for all $x, y \in a$,

$$x \leq_a y \Longleftrightarrow f(x) \leq_b f(y).$$

If $f: a \to b$ is order-preserving and f is bijective, then f is called an *isomorphism*, also written

$$(a, \leq_a) \overset{f}{\cong} (b, \leq_b).$$

$(a, \leq_a)$ and $(b, \leq_b)$ are called *isomorphic* iff there is an isomorphism $f: a \to b$, written

$$(a, \leq_a) \cong (b, \leq_b).$$

The following concept plays a key role in set theory.

Definition 2.3 Let $(a, \leq)$ be a partial order. Then $(a, \leq)$ is called a well-ordering iff for every $b \subset a$ with $b \neq \emptyset$, $\min(b)$ exists.

The natural ordering on $\mathbb{N}$ is a well-ordering, but there are many other well-orderings on $\mathbb{N}$ (cf. Problem 2.7).

Lemma 2.4 *Let $(a, \leq)$ be a well-ordering. Then $\leq$ is total.*

Proof If $x, y \in a$, then $\min(\{x, y\}) \leq x$ and $\min(\{x, y\}) \leq y$. Hence if $\min(\{x, y\}) = x$, then $x \leq y$, and if $\min(\{x, y\}) = y$, then $y \leq x$. □

Lemma 2.5 *Let $(a, \leq)$ be a well-ordering, and let $f: a \to a$ be order-preserving. Then $f(x) \geq x$ for all $x \in a$.*

Proof If $\{x \in a : f(x) < x\} \neq \emptyset$, set

$$x_0 = \min(\{x \in a : f(x) < x\}).$$

Then $y_0 = f(x_0) < x_0$ and so $f(y_0) < f(x_0) = y_0$, as f is order-preserving. But this contradicts the choice of x_0. □

Lemma 2.6 *If $(a, \leq)$ is a well-ordering, and if $(a, \leq) \overset{f}{\cong} (a, \leq)$, then f is the identity.*

Proof By the previous lemma applied to f as well as to f^{-1}, we must have $f(x) \geq x$ as well as $f^{-1}(x) \geq x$, i.e., $f(x) = x$, for every $x \in a$. □

Lemma 2.7 *Suppose that $(a, \leq_a)$ and $(b, \leq_b)$ are both well-orderings such that $(a, \leq_a) \cong (b, \leq_b)$. Then there is a unique f with $(a, \leq_a) \overset{f}{\cong} (b, \leq_b)$.*

Proof If $(a, \leq_a) \overset{f}{\cong} (b, \leq_b)$ and $(a, \leq_a) \overset{g}{\cong} (b, \leq_b)$, then $(a, \leq_a) \overset{g^{-1} \circ f}{\cong} (a, \leq_a)$, so $g^{-1} \circ f$ is the identity, so $f = g$. □

If $(a, \leq)$ is a partially ordered set, and if $x \in a$, then we write $(a, \leq) \upharpoonright x$ for the partially ordered set

$$(\{y \in a : y < x\}, \leq \cap \{y \in a : y < x\}^2),$$

i.e., for the restriction of $(a, \leq)$ to the predecessors of x.

Lemma 2.8 *If $(a, \leq)$ is a well-ordering, and if $x \in a$, then $(a, \leq) \not\cong (a, \leq) \upharpoonright x$.*

Proof If $(a, \leq) \overset{f}{\cong} (a, \leq) \upharpoonright x$, then $f : a \to a$ is order-preserving with $f(x) < x$. This contradicts Lemma 2.5. □

Theorem 2.9 *Let $(a, \leq_a)$, $(b, \leq_b)$ be well-orderings. Then exactly one of the following statements holds true.*

(1) $(a, \leq_a) \cong (b, \leq_b)$
(2) $\exists x \in b \, (a, \leq_a) \cong (b, \leq_b) \upharpoonright x$
(3) $\exists x \in a \, (a, \leq_a) \upharpoonright x \cong (b, \leq_b)$.

Proof Let us define $r \subset a \times b$ by

$$(x, y) \in r \Longleftrightarrow (a, \leq_a) \upharpoonright x \cong (b, \leq_b) \upharpoonright y.$$

By the previous lemma, for each $x \in a$ there is at most one $y \in b$ such that $(x, y) \in r$ and vice versa. Therefore, r is an injective function from a subset of a to b. We have that r is order-preserving, because, if $x <_a x'$ and

$$(a, \leq_a) \upharpoonright x' \overset{f}{\cong} (b, \leq_b) \upharpoonright y,$$

then

$$(a, \leq_a) \upharpoonright x \overset{f \upharpoonright \{y \in a : y < x\}}{\cong} (b, \leq_b) \upharpoonright f(x),$$

so that $r(x) = f(x) < y = r(x')$.

If both $a \setminus \operatorname{dom}(r)$ as well as $b \setminus \operatorname{ran}(r)$ were nonempty, say $x = \min(a \setminus \operatorname{dom}(r))$ and $y = \min(b \setminus \operatorname{dom}(r))$, then

$$(a, \leq_a) \upharpoonright x \overset{r}{\cong} (b, \leq_b) \upharpoonright y,$$

so that $(x, y) \in r$ after all. Contradiction! □

The following Theorem is usually called ZORN's *Lemma*. The reader will gladly verify that its proof is performed in the theory ZC.

Theorem 2.10 (Zorn) Let $(a, \leq)$ be a partial ordering, $a \neq \emptyset$, such that for all $b \subset a, b \neq \emptyset$, if $\forall x \in b \forall y \in b(x \leq y \vee y \leq x)$, then b has an upper bound. Then a has a maximal element.

Proof Fix $(a, \leq)$ as in the hypothesis. Let

$$A = \{\{(b, x) : x \in b\} : b \subset a, b \neq \emptyset\}.$$

Notice that A exists, as it can be separated from $\mathscr{P}(\mathscr{P}(a) \times \bigcup \mathscr{P}(a))$. (AC), the axiom of choice, gives us some set f such that for all $y \in A$ there is some z with $y \cap f = \{z\}$, which means that for all $b \subset a$, $b \neq \emptyset$, there is some unique $x \in b$ such that $(b, x) \in f$. Therefore, f is a function from $\mathscr{P}(a)\backslash\{\emptyset\}$ to a such that $f(b) \in b$ for every $b \in \mathscr{P}(a)\backslash\{\emptyset\}$.

Let us now define a binary relation $\leq^*$ on a as follows.

We let W denote the set of all well-orderings $\leq'$ of subsets b of a such that for all $u, v \in b$, if $u \leq' v$, then $u \leq v$, and for all $u \in b$, writing

$$B_u^{\leq'} = \{w \in a : w \text{ is a } \leq\text{–upper bound of } \{v \in b : v <' u\}\},$$

$B_u^{\leq'} \neq \emptyset$ and $u = f(B_u^{\leq'})$. Notice that W may be separated from $\mathscr{P}(a^2)$.

Let us show that if $\leq', \leq'' \in W$, then $\leq' \subset \leq''$ or else $\leq'' \subset \leq'$. Let $\leq' \in W$ be a well-ordering of $b \subset a$, and let $\leq'' \in W$ be a well-ordering of $c \subset a$.

By Theorem 2.9, we may assume by symmetry that either $(b, \leq') \cong (c, \leq'')$ or else there is some $v \in c$ such that $(b, \leq') \cong (c, \leq'') \upharpoonright v$. Let $g : b \to c$ be such that

$$(b, \leq') \overset{g}{\cong} (c, \leq'') \text{ or } (b, \leq') \overset{g}{\cong} (c, \leq'') \upharpoonright v.$$

We aim to see that g is the identity on b.

Suppose not, and let $u_0 \in b$ be $\leq'$-minimal in

$$\{w \in b : g(w) \neq w\}.$$

Writing $\bar{g} = g \upharpoonright \{w \in b : w <' u_0\}$,

$$(b, \leq') \upharpoonright u_0 \overset{\bar{g}}{\cong} (c, \leq'') \upharpoonright g(u_0),$$

and $\bar{g}$ is in fact the identity on $\{w \in b : w <' u_0\}$, so that

$$\{w \in b : w <' u_0\} = \{w \in c : w <'' g(u_0)\}.$$

But then $B_{u_0}^{\leq'} = B_{g(u_0)}^{\leq''} \neq \emptyset$ and thus

$$u_0 = f(B_{u_0}^{\leq'}) = f(B_{g(u_0)}^{\leq''}) = g(u_0).$$

Contradiction!

We have shown that if $\leq', \leq'' \in W$, then $\leq' \subset \leq''$ or $\leq'' \subset \leq'$.

But now $\bigcup W$, call it $\leq^*$, is easily seen to be a well-ordering of a subset b of a. Setting

$$B = \{w \in a : w \text{ is a } \leq\text{–upper bound of } b\},$$

our hypothesis on $\leq$ gives us that $B \neq \emptyset$. Suppose that b does have a maximum with respect to $\leq$. We must then have $B \cap b = \emptyset$, and if we set

$$u_0 = f(B)$$

and $\leq^{**} = \leq^* \cup \{(u, u_0) : u \in b\} \cup \{(u_0, u_0)\}$, then $B = B_{u_0}^{\leq^{**}}$. It is thus easy to see that $\leq^{**} \in W$. This gives $u_0 \in b$, a contradiction!

Thus b has a maximum with respect to $\leq$. ZORN's Lemma is shown. □

The following is a special case of ZORN's lemma (cf. Problem 3.10).

Corollary 2.11 (Hausdorff Maximality Principle) *Let $a \neq \emptyset$, and let $A \subset \mathscr{P}(a)$ be such that for all $B \subset A$, if $x \subset y \vee y \subset x$ for all $x, y \in B$, then there is some $z \in A$ such that $x \subset z$ for all $x \in B$. Then A contains an $\subset$-maximal element.*

In the next chapter, we shall use the HAUSDORFF Maximality Principle to show that every set can be well-ordered (cf. Theorem 3.23).

It is not hard to show that in the theory ZF, (AC) is in fact equivalent with ZORN's Lemma, with the HAUSDORFF Maximality Principle, as well as with the assertion that every set can be well-ordered, i.e., that for every set x there is some well-order $<$ on x (cf. Problem 3.10).

2.2 Gödel–Bernays Class Theory

There is another axiomatization of set theory, BGC, which is often more convenient to use. Its language is the same one as $\mathscr{L}_\in$, except that in addition there is a second type of variables. The variables $x, y, z, \ldots, a, b, \ldots$ of $\mathscr{L}_\in$ are supposed to range over *sets*, whereas the new variables, $X, Y, Z, \ldots, A, B, \ldots$ are supposed to range over *classes*. Each set is a class, and a given class is a set iff it is a member of some class (equivalently, of some set). Classes which are not sets are called *proper classes*. Functions may now be proper classes. The axioms of the BERNAYS–GÖDEL *class theory* BG are (Ext), (Fund), (Pair), (Union), (Pow), (Inf) exactly as before together with the following ones:

$$\forall X \forall Y \forall x((x \in X \leftrightarrow x \in Y) \rightarrow X = Y) \tag{2.1}$$

$$\forall x \exists X\, x = X \tag{2.2}$$

$$\forall X\, (\exists Y\, X \in Y \leftrightarrow \exists x\, x = X) \tag{2.3}$$

If F is a (class) function, then $F''a$ is a set for each set a, (Rep*) and for all φ such that φ is a formula of the language of BG, which contains exactly $x, X_1, \ldots, X_k$ (but not Y) as its free variables and which does not have quantifiers ranging over classes (in other words, φ results from a formula φ' of the language of ZF by replacing free occurences of set variables by class variables), then

$$\forall X_1 \ldots X_k \exists Y \forall x (x \in Y \leftrightarrow \varphi). \qquad (\text{Comp}_\varphi)$$

(Comp_φ) is called the *comprehension axiom* for φ, and the collection of all (Comp_φ) is called the *comprehension schema*. The BERNAYS–GÖDEL *class theory with choice*, BGC, is the theory BG plus the following version of the axiom of choice:

$$\text{There is a (class) function } F \text{ such that } \forall x(x \neq \emptyset \to F(x) \in x). \qquad (\text{AC})$$

It can be shown that ZFC and BGC prove the same theorems in their common language $\mathscr{L}_\in$ (i.e., BGC is conservative over ZFC).

If φ is a formula as in (Comp_φ), then we shall write $\{x : \varphi\}$ for the class given by (Comp_φ). (Rep*) says that for all class functions F and for all sets a, $F''a = \{y : \exists x\, (x, y) \in F\}$ is a set.

We shall write V for the *universe of all sets*, i.e., for $\{x : x = x\}$. V cannot be a set, because otherwise

$$R = \{x \in V : x \notin x\}$$

would be a set, and then $R \in R$ iff $R \notin R$. This is another instantiation of RUSSELL's antinomy.

If A is a class, then we write

$$\bigcup A = \{x : \exists y \in A\, x \in y\}$$

and

$$\bigcap A = \{x : \forall y \in A\, x \in y\}.$$

$\bigcup A$ and $\bigcap A$ always exist, and $\bigcup \emptyset = \emptyset$ and $\bigcap \emptyset = V$.

It may be shown that in contrast to ZFC, BGC can be *finitely* axiomatized. BGC will be the theory used in this book.

The books [15, 18, 23] present introductions to axiomatic set theory.

2.3 Problems

2.1. Let $k \in \mathbb{N}$. Show that there cannot be sets $x_1, x_2, \ldots, x_k$ such that $x_1 \in x_2 \in \ldots \in x_k \in x_1$.

2.2. Show that for all x, y, (x, y) exists. Show that if $(x, y) = (x', y')$, then $x = x'$ and $y = y'$. Show that for all a, b, $a \times b$ exists (cf. Lemma 2.1). Show that for all $d, b, {}^d b$ exists (cf. Lemma 2.2). Which axioms of **ZF** do you need in each case? Show that (Pair) may be derived from the rest of the axioms of **ZF** (from which ones?).

2.3. Show that neither in (Aus$_\varphi$) nor in (Rep$_\varphi$), as formulated on p. 11, we could have allowed b to occur freely in φ. Show that the separation schema (Aus) can be derived from the rest of the axioms of **ZF** augmented by the statement $\exists x\, x = \emptyset$.

2.4. Show that the following "version" of (**AC**) is simply false:

$$\forall x (\forall y \in x\; y \neq \emptyset) \to \exists z \forall y \in x \exists u\; z \cap y = \{u\}).$$

2.5. Show tht every partial order can be extended to a linear order. More precisely: Let a be any set. Show that for any partial order $\leq$ on a there is a linear order $\leq'$ on a with $\leq \subset \leq'$.

2.6. Show that in the theory **ZF**, the following statements are equivalent.

(i) (**AC**).

(ii) For every x such that $y \neq \emptyset$ for every $y \in x$ there is a *choice function*, i.e., some $f : x \to \bigcup x$ such that $f(y) \in y$ for all $y \in x$.

2.7. (a) Let $\leq$ denote the natural ordering on $\mathbb{N}$, and let $m \in \mathbb{N}$, $m \geq 2$. Let the ordering $\leq^m$ on $\mathbb{N}$ be defined as follows. $n \leq^m n'$ iff either $n \equiv n' (\text{mod } m)$ and $n \leq n'$, or else if $k < m$, $k \in \mathbb{N}$, is least such that $n \equiv k (\text{mod } m)$ and $k' < m$, $k' \in \mathbb{N}$, is least such that $n' \equiv k' (\text{mod } m)$, then $k < k'$. Show that $\leq^m$ is a well-ordering on $\mathbb{N}$.
(b) Let, for $m \in \mathbb{N}$, $\leq^m$ be *any* well-ordering of $\mathbb{N}$, and let $\varphi : \mathbb{N} \to \mathbb{N} \times \mathbb{N}$ be a bijection. Let us define $\leq'$ on $\mathbb{N}$ by $n \leq' n'$ iff, letting $(m, q) = \varphi(n)$ and $(m', q') = \varphi(n')$, $m < m'$ or else $m = m'$ and $q \leq^m q'$. Show that $\leq'$ is a well-ordering of $\mathbb{N}$.

2.8. (**Cantor**) Let $(a, <)$ be a linear order. $(a, <)$ is called *dense* iff for all $x, y \in a$ with $x < y$ there is some $z \in a$ with $x < z < y$. Show that if $(a, <)$ is dense (and a has more than one element), then $<$ is not a well-ordering on a. $(a, <)$ is said to have *no endpoints* iff for all $x \in a$ there are $y, z \in a$ with $y < x < z$. Let $(a, <_a)$ and $(b, <_b)$ be two dense linear orders with no endpoints such that both a and b are countable. Show that $(a, <_a)$ is isomorphic to $(b, <_b)$. [Hint. Write $a = \{x_n : n \in \mathbb{N}\}$ and $b = \{y_n : n \in \mathbb{N}\}$, and construct $f : a \to b$ by recursively choosing $f(x_0)$, $f^{-1}(y_0)$, $f(x_1)$, $f^{-1}(y_1)$, etc.]

2.9. Show that there is a set A of pairwise non-isomorphic linear orders on $\mathbb{N}$ such that $A \sim \mathbb{R}$.

2.10. Show that every axiom of **ZFC** is provable in **BGC**.
Let us introduce ACKERMANN's *set theory*, **AST**. The language of **AST** arises from $\mathscr{L}_\in$ by adding a single constant, say $\dot{v}$. The axioms of **AST** are (Ext), (Fund), (Aus), as well as (Str) and (Refl) which are formulated as follows.

$$\forall x \in \dot{v}\, \forall y\, ((y \in x \vee y \subset x) \rightarrow y \in \dot{v}). \tag{Str}$$

Let φ be any formula of $\mathscr{L}_\in$ in which exactly $v_1, \ldots, v_k$ occur freely. Then $\varphi^{\dot{v}}$ results fromφ by replacing every occurence of $\forall x$ by $\forall x \in \dot{v}$ and every occurence of $\exists x$ by $\exists x \in \dot{v}$. Then

$$\forall v_1 \in \dot{v} \ldots \forall v_k \in \dot{v}\, (\varphi^{\dot{v}} \longleftrightarrow \varphi). \tag{Refl$_\varphi$}$$

(Refl) is the schema of all (Refl$_\varphi$), where φ is a formula of $\mathscr{L}_\in$ (in which $\dot{v}$ does not occur). (Str) states that $\dot{v}$ is "supertransitive," and (Refl) states (as a schema) that $\dot{v}$ is a fully elementary submodel of V, the universe of all sets.

2.11. **(W. Reinhardt)** Show that every axiom of ZF is provable in AST.
AST is also conservative over ZF, cf. Problem 5.15.

Chapter 3
Ordinals

3.1 Ordinal Numbers

The axiom of infinity (Inf) states there is an inductive set. Recall that a set x is called *inductive* iff $\emptyset \in x$ and for each $y \in x$, $y \cup \{y\} \in x$. Let us write 0 for $\emptyset$ and $y + 1$ for $y \cup \{y\}$. The axiom of infinity then says that there is a set x such that $0 \in x$ and for each $y \in x$, $y + 1 \in x$. We shall also write 1 for $0 + 1$, 2 for $(0 + 1) + 1$, etc. Each inductive set therefore contains 0, 1, 2, etc. We shall write ω for

$$\bigcap\{x : x \text{ is inductive}\}.$$

This set exists by (Inf) plus the separation scheme: if x_0 inductive, then

$$\omega = \{y \in x_0 : \forall x(x \text{ is inductive } \rightarrow y \in x)\}.$$

Clearly, ω is inductive. Intuitively, the set ω contains exactly 0, 1, 2, etc.

We have the following "principle of induction".

Lemma 3.1 *Let $A \subset \omega$ be such that $0 \in A$ and for each $y \in A$, $y + 1 \in A$. Then $A = \omega$.*

Proof A is inductive, hence $\omega \subset A$, and thus $A = \omega$. □

In particular, if φ is a statement such that $\varphi(0)$ and $\forall y \in \omega(\varphi(y) \rightarrow \varphi(y + 1))$ both hold true, then $\forall y \in \omega\ \varphi(y)$ holds true as well. We shall call elements of ω *natural numbers* and ω itself the *set of natural numbers*. All natural numbers as well as ω will be ordinals according to Definition 3.3.

Definition 3.2 A set x is *transitive* iff for each $y \in x$, $y \subset x$.

We shall see later (cf. Lemma 3.14) that every set is contained in a transitive set.

R. Schindler, *Set Theory*, Universitext, DOI: 10.1007/978-3-319-06725-4_3,

The following concept of an "ordinal" was isolated by JÁNOS NEUMANN (1903–1957) which is why ordinals are sometimes called NEUMANN *ordinals*.

Definition 3.3 A set x is called an *ordinal number*, or just an *ordinal*, iff x is transitive and for all $y, z \in x$ we have $y \in z \vee y = z \vee z \in y$.

Ordinals will typically be denoted by $\alpha, \beta, \gamma, \ldots, i, j, \ldots$ We shall write OR for the class $\{\alpha \colon \alpha \text{ is an ordinal}\}$ of all ordinals.

By (Fund), if α is an ordinal, then $\in \restriction \alpha = \{(x, y) \in \alpha^2 \colon x \in y\}$ is a well-order of α.

Lemma 3.4 *Each natural number is an ordinal.*

Proof by induction, i.e., by using Lemma 3.1: 0 is trivially an ordinal. Now let α be an ordinal. We have to see that $\alpha + 1$ is an ordinal. $\alpha + 1 = \alpha \cup \{\alpha\}$ is transitive: let $y \in \alpha \cup \{\alpha\}$; then either $y \in \alpha$ and hence $y \subset \alpha \subset \alpha \cup \{\alpha\}$ because α is transitive, or else $y = \alpha$ and hence $y \subset \alpha \cup \{\alpha\}$. Now let $y, z \in \alpha + 1 = \alpha \cup \{\alpha\}$. We have to see that $y \in z \vee y = z \vee z \in y$. If $y, z \in \alpha$, then this follows from the fact that α is an ordinal; if $y, z \in \{\alpha\}$, then this is trivial; but it is also trivial if $y \in \alpha$ and $z \in \{\alpha\}$ or vice versa. □

Lemma 3.5 *ω is an ordinal.*

Proof We first show $\forall y \in \omega \; y \subset \omega$ by induction. This is trivial for $y = 0$. Now fix $y \in \omega$ with $y \subset \omega$. Then $\{y\} \subset \omega$, hence $y + 1 = y \cup \{y\} \subset \omega$.

We now show $\forall y \in \omega \forall z \in \omega (y \in z \vee y = z \vee z \in y)$ by a "nested induction." Let us write $\varphi(y, z)$ for $y \in z \vee y = z \vee z \in y$. In order to prove $\forall y \in \omega \forall z \in \omega \, \varphi(y, z)$ it obviously suffices to show the conjunction of the following three statements:

(a) $\varphi(0, 0)$,
(b) $\forall z \in \omega (\varphi(0, z) \rightarrow \varphi(0, z + 1))$,
(c) $\forall y \in \omega (\forall z' \in \omega \, \varphi(y, z') \rightarrow \forall z \in \omega \, \varphi(y + 1, z))$

This is because if (a) and (b) hold true, then $\forall z \in \omega \, \varphi(0, z)$ holds true by induction. This, together with (c), yields $\forall y \in \omega \forall z \in \omega \varphi(y, z)$ again by induction.

(a) and (b) are trivial.

As to (c), let us fix $y \in \omega$, and let us suppose that $\forall z' \in \omega \, \varphi(y, z')$. We aim to show $\forall z \in \omega \, \varphi(y + 1, z)$, and we will do so by induction. We already know that $\forall z \in \omega \varphi(0, z)$, which in particular gives $\varphi(0, y + 1)$ and thus also $\varphi(y + 1, 0)$ by symmetry. Let us assume that $\varphi(y + 1, z)$ holds true to deduce that $\varphi(y + 1, z + 1)$ holds true as well.

We have that $y + 1 \in z \vee y + 1 = z \vee z \in y + 1$ by hypothesis. If $y + 1 \in z$, then $y + 1 \in z + 1 = z \cup \{z\}$. If $y + 1 = z$, then $y + 1 \in z + 1 = z \cup \{z\}$ as well. Now let $z \in y + 1 = y \cup \{y\}$. If $z \in \{y\}$, then $y + 1 = z + 1$. So suppose that $z \in y$. We have that $y \in z + 1 \vee y = z + 1 \vee z + 1 \in y$ by our hypothesis $\forall z' \in \omega \, \varphi(y, z')$. But $z \in y \in z + 1 = z \cup \{z\}$ contradicts the axiom of foundation (consider $\{z, y\}$). Therefore, $z \in y$ implies $y = z + 1 \vee z + 1 \in y$, and therefore $z + 1 \in y \cup \{y\} = y + 1$ as desired. □

Lemma 3.6 *The following statements are true.*

(a) 0 *is an ordinal, and if* α *is an ordinal, then so is* $\alpha + 1$.
(b) *If* α *is an ordinal and* $x \in \alpha$, *then* x *is an ordinal.*
(c) *If* α, β *are ordinals, and* $\alpha \subsetneq \beta$, *then* $\alpha \in \beta$.
(d) *If* α, β *are ordinals, then* $\alpha \subset \beta$ *or* $\beta \subset \alpha$ *(and hence* $\alpha \in \beta \vee \alpha = \beta \vee \beta \in \alpha$*).*

Proof (a) is given by the proof of Lemma 3.4 above.

(b) is easy.

To show (c), let α be a proper subset of β. Let $\gamma \in \beta \backslash \alpha$ such that $\gamma \cap (\beta \backslash \alpha) = \emptyset$. (There is such a γ by the axiom of foundation.) If $\xi \in \gamma$, then $\xi \in \beta$ by the transitivity of β, so $\xi \in \alpha$, as otherwise $\xi \in \gamma \cap (\beta \backslash \alpha)$. If $\xi \in \alpha \subset \beta$, then $\xi \in \gamma \vee \xi = \gamma \vee \gamma \in \xi$, because β is an ordinal. But $\xi = \gamma \vee \gamma \in \xi$ implies $\gamma \in \alpha$, because $\xi \in \alpha$ and α is an ordinal; however, $\gamma \in \beta \backslash \alpha$. Therefore if $\xi \in \alpha$, then $\xi \in \gamma$. We have shown that $\gamma = \alpha$. Hence $\alpha \in \beta$.

(d): Suppose not. Let $\alpha \in \mathrm{OR}$ be such that there is some $\beta \in \mathrm{OR}$ with $\neg(\alpha \subset \beta \vee \beta \subset \alpha)$. Let α_0 be $\in$–minimal in $\alpha + 1 = \alpha \cup \{\alpha\}$ such that there is some $\beta \in \mathrm{OR}$ with $\neg(\alpha_0 \subset \beta \vee \beta \subset \alpha_0)$, and let $\beta_0 \in \mathrm{OR}$ be such that $\neg(\alpha_0 \subset \beta_0 \vee \beta_0 \subset \alpha_0)$. Clearly, $\alpha_0 \cup \beta_0$ is transitive, and if $\delta, \delta' \in \alpha_0 \cup \beta_0$, then $\delta \subset \delta'$ or $\delta' \subset \delta$ by the choice of α_0, so that $\delta \in \delta' \vee \delta = \delta' \vee \delta' \in \delta$ by (b) and (c). Hence $\alpha_0 \cup \beta_0$ is an ordinal, call it γ_0. We claim that $\gamma_0 = \alpha_0$ or $\gamma_0 = \beta_0$. Otherwise by (c), $\alpha_0 \in \gamma_0$ and $\beta_0 \in \gamma_0$, so that one of $\alpha_0 \in \alpha_0, \beta_0 \in \beta_0, \alpha_0 \in \beta_0 \in \alpha_0$ holds true, which contradicts the axiom of foundation. We have shown that $\alpha_0 \subset \beta_0$ or $\beta_0 \subset \alpha_0$ which contradicts the choice of α_0 and β_0. □

By Lemma 3.6 (b) and (d), OR cannot be a set, as otherwise $\mathrm{OR} \in \mathrm{OR}$.

If $\alpha, \beta \in \mathrm{OR}$, then we shall often write $\alpha \leq \beta$ instead of $\alpha \subset \beta$ (equivalently, $\alpha \in \beta \vee \alpha = \beta$) and $\alpha < \beta$ instead of $\alpha \in \beta$. We shall also write (α, β), $[\alpha, \beta)$, $(\alpha, \beta]$, and $[\alpha, \beta]$ for the sets $\{\gamma : \alpha < \gamma < \beta\}$, $\{\gamma : \alpha \leq \gamma < \beta\}$, $\{\gamma : \alpha < \gamma \leq \beta\}$, and $\{\gamma : \alpha \leq \gamma \leq \beta\}$, respectively.

Lemma 3.7 *The following statements are true.*

(a) *If* $X \neq \emptyset$ *is a set of ordinals, then* $\bigcap X$ *is the minimal element of* X.
(b) *If* X *is a set of ordinals, then* $\bigcup X$ *is also an ordinal.*

Proof To show (a), notice that $\bigcap X$ is certainly an ordinal. If $\bigcap X$ is a proper subset of every element of X, then $\bigcap X \in \bigcap X$. Contradiction!

(b) is easy by the previous lemma. □

If X is a set of ordinals, then we also write $\min(X)$ for $\bigcap X$ (provided that $X \neq \emptyset$) and $\sup(X)$ for $\bigcup X$.

Definition 3.8 An ordinal α is called a *successor ordinal* iff there is some ordinal β such that $\alpha = \beta + 1$. An ordinal α is a *limit ordinal* iff α is not a successor ordinal.

3.2 Induction and Recursion

Definition 3.9 A binary relation $R \subset B \times B$ on a set or class B is called well–founded iff every nonempty $b \subset B$ has an R-least element, i.e., there is $x \in b$ such that for all $y \in b$, $\neg yRx$. If R is not well–founded, then we say that R is *ill–founded.*

We have the following "principle of induction" for well–founded relations:

Lemma 3.10 *Let $R \subset B \times B$ be well-founded, where B is a set. Let $A \subset B$ be such that for all $x \in B$, if $\{y \in B : yRx\} \subset A$, then $x \in A$. Then $A = B$.*

Proof Suppose that $B \backslash A \neq \emptyset$. Let $x \in B \backslash A$ be R–least, i.e., for all $y \in B \backslash A$, $\neg yRx$. In other words, if yRx, then $y \in A$. Then $x \in A$ by hypothesis. Contradiction! □

If B is a set, then $\in\restriction B = \{(x, y) \in B \times B : x \in y\}$ is well–founded by the axiom of foundation.

Lemma 3.11 *$R \subset B \times B$ is well–founded iff there is no $f : \omega \to B$ such that $f(n + 1)Rf(n)$ for all $n \in \omega$.*

Proof Suppose there is an $f : \omega \to B$ such that $f(n + 1)Rf(n)$ for all $n \in \omega$. Then $\text{ran}(f) \subset B$ doesn't have an R–least element.

Now suppose that R is not well–founded. Pick $b \subset B, b \neq \emptyset$ with no R–least element; i.e.; for all $x \in b, \{y \in b : yRx\} \neq \emptyset$. Apply the axiom of choice to the set $\{\{(y, x) : y \in b \wedge yRx\} : x \in b\}$ to get a set u such that for all $x \in b$, $u \cap \{(y, x) : y \in b \wedge yRx\} = \{(y', x)\}$ for some y'; write y_x for this unique y'. We may now define $f : \omega \to B$ as follows. Pick $x_0 \in b$, and set $f(0) = x_0$. Set $f(n) = y$ iff there is some $g : n+1 \to b$ such that $g(n) = y$, $g(0) = x_0$, and for all $i \in n$, $g(i + 1) = y_{g(i)}$.

Obviously, for each $n \in \omega$ there is at most one such g, and an easy induction yields that for each $n \in \omega$, there is at least one such g. But then f is well–defined, and of course $f(n + 1)Rf(n)$ for all $n \in \omega$. □

If $R \subset B \times B$, then the *well–founded part* $\text{wfp}(R)$ of B is the class of all $x \in B$ such that there is no infinite sequence $(x_n : n < \omega)$ such that $x_0 = x$ and $(x_{n+1}, x_n) \in R$ for all $n < \omega$.

The previous proof gave an example of a recursive definition. There is a general "recursion theorem." We state the NBG version of it which extends the ZFC version.

Definition 3.12 Let $R \subset B \times B$, where B is a class. R is then called *set–like* iff $\{x : (x, y) \in R\}$ is a set for all $y \in B$.

Theorem 3.13 *(Recursion) Let $R \subset B \times B$ be well–founded and set–like, where B is a class. Let $\mathbf{p}$ be a set,*[1] *and let $\varphi(v_0, v_1, v_2, \mathbf{p})$ be such that for all sets u and x*

[1] $\mathbf{p}$ will play the role of a parameter in what follows.

there is exactly one set y with $\varphi(u, x, y, \mathbf{p})$. There is then a (class) function F with domain B such that for all x in B, $F(x)$ is the unique y with

$$\varphi(F \restriction \{\bar{y} \in B : \bar{y} R x\}, x, y, \mathbf{p}).$$

Proof Let us call a (set or class) function F *good for* x iff

(a) $x \in \text{dom}(F) \subset B$,
(b) $\forall x' \in \text{dom}(F) \forall \bar{y} \in B\ (\bar{y} R x' \rightarrow \bar{y} \in \text{dom}(F))$, and
(c) $\forall x' \in \text{dom}(F)\ (F(x')$ is the unique y with $\varphi(F \restriction \{\bar{y} \in B : \bar{y} R x'\}, x', y, \mathbf{p})$.

If F, F' are both good for x, then $F(x) = F'(x)$, as we may otherwise consider the R–least $x_0 \in \text{dom}(F) \cap \text{dom}(F')$ with $F(x_0) \neq F'(x_0)$ and get an immediate contradiction. For all x for which there is a *set* function $f \in V$ which is good for x,

$$\bigcap \{f \in V : f \text{ is good for } x\},$$

which we shall ad hoc denote by f^x, is then easily seen to be good for x.

We claim that for each $x \in B$, there is some set function $g \in V$ which is good for x. Suppose not, and let x_0 be R–least in the class of all x such that there is no set function $g \in V$ which is good for x. Then g^x exists for all $x R x_0$, and we may consider

$$g = \bigcup \{g^x : x R x_0\}.$$

As R is set-like, g is a set by the appropriate axiom of replacement. Moreover g is a function which is good for each $x \in B$ with $x R x_0$. Now let y be unique such that

$$\varphi(g \restriction \{x \in B : x R x_0\}, x_0, y, \mathbf{p}),$$

and set $g^* = g \cup \{(x_0, y)\}$. Then $g^* \in V$ is good for x_0. Contradiction!

We may now simply let

$$F = \bigcup \{f^x : x \in B\}.$$

Then F is a (class) function which is good for all $x \in B$. □

Lemma 3.14 *For every set x there is a transitive set y such that $x \in y$ and $y \subset y'$ for all transitive sets y' with $x \in y'$.*

Proof We use the recursion Theorem 3.13 to construct a function with domain ω such that $f(0) = \{x\}$ and $f(n+1) = \bigcup f(n)$ for $n < \omega$, and we consider $\bigcup \text{ran}(f)$. □

Definition 3.15 Let x be a set, and let y be as in Lemma 3.14. Then y is called the *transitive closure of* $\{x\}$, denoted by $\text{TC}(\{x\})$.

The following Lemma says that the $\in$–relation, restricted to any class, is well–founded.

Lemma 3.16 *Let A be a non-empty class. Then A has an $\in$-minimal member, i.e., there is some $a \in A$ with $a \cap A = \emptyset$.*

Proof Let $x \in A$ be arbitrary, and let y be a transitive set with $x \in y$. As $y \cap A \neq \emptyset$ is a set, the axiom of foundation gives us some $a \in y \cap A$ which is $\in$–minimal, i.e., $a \cap (y \cap A) = \emptyset$. Then $a \in A$, and if $z \in a$, then $z \in y$ (as y is transitive), so $z \notin A$. That is, $a \cap A = \emptyset$. □

Lemma 3.17 *Let B be a class, and let $R \subset B \times B$ be set-like. Then R is well–founded if and only if there is some (unique) α which is either an ordinal or else $\alpha = \mathrm{OR}$ and some (unique) $\rho\colon B \to \alpha$ such that $\rho(x) = \sup(\{\rho(y)+1\colon yRx\})$ for all $x \in B$.*

Proof Let us first suppose that R is well–founded. We may then apply the recursion theorem 3.13 to the formula $\varphi(u, x, y)$ which says that $y = \sup(\{u(y) + 1\colon y \in \mathrm{dom}(u)\})$ if u is a function whose range is contained in OR and $y = \emptyset$ otherwise. We then get an α and a function ρ as desired.

On the other hand, if $\rho\colon B \to \alpha$ is such that $\rho(x) = \sup(\{\rho(y) + 1\colon yRx\})$ for all $x \in B$, then in particular yRx implies that $\rho(y) < \rho(x)$, so that R must be well–founded. □

Definition 3.18 If $R \subset B \times B$ is well–founded and set like, and if α and $\rho\colon B \to \alpha$ are as in Lemma 3.17, then $\rho(x)$ is called the *R–rank of* $x \in B$, written $\mathrm{rk}_R(x)$ or $||x||_R$, and α is called the *rank of* R, written $||R||$.

Definition 3.19 The hierarchy $(V_\alpha\colon \alpha \in \mathrm{OR})$ is recursively defined by

$$V_\alpha = \bigcup\{\mathscr{P}(V_\beta)\colon \beta < \alpha\}. \tag{3.1}$$

We call V_α a *rank initial segment of* V.

Cf. Problem 3.1.

Definition 3.20 A binary relation $R \subset B \times B$ on a class B is called *extensional* iff for all $x, y \in B$,

$$\{z \in B\colon zRx\} = \{z \in B\colon zRy\} \Longleftrightarrow x = y.$$

By the axiom(s) of extensionality (and foundation), $\in\restriction B$ is (well–founded and) extensional for every set B.

The function π_R as in the following theorem is often called the "transitive collapse."

Theorem 3.21 (Mostowski Collapse) *Let B be a class. Let $R \subset B \times B$ be well–founded, extensional, and set–like. There is then a unique pair (X_R, π_R) such that X_R is transitive, $\pi_R : X_R \to B$ is bijective, and for all $x, y \in X_R$, $x \in y \Longleftrightarrow \pi_R(x) R \pi_R(y)$.*

Proof Apply the recursion theorem to the formula $\varphi(u, x, y) \equiv y = \text{ran}(u)$. We then get a function F with domain B such that for all $x \in B$, $F(x) = \{F(\bar{y}) : \bar{y}Rx\}$. Notice that F is injective, because R is extensional. We may then set $X_R = \text{ran}(F)$ and $\pi_R = F^{-1}$. □

In particular, we get that well–orderings are well–founded relations whose transitive collapse is an ordinal. Notice that if $R \subset B \times B$ is a well–ordering, then R is automatically extensional, so that we may indeed apply MOSTOWSKI's theorem to R. The reason is that if $\{z \in B : zRx\} = \{z \in B : zRy\}$ and $x \neq y$, then xRy, say, and so xRx; but then R would not be well–founded.

Definition 3.22 If R is a well–ordering on B, then the unique ordinal α such that there is some isomorphism $(\alpha; <\restriction \alpha) \overset{\pi}{\cong} (B; R)$ is called the *length* or the *order type of* R, denoted by $\text{otp}(R)$. If A is a set of ordinals, then we also denote by $\text{otp}(A)$ the order type of $<\restriction A$ and call it the *order type of* A. The isomorphism $(\text{otp}(A); <\restriction \text{otp}(A)) \overset{\pi}{\cong} (A; <\restriction A)$ is also called the *monotone enumeration of* A.

Theorem 3.23 (Zermelo) *Let A be any set. There is then a well–ordering on A. There is even an ordinal α and some bijection $\pi : \alpha \to A$.*

Proof We use the HAUSDORFF Maximality Principle 2.11 to show that there is a bijection $\pi : \alpha \to A$ for some ordinal α. Let F be the set of all injections $\sigma : \beta \to A$, where β is an ordinal.

F is indeed a set by the following argument. For each $\sigma : \beta \to A$, $R \subset A \times A$ is a well–ordering on $\text{ran}(\sigma)$, where we define $xRy \iff \sigma^{-1}(x) \in \sigma^{-1}(y)$ for $x, y \in \text{ran}(\sigma)$; but any such well–ordering is in $\mathscr{P}(A \times A)$. Conversely, any well–ordering R on a subset B of A induces a unique injection $\sigma : \beta \to A$ with $B = \text{ran}(\sigma)$ and $xRy \iff \sigma^{-1}(x) \in \sigma^{-1}(y)$ for $x, y \in \text{ran}(\sigma)$ by MOSTOWSKI's Theorem 3.21. Therefore, as $\mathscr{P}(A \times A)$ is a set, F is a set by the appropriate axiom of replacement.

Let $K \subset F$ be such that $\sigma \subset \tau$ or $\tau \subset \sigma$ (i.e., $\sigma \restriction \text{dom}(\tau) = \tau$ or $\tau \restriction \text{dom}(\sigma) = \sigma$) whenever $\sigma, \tau \in K$. Then $\bigcup K \in F$, as $\bigcup K$ is a function, $\text{dom}(\bigcup K) = \bigcup\{\text{dom}(\sigma) : \sigma \in K\}$ is an ordinal, and $\bigcup K$ is injective. Hence F satisfies the hypothesis of the HAUSDORFF Maximality Principle, Corollary 2.11, and there is some $\pi \in F$ such that for no $\sigma \in F$, $\pi \subsetneq \sigma$.

But now we must have $\text{ran}(\pi) = A$. Otherwise let $x \in A \setminus \text{ran}(\pi)$, and set $\sigma = \pi \cup \{(\text{dom}(\pi), x)\}$. Then $\sigma \in F$ (with $\text{dom}(\sigma) = \text{dom}(\pi) + 1$), $\pi \subsetneq \sigma$. Contradiction! □

If $f : \alpha \to A$, where α is an ordinal (or $\alpha = \text{OR}$), then f is also called a *sequence* and we sometimes write $(f(\xi) : \xi < \alpha)$ instead of f.

3.3 Problems

3.1 Use the recursion theorem 3.13 to show that there is a sequence $(V_\alpha : \alpha \in \text{OR})$ which satisfies (3.1). Show that every V_α is transitive and that $V_\beta \subset V_\alpha$ for $\beta \leq \alpha$.

Show that $V_0 = \emptyset$, $V_{\alpha+1} = \mathscr{P}(V_\alpha)$ for every α, and $V_\lambda = \bigcup_{\alpha<\lambda} V_\alpha$ for every limit ordinal λ.

3.2 Show that for every set x there is some α with $x \subset V_\alpha$ (and thus $x \in V_{\alpha+1}$). For any set x, let $\mathrm{rk}_\in(x)$ be as in Definition 3.18 for $B = V$ and $R = \in = \{(x, y) \colon x \in y\}$. Show that for every set x, the least α such that $x \subset V_\alpha$ is equal to $\mathrm{rk}_\in(x)$.
$\mathrm{rk}_\in(x)$ is called the *(set, or $\in$–) rank of* x, also just written $\mathrm{rk}(x)$.

3.3 If M is transitive, then we may construe $(M; \in\restriction M)$ as a model of $\mathscr{L}_\in$. Which axioms of **ZFC** hold true in all $(V_\alpha; \in\restriction V_\alpha)$, where $\alpha > \omega$ is a limit ordinal? Which ones hold true in $(V_\omega; \in\restriction V_\omega)$?

3.4 For a formula φ, let (Fund_φ) be the following version of the axiom of foundation.

$$\forall \mathbf{p}(\exists x\, \varphi(x, \mathbf{p}) \to \exists x(\varphi(x, \mathbf{p}) \wedge \forall y \in x\, \neg\varphi(y, \mathbf{p}))). \qquad (\mathrm{Fund}_\varphi)$$

Show that every instance of (Fund_φ) is provable in **ZFC**.

3.5 Let φ be a formula of $\mathscr{L}_\in$ in which exactly the variables $x, y, v_1, \ldots, v_p$ (all different from b) occur freely. The *collection principle* corresponding to φ, (Coll_φ) runs as follows.

$$\forall v_1 \ldots \forall v_p\, ((\forall x \exists y\, \varphi) \to (\forall a \exists b \forall x \in a \exists y \in b\, \varphi)). \qquad (\mathrm{Coll}_\varphi)$$

The *collection principle* is the set of all (Coll_φ). Show that in the theory **Z**, the collection principle is equivalent to the replacement schema (Rep).

3.6 Let α be an ordinal. Use the recursion Theorem 3.13 to show that there are functions $\beta \mapsto \alpha + \beta$, $\beta \mapsto \alpha \cdot \beta$, and $\beta \mapsto \alpha^\beta$ with the following properties.

(a) $\alpha + 0 = \alpha$, $\alpha + (\beta + 1) = (\alpha + \beta) + 1$ for all β, and $\alpha + \lambda = \sup(\{\alpha + \beta \colon \beta < \lambda\})$ for λ a limit ordinal.
(b) $\alpha \cdot 0 = 0$, $\alpha \cdot (\beta + 1) = (\alpha \cdot \beta) + \alpha$ for all β, and $\alpha \cdot \lambda = \sup(\{\alpha \cdot \beta \colon \beta < \lambda\})$ for λ a limit ordinal.
(c) $\alpha^0 = 1$, $\alpha^{\beta+1} = (\alpha^\beta) \cdot \alpha$ for all β, and $\alpha^\lambda = \sup(\{\alpha^\beta \colon \beta < \lambda\})$ for λ a limit ordinal.

3.7 Show that $+$ and $\cdot$ are associative. Show also that $\omega = 1 + \omega \neq \omega + 1$ and $\omega = 2 \cdot \omega \neq \omega \cdot 2$. Show that if λ is a limit ordinal $\neq 0$, then $\alpha + \lambda$ is a limit ordinal. Show that if γ is a successor ordinal, then $\alpha + \gamma$ is a successor ordinal. Show that if λ is a limit ordinal $\neq 0$, then $\alpha \cdot \lambda$ and $\lambda \cdot \alpha$ are limit ordinals.

3.8 Show that if $\alpha > 0$ is an ordinal, then there are unique positive natural numbers k and $c_1, \ldots, c_k$ and ordinals $0 \leq \beta_1 < \ldots < \beta_k$ such that

$$\alpha = \omega^{\beta_k} \cdot c_k + \ldots + \omega^{\beta_1} \cdot c_1. \qquad (3.2)$$

The representation (3.2) is called Cantor *normal form* of α.

3.9 Use (AC) to show the following statement, called the principle of *dependent choice*, DC. Let R be a binary relation on a set a such that for every $x \in a$ there is some $y \in a$ such that $(y, x) \in R$. Show that there is some function $f : \omega \to a$ such that for every $n \in \omega$, $(f(n+1), f(n)) \in R$. Use DC to prove Lemma 3.11.

3.10 Show that in the theory ZF, the following statements are equivalent.

(i) (AC).
(ii) ZORN's Lemma, i.e., Theorem 2.10.
(iii) The HAUSDORFF Maximality Principle, i.e., Corollary 2.11.
(iv) ZERMELO's Well–Ordering Theorem 3.23.

3.11 Show in ZC that for every set a there is some r such that r is a well–ordering of a.

3.12 (**F. Hartogs**) Show in ZF that for every set x there is an ordinal α such that there is no injection $f : \alpha \to x$. [Hint. Consider the set W of all well–orders of subsets of a, and well–order W via Theorem 2.9.]

Chapter 4
Cardinals

4.1 Regular and Singular Cardinal Numbers

We know by ZERMELO's Theorem 3.23 that for each set x there is an ordinal α such that $x \sim \alpha$, i.e., there is a bijection $f : x \to \alpha$.

Definition 4.1 Let x be a set. The cardinality of x, abbreviated by $\overline{\overline{x}}$, or $\text{Card}(x)$, is the least ordinal α such that $x \sim \alpha$.

Notice that $\text{Card}(x)$ exists for every set x. Namely, let $x \sim \alpha$. Then either $\alpha = \text{Card}(x)$, or else $\text{Card}(\alpha)$ is the least $\beta < \alpha$ such that $x \sim \beta$.

To give a few examples, $\text{Card}(n) = n$ for every $n \in \omega$; $\text{Card}(\omega) = \omega = \text{Card}(\omega+1) = \text{Card}(\omega+2) = \ldots = \text{Card}(\omega+\omega) = \ldots = \text{Card}(\omega \cdot \omega) = \ldots = \text{Card}(\omega^{\omega}) = \ldots$[1] We shall see more examples later.

Definition 4.2 An ordinal α is called a cardinal iff $\alpha = \overline{\overline{\alpha}}$.

Obviously, α is a cardinal iff there is some set x such that $\alpha = \text{Card}(x)$. We shall typically use the letters $\kappa, \lambda, \mu, \ldots$ to denote cardinals.

By CANTOR's Theorem 1.3, if x is any set, then there is no surjection $f : x \to \mathscr{P}(x)$. Therefore, if κ is a cardinal, then there is a cardinal $\lambda > \kappa$, and there is thus also a *least* cardinal $\lambda > \kappa$ which may also be identified as the least cardinal λ with $\kappa < \lambda \leq \overline{\overline{\mathscr{P}(\kappa)}}$.

The *Pigeonhole Principle* says that if κ and λ are cardinals with $\lambda > \kappa$ and if $f : \lambda \to \kappa$, then f cannot be injective.

Definition 4.3 Let κ be a cardinal. The least cardinal $\lambda > \kappa$ is called the cardinal successor of κ, abbreviated by κ^+. A cardinal κ is called a successor cardinal iff there is some cardinal $\mu < \kappa$ with $\kappa = \mu^+$; otherwise κ is called a limit cardinal.

All positive natural numbers are therefore successor cardinals, ω is a limit cardinal, $\omega^+, \omega^{++}, \ldots$ are successor cardinals, etc.

---

[1] We here use the notation for ordinal arithmetic from Problem 3.6.

R. Schindler, *Set Theory*, Universitext, DOI: 10.1007/978-3-319-06725-4_4,

The cardinal successor κ^+ of a given cardinal κ my also be characterized as the set of all ordinals of at most the same size as κ (cf. Problem 4.2).

As κ^+ exists for each cardinal κ, there are arbitrarily large successor cardinals. But there are also arbitrarily large limit cardinals.

Lemma 4.4 *Let X be a set of cardinals. Then $\bigcup X$ is a cardinal.*

Proof By Lemma 3.7 (b) we already know that $\bigcup X$ is an ordinal. We have to show that there is no $\alpha < \bigcup X$ such that $\alpha \sim \bigcup X$. Well, if $\alpha < \bigcup X$, i.e., $\alpha \in \bigcup X$, then there is some $\kappa \in X$ with $\alpha \in \kappa$, i.e., $\alpha < \kappa$. As κ is a cardinal, there is no surjection from α onto κ. But $\kappa \in X$ gives $\kappa \subset \bigcup X$, so that there is also no surjection from α onto $\bigcup X$. □

In particular, if κ is any cardinal, then $X = \{\kappa, \kappa^+, \kappa^{++}, \ldots\}$ exists by the replacement schema and we have that $\bigcup X$ is a limit cardinal $> \kappa$. There are therefore arbitrarily large limit cardinals.

If $X \neq \emptyset$ is a set of ordinals (or cardinals), then we also write $\sup(X)$ for $\bigcup X$ and $\min(X)$ for the least element of X, i.e., $\min(X) = \bigcap(X)$.

We may now recursively, i.e., by exploiting Theorem 3.13, define the $\aleph$-sequence as follows. $\aleph_0 = \omega$, the least infinite cardinal, and for $\alpha > 0$, $\aleph_\alpha =$ the least cardinal κ such that $\kappa > \aleph_\beta$ for all $\beta < \alpha$.

The first infinite cardinals are therefore

$$\aleph_0, \aleph_1, \aleph_2, \ldots, \aleph_\omega, \aleph_{\omega+1}, \ldots, \aleph_{\omega\cdot 2}, \ldots, \aleph_{\omega^\omega},$$

etc.[2]

An easy induction shows that $\alpha \leq \aleph_\alpha$ for every ordinal α. In particular, if κ is an infinite cardinal, then $\kappa \leq \aleph_\kappa$, so that there is some ordinal $\alpha \leq \kappa$ with $\kappa = \aleph_\alpha$. Every infinite cardinal is thus of the form $\aleph_\alpha$, where α is an ordinal.

We define cardinal addition, multiplication, and exponentiation as follows. By tradition, these operations are denoted the same way as ordinal addition, multiplication, and exponentiation, respectively, (cf. Problem 3.6) but it is usually clear from the context which one we refer to.

Definition 4.5 Let κ, λ be cardinals. We set

$$\begin{aligned} \kappa + \lambda &= \text{Card}((\kappa \times \{0\}) \cup (\lambda \times \{1\})), \\ \kappa \cdot \lambda &= \text{Card}(\kappa \times \lambda), \text{ and} \\ \kappa^\lambda &= \text{Card}({}^\lambda\kappa) = \text{Card}(\{f : f \text{ is a function with } \text{dom}(f) = \lambda \text{ and} \\ &\qquad \text{ran}(f) \subset \kappa\}). \end{aligned}$$

It is easy to verify that $\kappa + \lambda = \text{Card}(X \cup Y)$, whenever X, Y are disjoint sets with $\overline{\overline{X}} = \kappa$ and $\overline{\overline{Y}} = \lambda$. It is also easy to verify that if $\kappa, \lambda \geq 2$, then $\kappa + \lambda \leq \kappa \cdot \lambda$. Cf. Problem 4.3. Moreover, we have the usual rules for addition, multiplication and exponentiation.

---

[2] Again, we use the notation from Problem 3.6.

If κ is a cardinal and A is a set, then we write $[A]^\kappa$ for $\{x \subset A\colon \text{Card}(x) = \kappa\}$, and we also write $[A]^{<\kappa} = \bigcup_{\mu<\kappa}[A]^\mu$ and $[A]^{\leq\kappa} = \bigcup_{\mu\leq\kappa}[A]^\mu$. Trivially, $[A]^{\leq \text{Card}(A)} = \mathscr{P}(A)$. It is not hard to verify that $\kappa^\lambda = \text{Card}([\kappa]^\lambda)$ for cardinals κ, λ (cf. Problem 4.4).

We now want to verify that $\aleph_\alpha \cdot \aleph_\alpha = \aleph_\alpha$ for every α (cf. Theorem 4.6). For this purpose we need the GÖDEL pairing function.

We define an ordering $\leq$ on OR $\times$ OR as follows. We set $(\delta, \varepsilon) \leq (\delta', \varepsilon')$ iff either

(a) $\max\{\delta, \varepsilon\} < \max\{\delta', \varepsilon'\}$, or else
(b) $\max\{\delta, \varepsilon\} = \max\{\delta', \varepsilon'\}$, and $\delta < \delta'$, or else
(c) $\max\{\delta, \varepsilon\} = \max\{\delta', \varepsilon'\}$, $\delta = \delta'$, and $\varepsilon \leq \varepsilon'$.

We claim that $\leq$ is a well-ordering on OR $\times$ OR. We need to see that each non-empty $X \subset$ OR $\times$ OR has a $\leq$-least element. Let $X \neq \emptyset$ be given, $X \subset$ OR $\times$ OR. We let $X^0 = \{(\delta, \varepsilon) \in X : \forall(\delta', \varepsilon') \in X \ \max\{\delta, \varepsilon\} \leq \max\{\delta', \varepsilon'\}\}$; we let $X^1 = \{(\delta, \varepsilon) \in X^0\colon \forall(\delta', \varepsilon') \in X^0 \ \delta \leq \delta'\}$; finally we let $X^2 = \{(\delta, \varepsilon) \in X^1 : \forall(\delta', \varepsilon') \in X^1 \ \varepsilon \leq \varepsilon'\}$. Obviously, X^2 contains exactly one element and it is $\leq$-least in X.

It is easy to see that $\leq$ is set-like. Using Theorem 3.21 (cf. also Definition 3.22), we may therefore let $\pi\colon$ OR $\times$ OR $\to$ OR be the transitive collapse of $\leq$. I.e., π is a bijection such that $(\delta, \varepsilon) \leq (\delta', \varepsilon')$ iff $\pi((\delta, \varepsilon)) \leq \pi((\delta', \varepsilon'))$ (where the latter "$\leq$" denotes the usual well-ordering on ordinals).

Notice that $\pi \restriction (\gamma \times \gamma)\colon \gamma \times \gamma \to \pi(\gamma)$ is bijective for every γ. It is easy to verify that $\pi(\gamma) \geq \gamma$ for every γ. In what follows, we shall sometimes write $\langle \delta, \varepsilon \rangle$ for $\pi((\delta, \varepsilon))$. The map $\delta, \varepsilon \mapsto \langle \delta, \varepsilon \rangle$ is called the GÖDEL *pairing function*.

Theorem 4.6 (Hessenberg) *For every α, $\aleph_\alpha \cdot \aleph_\alpha = \aleph_\alpha$.*

Proof We use the notation from the preceeding paragraphs. One easily shows that $\pi(\aleph_0) = \aleph_0$, so that $\pi \restriction (\aleph_0 \times \aleph_0)$ witnesses that $\aleph_0 \times \aleph_0 \sim \aleph_0$, i.e., $\aleph_0 \cdot \aleph_0 = \aleph_0$.

Now suppose that there is some α with $\aleph_\alpha \cdot \aleph_\alpha > \aleph_\alpha$, and let us fix the least such α. We then must have $\alpha > 0$ and $\pi(\aleph_\alpha) > \aleph_\alpha$. Say $\pi((\delta, \varepsilon)) = \aleph_\alpha$, where $\delta, \varepsilon < \aleph_\alpha$. Let $\rho < \aleph_\alpha$ be such that $\delta, \varepsilon < \rho$. Then $\text{ran}(\pi \restriction ((\rho + 1) \times (\rho + 1))) \supset \aleph_\alpha$, so that in particular there is a surjection $f\colon (\rho + 1) \times (\rho + 1) \to \aleph_\alpha$. Now $\rho + 1 < \aleph_\alpha$, say $\text{Card}(\rho + 1) = \aleph_\beta$, where $\beta < \alpha$. We have $\aleph_\beta \cdot \aleph_\beta = \aleph_\beta$ by the choice of α, so that there is a surjection $g : \aleph_\beta \to \aleph_\beta \times \aleph_\beta$, and hence also a surjection $g^*\colon \aleph_\beta \to (\rho + 1) \times (\rho + 1)$. But then $f \circ g^*\colon \aleph_\beta \to \aleph_\alpha$ is surjective, contradicting the fact that $\beta < \alpha$ and $\aleph_\alpha$ is a cardinal. □

HESSENBERG's Theorem 4.6 yields that cardinal addition and multiplication are trivial.

Corollary 4.7 *For all α, β, $\aleph_\alpha + \aleph_\beta = \aleph_\alpha \cdot \aleph_\beta = \aleph_{\max\{\alpha,\beta\}}$.*

Proof Assume without loss of generality that $\alpha \leq \beta$. Then

$$\aleph_\beta \leq \aleph_\alpha + \aleph_\beta \leq \aleph_\alpha \cdot \aleph_\beta \leq \aleph_\beta \cdot \aleph_\beta = \aleph_\beta,$$

the last equality being true by HESSENBERG's Theorem 4.6. □

Cardinal exponentiation is a different matter.

Lemma 4.8 *For all* κ, $2^\kappa = \overline{\overline{\mathscr{P}(\kappa)}}$.

Proof 2^κ = the cardinality of the set of all functions $f : \kappa \to 2 = \{0, 1\}$, which is the same as the cardinality of the set $\mathscr{P}(\kappa)$. □

Corollary 4.9 $\kappa^+ \leq 2^\kappa$.

CANTOR's *Continuum Hypothesis*, abbreviated by CH, may now be restated as $2^{\aleph_0} = \aleph_1 (= \aleph_0^+)$. The assertion that

$$\forall\alpha\ 2^{\aleph_\alpha} = \aleph_{\alpha+1} \tag{4.1}$$

is called the *Generalized Continuum Hypothesis* and is abbreviated by GCH. We shall see that GCH as well as ¬CH are consistent with ZFC (cf. Theorems 5.31 and 6.33).

Definition 4.10 Let α be an ordinal. A function $f : A \to \alpha$ is called cofinal in α iff for all $\beta < \alpha$ there is some $a \in A$ such that $f(a) \geq \beta$. The *cofinality* of α, written $\mathrm{cf}(\alpha)$, is the least $\beta \leq \alpha$ such that there is a cofinal $f : \beta \to \alpha$.

Notice that $\mathrm{cf}(\alpha)$ is defined for all α, as the identity on α is always cofinal in α. $\mathrm{cf}(\alpha + 1) = 1$ for all α, so that $\mathrm{cf}(\alpha)$ is only interesting for limit ordinals α.

For instance, $\mathrm{cf}(\omega) = \omega = \mathrm{cf}(\omega + \omega) = \mathrm{cf}(\omega \cdot 3) = \ldots = \mathrm{cf}(\omega \cdot \omega) = \ldots = \mathrm{cf}(\omega^\omega)$.

The fact that $\beta = \mathrm{cf}(\alpha)$ is witnessed by a monotone function as follows. Let $\beta = \mathrm{cf}(\alpha)$. Let $f : \beta \to \alpha$ be cofinal. Define $f^* : \beta \to \alpha$ as follows: $f^*(\xi) = \sup\{f(\eta) : \eta < \xi\}$ for $\xi < \beta$. Notice that in fact for every $\xi < \beta$, $f^*(\xi) < \alpha$, as otherwise $f \upharpoonright \xi$ would witness that $\mathrm{cf}(\alpha) \leq \xi < \beta = \mathrm{cf}(\alpha)$. f^* is thus well-defined and cofinal, and of course if $\xi \leq \xi'$, then $f^*(\xi) \leq f^*(\xi')$. If $\pi : \gamma \to \mathrm{ran}(f^*)$ is the monotone enumeration of $\mathrm{ran}(f^*)$, then $\gamma \leq \mathrm{cf}(\alpha)$ and thus $\gamma = \mathrm{cf}(\alpha)$. Of course, π is then strictly monotone. π is also *continuous* in the sense that for all limit ordinals $\gamma \in \mathrm{dom}(\pi)$, $\pi(\gamma) = \sup(\{\pi(\bar{\gamma}) : \bar{\gamma} < \gamma\})$.

Definition 4.11 Let α be an ordinal. α is called regular iff $\mathrm{cf}(\alpha) = \alpha$, and α is called singular iff $\mathrm{cf}(\alpha) < \alpha$.

Examples of singular cardinals are $\aleph_\omega$, $\aleph_{\omega+\omega}$, etc., or more generally all $\aleph_\lambda$ where λ is a limit ordinal with $\lambda < \aleph_\lambda$. However, $\lambda = \aleph_\lambda$ does not imply that λ is regular, cf. Problem 4.5.

Lemma 4.12 *For every ordinal* α, $\mathrm{cf}(\alpha)$ *is regular.*

Proof Let $\beta = \mathrm{cf}(\alpha)$. We need to see that $\mathrm{cf}(\beta) = \beta$. Let $f : \beta \to \alpha$ be cofinal, and let $g : \mathrm{cf}(\beta) \to \beta$ be cofinal. By the above observation, we may and shall assume that f is monotone. Consider $f \circ g : \mathrm{cf}(\beta) \to \alpha$. If $\xi < \alpha$, then there is some $\eta < \beta$ with $f(\eta) \geq \xi$, and then there is some $\zeta < \mathrm{cf}(\beta)$ with $g(\zeta) \geq \eta$. But then $f \circ g(\zeta) = f(g(\zeta)) \geq f(\eta) \geq \xi$ by the monotonicity of f. I.e., $f \circ g$ is cofinal, so that $\beta = \mathrm{cf}(\beta)$. □

Lemma 4.13 *Let α be regular; then α is a cardinal.*

Proof Every bijection (or just surjection) $f: \overline{\overline{\alpha}} \to \alpha$ is cofinal. □

Corollary 4.14 *For every ordinal α, $\mathrm{cf}(\alpha)$ is a regular cardinal.*

Lemma 4.15 *Let κ be an infinite successor cardinal. Then κ is regular.*

Proof Let $\kappa = \mu^+$. Suppose that $\mathrm{cf}(\kappa) < \kappa$, i.e., $\mathrm{cf}(\kappa) \leq \mu$ by Lemma 4.13. Let $f : \mu \to \kappa$ be cofinal. Let $(g_\xi : \xi < \mu)$ be such that for each $\xi < \mu$, $g_\xi : \mu \to f(\xi)$ is surjective. (Here we use AC, the axiom of choice.) Let $h : \mu \to \mu \times \mu$ be bijective (cf. Theorem 4.6). We may then define a surjection $F : \mu \to \kappa$ as follows. Let $\eta < \mu$. Let $(\alpha, \beta) = h(\eta)$, and set $F(\eta) = g_\alpha(\beta)$. But because $\mu < \kappa$ and κ is a cardinal, there can't be such a surjection. □

We thus get that $\aleph_0, \aleph_1, \aleph_2, \ldots$ are all regular, whereas $\mathrm{cf}(\aleph_\omega) = \omega < \aleph_\omega$. $\aleph_{\omega+1}$ is again regular, etc.

Felix Hausdorff (1868–1942) asked whether every limit cardinal is singular. This question leads to the concept of "large cardinals," which will be discussed in detail below and in later chapters, cf. Definitions 4.41, 4.42, 4.48, 4.49, 4.54, 4.60, 4.62, 4.68, and 10.76. They are ubiquitous is current day set theory.

We now want to look at $\kappa, \lambda \mapsto \kappa^\lambda$.

Notice that $1^\kappa = 1$, but for every infinite cardinal κ, $2^\kappa \leq \kappa^\kappa \leq (2^\kappa)^\kappa = 2^{\kappa \cdot \kappa} = 2^\kappa$, i.e., $2^\kappa = \kappa^\kappa$. Therefore, $\mu^\kappa = 2^\kappa$ for all infinite κ and $2 \leq \mu \leq \kappa$.

If κ is a limit cardinal, then we write $2^{<\kappa}$ for $\sup_{\mu<\kappa} 2^\mu$. More generally, we write $\lambda^{<\kappa}$ for $\sup_{\mu<\kappa} \lambda^\mu$.

Lemma 4.16 *If κ is a limit cardinal, then $2^\kappa = (2^{<\kappa})^{\mathrm{cf}(\kappa)}$. In particular, if κ is a limit cardinal with $2^{<\kappa} = \kappa$, then $2^\kappa = \kappa^{\mathrm{cf}(\kappa)}$.*

Proof Let κ be an arbitrary limit cardinal, let $f: \mathrm{cf}(\kappa) \to \kappa$ be cofinal, and let us write κ_i for $f(i)$, where $i < \mathrm{cf}(\kappa)$. For $i < \mathrm{cf}(\kappa)$, let $g_i: \mathscr{P}(\kappa_i) \to 2^{<\kappa}$ be an injection. We may define

$$\Phi: \mathscr{P}(\kappa) \to {}^{\mathrm{cf}(\kappa)}(2^{<\kappa})$$

by letting $\Phi(X)(i) = g_i(X \cap \kappa_i)$, where $X \subset \kappa$ and $i < \mathrm{cf}(\kappa)$. Obviously, Φ is injective. Therefore,

$$2^\kappa \leq (2^{<\kappa})^{\mathrm{cf}(\kappa)} \leq (2^\kappa)^{\mathrm{cf}(\kappa)} = 2^{\kappa \cdot \mathrm{cf}(\kappa)} = 2^\kappa,$$

so that in fact $2^\kappa = (2^{<\kappa})^{\mathrm{cf}(\kappa)}$. □

Corollary 4.17 *Let κ be a singular limit cardinal and assume that there are $\mu_0 < \kappa$ and λ such that $2^\mu = \lambda$ whenever $\mu_0 \leq \mu < \kappa$. Then $2^\kappa = \lambda$.*

Proof Let $\mu \geq \mu_0$ be such that $\mu \geq \mathrm{cf}(\kappa)$. Then, using Lemma 4.16, $\lambda \leq 2^\kappa = \lambda^{\mathrm{cf}(\kappa)} = (2^\mu)^{\mathrm{cf}(\kappa)} = 2^{\mu \cdot \mathrm{cf}(\kappa)} = 2^\mu = \lambda$. □

The expression $\kappa^{\mathrm{cf}(\kappa)}$ will reappear in the statement of the Singular Cardinals Hypothesis, cf. (4.2) below.

Definition 4.18 Let κ be an infinite cardinal. We then say that a set x is *hereditarily smaller than* κ iff $\overline{\overline{\mathrm{TC}(\{x\})}} < \kappa$. We let

$$H_\kappa = \{x\colon x \text{ is hereditarily smaller than } \kappa\}.$$

We also write HF ("*h*ereditarily *f*inite") instead of $H_{\aleph_0}$ and HC ("*h*ereditarily *c*ountable") instead of $H_{\aleph_1}$.

It is not hard to show that H_κ is a *set* for every infinite cardinal κ and that $\mathrm{Card}(H_\kappa) = 2^{<\kappa}$, cf. Poblem 4.10.

The following is often referred to as the HAUSDORFF *Formula.*

Theorem 4.19 (Hausdorff) *For all infinite cardinals* κ, λ, $(\kappa^+)^\lambda = \kappa^\lambda \cdot \kappa^+$.

Proof Suppose first that $\kappa^+ \leq \lambda$. Then $(\kappa^+)^\lambda = 2^\lambda = \kappa^\lambda \cdot \kappa^+$.

Let us now assume that $\kappa^+ > \lambda$. Then, as κ^+ is regular by Lemma 4.15, every $f\colon \lambda \to \kappa^+$ is bounded, i.e., there is some $\xi < \kappa^+$ with $\mathrm{ran}(f) \subset \xi$. Therefore $(\kappa^+)^\lambda = \mathrm{Card}({}^\lambda(\kappa^+)) = \mathrm{Card}(\bigcup_{\xi<\kappa^+} {}^\lambda\xi) = \kappa^\lambda \cdot \kappa^+$. □

We may define infinite sums and products as follows. Let f be a function with $\mathrm{dom}(f) = I$ (where I is any non-empty set) and such that $f(i)$ is a cardinal for every $i \in I$. Let us write κ_i for $f(i)$, where $i \in I$. We then define

$$\sum_{i\in I} \kappa_i = \mathrm{Card}\left(\bigcup_{i\in I}(\kappa_i \times \{i\})\right),$$

and

$$\prod_{i\in I} \kappa_i = \text{the cardinality of the set of all functions } g \text{ with } \mathrm{dom}(g) = I \text{ and } g(i) \in \kappa_i \text{ for all } i \in I.$$

This generalizes the earlier definitions of $\kappa + \lambda$ and $\kappa \cdot \lambda$.

It is not hard to verify that if κ is a limit cardinal, then $\mathrm{cf}(\kappa)$ may be characterized as the least λ such that there is a sequence $(\kappa_i\colon i < \lambda)$ of cardinals less than κ with $\kappa = \sum_{i<\lambda} \kappa_i$ (cf. Problem 4.6).

If A_i is a set for each $i \in I$ (where I is any "index" set), then we write $X_{i\in I} A_i$ for the set of all functions g with $\mathrm{dom}(g) = I$ and $g(i) \in A_i$ for all $i \in I$. The axiom of choice says that $X_{i\in I} A_i \neq \emptyset$ provided that $I \neq \emptyset$ and $A_i \neq \emptyset$ for all $i \in I$. We have that $\prod_{i\in I} \kappa_i = \mathrm{Card}(X_{i\in I}\kappa_i)$.

Theorem 4.20 (König) *Let* $I \neq \emptyset$, *and suppose that for every* $i \in I$, κ_i *and* λ_i *are cardinals such that* $\kappa_i < \lambda_i$. *Then* $\sum_{i\in I} \kappa_i < \prod_{i\in I} \lambda_i$.

Proof Let $f\colon \bigcup_{i\in I}(\kappa_i \times \{i\}) \to X_{i\in I}\lambda_i$. We need to see that f is not surjective.

Let $i \in I$. Look at

$$\{\xi \in \lambda_i\colon \neg\exists\alpha \in \kappa_i \; f(\alpha, i)(i) = \xi\}.$$

As $\kappa_i < \lambda_i$, this set must be non-empty, so that we may let ξ_i be the least ξ such that for all $\alpha \in \kappa_i$, $f(\alpha, i)(i) \neq \xi$.

Now let $g \in X_{i \in I} \lambda_i$ be defined by $g(i) = \xi_i$ for $i \in I$. If $i \in I$ and $\alpha \in \kappa_i$, then $f(\alpha, i)(i) \neq \xi_i = g(i)$, i.e., $f(\alpha, i) \neq g$. Therefore, $g \notin \text{ran}(f)$. □

Corollary 4.21 *For all infinite cardinals κ, $\text{cf}(2^\kappa) > \kappa$ and $\kappa^{\text{cf}(\kappa)} > \kappa$.*

Proof Let $I = \kappa$, and let $\kappa_i < 2^\kappa$ for all $i \in I$. In order to show that $\text{cf}(2^\kappa) > \kappa$, it suffices to show that $\sum_{i \in I} \kappa_i < 2^\kappa$. Set $\lambda_i = 2^\kappa$ for all $i \in I$. Then $\sum_{i \in I} \kappa_i < \prod_{i \in I} \lambda_i = \text{Card}({}^\kappa(2^\kappa)) = (2^\kappa)^\kappa = 2^\kappa$.

To see that $\kappa^{\text{cf}(\kappa)} > \kappa$, let $f \colon \text{cf}(\kappa) \to \kappa$ be cofinal, and write $I = \text{cf}(\kappa)$ and $\kappa_i = f(i)$ for $i \in I$. Set $\lambda_i = \kappa$ for all $i \in I$. Then $\kappa \leq \sum_{i \in I} \kappa_i < \prod_{i \in I} \lambda_i = \text{Card}({}^{\text{cf}(\kappa)}\kappa) = \kappa^{\text{cf}(\kappa)}$. □

The *Singular Cardinal Hypothesis*, abbreviated by SCH, is the statement that for all singular limit cardinals κ,

$$\kappa^{\text{cf}(\kappa)} = 2^{\text{cf}(\kappa)} \cdot \kappa^+. \tag{4.2}$$

Notice that $\kappa^{\text{cf}(\kappa)} \geq 2^{\text{cf}(\kappa)} \cdot \kappa^+$ holds for all infinite κ by Corollary 4.21, so that SCH says that $\kappa^{\text{cf}(\kappa)}$ has the minimal possible value. Moreover, if κ is regular, then $2^{\text{cf}(\kappa)} \cdot \kappa^+ = 2^\kappa \cdot \kappa^+ = 2^\kappa = \kappa^\kappa = \kappa^{\text{cf}(\kappa)}$, so that (4.2) is always true for regular κ. A deep theorem of R. JENSEN will say that the negation of SCH implies the existence of an object called $0^\#$, cf. Corollary 11.61.

If GCH holds true and κ is a singular limit cardinal, then $\kappa^{\text{cf}(\kappa)} = 2^\kappa = \kappa^+ = 2^{\text{cf}(\kappa)} \cdot \kappa^+$ using Lemma 4.16. Therefore, GCH implies SCH.

Lemma 4.22 *Let κ be a limit cardinal, and suppose that SCH holds below κ, i.e., $\mu^{\text{cf}(\mu)} = 2^{\text{cf}(\mu)} \cdot \mu^+$ for every (infinite) $\mu < \kappa$. Then for every (infinite) $\mu < \kappa$ and for every infinite λ,*

$$\mu^\lambda = \begin{cases} 2^\lambda & \text{if } \mu \leq 2^\lambda, \\ \mu^+ & \text{if } \mu > 2^\lambda \text{ is a limit cardinal of cofinality } \leq \lambda, \text{ and} \\ \mu & \text{if } \mu > 2^\lambda \text{ is a successor cardinal or a limit cardinal} \\ & \text{of cofinality } > \lambda. \end{cases}$$

Proof by induction on μ, fixing λ. If $\mu \leq 2^\lambda$, then $\mu^\lambda \leq (2^\lambda)^\lambda = 2^\lambda \leq \mu^\lambda$, and thus $\mu^\lambda = 2^\lambda$. If $\mu = \nu^+ > 2^\lambda$, $\mu < \kappa$, then $\mu^\lambda = (\nu^+)^\lambda = \nu^\lambda \cdot \nu^+ = \nu^+ = \mu$ by the HAUSDORFF Formula 4.19 and the inductive hypothesis.

Now let $\mu < \kappa$, $\mu > 2^\lambda$, be a limit cardinal, and let $(\alpha_i : i < \text{cf}(\mu))$ be cofinal in μ, where $\alpha_i > 2^\lambda$ for all $i < \text{cf}(\mu)$. By the inductive hypothesis (cf. also Problem 4.16) and as we assume SCH to hold below κ, we have that

$$\mu^\lambda \leq \Big(\prod_{i<\text{cf}(\mu)} \alpha_i\Big)^\lambda = \prod_{i<\text{cf}(\mu)} (\alpha_i^\lambda) \leq \prod_{i<\text{cf}(\mu)} \alpha_i^+ \leq \prod_{i<\text{cf}(\mu)} \mu = \mu^{\text{cf}(\mu)} = 2^{\text{cf}(\mu)} \cdot \mu^+.$$

Therefore, if $\mathrm{cf}(\mu) \leq \lambda$, then with the help of Corollary 4.21 we get that $\mu^+ \leq \mu^{\mathrm{cf}(\mu)} \leq \mu^\lambda \leq 2^\lambda \cdot \mu^+ = \mu^+$, so that $\mu^\lambda = \mu^+$. On the other hand, if $\lambda < \mathrm{cf}(\mu)$, then every $f : \lambda \to \mu$ is bounded, so that by the inductive hypothesis

$$\mu^\lambda \leq \sum_{i<\mathrm{cf}(\mu)} \alpha_i^\lambda \leq \sum_{i<\mathrm{cf}(\mu)} \alpha_i^+ \leq \mathrm{cf}(\mu) \cdot \mu = \mu \leq \mu^\lambda,$$

so that $\mu^\lambda = \mu$. □

In order to prove more powerful statements in cardinal arithmetic, we need the concept of a "stationary" set.

4.2 Stationary Sets

Definition 4.23 Let A be a set of ordinals. A is called *closed* iff for all ordinals α, $\sup(A \cap \alpha) \in A$. If γ is an ordinal, then A is unbounded in γ iff for all $\xi < \gamma$, $(A \cap \gamma) \setminus \xi \neq \emptyset$. A is called δ*-closed*, where δ is an infinite regular cardinal, iff for all ordinals α with $\mathrm{cf}(\alpha) = \delta$ and such that $A \cap \alpha$ is unbounded in α, $\alpha = \sup(A \cap \alpha) \in A$. A is called *club in* α iff $A \cup \{\alpha\}$ is closed and A is unbounded in α. A is called δ*-club in* α, where δ is an infinite regular cardinal, iff $A \cup \{\alpha\}$ is δ-closed and A is unbounded in α.

A set A of ordinals is closed iff it is δ-closed for every δ, iff it is a closed subset of $\sup(A)$ in the topology generated by the non-empty open intervals below α. For any set A of ordinals we usually denote by A' the set of limit points of A, where α is a *limit point of* A iff for all $\beta < \alpha$ there is some $\gamma \in A$ with $\beta < \gamma < \alpha$. A' is always closed. Also, e.g., A is closed iff $A' \subset A \cup \{\sup(A)\}$.

For any κ, the cofinality of κ is the least size of a subset of κ which is unbounded in κ (cf. Problem 4.7). If $C \subset \kappa$ is club, where $\mathrm{cf}(\kappa) > \omega$, then C' is also club in κ. If $\pi : \mathrm{cf}(\kappa) \to \kappa$ is strictly monotone, continuous, and cofinal, then $\mathrm{ran}(\pi)$ is club in κ, and if $\mathrm{cf}(\kappa) > \omega$, then the set of limit points of $\mathrm{ran}(\pi)$ is club in κ and consists of points of cofinality strictly less than $\mathrm{cf}(\kappa)$.

Definition 4.24 Let $X \neq \emptyset$ be a set. $F \subset \mathscr{P}(X)$ is called a filter on X iff

(1) $F \neq \emptyset$,
(2) $\forall a \forall b (a \in F \wedge b \in F \to a \cap b \in F)$, and
(3) $\forall a \forall b (a \in F \wedge a \subset b \subset X \to b \in F)$.

F is called non-trivial iff $\emptyset \notin F$. F is called an *ultrafilter* iff for every $a \subset X$, either $a \in F$ or else $X \setminus a \in F$.

Let μ be a cardinal. Then F is called $< \mu$-closed iff for all $\alpha < \mu$ and for all $\{X_i : i < \alpha\} \subset F$, $\bigcap \{X_i : i < \alpha\} \in F$.

Notice that every filter is $< \omega$-closed. If α is a limit ordinal, then

$$\{X \subset \alpha \colon \exists \beta < \alpha \; \alpha \setminus \beta \subset X\}$$

is a non-trivial $< \mathrm{cf}(\alpha)$-closed filter on α, called the FRÉCHET *filter* (*on* α). Every filter can be extended to an ultrafilter, cf. Problem 4.11.

Lemma 4.25 *Let α be an ordinal such that $\mathrm{cf}(\alpha) > \omega$, and let F_α be the set of all $A \subset \alpha$ such that there is some $B \subset A$ which is club in α. Then F_α is a (non-trivial) $< \mathrm{cf}(\alpha)$-closed filter on α.*

Proof We need to see that if $\beta < \mathrm{cf}(\alpha)$ and A_i is club in α for every $i \in \beta$, then $\bigcap_{i \in \beta} A_i$ is club in α. Well, $(\bigcap_{i \in \beta} A_i) \cup \{\alpha\}$ is certainly closed, so that it suffices to verify that $\bigcap_{i \in \beta} A_i$ is unbounded in α. Let $\xi < \alpha$. We define $f : \beta \cdot \omega \to \alpha$ as follows.[3] For $n \in \omega$ and $i \in \beta$ we let $f(\beta \cdot n + i)$ be the least element of A_i which is bigger than $\sup(\{f(\eta) : \eta < \beta \cdot n + i\} \cup \{\xi\})$. Notice that this is welldefined as $\beta \cdot \omega < \mathrm{cf}(\alpha)$ and each A_i is unbounded in α.

Let $\rho = \sup\{f(\eta) : \eta < \beta \cdot \omega\}$. We have that $\rho < \alpha$ and in fact $\rho \in \bigcap_{i \in \beta} A_i$, because every A_i is closed; notice that for each $i \in \beta$, $\rho = \sup\{f(\beta \cdot n + i) : n \in \omega\} = \sup(A_i \cap \rho)$. □

F_α as in this lemma is called the *club filter on* α.

Definition 4.26 Let α be regular, and let $X_\xi \subset \alpha$ for all $\xi < \alpha$. The diagonal intersection of X_ξ, $\xi < \alpha$, abbreviated by $\Delta_{\xi<\alpha} X_\xi$, is the set $\left\{\eta < \alpha \colon \eta \in \bigcap_{\xi<\eta} X_\xi\right\}$.

Definition 4.27 Let α be regular, and let F be a filter on α. F is called *normal* iff for all $\left\{X_\xi \colon \xi < \alpha\right\} \subset F$, $\Delta_{\xi<\alpha} X_\xi \in F$.

Lemma 4.28 *Let α be regular, and let F_α be the club filter on α. Then F_α is normal.*

Proof We need to see that if A_ξ is club in α for every $\xi < \alpha$, then $\Delta_{\xi<\alpha} A_\xi$ is club in α. By replacing A_ξ by $\bigcap_{\eta \leq \xi} A_\eta$ if necessary, we may and shall assume that $A_{\xi'} \subset A_\xi$ whenever $\xi \leq \xi'$. (Notice that every $\bigcap_{\eta \leq \xi} A_\eta$ is again club in α by Lemma 4.25.) In order to see that $(\Delta_{\xi<\alpha} A_\xi) \cup \{\alpha\}$ is closed, let $\delta < \alpha$ be a limit ordinal such that $\delta = \sup((\Delta_{\xi<\alpha} A_\xi) \cap \delta)$. We want to argue that $\delta \in \Delta_{\xi<\alpha} A_\xi$, i.e., $\delta \in A_\xi$ for all $\xi < \delta$. Well, if $\xi \leq \xi' < \delta$, then there is some η with $\xi' < \eta < \delta$ and $\eta \in \Delta_{\bar{\xi}<\alpha} A_{\bar{\xi}}$, i.e., $\eta \in \bigcap_{\bar{\xi}<\eta} A_{\bar{\xi}}$, in particular $\eta \in A_\xi$. This shows that $\delta = \sup(A_\xi \cap \delta)$ for all $\xi < \delta$; hence $\delta \in A_\xi$ for all $\xi < \delta$.

In order to see that $\Delta_{\xi<\alpha} A_\xi$ is unbounded in α, let $\eta < \alpha$. We construct a sequence η_n, $n \in \omega$, as follows. Let $\eta_0 = \eta$. If η_n is defined, then let η_{n+1} be the least $\eta > \eta_n$ such that $\eta \in A_{\eta_n}$. We claim that $\sup\{\eta_n \colon n \in \omega\} \in \Delta_{\xi<\alpha} A_\xi$. Set $\beta = \sup\{\eta_n \colon n \in \omega\}$. We need to see that $\beta \in A_\xi$ for all $\xi < \beta$. Let $\xi < \beta$. Then $\xi < \eta_n$ for some $n \in \omega$. We have that $\eta_{m+1} \in A_{\eta m} \subset A_{\eta_n}$ for all $m \geq n$, so that $\beta = \sup\{\eta_{m+1} \colon m \geq n\} \in A_{x\eta_n} \subset A_\xi$. □

Definition 4.29 Let α be an ordinal such that $\mathrm{cf}(\alpha) > \omega$. $A \subset \alpha$ is called *stationary* (*in* α) iff $A \cap C \neq \emptyset$ for all C which are club in α.

---

[3] Here, $\beta \cdot \omega$ denotes ordinal multiplication.

If μ, κ are infinite regular cardinals with $\mu < \kappa$, then the set

$$\{\alpha < \kappa\colon \mathrm{cf}(\alpha) = \mu\}$$

is stationary in κ (cf. Problem 4.12). This immediately implies that for any regular $\kappa \geq \aleph_2$, F_κ is not an ultrafilter. We shall prove a stronger statement below, cf. Theorem 4.33.

Definition 4.30 Let $X \neq \emptyset$, and let F be a filter on X. Then we write

$$F^+ = \{a \subset X\colon \forall b \in F\, b \cap a \neq \emptyset\}.$$

The elements of F^+ are called the *positive sets (with respect to F)*.

The stationary sets are therefore just the positive sets with respect to the club filter.

Lemma 4.31 *Let κ be regular, and let F be a filter on κ. The following statements are equivalent.*

(*a*) *F is normal.*
(*b*) *Let $f\colon \kappa \to \kappa$ be such that $Y = \{\xi < \kappa\colon f(\xi) < \xi\} \in F^+$. There is then some $\alpha < \kappa$ and some $X \in F^+$, $X \subset Y$, such that $f(\xi) = \alpha$ for all $\xi \in X$.*

Proof (a) $\Longrightarrow$ (b): Let f be as in (b). If there is no $\alpha < \kappa$ and $X \in F^+$, $X \subset Y$, such that $f(\xi) = \alpha$ for all $\xi \in X$, then for every $\alpha < \kappa$ we may pick some $X_\alpha \in F$ such that $f(\xi) \neq \alpha$ for all $\xi \in X_\alpha \cap Y$. (Here we use AC, the axiom of choice.) By definition

$$\xi \in Y \cap \Delta_{\alpha<\kappa} X_\alpha \Longrightarrow f(\xi) \geq \xi. \tag{4.3}$$

By (a), $\Delta_{\alpha<\kappa} X_\alpha \in F$, so that by $Y = \{\xi < \kappa\colon\ f(\xi) < \xi\} \in F^+$, we may pick some $\xi \in \Delta_{\alpha<\kappa} X_\alpha$ such that $f(\xi) < \xi$. This contradicts (4.3).

(b) $\Longrightarrow$ (a): Let $X_\alpha \in F$, $\alpha < \kappa$. If $\Delta_{\alpha<\kappa} X_\alpha \notin F$, then

$$Y = \left\{\xi < \kappa\colon\ \xi \notin \bigcap_{\alpha<\xi} X_\alpha\right\} \in F^+.$$

Let $f\colon\ \kappa \to \kappa$ be such that $f(\xi) < \xi$ and $\xi \notin X_{f(\xi)}$ for all $\xi \in Y$. By (b), there is then some $X \in F^+$, $X \subset Y$, and some $\alpha < \kappa$ such that $\xi \notin X_\alpha$ for all $\xi \in X$. But $X \cap X_\alpha \neq \emptyset$, as $X_\alpha \in F$. Contradiction! □

In the light of Lemma 4.28, Lemma 4.31, applied to the club filter, immediately gives the following.

Theorem 4.32 (Fodor) *Let α be regular and uncountable, and let $S \subset \alpha$ be stationary. Let $f\colon S \to \alpha$ be* regressive *in the sense that $f(\xi) < \xi$ for all $\xi \in S$. Then there is a stationary $T \subset S$ such that $f \restriction T$ is constant.*

The following theorem is a strong from of saying that the club filter is not an ultrafilter, i.e. for any regular uncountable κ there are $X \subset \kappa^+$ which neither contain nor are disjoint from a club.

Theorem 4.33 (Solovay) *Let κ be a regular uncountable cardinal, and let $S \subset \kappa$ be stationary. Then S may be written as a disjoint union of κ stationary sets, i.e., there is $(S_\xi \colon \xi < \kappa)$ such that $S_\xi \subset S$ is stationary in κ for every $\xi < \kappa$, $S_\xi \cap S_{\xi'} = \emptyset$ for all $\xi, \xi' < \kappa$ with $\xi \neq \xi'$, and $S = \bigcup_{\xi<\kappa} S_\xi$.*

Proof Let us first write $S = S_0 \cup S_1$, where $S_0 = \{\alpha \in S \colon \mathrm{cf}(\alpha) < \alpha\}$ and $S_1 = \{\alpha \in S \colon \mathrm{cf}(\alpha) = \alpha\}$. At least one of S_0, S_1 must be stationary.

Claim 4.34 *There is some stationary $\bar{S} \subset S$ and some sequence $(A_\alpha \colon \alpha \in \bar{S})$ such that $A_\alpha \subset \alpha$ is club in α and $A_\alpha \cap \bar{S} = \emptyset$ for every $\alpha \in \bar{S}$.*

Proof Suppose first that S_0 is stationary. By Theorem 4.32, there is then some stationary $\bar{S} \subset S_0$ and some regular $\lambda < \kappa$ such that $\mathrm{cf}(\alpha) = \lambda$ for all $\alpha \in \bar{S}$. Let us pick $(A_\alpha \colon \alpha \in \bar{S})$, where A_α is club in α, $\mathrm{otp}(A_\alpha) = \lambda$, and $\mathrm{cf}(\gamma) < \lambda$ for all $\gamma \in A_\alpha$. Notice $A_\alpha \cap \bar{S} = \emptyset$ for each $\alpha \in \bar{S}$.

Now suppose S_0 to be non-stationary, so that S_1 is stationary. Let

$$\bar{S} = \{\alpha \in S_1 \colon S_1 \cap \alpha \text{ is non-stationary}\}.$$

We must have that $\bar{S}$ is stationary. To see this, let $C \subset (\kappa \setminus (\omega + 1))$ be club, and let $C' \subset \kappa$ be the club of all limit points of C. If $\alpha = \min(S_1 \cap C')$, then $C \cap \alpha$ is club in α, so that $C' \cap \alpha$ is still club in α, as $\alpha > \omega$ is regular; but $(C' \cap \alpha) \cap S_1 = \emptyset$, so that $\alpha \in C \cap \bar{S}$. We may thus pick $(A_\alpha \colon \alpha \in \bar{S})$, where A_α is club in α and $A_\alpha \cap \bar{S} \subset A_\alpha \cap S_1 = \emptyset$ for each $\alpha \in \bar{S}$. □

For $\alpha \in \bar{S}$, let $(\gamma_i^\alpha \colon i < \mathrm{otp}(A_\alpha))$ be the monotone enumeration of A_α.

Claim 4.35 *There is some $i < \kappa$ such that for all $\beta < \kappa$,*

$$\{\alpha \in \bar{S} \colon i < \mathrm{cf}(\alpha) \wedge \gamma_i^\alpha > \beta\}$$

is stationary in κ.

Proof Suppose first that $\bar{S} \subset S_0$, and let again $\lambda < \kappa$ be such that $\mathrm{cf}(\alpha) = \lambda$ for all $\alpha \in \bar{S}$. If Claim 4.35 fails, then for every $i < \lambda$ there is some $\beta_i < \kappa$ and some club $C_i \subset \kappa$ such that for all $\alpha \in \bar{S} \cap C_i$, $\gamma_i^\alpha \leq \beta_i$. But then if $\tilde{\beta} = \sup(\{\beta_i \colon i < \lambda\}) < \kappa$ and $\alpha \in (\bar{S} \cap \bigcap_{i<\lambda} C_i) \setminus (\tilde{\beta} + 1)$, then $\gamma_i^\alpha \leq \tilde{\beta}$ for all $i < \lambda$, so that A_α would be bounded in α. Contradiction!

Now suppose that $\bar{S} \subset S_1$, so that $\mathrm{cf}(\alpha) = \alpha$ for every $\alpha \in \bar{S}$. If Claim 4.35 fails, then for every $i < \kappa$ there is some $\beta_i < \kappa$ and some club $C_i \subset \kappa$ such that for all $\alpha \in \bar{S} \cap C_i$, either $i \geq \mathrm{cf}(\alpha)$ or else $\gamma_i^\alpha \leq \beta_i$. Let $D \subset \kappa$ be the club of all $\beta < \kappa$ such that $i < \beta$ implies $\beta_i < \beta$. By Lemma 4.28 we may pick

$$\alpha, \alpha' \in \bar{S} \cap \Delta_{i<\kappa} C_i \cap D,$$

with $\alpha < \alpha'$. If $i < \alpha$, then $\alpha' \in \bar{S} \cap C_i$, so that $\gamma_i^{\alpha'} \leq \beta_i$; but $\alpha \in D$, so that $\beta_i < \alpha$. I.e., $\gamma_i^{\alpha'} < \alpha$ for all $i < \alpha$. This yields $\gamma_\alpha^{\alpha'} \leq \alpha$, as $A_{\alpha'}$ is club; however, clearly $\gamma_i^{\alpha'} \geq i$ for every $i < \alpha'$, so that in fact $\gamma_\alpha^{\alpha'} = \alpha$. But then $\gamma_\alpha^{\alpha'} \bar{S}$. Contradiction! □

Now fix $i_0 < \kappa$ such that for all $\beta < \kappa$,

$$\{\alpha \in \bar{S}\colon\ i_0 < \mathrm{cf}(\alpha) \wedge \gamma_{i_0}^\alpha > \beta\}$$

is stationary in κ. Let us recursively define stationary sets S_ξ and ordinals β_ξ for $\xi < \kappa$ as follows.

Fix $\xi < \kappa$, and suppose that $S_{\bar{\xi}}$ and $\beta_{\bar{\xi}}$ have already been chosen for all $\bar{\xi} < \xi$. The set

$$\left\{\alpha \in \bar{S}\colon\ \gamma_{i_0}^\alpha > \sup_{\bar{\xi}<\xi} \beta_{\bar{\xi}}\right\}$$

is stationary in κ, and we may thus use FODOR's Theorem 4.32 to pick some $\beta > \sup_{\bar{\xi}<\xi} \beta_{\bar{\xi}}$ and some stationary $S^* \subset \bar{S}$ such that for all $\alpha \in S^*$, $\gamma_{i_0}^\alpha = \beta$. Let us set $S_\xi = S^*$ and $\beta_\xi = \beta$.

The rest is straightforward. □

The Singular Cardinals Hypothesis SCH cannot first fail at a singular cardinal of uncountable cofinality:

Theorem 4.36 (Silver) *Let κ be a singular cardinal of uncountable cofinality. If* SCH *holds below κ, then it holds at κ, i.e., if $\mu^{\mathrm{cf}(\mu)} = 2^{\mathrm{cf}(\mu)} \cdot \mu^+$ for every $\mu < \kappa$, then $\kappa^{\mathrm{cf}(\kappa)} = 2^{\mathrm{cf}(\kappa)} \cdot \kappa^+$.*

Proof Suppose first that $\kappa \leq 2^{\mathrm{cf}(\kappa)}$. Then $\kappa < 2^{\mathrm{cf}(\kappa)}$, as $\mathrm{cf}(2^{\mathrm{cf}(\kappa)}) > \mathrm{cf}(\kappa)$ by Corollary 4.21. I.e., $\kappa^+ \leq 2^{\mathrm{cf}(\kappa)}$, and hence $\kappa^{\mathrm{cf}(\kappa)} \leq (2^{\mathrm{cf}(\kappa)})^{\mathrm{cf}(\kappa)} = 2^{\mathrm{cf}(\kappa)} = 2^{\mathrm{cf}(\kappa)} \cdot \kappa^+ \leq \kappa^{\mathrm{cf}(\kappa)}$. Therefore, SCH holds at κ.

We may thus assume that $2^{\mathrm{cf}(\kappa)} < \kappa$. Let $C \subset \kappa$ be club in κ with $\mathrm{otp}(C) = \mathrm{cf}(\kappa)$, let C' be the set of all limit points of C. Let $(\mu_i : i < \mathrm{cf}(\kappa))$ be the monotone enumeration of $C' \setminus (2^{\mathrm{cf}(\kappa)})^+$. As $\mathrm{cf}(\mu_i) < \mathrm{cf}(\kappa)$ for every $i < \mathrm{cf}(\kappa)$ (as being witnessed by $C \cap \mu_i$), Lemma 4.22 gives that $\mu_i^{\mathrm{cf}(\kappa)} = \mu_i^+$ for each $i < \mathrm{cf}(\kappa)$. So for each $i < \mathrm{cf}(\kappa)$, we may pick a bijection $g_i : [\mu_i]^{\mathrm{cf}(\kappa)} \to \mu_i^+$.

We now have to count $[\kappa]^{\mathrm{cf}(\kappa)}$. To each $X \in [\kappa]^{\mathrm{cf}(\kappa)}$ we may associate a function $f_X : \mathrm{cf}(\kappa) \to \kappa$ by setting $f_X(i) = g_i(X \cap \mu_i)$. Obviously, $X \mapsto f_X$ is injective.

If $X, Y \in [\kappa]^{\mathrm{cf}(\kappa)}$, then we shall write $X \leq Y$ iff $\{i\colon f_X(i) \leq f_Y(i)\}$ is stationary. We must have $X \leq Y$ or $Y \leq X$ for any two $X, Y \in [\kappa]^{\mathrm{cf}(\kappa)}$.

Claim 4.37 *Let $X \in [\kappa]^{\mathrm{cf}(\kappa)}$. Then* $\mathrm{Card}(\{Y \in [\kappa]^{\mathrm{cf}(\kappa)}\colon Y \leq X\}) \leq \kappa$.

Proof Let us fix $X \in [\kappa]^{\mathrm{cf}(\kappa)}$ for a moment. For each $i < \mathrm{cf}(\kappa)$, let us pick an injection $g_i : f_X(i) + 1 \to \mu_i$. If $Y \leq X$, then the set $S_Y^X = \{i < \mathrm{cf}(\kappa)\colon f_Y(i) \leq f_X(i)\}$ is stationary, and we may look at

$$F_Y^X\colon\ S_Y^X \to \kappa,$$

as being defined by $F_Y^X(i) = g_i(f_Y(i))$ for $i < \mathrm{cf}(\kappa)$. In particular, $F_Y^X(i) < \mu_i$ for every $i \in S_Y^X$. Let D be the club of limit ordinals in below $\mathrm{cf}(\kappa)$. Then the map which sends $i \in S_Y^X \cap D$ to the least $j < \mathrm{cf}(\kappa)$ with $F_Y^X(i) < \mu_j$ is regressive. As $S_Y^X \cap D$ is still stationary, by FODOR's Theorem 4.32 there is a stationary set $\bar{S}_Y^X \subset S_Y^X$ and some $i_Y^X < \mathrm{cf}(\kappa)$ such that $F_Y^X(i) < \mu_{i_Y^X}$ for all $i \in \bar{S}_Y^X$.

If $Y, Z \leq X$, $\bar{S}_Y^X = \bar{S}_Z^X$, $i_Y^X = i_Z^X$, and $F_Y^X \upharpoonright \bar{S}_Y^X = F_Z^X \upharpoonright \bar{S}_Z^X$, then $Y = Z$. But there are only

$$\leq 2^{\mathrm{cf}(\kappa)} \cdot \mathrm{cf}(\kappa) \cdot (\sup_{\mu<\kappa} \mu^{\mathrm{cf}(\kappa)}) = \kappa$$

many possible triples $(\bar{S}_Y^X, i_Y^X, F_Y^X \upharpoonright \bar{S}_Y)$, so that there are at most κ many $Y \leq X$. □

In order to finish the proof of the Theorem, it thus remains to be shown that there is some $A \subset \mathscr{P}(\kappa)$ of cardinality κ^+ such that $\mathscr{P}(\kappa) = \{Y \subset \kappa \colon \exists X \in A\ Y \leq X\}$.

Let us recursively construct X_α for $\alpha < \kappa^+$ as follows. Given $\alpha < \kappa^+$, having constructed X_β for all $\beta < \alpha$, we pick X_α such that for no $\beta < \alpha$, $X_\alpha \leq X_\beta$. Notice that this choice is possible, as $\{Y \subset \kappa \colon \exists \beta < \alpha\ Y \leq X_\beta\}$ has size at most κ by Claim 4.37. Set $A = \{X_\alpha \colon \alpha < \kappa^+\}$. We must have that $\mathscr{P}(\kappa) = \{Y \subset \kappa \colon \exists X \in A\ Y \leq X\}$, as otherwise there would be some $Y \subset \kappa$ with $X_\alpha \leq Y$ for all $\alpha < \kappa^+$; but $X_\alpha \neq X_\beta$ for $\alpha \neq \beta$, so this is impossible by Claim 4.37. □

Corollary 4.38 *Let κ be a singular cardinal of uncountable cofinality. If* GCH *holds below κ, then it holds at κ, i.e., if $2^\mu = \mu^+$ for every $\mu < \kappa$, then $2^\kappa = \kappa^+$.*

Proof If $2^\mu = \mu^+$, then $\mu^{\mathrm{cf}(\mu)} = \mu^+$, so that $\kappa^{\mathrm{cf}(\kappa)} = 2^{\mathrm{cf}(\kappa)} \cdot \kappa^+$ by SILVER's Theorem 4.36. But $2^{\mathrm{cf}(\kappa)} < \kappa$, so that $\kappa^{\mathrm{cf}(\kappa)} = \kappa^+$. By Lemma 4.16, $2^\kappa = \kappa^{\mathrm{cf}(\kappa)}$, which gives $2^\kappa = \kappa^+$. □

Problems 4.17 and 4.18 produce generalizations of Theorem 4.36 and Corollary 4.38.

There is a generalization of stationarity which we shall now briefly discuss.

Definition 4.39 We say that $\mathscr{S} \subset [\theta]^\kappa$ is *stationary in* $[\theta]^\kappa$ iff for every $A \supset \theta$ and for every algebra $\mathfrak{A} = (A; (f_i \colon i < \bar{\kappa}))$ on A with at most $\bar{\kappa} \leq \kappa$ many functions $f_i, i < \bar{\kappa}$, there is some $X \in \mathscr{S}$ which is closed under all the $f_i, i < \bar{\kappa}$, from $\mathfrak{A}$, i.e., $f_i''[X]^{<\omega} \subset X$ for all $i < \bar{\kappa}$.

If $S \subset \kappa^+$, then $S \setminus \kappa \subset [\kappa^+]^\kappa$. It is easy to verify that if $S \subset \kappa^+$ is stationary in κ^+ in the sense of Definition 4.29, then $S \setminus \kappa$ is stationary in $[\kappa^+]^\kappa$ in the sense of Definition 4.39.

We may call a set $\mathscr{X} \subset [\theta]^{<\kappa}$ *unbounded in* $[\theta]^{<\kappa}$ iff for all $Y \in [\theta]^{<\kappa}$ there is some $X \in \mathscr{X}$ with $X \supset Y$, and we may call $\mathscr{X} \subset [\theta]^{<\kappa}$ *closed in* $[\theta]^{<\kappa}$ iff for all $(X_i \colon i < \bar{\kappa})$ such that $\bar{\kappa} < \kappa$ and $X_i \in \mathscr{X}$ and $X_i \subset X_j$ for $i < j < \bar{\kappa}$, $\bigcup_{i<\bar{\kappa}} X_i \in \mathscr{X}$. We may then call $\mathscr{C} \subset [\theta]^{<\kappa}$ *closed and unbounded* (*club*) *in* $[\theta]^{<\kappa}$ iff $\mathscr{C}$ is both unbounded and closed in $[\theta]^{<\kappa}$.

A set $\mathscr{X} \subset [\theta]^{<\kappa}$ is *stationary in* $[\theta]^{<\kappa}$ iff $\mathscr{X} \cap \mathscr{C} \neq \emptyset$ for all $\mathscr{C}$ which are club in $[\theta]^{<\kappa}$. Then $\mathscr{S} \subset [\theta]^\kappa$ is stationary in $[\theta]^{<\kappa^+}$ iff $\mathscr{S}$ is stationary in the sense of Definition 4.39, cf. Problem 4.15.

4.3 Large Cardinals

Definition 4.40 A cardinal κ is called a *strong limit cardinal* iff for all cardinals $\mu < \kappa$, $2^\mu < \kappa$.

Trivially, every strong limit cardinal is a limit cardinal. $\aleph_0$ is a strong limit cardinal, and if κ is an arbitray cardinal, then

$$\sup\left(\left\{\kappa, 2^\kappa, 2^{(2^\kappa)}, \ldots\right\}\right)$$

is a strong limit cardinal above κ. There are thus arbitrarily large strong limit cardinals. Also, any limit of strong limit cardinals is a strong limit cardinal.

Definition 4.41 A cardinal κ is called *weakly inaccessible* iff κ is an uncountable regular limit cardinal. A cardinal κ is called *(strongly) inaccessible* iff κ is an uncountable regular strong limit cardinal.

Trivially, every (strongly) inaccessible cardinal is weakly inaccessible. It can be shown that there may be weakly inaccessible cardinals which are not strongly inaccessible (cf. Problem 6.13). Moreover, the existence of weakly inaccessible cardinals cannot be proven in ZFC (cf. Problem 5.16). HAUSDORFF's question (cf. p. 35) as to whether every uncountable limit cardinal is singular thus does not have an answer in ZFC.

Large cardinals may be used to *prove* true statements which are unprovable in ZFC, cf. e.g. Theorems 12.20 and 13.7. They may also be used for showing that certain statements are *consistent with* ZFC, cf. e.g. Theorem 8.23.

Definition 4.42 A cardinal κ is called *weakly* MAHLO iff κ is weakly inaccessible and the set

$$\{\mu < \kappa : \ \mu \text{is regular}\}$$

is stationary. A cardinal κ is called *(strongly)* MAHLO iff κ is (strongly) inaccessible and the set

$$\{\mu < \kappa : \ \mu \text{is regular}\}$$

is stationary.

Again, every (strongly) MAHLO cardinal is weakly MAHLO, and there may be weakly MAHLO cardinals which are not (strongly) MAHLO. If κ is weakly/strongly MAHLO, then there are κ weakly/strongly inaccessible cardinals below κ (cf. Problem 4.21).

In the proof of Claim 1 of the proof of SOLOVAY's Theorem 4.33, S_1 can only be stationary if κ is weakly MAHLO.

Definition 4.43 Let κ be an infinite cardinal. A partially ordered set $(T, <_T)$ is called a *tree* iff for all $s \in T$, $\{t \in T : \ t <_T s\}$ is well-ordered by $<_T$. In this case, we write $\text{lv}_T(s)$ for the order-type of $\{t \in T : t <_T s\}$ and call it the *level of* s *in* T. We also write $\text{ht}(T)$ for $\sup(\{\text{lv}_T(s) + 1 : s \in T\})$ and call it the *height of* T.

A set $c \subset T$ is a *chain in T* iff for all $s, t \in c$, $s <_T t$ or $s = t$ or $t <_T s$. A chain $c \subset T$ is called a *branch through T* iff for all $s \in c$ and $t <_T s$, $t \in c$. A branch b through T is called *maximal* iff there is no branch $b' \supsetneq b$ through T, and a branch b through T is called *cofinal* iff for all $\alpha < \text{ht}(T)$ there is some $s \in b$ with $\text{lv}_T(s) = \alpha$.

A set $a \subset T$ is an *antichain* iff for all $s, t \in a$ with $s \neq t$ neither $s <_T t$ nor $t <_T s$.

Definition 4.44 Let κ be an infinite cardinal, and let $(T, <_T)$ be a tree. We call $(T, <_T)$ a κ-tree iff the following hold true.

(1) $\text{ht}(T) = \kappa$,
(2) there is a unique $r \in T$ with $\text{lv}_T(r) = 0$ (the root of T),
(3) for every $s \in T$ and every $\alpha > \text{lv}_T(s)$, $\alpha < \kappa$, there is some $t \in T$ with $s <_T t$ and $\text{ht}(T) = \alpha$,
(4) for every $s \in T$ there are $t, t' \in T$, $t \neq t'$, with $s <_T t$, $s <_T t'$ and $\text{lv}_T(t) = \text{lv}_T(t') = \text{lv}_T(s) + 1$, and
(5) for every $\alpha < \kappa$, $\text{Card}(\{s \in T\colon\ \text{lv}_T(s) = \alpha\}) < \kappa$.

A κ-tree $(T, <_T)$ is called κ-ARONSZAJN iff there is no cofinal branch through T.
A κ-tree $(T, <_T)$ is called κ-SOUSLIN iff T has no antichain of size κ.
A κ-tree $(T, <_T)$ is called κ-KUREPA iff T has at least κ^+ cofinal branches.

Notice that if $(T, <_T)$ is κ-SOUSLIN, then $(T, <_T)$ is κ-ARONSZAJN. (Cf. Problem 4.22.)

We may turn any tree $(T, <_T)$ with properties (1) and (5) from Definition 4.44 into a κ-tree without adding cofinal branches or antichains of size κ as follows, provided that κ be regular.

Lemma 4.45 *Let $\kappa \geq \aleph_0$ be a regular cardinal. If there is a tree with properties* (1) *and* (5) *from Definition* 4.44 *which has no cofinal branch and no antichain of size κ, then there is a κ-SOUSLIN tree.*

Proof Let $(T, <_T)$ be a tree with properties (1) and (5) from Definition 4.44. Let

$$T_0 = \{s \in T\colon\ \forall \alpha < \kappa(\alpha > \text{lv}_T(s) \rightarrow \exists t \in T(\text{lv}_T(t) = \alpha \wedge s <_T t))\}.$$

Then $T_0 = (T_0, <_T \upharpoonright T_0)$ is a tree with property (5) of Definition 4.44. Of course, $T_0 \subset T$ and $\text{lv}_{T_0}(s) = \text{lv}_T(s)$ for all $s \in T_0$.

Suppose that (1) of Definition 4.44 failed, and let $\alpha = \text{ht}(T_0) < \kappa$. For each $s \in T$ with $\text{lv}_T(s) = \alpha$ we must then have that $\rho(s) = \sup(\{\text{lv}_T(t)\colon s <_T t\}) < \kappa$, so that $\sup(\{\rho(s)\colon s \in T \wedge \text{lv}_T(s) = \alpha\}) < \kappa$, as κ is regular and T satisfies (5). But then T cannot have satisfied (1). Contradiction!

Therefore, T_0 satisfies (1) and (5). Let us show that T_0 satisfies (3). Let $s \in T_0$ and $\alpha > \text{lv}_{T_0}(s)$, $\alpha < \kappa$. As $s \in T_0$, for every $\beta > \alpha$ we may pick some $t_\beta \in T$, $s <_T t_\beta$ with $\text{lv}_T(t_\beta) = \beta$, and we may let $u_\beta \in T$ be unique such that $s <_T u_\beta <_T t_\beta$, $\text{lv}_T(u_\beta) = \alpha$. As κ is regular and T satisfies (5), there is some cofinal $X \subset \kappa$ such that $u_\beta = u_{\beta'}$ for all $\beta, \beta' \in X$. Write $u = u_\beta$, where $\beta \in X$. We must then have that $u \in T_0$, where $s <_T u$ and $\text{lv}_T(u) = \alpha$. This shows that T_0 satisfies (3).

Now pick $r \in T_0$ with $\mathrm{rk}_{T_0}(r) = 0$, and let

$$T_1 = \{s \in T_0 \colon\ r <_T s\}.$$

Then $T_1 = (T_1, <_T \restriction T_1)$ is a tree which satisfies (1), (2), (3), and (5) from Definition 4.44. We are left with having to arrange (2).

Now let us set

$$T_2 = \{s \in T_1 \colon\ \neg\exists t <_T s\, \forall r \in T_1\, (t \leq_T r \wedge \mathrm{lv}_{T_1}(r) \leq \mathrm{lv}_{T_1}(s) \longrightarrow r \leq_T s)\}.$$

Then $(T_2, <_T \restriction T_2)$ is again a tree.

T_2 trivially satisfies (2). As for (5), let $\alpha < \kappa$. If $L = \{s \in T_2 \colon\ \mathrm{lv}_{T_2}(s) = \alpha\}$ has size $\geq \kappa$, then L cannot be an antichain in T, so that there are $s, t \in L$ with $s <_T t$ or $t <_T s$. But then $\mathrm{lv}_{T_2}(s) \neq \mathrm{lv}_{T_2}(t)$. Contradiction! So T_2 satisfies (5).

As for (1) and (3), let us fix $s \in T_2$. For each $\alpha > \mathrm{lv}_{T_1}(s)$, let us pick some $t_\alpha \in T_1$ such that $\mathrm{lv}_{T_1}(t_\alpha) = \alpha$. Let $u_\alpha \leq_T t_\alpha$ be such that $u_\alpha \in T_2$ and

$$\forall r \in T_1\ \left(u_\alpha \leq_T r \wedge \mathrm{lv}_{T_1}(r) \leq \mathrm{lv}_{T_1}(t_\alpha) \longrightarrow r \leq_T t_\alpha\right).$$

If $\left\{\mathrm{lv}_{T_2}(u_\alpha) \colon\ \alpha < \kappa\right\}$ were bounded in κ, then by (5) for T_2 there would be some $X \subset \kappa$ of size κ such that $u_\alpha = u_{\alpha'}$ for all $\alpha, \alpha' \in X$. But then $\{s \in T \colon\ \exists \alpha \in X\, s \leq_T t_\alpha\}$ would be a cofinal branch through T. Contradiction! Hence $\{\mathrm{lv}_{T_2}(u_\alpha) \colon\ \alpha < \kappa\}$ is unbounded in κ, and (1) and (3) are shown for T_2.

(4) is clear by construction. □

Lemma 4.46 (König) *There is no* $\aleph_0$-ARONSZAJN *tree.*

Proof Cf. Problem 4.20. □

The following Lemma is in some sense a special case of Lemma 11.68.

Theorem 4.47 (Aronszajn) *Let* κ *be an infinite cardinal with* $\kappa^{<\kappa} = \kappa$. *There is then a* κ^+-ARONSZAJN *tree.*

Notice that $\aleph_0^{<\aleph_0} = \aleph_0$, so that Theorem 4.47 yields the existence of an $\aleph_1$-ARONSZAJN tree. As $\kappa^{\mathrm{cf}(\kappa)} > \kappa$ by Corollary 4.21, $\kappa^{<\kappa} = \kappa$ implies that κ is a regular cardinal. By Theorem 4.47, if there is no $\aleph_2$-ARONSZAJN tree, then CH fails.

Proof of Theorem 4.47. Let

$$U = \left\{s \in {}^\alpha\kappa \colon \alpha < \kappa^+ \wedge s \text{ is injective } \wedge \overline{\overline{\kappa \setminus \mathrm{ran}(s)}} = \kappa\right\}$$

We construe U as a tree by having it ordered under end-extension. The tree T which we are about to construct will be a subtree of U and also closed under initial segments and ordered by end-extension. Notice that U cannot have any cofinal branch, as this would yield an injection from κ^+ into κ, so that the tree T we are about to construct cannot have a cofinal branch either.

Because $T \subset U$ will be closed under initial segments and ordered by end-extension, we will have that $\mathrm{lv}_T(s) = \mathrm{dom}(s)$ for every $s \in T$. We shall construct

$$T_\alpha = \{s \in T : \mathrm{lv}_T(s) = \mathrm{dom}(s) < \alpha\}$$

by induction on α. We maintain the following conditions.

(1) $\overline{\overline{\kappa \setminus \mathrm{ran}(s)}} = \kappa$ for all $s \in T$.
(2) If $s \in T$, $\kappa \setminus \mathrm{ran}(s) = A \cup B$, where $A \cap B = \emptyset$ and $\overline{\overline{A}} = \overline{\overline{B}} = \kappa$, and if $\mathrm{lv}_T(s) < \beta < \kappa^+$, then there is some $t \in T$ with $\mathrm{lv}_T(t) = \beta$, $s \subset t$, and $\mathrm{ran}(t) \subset \mathrm{ran}(s) \cup A$.
(3) $\overline{\overline{T_\alpha}} \leq \kappa$ for all $\alpha < \kappa^+$.
(4) Let $\lambda < \kappa^+$ be a limit ordinal with $\mathrm{cf}(\lambda) < \kappa$. Let $C \subset \lambda$ be club in λ with $\mathrm{otp}(C) = \mathrm{cf}(\lambda)$, and let $(\lambda_i : i < \mathrm{cf}(\lambda))$ be the monotone enumeration of C. Let $\kappa = \bigcup_{i<\mathrm{cf}(\lambda)} A_i \cup B$, where $A_i \cap A_j = \emptyset$, $A_i \cap B = \emptyset$, and $\overline{\overline{B}} = \overline{\overline{A_i}} = \kappa$ for $i \neq j, i, j \leq \mathrm{cf}(\lambda)$. Let $s : \lambda \to \kappa$ be such that

$$s \upharpoonright \lambda_i \in T_{\lambda_i+1} \wedge s''\lambda_i \subset \bigcup_{j \leq i+1} A_j$$

for every $i < \mathrm{cf}(\lambda)$. Then $s \in T_{\lambda+1}$.

Well, $T_1 = \{\emptyset\}$, and $T_\lambda = \bigcup_{\alpha<\lambda} T_\alpha$ for limit ordinals $\lambda < \kappa^+$. If $\alpha = \beta + 1 < \kappa^+$, where $\beta = \gamma + 1$ is a successor ordinal, then we just let T_α consist of *all* injections $s \in {}^\beta\kappa$ such that $s \upharpoonright \gamma \in T_\beta$. Notice that (1) through (4) for the tree constructed so far follows from (1) through (4) for the ealier levels of the tree.

Now suppose that $\alpha = \lambda + 1 < \kappa^+$, where λ is a limit ordinal, and T_λ already has been constructed.

Let us first assume that $\mathrm{cf}(\lambda) < \kappa$. We then let $T_{\lambda+1}$ consist of all $s \in {}^\lambda\kappa$ which need to be there in order to satisfy (4). Notice that there are $\lambda^{\mathrm{cf}(\lambda)} \leq \kappa$ many clubs in λ of order type $\mathrm{cf}(\lambda)$, and for each such club C there are $\leq \kappa^{\mathrm{cf}(\lambda)} = \kappa$ possible choices of $s \in {}^\lambda\kappa$ such that for all $\xi \in C$, $s \upharpoonright \xi \in T_{\xi+1}$. Hence $\overline{\overline{T_{\lambda+1}}} \leq \kappa$, and (3) is maintained. (1) is ensured by the fact that for $s \in T_{\lambda+1}$ as being given by (4), $\mathrm{ran}(s) \cap B = \emptyset$.

Let us now assume that $\mathrm{cf}(\lambda) = \kappa$. Let us fix $C \subset \lambda$ club in λ with $\mathrm{otp}(C) = \lambda$, and let $(\lambda_i : i < \kappa)$ be the monotone enumeration of C. To each $s \in T_\lambda$ we shall assign some $t(s) \in U$ with $\mathrm{dom}(t(s)) = \lambda$ as follows. By (1), $\kappa \setminus \mathrm{ran}(s)$ has size κ, and we may hence pick sets A_i, $i < \kappa$, and B such that $\mathrm{ran}(s) \subset A_0$, $\kappa = \bigcup_{i<\kappa} A_i \cup B$, $A_i \cap A_j = \emptyset$, $A_i \cap B = \emptyset$, and $\overline{\overline{B}} = \overline{\overline{A_i}} = \kappa$ for $i, j < \kappa, i \neq j$. Using (2) and (4), we may construct some $t : \lambda \to \kappa$ extending s such that for every $i < \kappa$,

$$t \upharpoonright \lambda_i \in T_{\lambda_i+1} \wedge t''\lambda_i \subset \bigcup_{j \leq i+1} A_j.$$

Let us write $t(s)$ for this t. We may then set

$$T_{\lambda+1} = \{t(s) \colon \ s \in T_\lambda\}.$$

It is easy to verify that (1) through (4) remain true. □

It is much harder to construct a κ^+-Souslin tree. For $\kappa = \aleph_0$, this will be done from a principle called $\diamondsuit$ (cf. Lemma 5.36), and for $\kappa \geq \aleph_1$ we shall need $\diamondsuit_{\kappa^+}$ and $\square_\kappa$ (cf. Lemma 11.68).

Definition 4.48 A cardinal κ is said to have the tree property iff there is no κ-ARONSZAJN tree. A cardinal κ is called *weakly compact* iff κ is inaccessible and κ has the tree property.

The following large cardinal concept will be needed for the analysis of the combinatorial principle $\diamondsuit^*_\kappa$, cf. Definition 5.37.

Definition 4.49 Let κ be a regular uncountable cardinal. Then $R \subset \kappa$ is called *ineffable* iff for every sequence $(A_\xi \colon \ \xi \in R)$ such that $A_\xi \subset \xi$ for every $\xi \in R$ there is some $S \subset R$ which is stationary in κ such that $A_\xi = A_{\xi'} \cap \xi$ whenever ξ, $\xi' \in S, \xi \leq \xi'$.

Trivially, if $R \subset \kappa$ is ineffable, then R is stationary. On the other hand, if $\lambda < \kappa$ is an infinite regular cardinal, then $\{\xi < \kappa \colon \ \mathrm{cf}(\xi) = \lambda\}$ is stationary but *not* ineffable.

If κ itself is ineffable, then κ is weakly compact and the set

$$\{\mu < \kappa \colon \ \mu \text{ is weakly compact }\}$$

is stationary in κ (cf. Problem 4.24). On the other hand, every measurable cadinal (cf. Definition 4.54 below) is ineffable, cf. Lemma 4.58.

The study of (non-trivial) elementary embeddings between transitive structures plays a key role in set theory.

Definition 4.50 Let M, N be transitive sets or classes. We say that $\pi \colon M \to N$ is an *elementary embedding from M to N* iff $(\mathrm{ran}(\pi); \in)$ is an elementary substructure of $(N; \in)$, i.e., if for all formulae φ of the language of set theory and for all $a_1, \ldots,$ $a_k \in M$,

$$(M; \in) \models \varphi(a_1, \ldots, a_k) \Longleftrightarrow (N; \in) \models \varphi(\pi(a_1), \ldots, \pi(a_k)). \tag{4.4}$$

The elementary embedding π is called *non-trivial* iff there is some $x \in M$ with $\pi(x) \neq x$. The least ordinal α with $\pi(\alpha) \neq \alpha$ (if it exists) is called the *critical point of π*, abbreviated as $\mathrm{crit}(\pi)$.

Notice that if $\pi \colon \ M \to N$ is an elementary embedding between transitive sets or classes, then $\pi(\alpha) \geq \alpha$ for all $\alpha \in M$. This is because otherwise there would be a *least* α with $\pi(\alpha) < \alpha$; but then $\pi(\pi(\alpha)) < \pi(\alpha)$, a contradiction! Therefore, if $\mathrm{crit}(\pi)$ exists, then it is the least $\alpha \in M$ with $\pi(\alpha) > \alpha$.

We will mostly be concerned with non-trivial elementary embeddings $\pi\colon V \to M$ from V to M, where M is some transitive class. In this situation, M is of course also a model of ZFC, as by (4.4), the validity of any given axiom of ZFC is moved up from V to M.

Definition 4.51 An inner model is a transitive proper class model of ZFC.

Let M be a transitive model of (some fragment of) ZFC. For terms like V_α, $\text{rk}(x)$, λ^+, etc., we shall denote by $(V_\alpha)^M$, $\text{rk}^M(x)$, λ^{+M}, etc., the respective objects as defined in M, e.g. λ^{+M} = the unique ξ such that

$$(M; \in) \models \xi \text{ is the cardinal successor of } \lambda,$$

etc.

Lemma 4.52 *Let $\pi\colon V \to M$ be a non-trivial elementary embedding, where M is a transitive class. The following hold true.*

(a) $\pi(V_\alpha) = (V_{\pi(\alpha)})^M$ and $\pi(\text{rk}(x)) = \text{rk}(\pi(x))$ for all α and all x.
(b) There is some ordinal ξ with $\pi(\xi) \neq \xi$.

Let κ be the least ordinal ξ with $\pi(\xi) \neq \xi$. The following hold true.

(c) π is continuous at every ordinal of cofinality less than κ, i.e., if α is a limit ordinal and $\text{cf}(\alpha) < \kappa$, then $\pi(\alpha) = \sup(\pi''\alpha)$.
(d) κ is regular and uncountable.
(e) $(V_{\kappa+1})^M = V_{\kappa+1}$.
(f) κ is an inaccessible cardinal.
(g) κ is a MAHLO *cardinal.*
(h) κ is weakly compact.

Proof (a) This is easy.

(b) Let ξ be least such that there is some x with $\text{rk}(x) = \xi$ and $\pi(x) \neq x$. We show that $\pi(\xi) \neq \xi$.

Suppose that $\pi(\xi) = \xi$, i.e., $\xi = \text{rk}(x) = \text{rk}(\pi(x))$. Then by the choice of ξ, $\pi(y) = y$ for all $y \in x \cup \pi(x)$. This means that

$$y \in x \Longleftrightarrow y = \pi(y) \in \pi(x)$$

for all $y \in x \cup \pi(x)$, so that $\pi(x) = x$. Contradiction!

(c) Let $f\colon \text{cf}(\alpha) \to \alpha$ be cofinal in α, where $\text{cf}(\alpha) < \kappa$. Then $\pi(f)\colon \text{cf}(\alpha) \to \pi(\alpha)$ is cofinal in $\pi(\alpha)$ by the elementarity of π. But $\text{ran}(\pi(f)) \subset \pi''\alpha$, so that $\pi(\alpha) = \sup(\pi''\alpha)$.

(d) If κ were singular, then $\pi(\kappa) = \kappa$ using (c). It is easy to see that $\kappa > \aleph_0$.

(e) $(V_\kappa)^M = V_\kappa$ follows from (a) and the choice of κ. Therefore, if $X \subset V_\kappa$, then $X = \pi(X) \cap V_\kappa \in M$, so that $V_{\kappa+1} \subset (V_{\kappa+1})^M$. Trivially, $(V_{\kappa+1})^M \subset V_{\kappa+1}$.

(f) If κ is not inaccessible, then by (d) we may choose $\lambda < \kappa$ and a surjective $f\colon \mathscr{P}(\lambda) \to \kappa$. By (e) and the elementarity of π, $\pi(f)\colon \mathscr{P}(\lambda) \to \pi(\kappa)$ is surjective. However, for all $X \subset \lambda$, $f(X) < \kappa$ and thus $f(X) = \pi(f(X)) = \pi(f)(\pi(X)) = \pi(f)(X)$. This shows that in fact $\pi(f) = f$, so that $\pi(f)$ cannot be a surjection onto $\pi(\kappa) > \kappa$ after all. Contradiction!

(g) Let $C \subset \kappa$ be club in κ. Then $\pi(C)$ is club in $\pi(\kappa)$ by the elementarity of π. Also, $\pi(C) \cap \kappa = C$. Therefore, $\kappa \in \pi(C)$. By (e),

$$(M; \in) \models \kappa \text{ is inaccessible,}$$

so that

$$(M; \in) \models \exists \xi \ (\xi \text{ is inaccessible and } \xi \in \pi(C)).$$

By the elementarity of π,

$$(V; \in) \models \exists \xi \ (\xi \text{ is inaccessible and } \xi \in C).$$

As C was arbitrary, this shows that κ is a MAHLO cardinal.

(h): Let $(T, <_T)$ be a κ-tree. Writing $(\pi(T), <_{\pi(T)}) = \pi((T, <_T))$, we have that

$$(M; \in) \models (\pi(T), <_{\pi(T)}) \text{ is a } \pi(\kappa)\text{-tree.}$$

Let $s \in \pi(T)$ be such that $(M; \in) \models \mathrm{rk}_{\pi(T)}(s) = \kappa$, and set

$$b = \left\{t \in \pi(T) \colon \ t <_{\pi(T)} s\right\}.$$

We may assume without loss of generality that $(T, <_T) \in V_{\kappa+1}$, so that $\pi(T) \cap V_\kappa = T$ and $<_{\pi(T)} \restriction T = <_T$. But then $b \in M \subset V$ is a cofinal branch through T. □

A cardinal κ is called REINHARDT iff there is a non-trivial elementary embedding $\pi : V \to V$ with $\kappa = \mathrm{crit}(\pi)$. The following Theorem shows that there are no REINHARDT cardinals (in ZFC).

Theorem 4.53 (K. Kunen) *There is no non-trivial elementary embedding* $\pi \colon V \to V$.

Proof Let $\kappa_0 = \mathrm{crit}(\pi)$, and recursively define $\kappa_{n+1} = \pi(\kappa_n)$. Set $\lambda = \sup_{n<\omega}\kappa_n$. By Lemma 4.52 (c), $\pi(\lambda) = \lambda$, and therefore also $\pi(\lambda^+) = \pi(\lambda)^+ = \lambda^+$.

Let $S = \{\alpha < \lambda^+ \colon \ \mathrm{cf}(\alpha) = \omega\}$. Because S is a stationary subset of λ^+, S may be partitioned into κ_0 stationary sets by Theorem 4.33, i.e., we may choose $(S_i : i < \kappa_0)$ such that $S = \bigcup_{i<\kappa_0} S_i$, $S_i \cap S_j = \emptyset$ for $i \neq j$, i, $j < \kappa_0$, and each S_i, $i < \kappa_0$, is stationary in λ^+.

Set $(T_i \colon \ i < \kappa_1) = \pi((S_i \colon \ i < \kappa_0))$. We have that $T_i \cap T_j = \emptyset$ for $i \neq j$, i, $j < \kappa_1$, and each T_i, $i < \kappa_1$, is stationary in λ^+. Then T_{κ_0} is a stationary subset of λ^+. By Lemma 4.52 (c) and $\pi''\lambda^+ \subset \lambda^+$,

$$C = \left\{\alpha < \lambda^+ \colon \ \mathrm{cf}(\alpha) = \omega \wedge \pi(\alpha) = \alpha\right\}$$

is an ω-club in λ^+. There is hence some $\alpha \in T_{\kappa_0} \cap C$ (cf. Problem 4.12). As $C \subset S = \bigcup_{i<\kappa_0} S_i$, there must be some $i < \kappa_0$ with $\alpha \in S_i$. But then $\alpha = \pi(\alpha) \in \pi(S_i) = T_{\pi(i)}$, so that $T_{\pi(i)} \cap T_{\kappa_0} \neq \emptyset$. But $\kappa_0 = \text{crit}(\pi)$, so that κ_0 is not in the range of π and therefore $\pi(i) \neq \kappa_0$. Contradiction! □

The proof of Theorem 4.53 in fact shows that there can be no non-trivial elementary embedding $\pi: V_{\lambda+2} \to V_{\lambda+2}$ with $\text{crit}(\pi) < \lambda$. We remark that the proof of Theorem 4.53 uses Theorem 4.33 which in turn makes use of the Axiom of Choice. It is open whether Theorem 4.53 can be proven in **ZF** alone; this question leads to WOODIN's **HOD**-conjecture, cf. [45, Section 7].

Large cardinal theory studies the question which "fragments" of REINHARDT cardinals are consistent with **ZFC**.

Definition 4.54 Let κ be a cardinal. A filter F on κ is called *uniform* iff $\overline{\overline{X}} = \kappa$ for every $X \in F$. An uncountable cardinal κ is called *measurable* iff there is a $< \kappa$-closed uniform ultrafilter on κ. Such a filter is also called a *measure on* κ

It is easy to see that if U is a $< \kappa$-closed ultrafilter on κ, then U is uniform iff for no $\xi < \kappa$, $\{\xi\} \in U$, i.e., iff U is not generated by a singleton (cf. Problem 4.26). If we didn't require a measurable cardinal to be uncountable, then $\aleph_0$ would be a measurable cardinal.

Theorem 4.55 *Let κ be a cardinal. The following are equivalent.*

(1) *κ is measurable.*
(2) *There is a* normal *$< \kappa$-closed uniform ultrafilter on κ.*
(3) *There is an inner model M and an elementary embedding $\pi: V \to M$ with critical point κ.*

Proof (3) $\Longrightarrow$ (2): Let $\pi : V \to M$ be an elementary embedding with critical point κ. Let us set

$$U = U_\pi = \{X \subset \kappa : \kappa \in \pi(X)\}. \tag{4.5}$$

We claim that U is a normal $< \kappa$-closed uniform ultrafilter on κ. Well, $\{\xi\} \notin U$ for any $\xi < \kappa$, and $\kappa \in U$, as κ is the critical point of π. U is easily seen to be an ultrafilter, as $\pi(X \cap Y) = \pi(X) \cap \pi(Y)$ for all sets X, Y, $\pi(X) \subset \pi(Y)$ for all $X \subset Y$, and $\kappa \in \pi(\kappa) = \pi(X) \cup \pi(\kappa \setminus X)$ for all $X \subset \kappa$. Moreover, if $\alpha < \kappa$, then $\pi(\{X_i : i < \alpha\}) = \{\pi(X_i) : i < \alpha\}$ for all $\{X_i : i < \alpha\}$, which yields that U is $< \kappa$-closed. Hence U witnesses that κ is a measurable cardinal.

It remains to be shown that U is normal. Let $(X_i : i < \kappa)$ be such that $X_i \in U$ for all $i < \kappa$. We need to see that $\Delta_{i<\kappa} X_i \in U$, i.e., $\kappa \in \pi(\Delta_{i<\kappa} X_i)$. Writing $(Y_i : i < \pi(\kappa)) = \pi((X_i : i < \kappa))$, we have that $Y_i = \pi(X_i)$ for every $i < \kappa$, so that $\kappa \in \bigcap_{i<\kappa} Y_i$. This just means that $\kappa \in \pi(\Delta_{i<\kappa} X_i)$.

(2) $\Longrightarrow$ (1) is trivial.

(1) $\Longrightarrow$ (3): This will be shown by an ultrapower construction which is well-known from model theory and which will be refined later (cf. the proof of Theorem 10.48).

Let U be a $< \kappa$-closed uniform ultrafilter on κ. We aim to construct an inner model M and an elementary embedding $\pi : V \to M$ with critical point κ. We shall first construct M.

If $f, g \in {}^{\kappa}V$, we write $f \sim g$ iff $\{\xi < \kappa : f(\xi) = g(\xi)\} \in U$. It is easy to verify that $\sim$ is an equivalence relation. For any $f \in {}^{\kappa}V$, we write $[f]$ for the $\sim$-equivalence class of f, massaged by SCOTT's trick, i.e., for the set $\{g \in {}^{\kappa}V : g \sim f \wedge \forall h \in {}^{\kappa}V (h \sim f \to \text{rk}(h) \geq \text{rk}(g)\}$. If $f, g \in {}^{\kappa}V$, then we write $[f]E[g]$ iff $\{\xi < \kappa : f(x) \in g(x)\} \in U$. It is easy to check that E is extensional. Also, for all $f \in {}^{\kappa}V$, $\{[g] : [g]E[f]\}$ is a *set*.

Let us write $\text{ult}(V; U)$ for the structure $(\{[f] : f \in {}^{\kappa}V\}; E)$. We may define a map $\bar{\pi}$ from V into $\text{ult}(V; U)$ by setting

$$\bar{\pi}(x) = [c_x],$$

where $c_x : \kappa \to \{x\}$ is the constant function with value x. The following statement shows that $\bar{\pi}$ is elementary and it is referred to as the Łoś *Theorem*.

Claim 4.56 (Łoś Theorem) *Let $\varphi(v_1, \ldots, v_k)$ be a formula, and let $f_1, \ldots, f_k \in {}^{\kappa}V$. Then*

$$\text{ult}(V; U) \models \varphi([f_1], \ldots, [f_k]) \Longleftrightarrow$$
$$\{\xi < \kappa : V \models \varphi(f_1(\xi), \ldots, f_k(\xi))\} \in U.$$

Proof of Claim 4.56 by induction on the complexity of φ: The atomic case is immediate from the definition, as for $f, g \in {}^{\kappa}V$ we have that $\text{ult}(V; U) \models [f] \in [g]$ iff $[f]E[g]$ iff $\{\xi < \kappa : f(x) \in g(x)\} \in U$ and $\text{ult}(V; U) \models [f] = [g]$ iff $[f]E[g]$ iff $\{\xi < \kappa : f(x) = g(x)\} \in U$.

As for the sentential connectives, it suffices to discuss $\wedge$ and $\neg$.

As for $\wedge$, if $\varphi(v_1, \ldots, v_k)$ and $\psi(v_1, \ldots, v_\ell)$ are formulae, if $f_1, \ldots, f_k$, $g_1, \ldots, g_\ell \in {}^{\kappa}V$, and if the Claim holds for φ and ψ, then $\text{ult}(V; U) \models (\varphi([f_1], \ldots, [f_k]) \wedge \psi([g_1], \cdots, [g_\ell]))$ iff $\text{ult}(V; U) \models \varphi([f_1], \ldots, [f_k])$ *and* $\text{ult}(V; U) \models \psi([g_1], \ldots, [g_\ell])$ iff $\{\xi < \kappa : V \models \varphi(f_1(\xi), \ldots, f_k(\xi))\} \in U$ *and* $\{\xi < \kappa : V \models \psi(g_1(\xi), \ldots, g_\ell(\xi))\} \in U$ iff $\{\xi < \kappa : V \models (\varphi(f_1(\xi), \ldots, f_k(\xi)) \wedge \psi(g_1(\xi), \ldots, g_\ell(\xi)))\} \in U$, as U is a filter.

As for $\neg$, if $\varphi(v_1, \ldots, v_k)$ is a formula, if $f_1, \ldots, f_k \in {}^{\kappa}V$, and if the Claim holds for φ, then $\text{ult}(V; U) \models \neg\varphi([f_1], \ldots, [f_k])$ iff $\text{ult}(V; U)$ is *not* a model of $\varphi([f_1], \ldots, [f_k])$ iff $\{\xi < \kappa ; V \models \varphi(f_1(\xi), \cdots, f_k(\xi))\} \notin U$ iff $\{\xi < \kappa : V \models \neg\varphi(f_1(\xi), \ldots, f_k(\xi))\} \in U$, as U is an *ultra*filter.

Let us finally suppose that $\varphi(v_1, \ldots, v_k) \equiv \exists v_0 \, \psi(v_0, v_1, \ldots, v_k)$ for some formula ψ for which the Claim holds true. Let $f_1, \ldots, f_k \in {}^{\kappa}V$. If $\text{ult}(V; U) \models \exists v_0 \, \psi(v_0, [f_1], \ldots, [f_k])$, then there is some $f_0 \in {}^{\kappa}V$ such that

$$\text{ult}(V; U) \models \psi([f_0], [f_1], \ldots, [f_k]).$$

By induction, $\{\xi < \kappa \colon\ V \models \psi(f_0(\xi), f_1(\xi), \ldots, f_k(\xi))\} \in U$, which also gives that $\{\xi < \kappa \colon\ V \models \exists v_0\, \psi(v_0, f_1(\xi), \ldots, f_k(\xi))\} \in U$, as U is a filter.

Conversely, let us assume that $\{\xi < \kappa \colon\ V \models \exists v_0\, \psi(v_0, f_1(\xi), \ldots, f_k(\xi))\} \in U$. By the replacement schema in V, there is a set a such that for all $\xi < \kappa$, if there is some x with $V \models \psi(x, f_1(\xi), \ldots, f_k(\xi))$, then there is some $x \in a$ with $V \models \psi(x, f_1(\xi), \ldots, f_k(\xi))$. Let $<_a$ be a well-ordering of a. Let us define $f_0 \colon\ \kappa \to V$ as follows.

$$f_0(\xi) = \begin{cases} \text{the } <_a\text{-smallest } x \in a \text{ with} & \\ V \models \psi(x, f_1(\xi), \ldots, f_k(\xi)) & \text{if some such } x \text{ exists,} \\ \emptyset & \text{otherwise.} \end{cases}$$

By the choice of a we now have that $\{\xi < \kappa \colon\ V \models \psi(f_0(\xi), f_1(\xi), \ldots, f_k(\xi))\} \in U$, which inductively implies that $\mathrm{ult}(V; U) \models \psi([f_0], [f_1], \ldots, [f_k])$, and hence that $\mathrm{ult}(V; U) \models \exists v_0\, \psi(v_0, [f_1], \ldots, [f_k])$.

This verifies the Claim. □

We now prove that E is well-founded, using Lemma 3.11. If $([f_n] \colon n < \omega)$ were a sequence such that $[f_{n+1}] E [f_n]$ for all $n < \omega$, then

$$\bigcap \{\xi < \kappa \colon\ f_{n+1}(\xi) \in f_n(\xi)\} \in U$$

because U is $< \aleph_1$-closed, and then if $\xi \in \bigcap \{\xi < \kappa \colon\ f_{n+1}(\xi) \in f_n(\xi)\}$,

$$\ldots \in f_2(\xi) \in f_1(\xi) \in f_0(\xi),$$

a contradiction.

Therefore, by Theorem 3.21 there is an inner model N and some σ such that $(N; \in) \overset{\sigma}{\cong} (\{[f] \colon\ f \in {}^\kappa V\}; E)$. By Łoś' Theorem, we have that

$$N \models \varphi(\sigma^{-1}([f_1]), \ldots, \sigma^{-1}([f_k]))$$

if and only if

$$\{\xi < \kappa \colon\ V \models \varphi(f_1(\xi), \ldots, f_k(\xi))\} \in U$$

for all formulae φ and $f_1, \ldots, f_k \in {}^\kappa V$. This implies that we may define an elementary embedding $\pi_U \colon\ V \to N$ by setting $\pi_U(x) = \sigma^{-1} \circ \bar{\pi}(x) = \sigma^{-1}([c_x])$, where again $c_x \colon\ \kappa \to \{x\}$ is the constant function with value x.

It remains to be shown that κ is the critical point of π_U. We first prove that $\pi_U(\eta) = \eta$ for all $\eta < \kappa$ by induction on η. Fix $\eta < \kappa$ and suppose that $\pi_U(\bar{\eta}) = \bar{\eta}$ for all $\bar{\eta} < \eta$. Let $\sigma^{-1}([f]) < \pi_U(\eta)$, i.e., $\{\xi < \kappa \colon\ f(\xi) < \eta\} \in U$. As U is $< \eta^+$-closed, there is then some $\eta_0 < \eta$ such that $\{\xi < \kappa \colon\ f(\xi) = \eta_0\} \in U$. But then $[f] = [c_{\eta_0}]$, so that by the inductive hypothesis $\eta_0 = \pi_U(\eta_0) = \sigma^{-1}([f])$. This shows that $\pi_U(\eta) \leq \eta$, so that in fact $\pi_U(\eta) = \eta$.

Finally, if id denotes the identity function on κ, then for all $\eta < \kappa$, $\eta = \pi_U(\eta) = \sigma^{-1}([c_\eta]) < \sigma^{-1}([\mathrm{id}])$ by the uniformity of U. Also $\sigma^{-1}([\mathrm{id}]) < \sigma^{-1}([c_\kappa]) = \pi_U(\kappa)$. Hence κ is indeed the critical point of π_U. □

In the situation of the proof of (1) $\Longrightarrow$ (3) of Thorem 4.55, we usually also write $\mathrm{ult}(V; U)$ for the inner model which was called N there. Let $\gamma = \sigma^{-1}([\mathrm{id}])$. We must have that

$$\sigma^{-1}([f]) = \pi_U(f)(\gamma) \text{ for all } f\colon \kappa \to V. \tag{4.6}$$

This is because $\pi_U(f)(\gamma)$ may be written in a cumbersome way as

$$\sigma^{-1}([c_f])\left(\sigma^{-1}([\mathrm{id}])\right),$$

which with the help of Łoś' Theorem is easily seen to be equal to $\sigma^{-1}([f])$. In particular,

$$N = \mathrm{ult}(V; U) = \{\pi_U(f)(\gamma) : f\colon \kappa \to V\}. \tag{4.7}$$

γ is often called the *generator* of U. It is also easy to see that for $X \subset \kappa$,

$$X \in U \Longleftrightarrow \gamma \in \pi(X). \tag{4.8}$$

Now let $\pi\colon V \to M$ be any elementary embedding with critical point κ, where M is an inner model. Let $U = U_\pi$ be derived as in (4.5) in the proof of (3) $\Longrightarrow$ (2) of Theorem 4.55, i.e., $U = \{X \subset \kappa : \kappa \in \pi(X)\}$. Let $\pi_U\colon V \to \mathrm{ult}(V; U)$ be as constructed in the proof of (1) $\Longrightarrow$ (3). We may then define a factor map

$$k\colon \mathrm{ult}(V : U) \to M$$

by setting $k(\pi_U(f)(\kappa)) = \pi(f)(\kappa)$ for $f\colon \kappa \to V$. This map k is well-defined and elementary because we have that

$$\begin{aligned}
\mathrm{ult}(V; U) \models \varphi(\pi_U(f_1)(\kappa), \ldots, \pi_U(f_k)(\kappa)) &\Longleftrightarrow \\
\{\xi < \kappa\colon V \models \varphi(f_1(\xi), \ldots, f_k(\xi))\} \in U &\Longleftrightarrow \\
\kappa \in \pi(\{\xi < \kappa\colon V \models \varphi(f_1(\xi), \ldots, f_k(\xi))\}) &= \\
\{\xi < \pi(\kappa)\colon M \models \varphi(\pi(f_1)(\xi), \ldots, \pi(f_k)(\xi))\} &\Longleftrightarrow \\
M \models \varphi(\pi(f_1)(\kappa), \ldots, \pi(f_k)(\kappa)).&
\end{aligned}$$

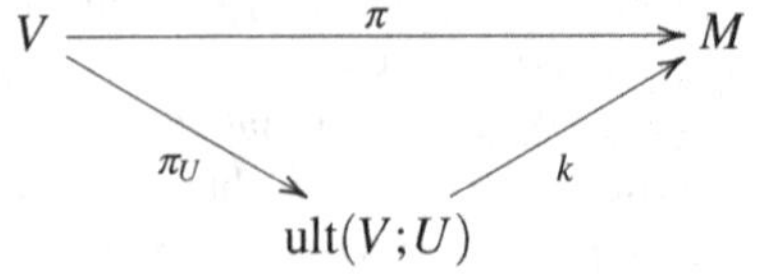

If $\pi: V \to M$ is itself an ultrapower map, i.e., $\pi = \pi_{U'}$ for some $< \kappa$-closed uniform ultrafilter U' on κ and if $\kappa = [g]$ in the sense of the ultrapower $\text{ult}(V; U)$, then

$$X \in U \Longleftrightarrow g^{-1\prime\prime} X \in U' \tag{4.9}$$

for every $X \subset \kappa$. We will have that U' is normal iff $U = U'$ iff $\{\xi < \kappa : g(\xi) = \xi\} \in U'$. (Cf. problem 4.26.)

We remark that M. Gitik has shown that AC is needed to prove (1) $\Longrightarrow$ (2) in the statement of Theorem 4.55 (cf. [7]).

Definition 4.57 Let M be a transitive model of ZFC, and let $M \models U$ is a measure. Then we shall ambiguously write $\text{ult}(M; U)$ for $\text{ult}(V; U)$ from the proof of Theorem 4.55 as defined inside M or for its transitive collapse. Also, we shall ambiguously write π_U^M for the map $\bar{\pi}$ from the proof of Theorem 4.55 or for the map π from the proof of Theorem 4.55. $\text{ult}(M; U)$ is called the *ultrapower of M by U*, and π_U^M is called the associated *ultrapower embedding*.

By (4.7), applied inside M, if U is a measure on κ, then

$$\text{ult}(M; U) = \left\{ \pi_U^M(f)(\gamma) : f : \kappa \to M \wedge f \in M \right\}, \tag{4.10}$$

where $\gamma = [id]$.

Lemma 4.58 *Let κ be a measurable cardinal, and let U be a normal $< \kappa$-closed ultrafilter on κ. If $R \in U$, then R is ineffable.*

Proof Let $(A_\xi : \xi \in R)$ be such that $A_\xi \subset \xi$ for every $\xi \in R$. Let

$$\pi = \pi_U^V : V \to M,$$

and let $(\tilde{A}_\xi : \xi \in \pi(R)) = \pi((A_\xi : \xi \in R))$. As $\kappa \in \pi(R)$, we may set $A = \tilde{A}_\kappa$.

By the Łoś Theorem, for every $\alpha < \kappa$ there is some $X_\alpha \in U$ such that for all $\xi \in X_\alpha$, $\alpha \in A$ iff $\alpha \in A_\xi$. Let $X = \Delta_{\alpha<\kappa} X_\alpha$. It is then easy to verify that for every $\xi \in X$, $A_\xi = A \cap \xi$. $\square$

Theorem 4.59 (Rowbottom) *Let κ be a measurable cardinal, and let U be a normal measure on κ. Let $\gamma < \kappa$, and let $F : [\kappa]^{<\omega} \to \gamma$. There is then some $X \in U$ such that for every $n < \omega$, $F \restriction [X]^n$ is constant.*

Proof Fix $\gamma < \kappa$. It suffices to show that for every $n < \omega$,

$$\forall F : [\kappa]^n \to \gamma \; \exists \xi < \gamma \; \exists X \in U \; F''[X]^n = \{\xi\}. \tag{4.11}$$

This is because if $F : [\kappa]^{<\omega} \to \gamma$ and for each $n < \omega$, $X_n \in U$ is such that $F \restriction [X_n]^n$ is constant, then $\bigcap_{n<\omega} X_n \in U$ is as desired.

We prove (4.11) by induction on n. For $n = 0$, (4.11) is trivial. So let us assume (4.11) for n and show it for $n + 1$.

Let $F\colon [\kappa]^{n+1} \to \gamma$ be given. Let

$$\pi\colon V \to_U M = \mathrm{ult}(V; U)$$

be the ultrapower map given by U, where M is an inner model, $\kappa = \mathrm{crit}(\pi)$, and for all $X \subset \kappa$, $X \in U$ iff $\kappa \in \pi(X)$. Let $F^*\colon [\kappa]^n \to \gamma$ be defined by $F^*(a) = \pi(F)(a \cup \{\kappa\})$ for $a \in [\kappa]^n$. By the inductive hypothesis, there is some $\xi < \gamma$ and some $X \in U$ such that $F^{*\prime\prime}[X]^n = \{\xi\}$. That is, for every $a \in [X]^n$, $\pi(F)(a \cup \{\kappa\}) = \xi$, or equivalently,

$$X_a = \{\eta < \kappa\colon \eta > \max(a) \wedge F(a \cup \{\eta\}) = \xi\} \in U. \tag{4.12}$$

By the normality of U, we may pick some $Y \in U$, $Y \subset X$, such that for every $\eta \in X$,

$$\eta \in \bigcap_{a \in [X \cap \eta]^n} X_a. \tag{4.13}$$

We then have for $a \cup \{\eta\} \in [Y]^{n+1}$ with $\eta > \max(a)$ that $\eta \in Y$, hence, as $Y \subset X$, $\eta \in X_a$ by (4.13), and so $F(a \cup \{\eta\}) = \xi$ by (4.12). We have shown that $F''[Y]^{n+1} = \{\xi\}$, where $Y \in U$. □

Definition 4.60 Let κ be a cardinal, and let $\alpha > \kappa$. Then κ is called *α-strong* iff there is some non-trivial elementary embedding $\pi\colon V \to M$, where M is an inner model and $\mathrm{crit}(\pi) = \kappa$, such that $V_\alpha \subset M$. κ is called *strong* iff κ is α-strong for all $\alpha > \kappa$.

Lemma 4.61 *If κ is measurable, then κ is $(\kappa + 1)$-strong. If κ is $(\kappa + 2)$-strong, then κ is measurable and there exists a measurable cardinal $\lambda < \kappa$.*

Proof The first part immediately follows from Lemma 4.52. As for the second part, let κ be $(\kappa + 2)$-strong, and let

$$\pi\colon V \to M$$

be an elementary embedding, where M is an inner model, $\mathrm{crit}(\pi) = \kappa$, and $V_{\kappa+2} \subset M$. By Lemma 4.55, κ is measurable, and there is hence some $< \kappa$-closed uniform ultrafilter U on κ. But $U \in V_{\kappa+2} \subset M$, which gives that

$$M \models \exists \lambda < \pi(\kappa) \exists U_0\ (U_0 \text{ is a } < \lambda\text{-closed uniform ultrafilter on } \lambda > \aleph_0).$$

By the elementarity of π, this gives that

$$V \models \exists \lambda < \kappa \exists U_0\ (U_0 \text{ is a } < \lambda\text{-closed uniform ultrafilter on } \lambda > \aleph_0).$$

There is thus a measurable cardinal $\lambda < \kappa$. □

Definition 4.62 Let κ be a cardinal, and let $\lambda \geq \kappa$ be a regular cardinal. Then κ is called λ-supercompact iff there is some non-trivial elementary embedding $\pi : V \to M$, where M is an inner model, $\text{crit}(\pi) = \kappa$, and $\pi(\kappa) > \lambda$, such that ${}^{\lambda}M \subset M$. κ is called *supercompact* iff κ is λ-supercompact for all regular $\lambda \geq \kappa$.

Measures cannot witness suprcompactess, cf. Problem 4.27. We will see later, cf. Lemmas 10.58 and 10.62, that "extenders" may be used to witness that κ is strong or supercompact in much the same way as a measure witnesses that a given cardinal is measurable.

Lemma 4.63 *If κ is measurable, then κ is κ-supercompact. If κ is 2^κ-supercompact, then κ is measurable and there is a measure U on κ such that*

$$\{\mu < \kappa : \mu \text{ is measurable }\} \in U.$$

Proof We use Theorem 4.55. Let κ be measurable, let U be a normal measure on κ, and let

$$\pi : V \to_U \text{ult}(V; U) = M$$

be the ultrapower embedding, where we assume M to be transitive. We need to see that ${}^{\kappa}M \subset M$. Let $(x_i : i < \kappa)$ be a sequence with $x_i \in M$ for all $i < \kappa$. Say $x_i = \pi(f_i)(\kappa)$, where $f_i : \kappa \to V$, for $i < \kappa$. Let us define $g : \kappa \to V$ as follows. For each $\xi < \kappa$, $g(\xi) : \xi \to V$ and for $i < \xi$, $g(\xi)(i) = f_i(\xi)$. We then get that $\pi(g)(\kappa) : \kappa \to M$ (here we use Problem 4.26), and for every $i < \kappa$,

$$\{\xi < \kappa : g(\xi)(i) = f_i(\xi)\} = \kappa \setminus (i+1) \in U,$$

so that $\pi(g)(\kappa)(i) = \pi(f_i)(\kappa) = x_i$. Thus $(x_i : i < \kappa) = \pi(g)(\kappa) \in M$, as desired.

If κ is 2^κ-supercompact, then we may pick some

$$\pi : V \to M,$$

where M is an inner model, $\text{crit}(\pi) = \kappa$, and ${}^{2^\kappa}M \subset M$. In particular, κ is measurable, and if U' is a measure on κ, then $U' \in M$. Therefore,

$$\kappa \in \pi(\{\xi < \kappa : V \models \xi \text{ is measurable}\}),$$

so that if U is the measure on κ derived from π as in the proof of Theorem 4.55, then $\{\mu < \kappa : \mu \text{ is measurable}\} \in U$ as desired. □

Definition 4.64 Let κ be a regular cardinal, and let F be a filter on κ. F is called *weakly normal* iff for all $f : \kappa \to \kappa$ with $\{\xi < \kappa : f(\xi) < \xi\} \in F^+$ there is some $\alpha < \kappa$ and some $X \in F^+$ such that $f(\xi) < \alpha$ for every $\xi \in X$.

By Lemma 4.31, every normal filter is weakly normal.

Lemma 4.65 *Let κ be λ-supercompact, where $\lambda \geq \kappa$ is a regular cardinal. There is then a $< \kappa$-closed uniform weakly normal ultrafilter on λ.*

Proof Let

$$\pi\colon\ V \to M$$

be an elementary embedding, where M is an inner model, $\text{crit}(\pi) = \kappa$, $\pi(\kappa) > \lambda$, and ${}^{\lambda}M \subset M$. Let us set

$$U = \{X \subset \lambda\colon\ \sup(\pi''\lambda) \in \pi(X)\}. \tag{4.14}$$

It is not hard to verify that U is a $< \kappa$-closed uniform ultrafilter on λ.

To show that U is weakly normal, let $f\colon\ \kappa \to \kappa$. As U is an ultrafilter, $U^+ = U$. If

$$\sup(\pi''\lambda) \in \pi(\{\xi < \kappa\colon\ f(\xi) < \xi\}) = \{\xi < \pi(\kappa)\colon \pi(f)(\xi) < \xi\},$$

then we may pick $\alpha < \lambda$ such that

$$\pi(f)(\sup(\pi''\lambda)) < \pi(\alpha),$$

so that

$$\sup(\pi''\lambda) \in \pi(\{\xi < \kappa\colon\ f(\xi) < \alpha\}).$$

U is thus weakly normal. □

Theorem 4.66 (Solovay) *Let κ be supercompact. Then $\lambda^{<\kappa} = \lambda$ for every regular cardinal $\lambda \geq \kappa$.*

Proof Let us fix λ, let again

$$\pi\colon\ V \to M$$

be an elementary embedding, where M is an inner model, $\text{crit}(\pi) = \kappa$, $\pi(\kappa) > \lambda$, and ${}^{\lambda}M \subset M$. Let U be the $< \kappa$-closed uniform weakly normal ultrafilter on λ which is given by (4.14).

Let us write

$$S = \{\alpha < \lambda\colon\ \text{cf}(\alpha) < \kappa\}.$$

Notice that $S \in U$. This is because ${}^{\lambda}M \subset M$ yields that $M \models \text{cf}(\sup \pi''\lambda) = \lambda$, which together with $\lambda < \pi(\kappa)$ then gives that $\sup \pi''\lambda \in \{\alpha < \pi(\lambda)\colon\ M \models \text{cf}(\alpha) < \pi(\kappa)\} = \pi(S)$.

For $\alpha \in S$, let us pick $C_\alpha \subset \alpha$ cofinal in α with $\text{otp}(C_\alpha) = \text{cf}(\alpha)$.

Let $\eta < \lambda$ be arbitrary. Because $S \in U$ and U is uniform,

$$\bar{S} = \{\alpha \in S\colon C_\alpha \setminus \eta \neq \emptyset\} \in U.$$

Let $f\colon\ \bar{S} \to \lambda$ be defined by

$$f(\alpha) = \text{ the least element of } C_\alpha \setminus \eta$$

for $\alpha \in \bar{S}$. Then f is a regressive function, and because U is weakly normal there is some $\eta' > \eta$, $\eta' < \lambda$, such that

$$\{\alpha \in \bar{S}\colon\ f(\alpha) < \eta'\} \in U.$$

We have shown that

$$\forall \eta < \lambda \exists \eta' < \lambda\ (\eta' > \eta \wedge \{\alpha \in S\colon\ C_\alpha \cap [\eta, \eta') \neq \emptyset\} \in U). \tag{4.15}$$

By (4.15), there is now a continuous and strictly increasing sequence $(\eta_\xi\colon\ \xi < \lambda)$ such that $\eta_0 = 0$ and

$$\forall \xi < \lambda\ \{\alpha \in S\colon\ C_\alpha \cap [\eta_\xi, \eta_{\xi+1}) \neq \emptyset\} \in U. \tag{4.16}$$

Let us write

$$I_\alpha = \{\xi < \lambda\colon\ C_\alpha \cap [\eta_\xi, \eta_{\xi+1}) \neq \emptyset\}$$

for $\alpha \in S$, so that $\overline{\overline{I_\alpha}} \leq \overline{\overline{C_\alpha}} = \mathrm{cf}(\alpha) < \kappa$ for every $\alpha \in S$.

Let $X \in [\lambda]^{<\kappa}$. For all $\xi \in X$, $\{\alpha \in S\colon\ \xi \in I_\alpha\} \in U$ by (4.16) and the definition of I_α. Therefore, as $\overline{\overline{X}} < \kappa$ and U is $<\kappa$-closed,

$$\{\alpha \in S\colon\ X \subset I_\alpha\} = \{\alpha \in S\colon\ \forall \xi \in X\ \xi \in I_\alpha\} \in U. \tag{4.17}$$

In particular, there is some $\alpha \in S$ such that $X \subset I_\alpha$.

We have shown that

$$[\lambda]^{<\kappa} \subset \bigcup_{\alpha \in S} \mathscr{P}(I_\alpha),$$

so that $\lambda^{<\kappa} \leq \lambda \cdot 2^{<\kappa} = \lambda \leq \lambda^{<\kappa}$, i.e., $\lambda^{<\kappa} = \lambda$. □

Corollary 4.67 *Let κ be supercompact. Then* SCH *holds above κ, i.e., if $\mu > \kappa$ is singular, then $\mu^{\mathrm{cf}(\mu)} = 2^{\mathrm{cf}(\mu)} \cdot \mu^+$.*

Proof By (the proof of) SILVER's Theorem 4.36 (cf. Problem 4.17 (1)) it suffices to prove that $\mu^{\aleph_0} = \mu^+$ for every $\mu > \kappa$ with $\mathrm{cf}(\mu) = \omega$. However, for every such μ, $\mu^{\aleph_0} \leq (\mu^+)^{\aleph_0} = \mu^+$ by Theorem 4.66. □

The following large cardinal will play a role for the failure of $\square_\kappa$, cf. Definition 11.62 and Lemma 11.69.

Definition 4.68 A cardinal κ is called *subcompact* iff for every $A \subset H_{\kappa^+}$ there is some $\lambda < \kappa$ and some $B \subset H_{\lambda^+}$ such that there is an elementary embedding $\sigma\colon (H_{\lambda^+}; \in, B) \to (H_{\kappa^+}; \in, A)$.

Notice that in the situation of Definition 4.68, $\sigma(\lambda) = \kappa$.

Lemma 4.69 *Suppose that κ is 2^κ-supercompact. Then κ is subcompact.*

Proof Let $A \subset H_{\kappa^+}$, and let

$$\pi: V \to M$$

be such that M is an inner model, $\text{crit}(\pi) = \kappa$, and ${}^{2^\kappa}M \subset M$. We have that $(H_{\kappa^+})^M = H_{\kappa^+} \in M$ and $\pi \restriction H_{\kappa^+} \in M$, and therefore

$M \models \exists \lambda < \pi(\kappa) \exists B \subset H_{\lambda^+} \exists \sigma (\sigma: (H_{\lambda^+}; \in, B) \to (H_{\pi(\kappa)^+}; \in, \pi(A))$ is elementary $\wedge \text{crit}(\sigma) = \lambda)$.

By the elementarity of π,

$V \models \exists \lambda < \kappa \exists B \subset H_{\lambda^+} \exists \sigma (\sigma: (H_{\lambda^+}; \in, B) \to (H_{\kappa^+}; \in, A)$ is elementary $\wedge \text{crit}(\sigma) = \lambda)$.

We have shown that κ is subcompact. □

Amazing results in cardinal arithmetic were obtained by SAHARON SHELAH via his pcf-theory, cf. e.g. [1]

4.4 Problems

4.1. Let $\alpha < \omega_1$. Show that there is some $X \subset \mathbb{Q}$ such that $(\alpha; <) \cong (X; <_{\mathbb{Q}} \restriction X)$. (Here, $<$ denotes the natural order on α and $<_{\mathbb{Q}}$ denotes the natural order on $\mathbb{Q}$.) [Hint. Use induction on α.]

4.2. Let κ be a cardinal. Let $y = \kappa \cup x$, where $x = \{\alpha : \alpha$ is an ordinal with $\alpha \sim \kappa\}$. Show that $y = \kappa^+$.

4.3. Let κ and λ be cardinals. Show that $\kappa + \lambda = \text{Card}(X \cup Y)$, whenever X, Y are disjoint sets with $\overline{\overline{X}} = \kappa$ and $\overline{\overline{Y}} = \lambda$. Also show that $\kappa + \lambda \leq \kappa \cdot \lambda$ whenever $\kappa, \lambda \geq 2$.

4.4. Show that if κ and λ are cardinals, then $\kappa^\lambda = \text{Card}([\kappa]^\lambda)$.

4.5. Show that the least κ with $\aleph_\kappa = \kappa$ is singular of cofinality ω. Show that for every regular cardinal λ there is some κ with $\aleph_\kappa = \kappa$ and $\text{cf}(\kappa) = \lambda$.

4.6. Show that if κ is a limit cardinal, then $\text{cf}(\kappa)$ may be characterized as the least λ such that there is a sequence $(\kappa_i : i < \lambda)$ of cardinals less than κ with $\kappa = \sum_{i<\lambda} \kappa_i$.

4.7. Let α be an ordinal. Show that the cofinality of α is the least size of a subset of α which is unbounded in α. Show also that there is a club $C \subset \alpha$ such that $\text{Card}(C) = \text{otp}(C) = \text{cf}(\alpha)$.

4.8. Show that $\aleph_\omega^{\aleph_1} = \aleph_\omega^{\aleph_0} \cdot 2^{\aleph_1}$. Show also that if $\omega \leq \alpha < \omega_1$, then $\aleph_\alpha^{\aleph_1} = \aleph_\alpha^{\aleph_0} \cdot 2^{\aleph_1}$.

4.9. Use the recursion theorem 3.13 to show that there is a sequence $(\beth_\alpha : \alpha \in \mathrm{OR})$ such that $\beth_0 = \aleph_0$, $\beth_{\alpha+1} = 2^{\beth_\alpha}$ for all α, and $\beth_\lambda = \sup_{\alpha<\lambda} \beth_\alpha$ for every limit ordinal λ. Show that $\mathrm{Card}(V_{\omega+\alpha}) = \beth_\alpha$ for every α.

4.10. (a) Let κ be an infinite cardinal. Show that H_κ is a set. [Hint. Show e.g. that $H_\kappa \subset V_\kappa$ by induction on κ.] Also show that $\mathrm{Card}(H_\kappa) = 2^{<\kappa}$.
(b) Show that $\mathrm{HF} = V_\omega$ and $(\mathrm{HF}; \in \restriction \mathrm{HF}) \models \mathsf{ZFC}^{-\infty}$.
(c) **(W. Ackermann)** Let us define $E_A \subset \omega \times \omega$ as follows. $n E_A m$ iff: if $m = \sum_i k_i \cdot 2^i$, where $k_i \in \{0, 1\}$ for all i, then $k_n = 1$. Show that $(\omega; E_A) \cong (\mathrm{HF}; \in)$.
(d) Show that if κ is uncountable and regular, then $(H_\kappa; \in \restriction H_\kappa) \models \mathsf{ZFC}^-$.

4.11. **(A. Tarski)** Let X be a set, and let F be a filter on X. Show that there is an ultrafilter U on X with $U \supset F$. [Hint. Use the HAUSDORFF Maximality Principle 2.11.]

4.12. Show that if μ, κ are infinite regular cardinals with $\mu < \kappa$, then the set

$$S = \{\alpha < \kappa : \mathrm{cf}(\alpha) = \mu\}$$

is stationary in κ. Show also that if $T \subset S$ is stationary in κ and $C \subset \kappa$ is μ-club in κ, then $T \cap C \neq \emptyset$.

4.13. Let $S \subset \omega_1$ be stationary, and let $\alpha < \omega_1$. Show that S has a *closed* subset of order type α.

4.14. Let κ be regular and uncountable, and let $R \subset \kappa$ be stationary. Let $(U; \in, A_1, \ldots, A_n)$ be a model such that U is transitive and $\kappa \subset U$. Show that there is some $X \prec (U; \in, A_1, \ldots, A_n)$ such that $X \cap \kappa \in R$. Show also that if $\mathrm{Card}(U) = \kappa$ and $f : \kappa \to U$ is surjective, then $\{\xi < \kappa : f''\xi \prec (U; \in, A_1, \ldots, A_n)\}$ is club in κ.

4.15. Show that set $\mathscr{X} \subset [\theta]^\kappa$ is stationary in $[\theta]^\kappa$ iff $\mathscr{X} \cap \mathscr{C} \neq \emptyset$ for all $\mathscr{C}$ which are club in $[\theta]^\kappa$ according to the definition on p. 44.

4.16. Show that $\left(\prod_{i\in I} \alpha_i\right)^\kappa = \prod_{i\in I} (\alpha_i^\kappa)$.

4.17. Let κ be a singular cardinal of uncountable cofinality.

(1) Suppose that there is some cardinal $\lambda < \kappa$ with $\lambda^{\mathrm{cf}(\kappa)} = \lambda$ and SCH holds for every singular $\mu \in (\lambda, \kappa)$, i.e., if μ is a singular cardinal, $\lambda < \mu < \kappa$, then $\mu^{\mathrm{cf}(\mu)} = 2^{\mathrm{cf}(\mu)} \cdot \mu^+$. Show that SCH holds at κ, i.e., $\kappa^{\mathrm{cf}(\kappa)} = 2^{\mathrm{cf}(\kappa)} \cdot \kappa^+$.
(2) Suppose that $\mu^{\mathrm{cf}(\kappa)} < \kappa$ for all $\mu < \kappa$, and SCH holds on a stationary set below κ, i.e., $\{\mu < \kappa : \mu^{\mathrm{cf}(\mu)} = 2^{\mathrm{cf}(\mu)} \cdot \mu^+\}$ is stationary in κ. Show that SCH holds at κ, i.e., $\kappa^{\mathrm{cf}(\kappa)} = 2^{\mathrm{cf}(\kappa)} \cdot \kappa^+$.

4.18. Let κ be a singular cardinal of uncountable cofinality. If $\{\mu < \kappa : 2^\mu = \mu^+\}$ is stationary in κ, then $2^\kappa = \kappa^+$.

4.19. Use the axiom of choice to show that there is some $A \subset \mathbb{R}$ such that neither A nor $\mathbb{R}$ contains a perfect subset. [*Hint.* Show that there is an enumeration $e\colon 2^{\aleph_0} \to C$, where C is the collection of perfect sets. If $<$ is a well-ordering of $\mathbb{R}$, then construct $(a_\xi, b_\xi : \xi < 2^{\aleph_0})$ by letting a_ξ be the $<$-least element of $e(\xi) \setminus (\{a_{\bar\xi} : \bar\xi < \xi\} \cup \{b_{\bar\xi} : \bar\xi < \xi\})$ and b_ξ be the $<$-least element of $e(\xi) \setminus (\{a_{\bar\xi} : \bar\xi \leq \xi\} \cup \{b_{\bar\xi} : \bar\xi < \xi\})$. Show that $A = \{a_\xi : \xi < 2^{\aleph_0}\}$ works.]

4.20. Prove Lemma 4.46.

4.21. Show that if κ is weakly MAHLO, then $\{\mu < \kappa : \mu$ is weakly inaccessible$\}$ is stationary. Also show that if κ is MAHLO, then $\{\mu < \kappa : \mu$ is inaccessible$\}$ is stationary.

4.22. Let $\kappa \geq \aleph_1$ be a cardinal. If $(T, <_T)$ is a κ-SOUSLIN tree, then $(T, <_T)$ is a κ-ARONSZAJN tree.

4.23. Let κ be a cardinal. Show that the following are equivalent.

(a) κ is weakly compact.
(b) If $X \subset \mathscr{P}(\kappa)$, $\mathrm{Card}(X) \leq \kappa$, then there are transitive models H and H^* with $X \subset H$, $\mathrm{Card}(H) = \kappa$, ${}^{\mu}H \subset H$ and ${}^{\mu}H^* \subset H^*$ for every $\mu < \kappa$ and there is some elementary embedding $\sigma : H \to H^*$ such that κ is the critical point of σ.

4.24. Let κ be a regular uncountable cardinal. Show:

(a) If $\lambda < \kappa$ is an infinite regular cardinal, then $\{\xi < \kappa : \mathrm{cf}(\xi) = \lambda\}$ is not ineffable.
(b) If κ itself is ineffable, then κ is weakly compact and the set

$$\{\mu < \kappa : \mu \text{ is weakly compact}\}$$

is stationary in κ.
(c) Let κ be a measurable cardinal, and let U be a normal $< \kappa$-closed ultrafilter on κ. Then $\{\mu < \kappa : \mu$ is ineffable $\} \in U$.

4.25. **(Jensen–Kunen)** Show that if κ is ineffable, then there is no κ-KUREPA tree.

4.26. Let κ be a cardinal. Show that if U is a $< \kappa$-closed ultrafilter on κ, then U is uniform iff for no $\xi < \kappa$, $\{\xi\} \in U$. Show also that if U is a $< \kappa$-closed uniform ultrafilter on κ, then U is normal iff $\kappa = \sigma^{-1}[\mathrm{id}]$, where id is the identity function and σ is the (inverse of) transitive collapse as in the proof of Theorem 4.55.
Let U be a $< \kappa$-closed normal ultrafilter on κ, and let $(X_s : s \in [\kappa]^{<\omega})$ be a family such that $X_s \in U$ for every $s \in [\kappa]^{<\omega}$. Let us define

$$\Delta_{s \in [\kappa]^{<\omega}} X_s = \{\xi \in \kappa : \xi \in \bigcap_{s \in [\xi]^{<\omega}} X_s\}.$$

Show that $\Delta_{s \in [\kappa]^{<\omega}} X_s \in U$.

4.27. Let κ be a measurable cardinal.

(a) Let U be a measure on κ. Show that $\kappa^{+\text{ult}(V;U)} = \kappa^{+V}$ and that $2^\kappa < \pi_U^V(\kappa) < ((2^\kappa)^+)^V$. [Hint: If $\alpha < \pi_U^V(\kappa)$, then α is represented by some $f: \kappa \to \kappa$, and there are 2^κ functions from κ to κ.] Conclude that $U \notin \text{ult}(V;U)$.

(b) Let U, U' be measures on κ. We define $U <_{\text{M}} U'$ iff $U \in \text{ult}(V;U')$. Show that $\pi_U^V(\kappa) = \pi_U^{\text{ult}(V;U')}(\kappa) < \pi_{U'}^V(\kappa)$. Conclude that $<_{\text{M}}$ is well-founded, and that the rank of $<_{\text{M}}$ is always less than or equal to $(2^\kappa)^+$. $<_{\text{M}}$ is called the MITCHELL *order*.

4.28. Let κ be a measurable cardinal, let U be a measure on κ, and let $\pi_U: V \to_U M$ be the ultrapower map, where M is an inner model. Let $\mu > \kappa$ be a cardinal, $\text{cf}(\mu) \neq \kappa$, and $\rho^\kappa < \mu$ for every $\rho < \mu$. Show that $\pi_U(\mu) = \mu$.

4.29. **(Magidor)** Show that if κ is supercompact, then for every $\alpha > \kappa$ there are $\mu < \beta < \kappa$ together with an elementary embedding $\sigma: V_\beta \to V_\alpha$ such that $\text{crit}(\sigma) = \mu$ and $\sigma(\mu) = \kappa$. [Hint: Let $\pi: V \to M$, where M is an inner model, $\text{crit}(\pi) = \kappa$, and ${}^{\overline{\overline{V_\alpha}}}M \subset M$. Show that in M, there is some $\sigma: V_\alpha \to (V_{\pi(\alpha)})^M$ such that $\text{crit}(\sigma) = \kappa$ and $\sigma(\kappa) = \pi(\kappa)$. Pull this statement back via π.]

4.30. Let κ be supercompact. Show that for every cardinal $\lambda \geq \kappa$ there is a $<\kappa$-closed ultrafilter U on $[\lambda]^{<\kappa}$ such that $\{a\} \notin U$ for all $a \in [\lambda]^{<\kappa}$, $\{a \in [\lambda]^{<\kappa} : \xi \in a\} \in U$ for all $\xi < \lambda$, and if $(A_\xi : \xi < \lambda)$ is such that $A_\xi \in U$ for all $\xi < \lambda$, then there is some $A \in U$ such that whenever $\xi \in a \in A$, $a \in A_\xi$. [Hint. Let U be derived from π morally as in (4.14).]
Problems 10.21 and 10.22 will show that the conclusions of Problems 4.29 and 4.30 actually both characterize the supercompactness of κ.

4.31. Use the necessary criterion for supercompactness provided by Problem 4.29 to show that if κ is supercompact, then κ is subcompact (cf. Lemma 4.69) and in fact there is a measure U on κ such that $\{\mu < \kappa : \mu \text{ is subcompact}\} \in U$.

Chapter 5
Constructibility

Models of the language $\mathscr{L}_\in$ of set theory are of the form $(M; E)$, where $M \neq \emptyset$ is a set and $E \subset M \times M$ interprets $\in$. We shall also consider "class models" $(M; E)$ of $\mathscr{L}_\in$ where M is a proper class rather than a set.

5.1 The Constructible Universe

Definition 5.1 A formula φ of $\mathscr{L}_\in$ is called Σ_0 (or Π_0, or Δ_0) iff φ is contained in each set Γ for which the following hold true.

(a) Every atomic formula is in Γ,
(b) if ψ_0, ψ_1 are in Γ then so are $\neg\psi_0$, $(\psi_0 \wedge \psi_1)$, $(\psi_0 \vee \psi_1)$, $(\psi_0 \rightarrow \psi_1)$, and $(\psi_0 \leftrightarrow \psi_1)$, and
(c) if ψ is in Γ and x, y are variables, then $\forall x(x \in y \rightarrow \psi)$ and $\exists x(x \in y \wedge \psi)$ are in Γ.

For $n \in \omega \setminus \{0\}$, a formula φ of $\mathscr{L}_\in$ is called Σ_n iff φ is of the form

$$\exists x_1 \ldots \exists x_k \psi,$$

where $x_1, \ldots, x_k$ are variables and ψ is Π_{n-1}, and φ is called Π_n iff φ is of the form

$$\forall x_1 \ldots \forall x_k \psi,$$

where $x_1, \ldots, x_k$ are variables and ψ is Σ_{n-1}.

If M is a transitive set or class and φ is a sentence of $\mathscr{L}_\in$, then we write $M \models \varphi$ for $(M; \in \restriction M) \models \varphi$ (where $\in \restriction M = \in \cap M^2$). If φ is a formula, φ being $\varphi(x_1, \ldots, x_k)$ with all free variables shown, and if $a_1, \ldots, a_k \in M$, then we write $M \models \varphi(a_1, \ldots, a_k)$ for the assertion that φ holds in $(M; \in \restriction M)$ for an assignment which sends v_l to a_l $(1 \leq l \leq k)$.

R. Schindler, *Set Theory*, Universitext, DOI: 10.1007/978-3-319-06725-4_5,

We have the following absoluteness properties.

Lemma 5.2 *Let M be transitive, let φ be a Σ_0-formula, and let $a_1, \ldots, a_k \in M$. Then*

$$M \models \varphi(a_1, \ldots, a_k) \Leftrightarrow V \models \varphi(a_1, \ldots, a_k).$$

Proof by induction on the complexity of φ. Let us only consider the case where φ is of the form $\exists x_0 \in x_1 \psi(x_0, x_1, \ldots, x_k)$.

"$\Longleftarrow$": If $V \models \varphi(a_1, \ldots, a_k)$, let $a \in a_1$ be such that $V \models \psi(a, a_1, \ldots, a_k)$. As M is transitive, $a \in a_1 \in M$ gives $a \in M$, and $V \models a \in a_1 \wedge \psi(a, a_1, \ldots, a_k)$ gives

$$M \models a \in a_1 \wedge \psi(a, a_1, \ldots, a_k)$$

by the inductive hypothesis, and so $M \models \varphi(a_1, \ldots, a_k)$.

"$\Longrightarrow$": If $M \models \varphi(a_1, \ldots, a_k)$, let $a \in M$ be such that $M \models a \in a_1 \wedge \psi(a, a_1, \ldots, a_k)$. Then

$$V \models a \in a_1 \wedge \psi(a, a_1, \ldots, a_k)$$

by the inductive hypothesis, and so $V \models \varphi(a_1, \ldots, a_k)$. □

This proof also shows the following.

Lemma 5.3 *Let M be transitive, let $\varphi(v_1, \ldots, v_k)$ be a Σ_1-formula, let $\psi(v_1, \ldots, v_k)$ be a Π_1-formula, and let $a_1, \ldots, a_k \in M$. Then φ is* upward absolute, *i.e.*

$$M \models \varphi(a_1, \ldots, a_k) \Longrightarrow V \models \varphi(a_1, \ldots, a_k),$$

and ψ is downward absolute, *i.e.*

$$V \models \psi(a_1, \ldots, a_k) \Longrightarrow M \models \psi(a_1, \ldots, a_k).$$

Definition 5.4 Let T be a theory in the language $\mathscr{L}_\in$, and let φ be a formula of $\mathscr{L}_\in$. Then φ is called Δ_1^T iff there are $\mathscr{L}_\in$-formulae ψ and ψ' such that

ψ is Σ_1,
ψ' is Π_1, and
$T \vdash \varphi \longleftrightarrow \psi \longleftrightarrow \psi'$

Lemma 5.3 immediately implies:

Lemma 5.5 *Let T be a theory in the language of $\mathscr{L}_\in$, and let φ be a formula of $\mathscr{L}_\in$ which is Δ_1^T. Let M be a transitive model of T. Then φ is* absolute between V and M, *i.e.,*

$$V \models \varphi(a_1, \ldots, a_k) \Longleftrightarrow M \models \varphi(a_1, \ldots, a_k)$$

for all $a_1, \ldots, a_k \in M$ making an assignment to all the free variables of φ.

The following Lemma expresses the "absoluteness of well-foundedness."

Lemma 5.6 *The statement*[1] *"R is a well-founded relation" is* $\Delta_1^{\mathsf{ZFC}^-}$. *In particular, if M is a transitive model of* ZFC^- *such that* $R \in M$ *is a binary relation, then*

$$V \models \text{"}R \text{ is well-founded"} \iff M \models \text{"}R \text{ is well-founded"}.$$

Proof The first part, that "R is a well-founded relation" be $\Delta_1^{\mathsf{ZFC}^-}$ follows directly from (the proofs of) Lemmas 3.11 and 3.17. The second part is then a consequence of Lemma 5.5. □

The following easy lemma will be used to verify that fragments of ZFC hold in a given transitive model.

Lemma 5.7 *Let M be a transitive set or class.*

(1) *M is a model of* (Ext), *the axiom of extensionality.*
(2) *M is a model of* (Fund), *the axiom of foundation.*
(3) *If* $\omega \in M$ *then M is a model of* (Inf), *the axiom of infinity.*
(4) *If M is closed under* $x, y \mapsto \{x, y\}$ *(i.e.,* $a, b \in M \Longrightarrow \{a, b\} \in M$*), then M is a model of* (Pair), *the pairing axiom.*
(5) *If M is closed under* $x \mapsto \bigcup x$ *(i.e.,* $a \in M \Longrightarrow \bigcup a \in M$*), then M is a model of* (Union), *the axiom of union.*

Proof We shall use Lemma 5.2.

(1) The axiom of extensionality holds in V and it is Π_1, as it says

$$\forall x \forall y ((\forall z \in x\, z \in y \wedge \forall z \in y\, z \in x) \rightarrow x = y).$$

(2) The axiom of foundation holds in V and is Π_1, as it says

$$\forall x (x \neq \emptyset \rightarrow \exists y \in x\; y \cap x = \emptyset).$$

Notice that $x \neq \emptyset$ can be written as $\exists y \in x\; y = y$, and $y \cap x = \emptyset$ can be written as $\neg \exists z \in y\; z \in x$.

(3) The axiom of infinity is Σ_1. It says that $\exists x\, \varphi(x)$, where $\varphi(x)$ is

$$\emptyset \in x \wedge \forall y \in x\, \exists z \in x\; z = y \cup \{y\}.$$

Here, $\emptyset \in x$ can be written as $\exists y \in x\; \neg \exists z \in y\; z = z$, and $z = y \cup \{y\}$ can be written as

$$\forall u \in z\, (u \in y \vee u = y) \wedge \forall u \in y\; u \in z \wedge y \in z.$$

Thus $\varphi(x)$ is Σ_0, $V \models \varphi(\omega)$ holds, and thus $M \models \varphi(\omega)$ holds provided that $\omega \in M$.

---

[1] Recall from p. 12 that ZFC^- is ZFC without the power set axiom.

(4) The pairing axiom is of the form

$$\forall x\, \forall y\, \exists z\; z = \{x, y\},$$

where $z = \{x, y\}$ can be written as $x \in z \wedge y \in z \wedge \forall u \in z(u = x \vee u = y)$.

(5) The union axiom is of the form

$$\forall x\, \exists y\; y = \bigcup x,$$

where $y = \bigcup x$ can be written as

$$\forall z \in y\, \exists u \in x\; z \in u \wedge \forall u \in x\, \forall z \in u\; z \in y.$$

Lemma 5.7 is shown. □

It is more delicate verifying that a given transitive model M is a model of (Pow), $(\mathrm{Aus})_\varphi$, $(\mathrm{Ers})_\varphi$, and (AC).

Recall (cf. Definition 4.51) that an inner model is a transitive proper class model of ZFC. If E is a set or a proper class, then $L[E]$ is the least inner model which is closed under the operation $x \mapsto E \cap x$. An important example will be $L = L[\emptyset]$, GÖDEL's *constructible universe*, which we shall study in detail.

In order to show that $L[E]$ indeed always exists, we need to define it in a way that is different from saying "the least inner model which is closed under the operation $x \mapsto E \cap x$." Any model of the form $L[E]$ may be stratified in two ways: into levels of the L-hierarchy and into levels of the J-hierarchy. The former approach was GÖDEL's original one, but it turned out that the latter one (which was discovered by RONALD B. JENSEN, cf. [16]) is much more useful.

In order to define the J-hierarchy we need the concept of rudimentary functions.

Definition 5.8 Let E be a set or a proper class. A function $f: V^k \to V$, where $k < \omega$, is called *rudimentary in* E (or, rud_E) if it is generated by the following schemata:

$$\begin{aligned}
f(x_1, \ldots, x_k) &= x_i \\
f(x_1, \ldots, x_k) &= x_i \backslash x_j \\
f(x_1, \ldots, x_k) &= \{x_i, x_j\} \\
f(x_1, \ldots, x_k) &= h(g_1(x_1^1, \ldots, x_{k_1}^1), \cdots, g_\ell(x_1^\ell, \ldots, x_{k_\ell}^\ell)) \\
f(x_1, \ldots, x_k) &= \bigcup_{y \in x_1} g(y, x_2, \ldots, x_k) \\
f(x) &= x \cap E
\end{aligned}$$

f is called *rudimentary* (or, rud) if f is $\mathrm{rud}_\emptyset$.

We often write $\mathbf{x}$ for $(x_1, \ldots, x_k)$ in what follows. It is easy to verify that for instance the following functions are rudimentary: $f(\mathbf{x}) = \bigcup x_i$, $f(\mathbf{x}) = x_i \cup x_j$,

$f(\mathbf{x}) = x_i \cap x_j$, $f(\mathbf{x}) = \{x_1, \ldots, x_k\}$, $f(\mathbf{x}) = (x_1, \ldots, x_k)$, and $f(\mathbf{x}) = x_1 \times \ldots \times x_k$. (Cf. Problem 5.5.) Lemma 5.10 below will provide more information.

If U is a set and E is a set or a proper class then we shall denote by $\mathrm{rud}_E(U)$ the rud_E closure of U, i.e., the set

$$U \cup \{f((x_1, ..., x_k));\ f \text{ is } \mathrm{rud}_E \text{ and } x_1, ..., x_k \in U\}.$$

It is not hard to verify that if U is transitive then so is $\mathrm{rud}_E(U \cup \{U\})$ (cf. Problem 5.7). We shall now be interested in $\mathscr{P}(U) \cap \mathrm{rud}_E(U \cup \{U\})$ (cf. Lemma 5.11 below).

Definition 5.9 Let E be a set or a proper class. A relation $R \subset V^k$, where $k < \omega$, is called *rudimentary in E* (or, rud_E) if there is a rud_E function $f\colon V^k \to V$ such that $R = \{\mathbf{x}\colon f(\mathbf{x}) \neq \emptyset\}$. R is called *rudimentary* (or, rud) if R is $\mathrm{rud}_\emptyset$.

Lemma 5.10 *Let E be a set or a proper class.*

(a) *The relation* $\notin$ *is* rud.

(b) *Let* f, R *be* rud_E. *Let* $g(\mathbf{x}) = f(\mathbf{x})$ *if* $R(\mathbf{x})$ *holds, and* $g(\mathbf{x}) = \emptyset$ *if not. Then* g *is* rud_E.

(c) *If* R, S *are* rud_E *then so is* $R \cap S$.

(d) *Membership in E is* rud_E.

(e) *If* R *is* rud_E, *then so is its characteristic function* χ_R.[2]

(f) R *is* rud_E *iff* $\neg R$ *is* rud_E.

(g) *Let* R *be* rud_E. *Let* $f(y, \mathbf{x}) = y \cap \{z\colon R(z, \mathbf{x})\}$. *Then* f *is* rud_E.

(h) *If* $R(y, \mathbf{x})$ *is* rud_E, *then so is* $\exists z \in y R(z, \mathbf{x})$.

Proof (a) $x \notin y$ iff $\{x\} \setminus y \neq \emptyset$.

(b) If $R(\mathbf{x}) \iff r(\mathbf{x}) \neq \emptyset$, where r is rud_E, then $g(\mathbf{x}) = \bigcup_{y \in r(\mathbf{x})} f(\mathbf{x})$.

(c) Let $R(\mathbf{x}) \iff f(\mathbf{x}) \neq \emptyset$, where f is rud_E. Let $g(\mathbf{x}) = f(\mathbf{x})$ if $S(\mathbf{x})$ holds, and $g(\mathbf{x}) = \emptyset$ if not. g is rud_E by (b), and thus g witnesses that $R \cap S$ is rud_E.

(d) $x \in E$ iff $\{x\} \cap E \neq \emptyset$.

(e): by (b).

(f) $\chi_{\neg R}(\mathbf{x}) = 1 \setminus \chi_R(\mathbf{x})$.

(g) Let $g(z, \mathbf{x}) = \{z\}$ if $R(z, \mathbf{x})$ holds, and $g(z, \mathbf{x}) = \emptyset$ if not. We have that g is rud_E by (b), and $f(y, \mathbf{x}) = \bigcup_{z \in y} g(z, \mathbf{x})$.

(h) Set $f(y, \mathbf{x}) = y \cap \{z;\ R(z, \mathbf{x})\}$. f is rud_E by (g), and thus f witnesses that $\exists z \in y R(z, \mathbf{x})$ is rud_E. □

We shall often be concerned with models of the form $(U; \in, A_1, \ldots, A_m)$, where $A_1, \ldots, A_m \subset U^{<\omega}$ and $\in$ stands for $\in\restriction U = \in \cap U^2$. Each such structure comes with a language $\mathscr{L}_{\in, A_1, ..., A_m}$ with predicates $\dot{\in}$, $\dot{A}_1, \ldots, \dot{A}_m$. We shall mostly restrict ourselves to discussing the cases where $m = 0$ (i.e., where there is no A_i around) or $m = 1$ or $m = 2$.

Let $\mathscr{U} = (U; \in, A_1, \ldots, A_m)$ be a model as above. The notions of Σ_n-and Π_n-formulae of the language of $\mathscr{L}_{\in, A_1, ..., A_m}$ is defined as in Definition 5.1, where $\mathbf{x} \in A_i$,

[2] I.e., $\chi_R(\mathbf{x}) = 1$ iff $R(\mathbf{x})$ and $= 0$ otherwise.

$0 < i \leq m$, count as atomic formulae. If $X \subset U$, and $n < \omega$, then we let $\Sigma_n^{\mathscr{U}}(X)$ denote the set of all relations which are Σ_n definable over $\mathscr{U}$ from parameters in X, i.e., the set of all R such that there are $k, l < \omega$, some Σ_n-formula $\varphi(x_1, \ldots, x_{k+l})$ of $\mathscr{L}_{\in, A_1, \ldots, A_m}$ and $a_1, \ldots, a_l \in X$ such that $R \subset U^k$ and for all $\mathbf{z} = (z_1, \ldots, z_k) \in U^k$,

$$\mathbf{z} \in R \Longleftrightarrow \mathscr{U} \models \varphi(z_1, \ldots, z_k, a_1, \ldots, a_l).$$

We shall also write $\underset{\sim}{\Sigma}{}_n^{\mathscr{U}}$ for $\Sigma_n^{\mathscr{U}}(U)$, and we shall write $\underset{\sim}{\Sigma}{}_\omega^{\mathscr{U}}$ for $\bigcup_{n<\omega} \underset{\sim}{\Sigma}{}_n^{\mathscr{U}}$. We shall also write $\Sigma_{n/\omega}^{\mathscr{U}}$ for $\Sigma_{n/\omega}^{\mathscr{U}}(\emptyset)$.

The notions $\Pi_n^{\mathscr{U}}(X)$, $\underset{\sim}{\Pi}{}_n^{\mathscr{U}}$, $\Pi_n^{\mathscr{U}}$, etc. are defined in an entirely analoguous fashion. A relation is $\Delta_n^{\mathscr{U}}$ iff it is both $\Sigma_n^{\mathscr{U}}$ and $\Pi_n^{\mathscr{U}}$.

Let $\mathscr{U} = (U; \in, A_1, \ldots, A_m)$ and $\mathscr{U}' = (U'; \in, A_1', \ldots, A_m')$ be models as above. Generalizing Definition 4.50, we say that $\pi: U \to U'$ is a Σ_n-*elementary embedding*, written $\pi: U \to_{\Sigma_n} U'$, where $n \in \omega \cup \{\omega\}$ iff $\operatorname{ran}(\pi)$ is a Σ_n-elementary substructure of $\mathscr{U}'$ in the common language $\mathscr{L}_{\in, A_1, \ldots, A_m}$, i.e., if for all Σ_n-formulae φ of the language $\mathscr{L}_{\in, A_1, \ldots, A_m}$ and for all $a_1, \ldots, a_k \in U$,

$$\mathscr{U} \models \varphi(a_1, \ldots, a_k) \Longleftrightarrow \mathscr{U}' \models \varphi(\pi(a_1), \ldots, \pi(a_k)). \tag{5.1}$$

If π is the identity on U, then we also write this as $\mathscr{U} \prec_{\Sigma_n} \mathscr{U}'$.

The following lemma says that $\operatorname{rud}_E(U \cup \{U\})$ is just the result of "stretching" $\underset{\sim}{\Sigma}{}_\omega^{(U;\in,E)}$ without introducing additional elements of $\mathscr{P}(U)$. By $(U; \in, E)$ we shall always mean the model $(U; \in\restriction U, E \cap U)$. A set U is rud_E *closed* iff $\operatorname{rud}_E(U) \subset U$.

Lemma 5.11 *Let U be a transitive set, and let $E \subset U$. Then $\mathscr{P}(U) \cap \operatorname{rud}_E(U \cup \{U\}) = \mathscr{P}(U) \cap \underset{\sim}{\Sigma}{}_\omega^{(U;\in,E)}$.*

Proof Notice that $\mathscr{P}(U) \cap \underset{\sim}{\Sigma}{}_\omega^{(U;\in,E)} = \mathscr{P}(U) \cap \underset{\sim}{\Sigma}{}_0^{(U\cup\{U\};\in,E)}$, so that we have to prove that

$$\mathscr{P}(U) \cap \operatorname{rud}_E(U \cup \{U\}) = \mathscr{P}(U) \cap \underset{\sim}{\Sigma}{}_0^{(U\cup\{U\};\in,E)}.$$

"$\supset$": By Lemma 5.10 (a) and (d), $\notin$ and membership in E are both rud_E. By Lemma 5.10 (f), (c), and (h), the collection of rud_E relations is closed under complement, intersection, and bounded quantification. Therefore we get inductively that every relation which is Σ_0 in the language $\mathscr{L}_{\in,E}$ with $\dot{\in}$ and $\dot{E}$ is also rud_E.

Now let $x \in \mathscr{P}(U) \cap \underset{\sim}{\Sigma}{}_0^{(U\cup\{U\};\in,E)}$. There is then some rud_E relation R and there are $x_1, \ldots, x_k \in U \cup \{U\}$ such that $y \in x$ iff $y \in U$ and $R(y, x_1, \ldots, x_k)$ holds. But then $x = U \cap \{y: R(y, x_1, \ldots, x_k)\} \in \operatorname{rud}_E(U \cup \{U\})$ by Lemma 5.10 (g).

"$\subset$": Call a function $f: V^k \to V$, where $k < \omega$, *simple* iff the following holds true: if $\varphi(v_0, v_1, \ldots, v_m)$ is Σ_0 in the language $\mathscr{L}_{\in,E}$, then $\varphi(f(v_1', \ldots, v_k'), v_1, \ldots, v_m)$ is equivalent over transitive rud_E closed structures to a Σ_0 formula in the same language. It is not hard to verify inductively that every rud_E function is simple (cf. Problem 5.8).

Now let $x \in \mathcal{P}(U) \cap \text{rud}_E(U \cup \{U\})$, say $x = f(x_1, \ldots, x_k)$, where x_1, ..., $x_k \in U \cup \{U\}$ and f is rud_E. Then, as f is simple, "$v_0 \in f(v_1, ..., v_k)$" is (equivalent over $\text{rud}_E(U \cup \{U\})$ to) a Σ_0 formula in the language $\mathcal{L}_{\in,E}$, and hence $x = \{y \in U: y \in f(x_1, ..., x_n)\}$ is in $\Sigma_0^{(U\cup\{U\},\in,E)}(\{x_1, ..., x_n\})$. □

Of course Lemma 5.11 also holds with $\mathcal{P}(U)$ being replaced by the set of all relations on U.

Let U be rud_E closed, and let $x \in U$ be transitive. Suppose that

$$B \in \Sigma_0^{(U;\in,E)}(\{x_1, ..., x_k\}),$$

where $x_1, \cdots, x_k \in x$. Then $B \cap x \in \underset{\sim}{\Sigma}_0^{(x;\in,E\cap x)}$, and hence $B \cap x \in \text{rud}_E(x \cup \{x\})$ by Lemma 5.11. But $\text{rud}_E(x \cup \{x\}) \subset U$, and therefore $B \cap x \in U$. We have shown the following.

Lemma 5.12 *Let U be a transitive set such that for every $x \in U$ there is some transitive $y \in U$ with $x \in y$, let E be a set or a proper class, and suppose that U is* rud_E *closed. Then $(U; \in, E)$ is a model of* Σ_0 comprehension *in the sense that if $B \in \underset{\sim}{\Sigma}_0^{(U;\in,E)}$ and $x \in U$, then $B \cap x \in U$.*

In the situation of Lemma 5.12, $(U; \in, E, B)$ is therefore "amenable" in the sense of the following definition.

Definition 5.13 A structure $(U; \in, A_1, \ldots, A_m)$, where U is transitive and $A_1, \ldots$, $A_m \subset U^{<\omega}$, is called *amenable* if and only if $A_i \cap x \in U$ whenever $0 < i \leq m$ and $x \in U$.

Later on, cf. Definition 11.4, we will study possible failures of Σ_1 comprehension in rud_E closed structues. Lemma 5.12 provides the key element for proving that (all but two of) the structures we are now about to define are models of "basic set theory," a theory which consists of Σ_0 comprehension together with extensionality, foundation, pairing, union, infinity, and the statement that Cartesian products and transitive closures exist.

We are ready to define the $J_\alpha[E]$ hierarchy as follows. For later purposes it is convenient to index this hierarchy by limit ordinals.

Definition 5.14 Let E be a set or a proper class.

$$J_0[E] = \emptyset$$
$$J_{\alpha+\omega}[E] = \text{rud}_E(J_\alpha[E] \cup \{J_\alpha[E]\})$$
$$J_{\omega\lambda}[E] = \bigcup_{\alpha<\lambda} J_{\omega\alpha}[E] \text{ for limit } \lambda$$
$$L[E] = \bigcup_{\alpha\in\text{OR}} J_{\omega\alpha}[E]$$

Obviously, every $J_\alpha[E]$ is rud_E closed and transitive. We shall also denote by $J_\alpha[E]$ the model $(J_\alpha[E]; \in\upharpoonright J_\alpha[E], E \cap J_\alpha[E])$.

An important special case is obtained by letting $E = \emptyset$ in Definition 5.14. We write J_α for $J_\alpha[\emptyset]$, and L for $L[\emptyset]$. L is GÖDEL's *Constructible Universe.* Other important examples which are studied in contemporary set theory are obtained by letting E be a set or proper class with certain condensation properties or by letting E code a (carefully chosen) sequence of *extenders* (cf. Definition 10.45 and also Problem 10.5).

The next lemma is an immediate consequence of Lemma 5.11.

Lemma 5.15 *Let E be a set or proper class. Assume that*[3] $E \subset \text{Lim} \times V$. *Let us write*

$$E_\alpha = \{x : (\alpha, x) \in E\}$$

and

$$E \upharpoonright \alpha = E \cap (\alpha \times V)$$

for limit ordinals α. Let us assume that $E_\alpha \subset J_\alpha[E]$ and that $(J_\alpha[E]; \in, E_\alpha)$ is amenable for every limit ordinal α.

Then

$$\mathscr{P}(J_\alpha[E]) \cap J_{\alpha+\omega}[E] = \mathscr{P}(J_\alpha[E]) \cap \underset{\sim}{\Sigma}{}_\omega^{(J_\alpha[E]; \in, E\upharpoonright\alpha, E_\alpha)}.$$

The hypothesis of Lemma 5.15 is satisfied for all present-day canonical inner models. (Cf. also Problem 5.13.) We will assume from now on, that E always satisfies the hypothesis of Lemma 5.15.

The following is easy to verify inductively (cf. Problem 5.9).

Lemma 5.16 *For every limit ordinal,* $J_\alpha[E] \cap \text{OR} = \alpha$, *and* $\text{Card}(J_\alpha[E]) = \text{Card}(\alpha)$.

The following can be easily proved by induction on α, with a subinduction on the rank according to Definition 5.8. We shall produce a much stronger statement later on, cf. (5.2).

Lemma 5.17 *Let α be a limit ordinal. If $x \in J_\alpha[E]$, then there is a transitive set $y \in J_\alpha[E]$ such that $x \in y$.*

Lemmas 5.7 and 5.17 (and Problem 5.5) immediately give the following.

Corollary 5.18 *Let E be a set or a proper class. Let α be a limit ordinal, $\alpha > \omega$. Then $J_\alpha[E]$ is a model of the following statements:* (Ext), (Fund), (Inf), (Pair), (Union), *the statement that every set is an element of a transitive set, and Σ_0-comprehension. Also, $J_\alpha[E]$ is a model of "$\forall x \forall y\ x \times y$ exists."*

Theorem 5.19 *Let E be a set or a proper class.* $L[E] \models \mathsf{ZF}$.

---

[3] Here, Lim denotes the class of all limit ordinals.

Proof We have to verify that the following axioms hold in $L[E]$: the power set axiom, the separation schema and the replacement schema.

Let us start with the power set axiom (Pow). Fix $a \in L[E]$. By the replacement schema in V, there is some α such that

$$\mathscr{P}(a) \cap L[E] \subset J_\alpha[E].$$

But then

$$\mathscr{P}(a) \cap L[E] = \{x \in J_\alpha[E] \colon J_\alpha[E] \models \forall y \in x \; y \in a\} \in J_\alpha[E],$$

as $J_\alpha[E]$ satisfies Σ_0 comprehension. This shows that $L[E] \models \exists z(z = \mathscr{P}(a))$.

In order to show that the separation schema (Sep) holds in $L[E]$, let $\varphi(x_1, \ldots, x_k)$ be a formula, and let $a, a_1, \ldots, a_k \in L[E]$. As $(J_\beta[E] \colon \beta \in OR)$ is a continuous cumulative hierarchy, we may pick some α with $a, a_1, \ldots, a_k \in J_\alpha[E]$ such that for all $b \in J_\alpha[E]$,

$$J_\alpha[E] \models \varphi(b, a_1, \ldots, a_k) \iff L[E] \models \varphi(b, a_1, \ldots, a_k).$$

(Cf. Problem 5.14.) But then

$$\begin{aligned} &\{b \in a \colon L[E] \models \varphi(b, a_1, \ldots, a_k)\} \\ &\quad = \{b \in a \colon J_\alpha[E] \models \varphi(b, a_1, \ldots, a_k)\} \in J_{\alpha+\omega}[E] \subset L[E], \end{aligned}$$

using Lemma 5.15. This verifies that the separation schema holds in $L[E]$.

That the replacement schema (Rep) holds in $L[E]$ can be shown similarily by using the replacement schema in V. □

It is often necessary to work with the auxiliary hierarchy $S_\alpha[E]$ which is defined as follows:

$$\begin{aligned} S_0[E] &= \emptyset \\ S_{\alpha+1}[E] &= \mathbf{S}^E(S_\alpha[E]) \\ S_\lambda[E] &= \bigcup_{\xi<\lambda} S_\xi[E] \quad \text{for limit } \lambda \end{aligned}$$

where $\mathbf{S}^E$ is an operator which, applied to a set U, adds images of members of $U \cup \{U\}$ under rud_E functions from a certain carefully chosen fixed *finite* list. We may set

$$\mathbf{S}^E(U) = \bigcup_{i \in \{3,4,5,16\}} F_i''(U \cup \{U\}) \cup \bigcup_{i=0,\, i \neq 3,4,5}^{15} F_i''(U \cup \{U\})^2,$$

where

$$\begin{aligned}
F_0(x, y) &= \{x, y\} \\
F_1(x, y) &= x \setminus y \\
F_2(x, y) &= x \times y \\
F_3(x) &= \bigcup x \\
F_4(x) &= \{a : \exists b\, (a, b) \in x\} \\
F_5(x) &= \in \cap (x \times x) = \{(b, a) : a, b \in x \wedge a \in b\} \\
F_6(x, y) &= \{\{b : (a, b) \in x\} : a \in y\} \\
F_7(x, y) &= \{(a, b, c) : a \in x \wedge (b, c) \in y\} \\
F_8(x, y) &= \{(a, c, b) : (a, b) \in x \wedge c \in y\} \\
F_9(x, y) &= (x, y) \\
F_{10}(x, y) &= \{b : (y, b) \in x\} \\
F_{11}(x, y) &= (x, (y)_0, (y)_1) \\
F_{12}(x, y) &= ((y)_0, x, (y)_1) \\
F_{13}(x, y) &= \{((y)_0, x), (y)_1\} \\
F_{14}(x, y) &= \{(x, (y)_0), (y)_0\} \\
F_{15}(x, y) &= \{(x, y)\} \\
F_{16}(x) &= E \cap x.
\end{aligned}$$

(Here, $(y)_0 = u$ and $(y)_1 = v$ if $y = (u, v)$ and $(y)_0 = 0 = (y)_1$ if y is not an ordered pair.) It is not difficult to show that each F_i, $0 \leq i \leq 15$, is rud_E, and that $\mathbf{S}^E$ is rud_E as well (cf. Problem 5.5).

Lemma 5.20 *The ten functions $F_0, \ldots, F_8$, and F_{16} form a* basis *for the* rud_E *functions in the sense that every* rud_E *function can be generated as a composition of $F_0, \ldots, F_8$, and F_{16}.*

Proof (Cf. also Problem 5.6.) It obviously suffices to prove that the nine functions $F_0, \ldots, F_8$ form a basis for the rud functions. Let us write $\mathfrak{C}$ for the class of all functions which can be obtained from $F_0, \ldots, F_8$ via composition. We aim to see that every rud function is in $\mathfrak{C}$.

If $\varphi(v_1, \ldots, v_k)$ is a formula of $\mathscr{L}_\in$ with the free variables among $v_1, \ldots, v_k$, then we write[4]

$$g_{\varphi,k}(x) = \{(y_1, \ldots, y_k) \in x^k : (x; \in \cap x^2) \models \varphi(y_1, \ldots, y_k)\}.$$

[4] Here, x is *not* assumed to be transitive.

Claim 5.21 *$g_{\varphi,k} \in \mathfrak{C}$ for every $\varphi(v_1, \ldots, v_k)$.*

Proof First let $\varphi(v_1, \ldots, v_k) \equiv v_j \in v_i$, where $1 \leq i < j \leq k$. If $k = 2$, then we simply have $g_{\varphi,k}(x) = F_5(x)$, but in general we also need F_2, as well as F_7 and F_8 for "reshuffling." Let us write $X^1(z, x) = z$ and $X^{n+1}(z, x) = F_2(X^n(z, x), x) = X^n(z, x) \times x$ for $n \geq 1$, and let us also write $F_8^1(z, x) = F_8(z, x)$ and $F_8^{n+1}(z, x) = F_8(F_8^n(z, x), x)$ for $n \geq 1$. Say $2 \leq i < i + 1 < j \leq k$; then

$$g_{\varphi,k}(x) = X^{k-j+1}(F_8^{j-i-1}(F_7(X^{i-1}(x, x), F_5(x)), x), x).$$

The case $i = 1$ or $j = i + 1$ is similar.

Next, notice that

$$g_{\neg\varphi,k}(x) = F_1(X^k(x, x), g_{\varphi,k}(x))$$

and[5]

$$g_{\varphi\wedge\psi,k}(x) = F_1(g_{\varphi,k}(x), F_1(g_{\varphi,k}(x), g_{\psi,k}(x))).$$

Also

$$g_{\exists v_k\, \varphi,k}(x) = F_4(g_{\varphi,k+1}(x)).$$

Finally, if $\varphi(v_1, \ldots, v_k) \equiv v_i = v_j$, where $1 \leq i, j \leq k, i \neq j$, then

$$g_{\varphi,k}(x) = F_4(g_{\forall v_{k+1}\,(v_{k+1} \in v_i \leftrightarrow v_{k+1} \in v_j),k+1}(x \cup \bigcup x)) \cap X^k(x),$$

which may be generated with the additional help of F_0 and F_3, and if $\varphi(v_1, \ldots, v_k) \equiv v_i \in v_j$, where $1 \leq i < j \leq k$, then

$$g_{\varphi,k}(x) = F_4(F_4(g_{\exists v_{k+2} \exists v_{k+1} (v_{k+2} = v_i \wedge v_{k+1} = v_j \wedge v_{k+2} \in v_{k+1}),k+2}(x))).$$

Also,

$$g_{\exists v_i \varphi,k}(x) = F_4(g_{\exists v_{k+1} (v_i = v_{k+1} \wedge \varphi'),k+1}(x)),$$

where φ' results from φ by replacing each (free) occurence of v_i by v_{k+1}.

We have shown Claim 5.21. □

The proof of Claim 5.21 made use of all of $F_0, \ldots, F_8$ except for F_6. The role of F_6 is to verify the following.

Claim 5.22 *If f is* rud *and k-ary, then the function*

$$h_{f,k}(x) = f''x^k = \{z : \exists \mathbf{y} \in x^k\, f(\mathbf{y}) = z\}$$

is in $\mathfrak{C}$.

---

[5] $a \cap b = a \setminus (a \setminus b)$.

Proof We use the obvious induction along the schemata from Definition 5.8.

Let $f(\mathbf{x}) = x_i \backslash x_j$. Let $\varphi(v_1, v_2, v_3) \equiv v_3 \in v_1 \backslash v_2$. Then

$$h_{f,k}(x) = F_6(g_{\varphi,3}(x \cup \bigcup x) \cap (x \times x \times \bigcup x), x \times x).$$

If $f(\mathbf{x}) = \{x_i, x_j\}$, then $h_{f,k}(x) = \bigcup(x \times x)$.

Let $f(\mathbf{x}) = g_0(g_1(\mathbf{x}), \dots, g_\ell(\mathbf{x}))$. As every rud function is simple (cf. the proof of Lemma 5.11), we may let $\varphi(v_0, v)$ be a formula expressing that[6]

$$\exists(w_1, \dots, w_\ell)\exists v_1 \dots \exists v_k\, (v = (v_1, \dots, v_k) \wedge \\ w_1 = g_1(v_1, \dots, v_k) \wedge \dots \wedge w_\ell = g_\ell(v_1, \dots, v_k) \wedge v_0 = g_0(w_1, \dots, w_\ell)).$$

Let us write $H_1 = h_{g_1,k}(x) \cup \dots \cup h_{g_\ell,k}(x)$, $H_2 = h_{g_0,\ell}(H_1)$, and $H_3 = H_1 \cup (H_1)^2 \cup \dots \cup (H_1)^\ell \cup H_2 \cup x \cup x^2 \cup \dots \cup x^k$. Then

$$h_{f,k}(x) = F_4(g_{\varphi,2}(H_3) \cap (H_2 \times x^k)).$$

Finally, if $f(\mathbf{x}) = \bigcup_{y \in x_1} g(y, x_2, \dots, x_k)$, then we may argue analogously. Claim 5.22 is thus proven.

Claim 5.22 now immediately implies that every rud function is in $\mathfrak{C}$. Let f be rud, say f is k-ary. Let $\tilde{f}$ be defined by

$$\tilde{f}(u) = \begin{cases} f(x_1, \dots, x_k), & \text{if } u = (x_1, \dots, x_k) \\ \emptyset & \text{otherwise.} \end{cases}$$

Then $\tilde{f}$ is rud, so that by Claim 5.22 the function $u \mapsto \tilde{f}''u$ is in $\mathfrak{C}$. But then

$$x_1, \dots, x_k \mapsto \bigcup \tilde{f}''\{(x_1, \dots, x_k)\} = f(x_1, \dots, x_k)$$

is in $\mathfrak{C}$ as well. □

It is now straightforward to verify that if U is transitive, then $\mathbf{S}^E(U)$ is transitive as well, cf. Problem 5.5.[7] We thus inductively get that every $S_\alpha[E]$ is transitive. Moreover, by Lemma 5.20 and the definition of the S-hierarchy,

$$S_\beta[E] \in J_\alpha[E] = S_\alpha[E] \tag{5.2}$$

for all limit ordinals α and all $\beta < \alpha$. It is easy to see that there is only a finite jump in $\in$-rank from $S_\alpha[E]$ to $S_{\alpha+1}[E]$.

Lemma 5.12 together with (5.2) readily gives the following.

[6] We assume w.l.o.g. that every g_i, $1 \le i \le \ell$, is k-ary.

[7] The reason why the functions F_9 through F_{15} were added to the above list is in fact to guarantee that if U is transitive, then $\mathbf{S}^E(U)$ is transitive as well.

Lemma 5.23 *Let E be a set or proper class, and let α be a limit ordinal. Let $B \in (\underset{\sim}{\Sigma}_0)^{J_\alpha[E]}$. Then $(J_\alpha[E]; \in, E, B)$ is amenable, i.e., $J_\alpha[E]$ is a model of Σ_0 comprehension in the language $\mathscr{L}_{\dot{E},\dot{B}}$ with $\dot{\in}$, $\dot{E}$ and $\dot{B}$.*

Definition 5.24 A *J-structure* is an amenable structure of the form $(J_\alpha[E], B)$ for a limit ordinal α and predicates E, B, where E satisfies the hypothesis of Lemma 5.15.

Here, $(J_\alpha[E], B)$ denotes the structure $(J_\alpha[E]; \in\restriction J_\alpha[E], E \cap J_\alpha[E], B \cap J_\alpha[E])$. Of course, every $J_\alpha[E]$ is a J-structure.

Lemma 5.25 *Let $J_\alpha[E]$ be a J-structure, where α is a limit ordinal.*

(1) *For all $\beta < \alpha$, $(S_\gamma[E] : \gamma < \beta) \in J_\alpha[E]$. In particular, $S_\beta[E] \in J_\alpha[E]$ for all $\beta < \alpha$.*
(2) *$(S_\gamma[E] : \gamma < \alpha)$ is uniformly $\Sigma_1^{J_\alpha[E]}$. I.e., "$x = S_\gamma[E]$" is Σ_1 over $J_\alpha[E]$, as being witnessed by a formula which does not depend on α.*

Proof (1) and (2) are shown simultaneously by induction on (α, β), ordered lexicographically. Fix a limit ordinal α and some $\beta < \alpha$. If β is a limit ordinal, then inductively by (2), $(S_\gamma[E] : \gamma < \beta)$ is $\Sigma_1^{J_\beta[E]}$, and hence $(S_\gamma[E] : \gamma < \beta) \in J_\alpha[E]$ by Lemma 5.15. If $\beta = \delta + 1$, then inductively by (1), $(S_\gamma[E] : \gamma < \delta) \in J_\alpha[E]$. If δ is a limit ordinal, then $S_\delta[E] = \bigcup_{\gamma<\delta} S_\gamma[E] \in J_\alpha[E]$; and if $\delta = \bar{\delta} + 1$ then $S_\delta[E] = \mathbf{S}^E(S_{\bar{\delta}}[E]) \in J_\alpha[E]$ as well, as $\mathbf{S}^E$ is rud$_E$ (cf. Problem 5.6). It follows that $(S_\gamma[E] : \gamma < \beta) = (S_\gamma[E] : \gamma < \delta) \cup \{(\delta, S_\delta[E])\} \in J_\alpha[E]$, which proves (1). (2) is then not hard to verify. □

In order to show that $L[E]$ satisfies the Axiom of Choice (even locally), we may inductively define a well-ordering $<_\beta^E$ of $S_\beta[E]$ as follows. If β is a limit ordinal then we let $<_\beta^E = \bigcup_{\gamma<\beta} <_\gamma^E$. Now suppose that $\beta = \bar{\beta} + 1$. The order $<_{\bar{\beta}}^E$ induces a lexicographical order, call it $<_{\bar{\beta},\mathrm{lex}}^E$, of $17 \times S_{\bar{\beta}}[E] \times S_{\bar{\beta}}[E]$. We may then set[8]

$$x <_\beta^E y \iff \begin{cases} x, y \in S_{\bar{\beta}}[E] \text{ and } x <_{\bar{\beta}}^E y, \text{ or else} \\ x \in S_{\bar{\beta}}[E] \wedge y \notin S_{\bar{\beta}}[E], \text{ or else} \\ x, y \notin S_{\bar{\beta}}[E] \text{ and } (i, u_x, v_x) <_{\bar{\beta},\mathrm{lex}}^E (j, u_y, v_y) \\ \quad \text{where } (i, u_x, v_x) \text{ is } <_{\bar{\beta},\mathrm{lex}}^E -\text{minimal with } x = F_i(u_x, v_x) \\ \quad \text{and } (j, u_y, v_y) \text{ is } <_{\bar{\beta},\mathrm{lex}}^E -\text{minimal with } y = F_j(u_y, v_y). \end{cases}$$

If $M = J_\alpha[E]$, then we shall also write $<_M$ for $<_\alpha^E$.

Using Lemma 5.25 and by a proof similar to the one for Lemma 5.25, one may show the following. (Cf. Problem 5.10.)

---

[8] we pretend that all F_i, $i < 17$, are binary.

Lemma 5.26 *Let $J_\alpha[E]$ be a J-structure.*

(1) *For all $\beta < \alpha$, $(<^E_\gamma : \gamma < \beta) \in J_\alpha[E]$. In particular, $<^E_\beta \in J_\alpha[E]$ for all $\beta < \alpha$.*
(2) *$(<^E_\gamma : \gamma < \alpha)$ is uniformly $\Sigma_1^{J_\alpha[E]}$. I.e., "$x = <^E_\gamma$" is Σ_1 over $J_\alpha[E]$, as being witnessed by a formula which does not depend on α.*

Theorem 5.27 (Gödel) *Let E be a set or a proper class. Then $L[E] \models \mathsf{ZFC}$.*

Proof This is an immediate consequence of Theorem 5.19 and Lemma 5.26. □

By "$V = L[\dot{E}]$" we abbreviate the $\mathscr{L}_{\in,\dot{E}}$-sentence

$$\forall x \exists y \exists \gamma (y = S_\gamma[\dot{E}] \wedge x \in y),$$

where "$y = S_\gamma[\dot{E}]$" stands for the Σ_1 formula given by Lemma 5.25 (2). By Lemma 5.25 (2), "$V = L[\dot{E}]$ is Π_2, uniformly over $J_\alpha[E]$ (including $\alpha = \infty$). As a special case, by "$V = L$" we abbreviate the $\mathscr{L}_\in$-sentence

$$\forall x \exists y \exists \gamma (y = S_\gamma[\emptyset] \wedge x \in y).$$

We then get:

Lemma 5.28 *Let M be a transitive model. Then $M \models V = L$ iff $M = J_\alpha$ for some α. (In particular, for any transitive model M, $L^M = J_\alpha$, where $\alpha = M \cap \mathrm{OR}$.) More generally, $M \models V = L[\dot{E}]$ iff $M = J_\alpha[\bar{E}]$ for some α and $\bar{E}$, where $x \in \bar{E} \Longleftrightarrow M \models x \in \dot{E}$ for all $x \in M$. (Here, $\alpha \in \mathrm{OR}$ if M is a set and $\alpha = \infty$ if M is a proper class.)*

Proof We prove the first assertion; the proof of the second one is basically identical. "$\Longleftarrow$" immediately follows from Lemma 5.25 (1). To see "$\Longrightarrow$", let $M \models V = L$. If $x \in M$, then $M \models \exists y \exists \gamma (y = S_\gamma[\emptyset] \wedge x \in y)$. But this is Σ_1, so that by Lemma 5.3, every $x \in M$ is really contained in some $S_\gamma[\emptyset]$ for $\gamma \in M$. Thus, setting $\alpha = M \cap \mathrm{OR}$, $M \subset \bigcup_{\gamma<\alpha} S_\gamma[\emptyset] = S_\alpha[\emptyset] = J_\alpha$. By the same reasoning, if $\gamma' < \alpha$, then $M \models \exists y \exists \gamma (y = S_\gamma[\emptyset] \wedge \gamma' \in y)$ and $J_\alpha = \bigcup_{\gamma<\alpha} S_\gamma[\emptyset] \subset M$. □

The *Condensation Lemma* for the constructible hierarchy is the following statement.

Theorem 5.29 *Let $M = (J_\alpha[E], B)$ be a J-structure, and let $\pi : \bar{M} \longrightarrow_{\Sigma_1} M$, where $\bar{M}$ is transitive. Then there are $\bar{\alpha} \leq \alpha$, $\bar{E}$, and $\bar{B}$ such that $\bar{M} = (J_{\bar{\alpha}}[\bar{E}], \bar{B})$ is a J-structure.*

Proof Set $\bar{\alpha} = \mathrm{OR} \cap \bar{M} \leq \alpha$, $\bar{E} = \pi^{-1}{}''E$, and $\bar{B} = \pi^{-1}{}''B$. The sentence "$V = L[\dot{E}]$" is Π_2, so that $\bar{M} \models V = L[\dot{E}]$ is inherited from $M \models V = L[\dot{E}]$, and Lemma 5.28 then immediately gives $\bar{M} = (J_{\bar{\alpha}}[\bar{E}], \bar{B})$. □

The Condensation Lemma 5.29 leads to the following natural concept.

Definition 5.30 Let E be a set or a proper class. Then E is said to *satisfy full condensation* iff for every limit ordinal α and for every $\pi: M = J_{\bar{\alpha}}[\bar{E}] \to_{\Sigma_1} J_\alpha[E]$, $\bar{E} \cap M = E \cap M$ (in particular, $M = J_{\bar{\alpha}}[E]$).

Trivially, $E = \emptyset$ (or more generally, $E \subset \omega$) satisfies full condensation. There are non-trivial examples, though (cf. Problem 8.9).

The following Theorem was shown by KURT GÖDEL (1906–1978).

Theorem 5.31 (Gödel) *Let E be a set or a proper class which satisfies full condensation. Then $L[E] \models$ GCH. In fact, if κ is an infinite cardinal in $L[E]$ and $\tau = \kappa^{+L[E]}$, then*

$$\mathscr{P}(\kappa) \cap L[E] \subset J_\tau[E].$$

Proof Let $x \in \mathscr{P}(\kappa) \cap L[E]$, and pick some α such that $x \in J_\alpha[E]$. Let us work in $L[E]$. Pick

$$\pi: M \longrightarrow_{\Sigma_1} J_\alpha[E],$$

where M is transitive, $(\kappa + 1) \cup \{x\} \subset \text{ran}(\pi)$, and $\text{Card}(M) = \kappa$. By the Condensation Lemma 5.29, $M = J_{\bar{\alpha}}[\bar{E}]$ for some $\bar{\alpha} \leq \alpha$ and some $\bar{E}$. We must have $\text{Card}(\bar{\alpha}) < \kappa^+$ by Lemma 5.16. Also, $\bar{E} \cap M = E \cap M$, as E satisfies full condensation, so that in fact $M = J_{\bar{\alpha}}[\bar{E}] = J_{\bar{\alpha}}[E]$, where $\bar{\alpha} < \tau = \kappa^+$. We have shown that $x = \pi^{-1}(x) \in J_\tau[E]$. As x was arbitrary, $\mathscr{P}(\kappa) \subset J_\tau[E]$. Finally, because $\text{Card}(J_\tau[E]) = \tau$ again by Lemma 5.16, $\text{Card}(\mathscr{P}(\kappa)) = \tau = \kappa^+$. Because we worked in $L[E]$, the Theorem is shown. □

Corollary 5.32 *If* ZFC *is consistent, then so are* ZFC + *"$V = L$" as well as* ZFC + GCH.

Proof Let $(M; E)$ be a model of ZFC. Construct L inside $(M; E)$. This yields a model of ZFC plus GCH which thinks that "$V = L$," by Theorems 5.27, 5.28, and 5.31. □

We aim to study refinements of Theorem 5.31. For one thing, we may "localize" condensation for $L[E]$, cf. Definition 5.33. We will obtain the combinatorial principle $\diamondsuit_\kappa$ (and its variants) to hold in $L[E]$, cf. Definitions 5.34 and 5.37 and Theorems 5.35 and 5.39. For another thing, we may "localize" GCH in $L[E]$; this will lead to the concept of "acceptability" and the fine structure theory of $L[E]$, cf. Definition 11.1.

Definition 5.33 Let E be a set or a proper class. E is said to *satisfy local condensation* iff for every limit ordinal α, for every $\pi: M = J_{\bar{\alpha}}[\bar{E}] \to_{\Sigma_1} J_\alpha[E]$ with critical point δ, and for all limit ordinals $\beta < \bar{\alpha}$, if

$$(\mathscr{P}(\delta) \cap J_{\beta+\omega}[\bar{E}]) \setminus J_\beta[\bar{E}] \neq \emptyset,$$

then $\bar{E} \cap J_{\beta+\omega}[\bar{E}] = E \cap J_{\beta+\omega}[\bar{E}]$ (in particular, $J_{\beta+\omega}[\bar{E}] = J_{\beta+\omega}[E]$).

Trivially, full condensation implies local condensation, and $E = \emptyset$ satisfies full condensation. Theorem 5.31 also holds if E just satisfies local condensation (cf. Problem 5.17).

The definitions of variants of $\diamondsuit$ as well as the results on them which we are about to prove are all due to RONALD JENSEN.

Definition 5.34 Let κ be a regular uncountable cardinal, and let $R \subset \kappa$. By $\diamondsuit_\kappa(R)$ we mean the following statement. There is a sequence $(A_\xi : \xi \in R)$ such that for all $\xi \in R$, $A_\xi \subset \xi$, and for every $A \subset \kappa$, the set

$$\{\xi \in R : A \cap \xi = A_\xi\}$$

is stationary in κ. We also write $\diamondsuit_\kappa$ for $\diamondsuit_\kappa(\kappa)$ and $\diamondsuit$ for $\diamondsuit_{\omega_1}$.

Trivially, $\diamondsuit_\kappa(R)$ implies that R be stationary. It is not hard to show that $\diamondsuit_{\kappa^+}$ implies that $2^\kappa = \kappa^+$ (cf. Problem 5.20).

Theorem 5.35 (Jensen) *Let E be a set or a proper class which satisfies local condensation. Then inside $L[E]$, for every regular uncountable cardinal κ and every stationary $R \subset \kappa$, $\diamondsuit_\kappa(R)$ holds true.*

In particular, inside L, for every regular uncountable cardinal κ and every stationary $R \subset \kappa$, $\diamondsuit_\kappa(R)$ holds true.

Proof Let us work inside $L[E]$, and let us fix a stationary set $R \subset \kappa$. Let us recursively construct $(B_\xi, C_\xi : \xi \in R)$ as follows. Having constructed $(B_{\bar\xi}, C_{\bar\xi} : \bar\xi \in R \cap \xi)$, where $\xi \in R$, let (B_ξ, C_ξ) be the $<^E_\kappa$-least pair $(B, C) \in J_\kappa[E]$ such that $B \subset \xi$, C is club in ξ, and

$$\{\bar\xi \in R \cap \xi : B_{\bar\xi} = B \cap \bar\xi\} \cap C = \emptyset,$$

provided such a pair exists; otherwise we set $(B_\xi, C_\xi) = (\emptyset, \emptyset)$.

We claim that $(B_\xi : \xi \in R)$ witnesses that $\diamondsuit_\kappa(R)$ holds true. If not, then let $(B, C) \in L[E]$ be the $<_{L[E]}$-least pair such that $B \subset \kappa$, C is club in κ, and

$$\{\xi \in R : B_\xi = B \cap \xi\} \cap C = \emptyset. \tag{5.3}$$

Let $(B, C) \in J_\alpha[E]$. We have that $R \cap C$ is stationary, so that we may pick some

$$\pi : J_{\bar\alpha}[\bar E] \to_{\Sigma_\omega} J_\alpha[E],$$

where $\xi = \pi^{-1}(\kappa)$ is the critical point of π, $\mathrm{Card}(J_{\bar\alpha}[\bar E]) < \kappa$, $\{R, B, C\} \subset \mathrm{ran}(\pi)$, and $\xi \in R \cap C$ (cf. Problem 4.14). Say $\bar R = \pi^{-1}(R)$, $\bar B = \pi^{-1}(B)$, and $\bar C = \pi^{-1}(C)$. We have that $\bar R = R \cap \xi$, $\bar B = B \cap \xi$, and $\bar C = C \cap \xi$ is club in ξ.

As $<^E_\alpha$ is an initial segment of $<_{L[E]}$, (B, C) is also the $<^E_\alpha$-least pair such that $B \subset \kappa$, C is club in κ, and (5.3) holds true. By the elementarity of π, $(B \cap \xi, C \cap \xi)$ is then the $<^{\bar E}_{\bar\alpha}$-least pair such that $B \cap \xi \subset \xi$, $C \cap \xi$ is club in ξ, and

$$\{\bar{\xi} \in R \cap \xi : B_{\bar{\xi}} = B \cap \bar{\xi}\} \cap (C \cap \xi) = \emptyset. \tag{5.4}$$

Let $\beta < \bar{\alpha}$ be least such that

$$\{R \cap \xi, B \cap \xi, C \cap \xi\} \subset J_{\beta+\omega}[\bar{E}].$$

As E satisfies local condensation, $\bar{E} \cap J_{\beta+\omega}[\bar{E}] = E \cap J_{\beta+\omega}[\bar{E}]$, and hence $J_{\beta+\omega}[\bar{E}] = J_{\beta+\omega}[E]$. This gives that $(B \cap \xi, C \cap \xi)$ is the $<^E_{\beta+\omega}$-least (and thus also the $<^E_\kappa$-least) pair such that $B \cap \xi \subset \xi$, $C \cap \xi$ is club in ξ, and (5.4) holds true.

Therefore, $(B_\xi, C_\xi) = (B \cap \xi, C \cap \xi)$ by the choice of (B_ξ, C_ξ). Therefore,

$$\xi \in \{\bar{\xi} \in R : B_{\bar{\xi}} = B \cap \bar{\xi}\} \cap C,$$

which contradicts (5.3). □

Lemma 5.36 (Jensen) *If $\diamondsuit$ holds true, then there is an $\aleph_1$-Souslin tree.*

We shall prove a more general statement, cf. Lemma 11.68, later.

We also discuss a strengthening of $\diamondsuit_\kappa$, called $\diamondsuit^*_\kappa$.

Definition 5.37 Let κ be a regular uncountable cardinal, and let $R \subset \kappa$. By $\diamondsuit^*_\kappa(R)$ we mean the following statement. There is a sequence $(\mathscr{A}_\xi : \xi \in R)$ such that for all $\xi \in R$, $\mathscr{A}_\xi \subset \mathscr{P}(\xi)$ and $\text{Card}(\mathscr{A}_\xi) \leq \overline{\overline{\xi}}$, and for every $A \subset \kappa$ there is some club $C \subset \kappa$ such that $A \cap \xi \in \mathscr{A}_\xi$ for every $\xi \in C \cap R$. We also write $\diamondsuit^*_\kappa$ for $\diamondsuit^*_\kappa(\kappa)$ and $\diamondsuit^*$ for $\diamondsuit^*_{\omega_1}$.

Lemma 5.38 (Kunen) *Let κ be a regular uncountable cardinal, and let $R \subset \kappa$ be stationary. Then $\diamondsuit^*_\kappa(R)$ implies $\diamondsuit_\kappa(R)$.*

Proof Let $(\mathscr{A}_\xi : \xi \in R)$ witness $\diamondsuit^*_\kappa(R)$. Say $\mathscr{A}_\xi = \{A_{\xi,i} : i < \xi\}$ for $\xi \in R$. If i, $j < \xi$, $\xi \in R$, then let us write

$$A^j_{\xi,i} = \{\alpha < \xi : \langle \alpha, j \rangle \in A_{\xi,i}\},$$

where $\alpha, j \mapsto \langle \alpha, j \rangle$ is the GÖDEL pairing function (cf. p. 33). We claim that there is some $i < \kappa$ such that $(A^i_{\xi,i} : \xi \in R)$ witnesses $\diamondsuit_\kappa(R)$.

Suppose not. Then for every $i < \kappa$ there is some $A_i \subset \kappa$ and some club $C_i \subset \kappa$ such that

$$\{\xi \in R : A_i \cap \xi = A^i_{\xi,i}\} \cap C_i = \emptyset. \tag{5.5}$$

Let $D = \Delta_{i<\kappa} C_i$, and let

$$A = \{\langle \alpha, i \rangle : \alpha \in A_i\}.$$

As $(\mathscr{A}_\xi : \xi \in R)$ witnesses $\diamondsuit^*_\kappa(R)$, there is some club $C \subset D$ such that $A \cap \xi \in \mathscr{A}_\xi$ for every $\xi \in C \cap R$.

Let $\xi \in C \cap R$, where we may assume without loss of generality that ξ is closed under the GÖDEL pairing function ($\eta \mapsto \langle \eta, \eta \rangle$ is incresing and continuous). There is then some $i_0 < \xi$ such that $A \cap \xi = A_{\xi, i_0}$, and we also have that $A \cap \xi = \{\langle \alpha, i \rangle : \alpha \in A_i \wedge \alpha, i < \xi\}$. This yields that

$$A^{i_0}_{\xi, i_0} = A_{i_0} \cap \xi. \tag{5.6}$$

But $i_0 < \xi \in D = \Delta_{i<\kappa} C_i$ implies that $\xi \in C_{i_0}$, so that (5.6) contradicts (5.5). □

We now aim to characterize for which $R \subset \kappa$ we have that $\diamondsuit^*_\kappa(R)$ holds true in models of the form $L[E]$, where E satisfies local condensation. It turns out that the notion of "ineffability" (cf. Definition 4.49) is the relevant concept.

Theorem 5.39 (Jensen) *Let E be a set or a proper class which satisfies local condensation. The following is true inside $L[E]$.*

If κ is regular and uncountable and if $R \subset \kappa$ is not ineffable, then $\diamondsuit^*_\kappa(R)$ holds true.

In particular, inside L, if κ is regular and uncountable and if $R \subset \kappa$ is not ineffable, then $\diamondsuit^*_\kappa(R)$ holds true.

Proof Let us work inside $L[E]$. Let $(A_\xi : \xi \in R)$ witness that R is not ineffable, i.e., $A_\xi \subset \xi$ for every $\xi \in R$ and whenever $S \subset R$ is stationary, then there are $\xi \leq \xi'$ with $\xi, \xi' \in S$ and $A_\xi \neq A_{\xi'} \cap \xi$. For every $\xi \in R$, let $\delta(\xi)$ be the least $\delta > \xi$ such that

$$A_\xi \in J_\delta[E],$$

and set

$$\mathscr{A}_\xi = \mathscr{P}(\xi) \cap J_{\delta(\xi)}[E].$$

We claim that $(\mathscr{A}_\xi : \xi \in R)$ witnesses $\diamondsuit^*_\kappa(R)$.

By Problem 5.17 and Lemma 5.16, $\text{Card}(\mathscr{A}_\xi) \leq \overline{\overline{\xi}}$ for every $\xi \in R$. Let us fix $B \subset \kappa$. We aim to find some club $C \subset \kappa$ such that $B \cap \xi \in \mathscr{A}_\xi$ for every $\xi \in C \cap R$.

Let $\beta > \kappa$ be least such that $B \in J_\beta[E]$. By Problem 5.17, $\beta < \kappa^+$. Using Lemma 5.16, let us pick some bijection $g : \kappa \to J_\beta[E]$. The set

$$D = \{\xi < \kappa : B \in g``\xi \wedge g``\xi \prec J_\beta[E] \wedge \xi = (g``\xi) \cap \kappa\}$$

is easily verified to be club in κ. (Cf. Problem 4.14.) For every $\xi \in D$ there is some $\bar{E}$, $\varepsilon(\xi)$ and some π_ξ with critical point ξ such that

$$J_{\varepsilon(\xi)}[\bar{E}] \overset{\pi_\xi}{\cong} g``\xi \prec J_\beta[E].$$

Notice that $\beta = \bar{\beta} + \omega$, some limit $\bar{\beta}$, and hence if $\varepsilon(\xi) = \bar{\varepsilon} + \omega$, where $\bar{\varepsilon}$ is a limit, then $B \cap \xi = \pi_\xi^{-1}(B) \in J_{\varepsilon(\xi)}[\bar{E}] \setminus J_{\bar{\varepsilon}}[\bar{E}]$. As E satisfies local condensation, we therefore have that $\bar{E} = E \cap J_{\varepsilon(\xi)}[\bar{E}]$ and $J_{\varepsilon(\xi)}[\bar{E}] = J_{\varepsilon(\xi)}[E]$. I.e.,

$$B \cap \xi = \pi_\xi^{-1}(B) \in J_{\varepsilon(\xi)}[E] \tag{5.7}$$

for all $\xi \in D$.

Suppose that there were no club $C \subset D$ such that $B \cap \xi \in \mathscr{A}_\xi$ for every $\xi \in C \cap R$. This means that

$$\{\xi \in D \cap R : B \cap \xi \notin \mathscr{A}_\xi\} \text{ is stationary.} \tag{5.8}$$

By (5.7), $B \cap \xi \notin \mathscr{A}_\xi = \mathscr{P}(\xi) \cap J_{\delta(\xi)}[E]$ implies that $\delta(\xi) < \varepsilon(\xi)$, which in turn gives $A_\xi \in J_{\delta(\xi)}[E] \subset J_{\varepsilon(\xi)}[E]$. Hence (5.8) yields that

$$S = \{\xi \in D \cap R : A_\xi \in J_{\varepsilon(\xi)}[E]\} \text{ is stationary.} \tag{5.9}$$

If $\xi \in S$, then $\pi_\xi(A_\xi) \in g''\xi$, so that there is some $f(\xi) < \xi$ with $\pi_\xi(A_\xi) = g(f(\xi))$. By Fodor's Theorem 4.32, there is some stationary $T \subset S$ and some $\eta < \kappa$ such that for all $\xi \in T$, $f(\xi) = \eta$.

Let us write $A = g(\eta)$. If $\xi \in T$, then $\pi_\xi(A_\xi) = g(\eta) = A$, which means that $A_\xi = A \cap \xi$. But then $A_\xi = A_{\xi'} \cap \xi$ whenever $\xi \leq \xi'$, $\xi, \xi' \in T$. This contradicts the fact that $(A_\xi : \xi \in R)$ witnesses that R is not ineffable. □

Lemma 5.40 *Let κ be an uncountable regular cardinal, and assume $R \subset \kappa$ to be ineffable. Then $\diamondsuit_\kappa^*(R)$ fails.*

Proof Suppose $(\mathscr{A}_\xi : \xi \in R)$ were to witness $\diamondsuit_\kappa^*(R)$. For $\xi \in R$, say $\mathscr{A}_\xi = \{A_{\xi,i} : i < \xi\}$, and set

$$A_\xi = \{\langle \alpha, i \rangle : \alpha \in A_{\xi,i}\},$$

where $\alpha, i \mapsto \langle \alpha, i \rangle$ again is the Gödel pairing function. Applying the ineffability of R to $(A_\xi \cap \xi : \xi \in R)$, we find some stationary $S \subset R$ such that for all $\xi \leq \xi'$ with $\xi, \xi' \in S$,

$$A_\xi \cap \xi = A_{\xi'} \cap \xi.$$

For $i < \kappa$, let us write

$$A^i = \{\alpha < \kappa : \exists \xi \in S \ \langle \alpha, i \rangle \in A_\xi \cap \xi\}.$$

By the properties of S,

$$A_{\xi,i} = A^i \cap \xi \tag{5.10}$$

for all $i < \kappa$ and $\xi \in S$.

Let us now pick any $A \in \mathscr{P}(\kappa) \setminus \{A^i : i < \kappa\}$. As $(\mathscr{A}_\xi : \xi \in R)$ is supposed to witness $\diamondsuit_\kappa^*(R)$, there is some club $C \subset \kappa$ such that $A \cap \xi \in \mathscr{A}_\xi$ for all $\xi \in R \cap C$. By Fodor's Theorem 4.32 there is some stationary $T \subset S \cap C$ and some $i_0 < \kappa$ such that for all $\xi \in T$,

$$A \cap \xi = A_{\xi, i_0}.$$

But this implies that $A = A^{i_0}$ by (5.10). Contradiction! □

There is a principle which is slightly stronger than $\diamondsuit_\kappa^*$ and which is called $\diamondsuit_\kappa^+$, cf. Problem 5.22.

5.2 Ordinal Definability

We also need to introduce HOD.

Definition 5.41 Let Y be a set or a proper class. We say that x is *hereditarily in* Y iff $TC(\{x\}) \subset Y$.

Definition 5.42 Let z be a set or a proper class. We write OD_z for the class of all x which are *ordinal definable from elements of* z, i.e., such that there is a formula φ, there are ordinals $\alpha_1, \ldots, \alpha_n$ and elements $y_1, \ldots, y_m$ of z such that for all u,

$$u \in x \iff \varphi(u, \alpha_1, \ldots, \alpha_n, y_1, \ldots, y_m).$$

We also write HOD_z for the class of all x which are hereditarily in OD_z, i.e.,

$$\mathsf{HOD}_z = \{x : TC(\{x\}) \subset \mathsf{OD}_z\}.$$

If $x \in \mathsf{HOD}_z$, then we say that x is *hereditarily ordinal definable from elements of* z. If $z = \emptyset$, then we write OD instead of $\mathsf{OD}_\emptyset$ and HOD instead of $\mathsf{HOD}_\emptyset$.

Lemma 5.43 *Let z be a set. Then $x \in \mathsf{OD}_z$ iff there is a formula φ, there are ordinals $\alpha_1, \ldots, \alpha_n, \alpha$ and elements $y_1, \ldots, y_m$ of z such that $\alpha_1, \ldots, \alpha_n, z \in V_\alpha$ and for all u,*

$$u \in x \iff V_\alpha \models \varphi(u, \alpha_1, \ldots, \alpha_n, y_1, \ldots, y_m).$$

Proof By the reflection principle, cf. Problem 5.14, given φ, $\alpha_1, \ldots, \alpha_n$ and $y_1, \ldots, y_m \in z$ there is some α with $\alpha_1, \ldots, \alpha_n, z \in V_\alpha$ and for all $u \in V_\alpha$,

$$\varphi(u, \alpha_1, \ldots, \alpha_n, y_1, \ldots, y_m) \iff V_\alpha \models \varphi(u, \alpha_1, \ldots, \alpha_n, y_1, \ldots, y_m).$$

The Lemma then follows using Problem 5.1. □

By Lemma 5.43 and Problem 5.3, $\mathsf{OD}_z = \{x : \varphi(x, z)\}$ for some formula φ which is is Σ_2. This implies that $\mathsf{HOD}_z = \{x : \psi(x, z)\}$ for some formula ψ which is is Σ_2.

Theorem 5.44 (Gödel) *Let z be a set such that $z \in \mathsf{OD}_z$. Then $\mathsf{HOD}_z \models \mathsf{ZF}$.*

Proof Notice that HOD_z is trivially transitive. (Ext) and (Fund) are therefore true in HOD_z. It is easy to see that OR $\subset \mathsf{HOD}_z$, so that (Inf) is also true in HOD_z. (Pair), (Union), and (Sep) are also straightforward.

Let us show that the Power Set Axiom (Pow) holds in HOD_z. We need to see that if $x \in \mathsf{HOD}_z$, then $\mathscr{P}(x) \cap \mathsf{OD}_z \in \mathsf{OD}_z$. Fix $x \in \mathsf{HOD}_z$. By Lemma 5.43 and Replacement in V, there is some α such that $y \in \mathscr{P}(x) \cap \mathsf{OD}_z$ iff there is a formula φ, there are $\alpha_1, \ldots, \alpha_n, z \in V_\alpha$ and there are $y_1, \ldots, y_m \in z$ such that for all u,

$$u \in y \Longleftrightarrow V_\alpha \models \varphi(u, \alpha_1, \ldots, \alpha_n, y_1, \ldots, y_m).$$

Because $z \in \mathsf{OD}_z$, this shows that $\mathscr{P}(x) \cap \mathsf{OD}_z \in \mathsf{OD}_z$.

In order to show the Replacement Schema (Rep) in HOD_z, it is easy to see that it suffices to prove $\mathsf{HOD}_z \cap V_\alpha \in \mathsf{HOD}_z$ for all α. However, $y \in \mathsf{HOD}_z \cap V_\alpha$ iff $y \in V_\alpha$ and for all $x \in TC(\{y\})$ there is a formula φ, there are ordinals $\alpha_1, \ldots, \alpha_n, \beta$ and elements $y_1, \ldots, y_m$ of z such that $\alpha_1, \ldots, \alpha_n, z \in V_\beta$ and for all u,

$$u \in x \Longleftrightarrow V_\beta \models \varphi(u, \alpha_1, \ldots, \alpha_n, y_1, \ldots, y_m).$$

This shows that $\mathsf{HOD}_z \cap V_\alpha \in \mathsf{OD}_z$. Trivially, $\mathsf{HOD}_z \cap V_\alpha \subset \mathsf{HOD}_z$, and therefore in fact $\mathsf{HOD}_z \cap V_\alpha \in \mathsf{HOD}_z$. □

Theorem 5.45 (Gödel) *Let z be a set with $z \in \mathsf{OD}_z$. If there is a well-order of z in OD_z, then $\mathsf{HOD}_z \models \mathsf{ZFC}$. In particular, $\mathsf{HOD} \models \mathsf{ZFC}$.*

Proof By Theorem 5.44, we are left with having to verify that $\mathsf{HOD}_z \models (\mathsf{AC})$. Let $\leq_z$ be a well-order of z which exists in OD_z.

We write $a \Delta b$ for the *symmetric difference* $(a \backslash b) \cup (b \backslash a)$ *of a and b*. For finite sets u, v of ordinals, i.e., $u, v \in \mathrm{OR}^{<\omega}$, let us write $u \leq^* v$ iff $u = v$ or else $\max(u \Delta v) \in v$. It is easy to show that $\leq^*$ is a well-ordering on $\mathrm{OR}^{<\omega}$, cf. Problem 5.19. The well-order $\leq_z$ induces a well-order $\leq_z^*$ of finite subsets of z in the same fashion: for $u, v \in [z]^{<\omega}$, let $u \leq^* v$ iff $u = v$ or else y is largest (in the sense of $\leq_z$) in $u \Delta v$, then $y \in v$. Notice that $\leq^*, \leq_z^* \in OD_z$. For formulae $\varphi \equiv \varphi(v_0, v_1, \ldots, v_n, v_1', \ldots, v_n')$, $\psi = \psi(v_0, v_1, \ldots, v_p, v_1', \ldots, v_q')$, ordinals α, β, finite sets $\boldsymbol{\alpha} = \{\alpha_1, \ldots, \alpha_n\}$, $\boldsymbol{\beta} = \{\beta_1, \ldots, \beta_p\}$ of ordinals, and finite vectors $\mathbf{y} = (y_1, \ldots, y_m)$, $\mathbf{w} = (w_1, \ldots, w_q)$ of elements of z, we may then set

$$(\varphi, \alpha, \boldsymbol{\alpha}, \mathbf{y}) \leq^{**} (\psi, \beta, \boldsymbol{\beta}, \mathbf{w})$$

iff (the GÖDEL no. of) φ is smaller than (the GÖDEL no. of) ψ, or else if $\alpha < \beta$, or else $\boldsymbol{\alpha} <^* \boldsymbol{\beta}$, or else if $\mathbf{y} \leq_z^* \mathbf{w}$. We have that $\leq^{**}$ is a well-order.

Now if $x \in OD_z$, we may let $(\varphi_x, \alpha_x, \boldsymbol{\alpha}_x, \mathbf{y}_x)$ be the $\leq^{**}$-least tuple $(\varphi, \alpha, \boldsymbol{\alpha}, \mathbf{y})$ such that if $\varphi \equiv \varphi(v_0, v_1, \ldots, v_n, v_1', \ldots, v_m')$, $\boldsymbol{\alpha} = (\alpha_1, \ldots, \alpha_n)$, and $\mathbf{y} = (y_1, \ldots, y_m)$ then for all u,

$$u \in x \Longleftrightarrow V_\alpha \models \varphi(u, \alpha_1, \ldots, \alpha_n, y_1, \ldots, y_n).$$

For $x, y \in \mathsf{OD}_z$ we may then set $x \leq^{***} y$ iff $(\varphi_x, \alpha_x, \boldsymbol{\alpha}_x, \mathbf{y}_x) \leq^{**} (\varphi_y, \alpha_y, \boldsymbol{\alpha}_y, \mathbf{y}_y)$. We have that $\leq^{**}$ is a well-order of OD_z. For any ordinal γ, the restriction of $\leq^{***}$

to sets in $\mathsf{OD}_z \cap V_\gamma$ is an element of OD_z. But this implies that for any ordinal γ, the restriction of $\leq^{***}$ to $\mathsf{HOD}_z \cap V_\gamma$ is in HOD_z. It follows that $\mathsf{HOD}_z \models$ (AC). □

We refer the reader to [39] for an outline of the status quo of current day inner model theory.

5.3 Problems

5.1. Show that the relation $(M; E) \models \varphi(x)$ is definable in the language $\mathcal{L}_\in$ by a Σ_1- as well as by a Π_1-formula, i.e., there is a Σ_1-formula ψ and a Π_1-formula ψ' such that for all models $(M; E)$ of $\mathcal{L}_\in$ and for all $x \in M$,

$$(M; E) \models \varphi(x) \Longleftrightarrow \psi(M, E, \ulcorner\varphi\urcorner, x) \Longleftrightarrow \psi'(M, E, \ulcorner\varphi\urcorner, x).$$

Here, $\ulcorner\varphi\urcorner$ is the GÖDEL number of φ. We shall produce a stronger statement in Section 11.1, cf. Lemma 10.14.

5.2. Let E_A be as in Problem 4.10 (c). Let $(\omega; E_A) \overset{\pi}{\cong} (\mathrm{HF}; \in)$. Show that the relation $R \subset \omega \times \mathrm{HF}$, where $(n, a) \in R \Longleftrightarrow \pi(n) = a$, is $\Delta_1^{\mathsf{ZFC}^{-\infty}}$.

5.3. Show that a formula $\varphi(v)$ of $\mathcal{L}_\in$ is Σ_2 iff there is a formula $\varphi'(v)$ of $\mathcal{L}_\in$ such that

$$\mathsf{ZFC} \vdash \forall v(\varphi(v) \longleftrightarrow \exists\alpha\, V_\alpha \models \varphi'(v)).$$

5.4. Let κ be an infinite cardinal, and let φ be a Σ_2-sentence. Show that if $H_\kappa \models \varphi$, then $V \models \varphi$.

5.5. Show that the following functions are rudimentary: $f(\mathbf{x}) = \bigcup x_i$, $f(\mathbf{x}) = x_i \cup x_j$, $f(\mathbf{x}) = \{x_1, ..., x_k\}$, $f(\mathbf{x}) = (x_1, ..., x_k)$, and $f(\mathbf{x}) = x_i \times x_j$.

5.6. Let E be a set or a proper class. Let F_i, $0 \leq i \leq 16$, be the collection of functions from p. 73 which produce the $S_\alpha[E]$ hierarchy.
Show that each F_i, $0 \leq i \leq 15$, is rud, and that $\mathbf{S}^E$ is rud_E as well. Also, fill in the details in the proof of Lemma 5.20. Show that if U is transitive, then $\mathbf{S}^E(U)$ is transitive as well.

5.7. Show that if U is a transitive set and E is a set or a proper class, then $\mathrm{rud}_E(U \cup \{U\})$ is transitive as well.

5.8. Show that every rud_E function is simple.

5.9. Prove Lemma 5.16!

5.10. Prove Lemma 5.26!

5.11. Show that for every $x \in V$ there is some A such that $x \in L[A]$. Show also that it need not be the case that $x \in L[x]$. Show in BGC that there is a class A of ordinals such that $V = L[A]$.

5.12. Let M and N be two transitive models of ZFC. Show that if for all sets x of ordinals, $x \in M \iff x \in N$, then $M = N$.

5.13. Let $A \subset \mathrm{OR}$. Set

$$E = \{(\alpha + \omega, \xi) : \xi \in A \cap [\alpha, \alpha + \omega)\}$$

for every limit ordinal α. Show that E satisfies the hypotheses of Lemma 5.15 and that if λ is a limit of limit ordinals, then $J_\lambda[E] = J_\lambda[A]$ (and hence $L[E] = L[A]$).

5.14. Let $(M_\alpha : \alpha \in \mathrm{OR})$ be a continuous cumulative hierarchy of transitive sets, i.e., every M_α is transitive, $M_\alpha \subset M_\beta$ for $\alpha \leq \beta$, and $M_\lambda = \bigcup_{\alpha<\lambda} M_\alpha$ for limit ordinals λ. Set $M = \bigcup_\alpha M_\alpha$. Let $\varphi(x_1, \ldots, x_k)$ be a formula, and let $a, a_1, \ldots, a_k \in M$.
Show that there is then some α with $a, a_1, \ldots, a_k \in M_\alpha$ such that for all $b \in M_\alpha$,

$$M_\alpha \models \varphi(b, a_1, \ldots, a_k) \Leftrightarrow M \models \varphi(b, a_1, \ldots, a_k).$$

In particular, the *Reflection Principle* holds true: If φ is a formula, then there is a club class of α such that for all $x_1, \ldots, x_k \in V_\alpha$,

$$\varphi(x_1, \ldots, x_k) \iff V_\alpha \models \varphi(x_1, \ldots, x_k).$$

5.15. **(A. Levy)** Use Problem 5.14 to show that ACKERMANN's set theory AST is conservative over ZF, i.e., if φ is a formula of $\mathscr{L}_\in$ which is provable in AST, then φ is provable in ZF. [Hint. Use the compactness theorem.]

5.16. Let κ be weakly inaccessible. Show that $J_\kappa \models$ ZFC. Conclude that the existence of weakly inaccessible cardinals cannot be proven in ZFC.

5.17. Show that the conclusion of Theorem 5.31 also holds if E is just assumed to satisfy local condensation. I.e., if E is a set or a proper class which satisfies local condensation, then $L[E] \models$ GCH, and in fact, if κ is an infinite cardinal in $L[E]$ and $\tau = \kappa^{+L[E]}$, then $\mathscr{P}(\kappa) \cap L[E] \subset J_\tau[E]$.

5.18. Let $M = J_\alpha[E]$ be a J-structure. Show that there is a (partial) surjection $h : \alpha \to [\alpha]^{<\omega}$ such that $h \in \underset{\sim}{\Sigma}_1^M$.

5.19. For $u, v \in \mathrm{OR}^{<\omega}$, let $u \leq^* v$ iff $\max(u \Delta v) \in v$. Show that $\leq^*$ is a well-ordering on $\mathrm{OR}^{<\omega}$

5.20. Let κ be an infinite cardinal such that $\diamondsuit_{\kappa^+}$ holds true. Show that $2^\kappa = \kappa^+$.

Let κ be a regular uncoutable cardinal. $R \subset \kappa$ is called *subtle* iff for every sequence $(A_\xi : \xi \in R)$ such that $A_\xi \subset \xi$ for every $\xi \in R$ and for every club $C \subset \kappa$ there are $\xi, \xi' \in R \cap C$ with $\xi < \xi'$ such that $A_\xi = A_{\xi'} \cap \xi$.

5.21. **(K. Kunen)** Let κ be a regular uncoutable cardinal, and let $R \subset \kappa$. Show that if R is ineffable, then R is subtle. Show also that if R is subtle, then $\diamondsuit_\kappa(R)$ holds true. [Hint. Follow the proof of Theorem 5.35.]

Let κ be a regular uncountable cardinal, and let $R \subset \kappa$. By $\diamondsuit_\kappa^+(R)$ we mean the following statement. There is a sequence $(\mathscr{A}_\xi : \xi \in R)$ such that for all $\xi \in R$, $\mathscr{A}_\xi \subset \mathscr{P}(\xi)$ and $\text{Card}(\mathscr{A}_\xi) \leq \overline{\overline{\xi}}$, and for every $A \subset \kappa$ there is some club $C \subset \kappa$ such that for every $\xi \in C \cap R$, $\{A \cap \xi, C \cap \xi\} \subset \mathscr{A}_\xi$. $\diamondsuit_\kappa^+$ is $\diamondsuit_\kappa^+(\kappa)$.

5.22. **(R. Jensen)** Assume $V = L[E]$, where E satisfies local condensation. Let κ be a regular uncountable cardinal, and let $R \subset \kappa$. Show that if R is not ineffable, then $\diamondsuit_\kappa^+(R)$ holds true. [Hint. Follow the proof of Theorem 5.39. Pick β such that $R \in J_\beta$, say $R = g(0)$. Towards the end, after (5.7), let $\bar{C} = \{\varepsilon(\xi) : \xi \in D\}$. Verify as follows that if $\xi_0 \in R$ is a limit point of $\bar{C}$, then $B \cap \xi_0, \bar{C} \cap \xi_0 \in \mathscr{A}_{\xi_0}$. Otherwise $A_{\xi_0} \in J_{\varepsilon(\xi_0)}$. Set $S = \{\bar{\xi} \in R \cap \xi_0 : A_{\bar{\xi}} = A_{\xi_0} \cap \bar{\xi}\}$. By the choice of $(A_\xi : \xi \in R)$ and the elementarity of π_{ξ_0}, S can't be stationary in $J_{\varepsilon(\xi_0)}$, so that $J_{\varepsilon(\xi_0)}$ has a club $I \subset \xi$ disjoint from S. But setting $I^* = \pi_{\xi_0}(I)$ and $A^* = \pi_{\xi_0}(A_{\xi_0})$, $\xi_0 \in R \cap I^*$ and $A_{\xi_0} = A^* \cap \xi_0$. Contradiction!]

KUREPA'*s Hypothesis at* κ, KH_κ, is the statement that there is some set $B \subset \mathscr{P}(\kappa)$ of size κ^+ such that for all $\xi < \kappa$, $\{X \cap \xi : X \in B\}$ has size at most $\text{Card}(\xi)$.

5.23. **(R. Jensen)** Let κ be regular and uncoutable. Then $\diamondsuit_\kappa^+$ implies KH_κ. [Hint. Let $(\mathscr{A}_\xi : \xi < \kappa)$ witness $\diamondsuit_\kappa^+$. For $\omega \leq \xi < \kappa$, let M_ξ be a transitive model of ZFC^- of size $\text{Card}(\xi)$ such that $\mathscr{A}_\xi \subset M_\xi$. Let $B = \{X \subset \kappa : \forall \xi < \kappa\ X \cap \xi \in M_\xi\}$.]

5.24. **(R. Jensen, K. Kunen)** Show that if κ is ineffable, then KH_κ fails. Conclude that if κ is ineffable, then $\diamondsuit_\kappa^+$ fails.

KRIPKE- PLATEK *set theory*, KP, for short, is the theory in the language of $\mathscr{L}_\in$ which has the following axioms. (Ext), (Fund_φ) for every formula φ (cf. Problem 3.4), (Pair), (Union), (Inf), (Aus_φ) for all Σ_0-formulae φ, and (Coll_φ) for all Σ_0-formulae φ (cf. Problem 3.5).

5.25. (a) Show that there is a proof of Lemma 2.1 in KP, i.e., "for all a, b, $a \times b$ exists" is provable in KP.

(b) Show also that KP proves (Coll_φ) for all Σ_1-formulae φ.

(c) Also show that if φ and ψ are both Σ_1 and such that KP proves $\varphi \leftrightarrow \neg\psi$, then KP proves (Aus_φ).

5.26. Let $\alpha > \omega$ be a limit ordinal. Show that the following statements are equivalent.

(a) $J_\alpha \models \mathsf{KP}$.
(b) $J_\alpha \models (\text{Coll}_\varphi)$ for all Σ_0-formulae φ.

(c) if φ and ψ are both Σ_1 and such that $J_\alpha \models \forall x \forall \mathbf{v}(\varphi \leftrightarrow \neg\psi)$, then $J_\alpha \models (\text{Aus}_\varphi)$.

(d) there is no total $f: a \to J_\alpha$, where $a \in J_\alpha$ and $f \in \underset{\sim}{\Sigma}_1^{J_\alpha}$.

Let $(M; E)$ be a model of KP, and let $(N; \in) \cong (\text{wfp}(E); E \restriction \text{wfp}(M))$, i.e. N is the transitive collapse of the well-founded part of $(M; E)$. Then $(M; E)$ is called an *ω-model* iff $\omega \in N$. A set N is called *admissible* iff N is transitive and $(N; \in\restriction N) \models \mathsf{KP}$.

5.27. **(Ville's Lemma)** Let $(M; E) \models \mathsf{KP}$ be an ω-model. Show that if N is the transitive collapse of the well-founded part of $(M; E)$, then N is admissible.

5.28. Show in KP that if $R \subset M \times M$ is well-founded, where M is a set, then the function $\rho: R \to \text{OR}$ is a set, where $\rho(x) = \{\rho(y): yRx\}$, i.e., $\rho(x)$ is the R-rank of x for every $x \in R$. [Hint. The relevant φ in the Recursion Theorem 3.13 is Σ_1.] Show also in KP that if $R \subset M \times M$ is well-founded and extensional, where M is a set, then the transitive collapse as defined in Theorem 3.21 is a set.

Conclude the following. Let N be admissible. Let $R \in N$ be a binary relation which is well-founded in V. Then $||R|| < N \cap \text{OR}$. If $R \in N$ is well-founded and extensional in V, then the transitive collapse of R is an element of N. Also, if $R \in N$ is *any* relation, then $||\text{wfp}(R)|| \leq N \cap \text{OR}$.

Let $z \in {}^\omega\omega$. We call α *z-admissible* iff $J_\alpha[z]$ is an admissible set. We write ω_1^z for the least z-admissible ordinal. We also write ω_1^{CK} for ω_1^0 ($0 =$ the constant function wth value 0, CK = Church–Kleene).

Chapter 6
Forcing

The method of *forcing* was invented by PAUL COHEN (1934–2007) to show the independence of the Continuum Hypothesis from the axioms of ZFC, using COHEN forcing (cf. Definition 6.5 and Theorem 6.33).

6.1 The General Theory of Forcing

Recall that $\mathbb{P} = (\mathbb{P}; \leq)$ is a *partial order* iff $\leq$ is reflexive, symmetric, and transitive (cf. p. 14). In what follows, we shall always assume that $\mathbb{P} \neq \emptyset$. As before, we write $p < q$ for $p \leq q \wedge q \not\leq p$ (which, by symmetry, is equivalent to $p \leq q \wedge p \neq q$).

Definition 6.1 Let $\mathbb{P} = (\mathbb{P}; \leq)$ be a partial order. We also call $\mathbb{P}$ a *notion of forcing* and the elements of $\mathbb{P}$ the *forcing conditions*. For $p, q \in \mathbb{P}$ we say that p is *stronger than* q iff $p \leq q$, and we say that p is *strictly stronger than* q iff $p < q$.

Definition 6.2 Let $\mathbb{P} = (\mathbb{P}; \leq)$ be a partial order. A set $D \subset \mathbb{P}$ is called dense (in $\mathbb{P}$) iff $\forall p \in \mathbb{P}\, \exists q \in D\; q \leq p$. If $p \in \mathbb{P}$, then $D \subset \mathbb{P}$ is called *dense below* p iff $\forall p' \leq p\, \exists q \in D\; q \leq p'$. A set $G \subset \mathbb{P}$ is called a filter iff (a) if $p, q \in G$, then there is some $r \in G$ with $r \leq p \wedge r \leq q$, and (b) $\forall p \in G\, \forall q \in \mathbb{P}(p \leq q \to q \in G)$.

If $\mathbb{P} = (\mathbb{P}; \leq)$ is a partial order, and if $p, q \in \mathbb{P}$, then we write $p \parallel q$ for $\exists r \in \mathbb{P}(r \leq p \wedge r \leq q)$, in which case p, q are called *compatible*, and we write $p \perp q$ for $\neg p \parallel q$, in which case p, q are called *incompatible*. If G is a filter, then any two $p, q \in G$ are compatible (as being witnessed by an element of the filter).

Definition 6.3 Let $\mathbb{P} = (\mathbb{P}; \leq)$ be a partial order, and let $\mathscr{D}$ be a family of dense sets. A filter $G \subset \mathbb{P}$ is called $\mathscr{D}$-generic iff $G \cap D \neq \emptyset$ for all $D \in \mathscr{D}$.

Lemma 6.4 *Let $\mathbb{P} = (\mathbb{P}; \leq)$ be a partial order, and let $\mathscr{D}$ be family of dense sets such that $\mathscr{D}$ is at most countable. Then for every $p \in \mathbb{P}$ there is a $\mathscr{D}$-generic filter G with $p \in G$.*

R. Schindler, *Set Theory*, Universitext, DOI: 10.1007/978-3-319-06725-4_6,

Proof Say $\mathscr{D} = \{D_n : n < \omega\}$. Let $p \in \mathbb{P}$ be given, and recursively construct $(p_n : n < \omega)$ as follows. Set $p_0 = p$. If p_n is constructed, where $n < \omega$, then pick some $q \leq p_n$ with $q \in D_n$ (this is possible because D_n is dense in $\mathbb{P}$), and set $p_{n+1} = q$. It is then easy to see that

$$G = \{r \in \mathbb{P} : \exists n \; p_n \leq r\}$$

is a $\mathscr{D}$-generic filter. □

This lemma produces a "forcing proof" of CANTOR's Theorem 1.1 as follows.

Definition 6.5 Let $\mathbb{C} = {}^{<\omega}\omega$, i.e., the set of all finite sequences of natural numbers. For $p, q \in \mathbb{C}$, let $p \leq q$ iff $p \supset q$ (iff $\exists n \; p \upharpoonright n = q$). The partial order $\mathbb{C} = (\mathbb{C}; \leq)$ is called COHEN *forcing*.

Now let X be a countable subset of ${}^{\omega}\omega$, the set of all infinite sequences of natural numbers. Say $X = \{x_n : n < \omega\}$. Set

$$D_n = \{p \in \mathbb{C} : p \neq x_n \upharpoonright \mathrm{dom}(p)\},$$

and

$$D_n^* = \{p \in \mathbb{C} : n \in \mathrm{dom}(p)\}.$$

It is easy to verify that each D_n as well as each D_n^* is dense in $\mathbb{C}$.

Set $\mathscr{D} = \{D_n : n < \omega\} \cup \{D_n^* : n < \omega\}$, and let G be $\mathscr{D}$-generic, via Lemma 6.4.

If $n < \omega$, then, as D_n^* is dense, there is some $p \in G$ with $n \in \mathrm{dom}(p)$. Therefore, as G is a filter, $\bigcup G \in {}^{\omega}\omega$. Also, if $n < \omega$, then, as D_n is dense, there is some $p \in G$ with $p \neq x_n \upharpoonright \mathrm{dom}(p)$, so that $\bigcup G \neq x_n$. We have seen that $\bigcup G \in {}^{\omega}\omega \setminus X$. In particular, ${}^{\omega}\omega \setminus X \neq \emptyset$. We have shown that $2^{\aleph_0} > \aleph_0$.

In what follows, we aim to produce "generic extensions" $M[G]$ of given (countable) transitive models M of ZFC.

Lemma 6.6 *Let* $\mathbb{P} = (\mathbb{P}; \leq) \in M$*, where* M *is a transitive model of* ZFC*. Then* $\mathbb{P}$ *is a partial order* $\Longleftrightarrow M \models$ *"*$\mathbb{P}$ *is a partial order". Also, if* $D \subset \mathbb{P}$*, where* $D \in M$*, and if* $p \in \mathbb{P}$*, then* D *is dense in* $\mathbb{P} \Longleftrightarrow M \models$ *"*D *is dense in* $\mathbb{P}$*" and* D *is dense below* $p \Longleftrightarrow M \models$ *"*D *is dense below* p*."*

Proof "$(\mathbb{P}; \leq)$ is a partial order," "D is dense in $\mathbb{P}$," and "D is dense below p," may all be written as Σ_0-formulae. □

Definition 6.7 Let M be a transitive model of ZFC, and let $\mathbb{P} = (\mathbb{P}; \leq) \in M$ be a partial order. A filter $G \subset \mathbb{P}$ is called $\mathbb{P}$-generic over M (or, M-generic for $\mathbb{P}$) iff G is $\mathscr{D}$-generic, where $\mathscr{D} = \{D \in M : D \text{ is dense in } \mathbb{P}\}$.

By Lemma 6.4, if M is a *countable* transitive model of ZFC and $\mathbb{P} \in M$ is a partial order, then for each $p \in \mathbb{P}$ there is a $\mathbb{P}$-generic filter G over M with $p \in G$.

Lemma 6.8 *Let M be a transitive model of* ZFC, *let $\mathbb{P} = (\mathbb{P}; \leq) \in M$ be a partial order, and let G be $\mathbb{P}$-generic over M. If $p \in G$, and if $D \subset \mathbb{P}$, $D \in M$, is dense below p, then $G \cap D \neq \emptyset$.*

Proof Set

$$D' = \{q \in \mathbb{P} \colon \exists r \in D\ q \leq r\} \cup \{q \in \mathbb{P} \colon \forall r \in D\ q \perp r\}.$$

Then $D' \in M$, and D' is easily be seen to be dense. Let $q \in G \cap D'$, and let $s \in G$ be such that $s \leq p$ and $s \leq q$. As D is dense below p, there is some $t \leq s$ with $t \in D$, so that in particular $q \| t$. But $q \in D'$, so that we must now have some $r \in D$ with $q \leq r$, which gives $r \in D \cap G$ as desired. □

Definition 6.9 Let $\mathbb{P} = (\mathbb{P}; \leq)$ be a partial order. $A \subset \mathbb{P}$ is called an *antichain* iff for all $p, q \in A$, if $p \neq q$, then $p \perp q$. $A \subset \mathbb{P}$ is called a *maximal antichain* iff A is an antichain and for all $p \in \mathbb{P}$ there is some $q \in A$ with $p \parallel q$. $D \subset \mathbb{P}$ is called *open* iff for all $p \in D$, if $q \leq p$, then $q \in D$.

The HAUSDORFF Maximality Principle 2.11 easily gives that every antichain is contained in a maximal antichain.

Lemma 6.10 *Let M be a transitive model of ZFC, and let $\mathbb{P} = (\mathbb{P}; \leq) \in M$ be a partial order. Let $G \subset \mathbb{P}$ be a filter. The following are equivalent.*

(1) *G is $\mathbb{P}$-generic over M.*
(2) *$G \cap A \neq \emptyset$ for every maximal antichain $A \subset \mathbb{P}$, $A \in M$.*
(3) *$G \cap D \neq \emptyset$ for every open dense set $D \subset \mathbb{P}$, $D \in M$.*

Proof (1) $\Longrightarrow$ (2): Let $A \subset \mathbb{P}$, $A \in M$, be a maximal antichain. Let $D = \{p \in \mathbb{P} : \exists q \in A\ p \leq q\}$. D is easily seen to be dense, and of course $D \in M$. Let $p \in G \cap D$. There is some $q \in A$ with $p \leq q$. But then $q \in G \cap A$, as G is a filter.

(2) $\Longrightarrow$ (1): Let $D \subset \mathbb{P}$, $D \in M$ be dense. Working in M, let $A \subset D$ be an antichain such that for every $p \in D$ there is some $q \in A$ with $p \parallel q$. It is easy to see that A is then a maximal antichain. (Cf. Problem 6.1.) But then $p \in A \cap G$ implies $p \in D \cap G$.

(1) $\Longrightarrow$ (3): This is trivial.

(3) $\Longrightarrow$ (1): Let $D \subset \mathbb{P}$, $D \in M$ be dense. Let $D^* = \{p : \exists q \in D\ p \leq q\}$. D^* is then open dense, and of course $D^* \in M$. Let $p \in D^* \cap G$. There is then some $q \in D$ with $p \leq q$. But then $q \in D \cap G$, as G is a filter. □

$\mathbb{P}$ is called *atomless* iff $\forall p \in \mathbb{P}\ \exists q \in \mathbb{P}\ \exists r \in \mathbb{P}\ (q \leq p \wedge r \leq p \wedge q \perp r)$. COHEN forcing is atomless as are all the other forcings considered in this book.

Lemma 6.11 *Let M be a transitive model of* ZFC, *let $\mathbb{P} = (\mathbb{P}; \leq)$ be an atomless partial order, and let $G \subset \mathbb{P}$ be $\mathbb{P}$-generic over M. Then $G \notin M$.*

Proof Suppose that $G \in M$. Then $D = \mathbb{P} \setminus G \in M$, and D is dense: if $p \in \mathbb{P}$ and if $q \leq p$ and $r \leq p$ are incompatible, then at most one of q, r can be in G, i.e., at least one of q, r must be in D. But then $D \cap G \neq \emptyset$, which is nonsense. □

Definition 6.12 Let M be a transitive model of ZFC, and let $\mathbb{P} \in M$ be a partial order. For $\alpha < M \cap \mathrm{OR}$, we recursively define the sets $M^{\mathbb{P}}_{\alpha}$ of $\mathbb{P}$-names (of M) of rank $\leq \alpha$ as follows. Set

$$M^{\mathbb{P}}_{\alpha} = \{\tau \in M : \tau \text{ is a binary relation and } \\ \forall (\sigma, p) \in \tau \, (p \in \mathbb{P} \wedge \exists \beta < \alpha \, \sigma \in M^{\mathbb{P}}_{\beta})\}$$

We also write $M^{\mathbb{P}} = \bigcup_{\alpha < M \cap \mathrm{OR}} M^{\mathbb{P}}_{\alpha}$ for the class of $\mathbb{P}$-names (of M).

Of course $M^{\mathbb{P}} \subset M$, so that if M is a set, then $M^{\mathbb{P}}$ is a set as well.

Definition 6.13 Let M be a transitive model of ZFC, and let $\mathbb{P} \in M$ be a partial order. Let $G \subset \mathbb{P}$ be $\mathbb{P}$-generic over M. For $\tau \in M^{\mathbb{P}}$ we write τ^G for the *G-interpretation of τ*, which is defined to be

$$\{\sigma^G : \exists p \in G \, (\sigma, p) \in \tau\}.$$

We also write $M[G] = \{\tau^G : \tau \in M^{\mathbb{P}}\}$ and call it a (the) *generic extension of M (via $\mathbb{P}$, G)*.

Of course, the definition of τ^G is by recursion on the rank of τ in the sense of Definition 6.12. We aim to prove that any generic extension of a transitive model of ZFC is again a transitive model of ZFC.

In what follows, we want to assume that our partial order $\mathbb{P}$ always has a "least" ("weakest") element $1 = 1_{\mathbb{P}}$, i.e., $p \leq 1$ for all $p \in \mathbb{P}$. (Most often, $1_{\mathbb{P}} = \emptyset$.)

By recursion on the $\in$-rank of $x \in M$, we define $\check{x} = \{(\check{y}, 1) : y \in x\}$ for every $x \in M$. A trivial induction shows $\check{x} \in M^{\mathbb{P}}$ for every $x \in M$. We also define $\dot{G}$ to be the $\mathbb{P}$-name $\{(\check{p}, p) : p \in \mathbb{P}\}$; obviously, $\dot{G} \in M^{\mathbb{P}}$.

Lemma 6.14 *Let M be a transitive model of ZFC, let $\mathbb{P} \in M$ be a partial order, and let $G \subset \mathbb{P}$ be $\mathbb{P}$-generic over M. $M[G]$ is transitive, and $M \cup \{G\} \subset M[G]$.*

Proof The transitivity of $M[G]$ is trivial. In order to verify $M \subset M[G]$, we show $\check{x}^G = x$ for every $x \in M$ by induction on the rank of x. We have $\check{x}^G = \{\sigma^G : \exists p \in G \, (\sigma, p) \in \check{x}\} = \{\check{y}^G : y \in x\}$ (notice that $1 \in G$) $= x$ by the inductive hypothesis. In order to verify $G \in M[G]$, we show that $\dot{G}^G = G$. Well, $\dot{G}^G = \{\sigma^G : \exists p \in G \, (\sigma, p) \in \dot{G}\} = \{\check{p}^G : p \in G\} = G$, as $\check{p}^G = p$ for all $p \in \mathbb{P}$. □

It is easy to verify that if N is a transitive model of ZFC with $M \cup \{G\} \subset N$, then $M[G] \subset N$. Therefore, once we showed that $M[G]$ is indeed a model of ZFC, we know that it is the smallest ZFC-model which contains $M \cup \{G\}$. To begin, the ordinal height of $M[G]$ is the same as the one of M:

Lemma 6.15 *Let M be a transitive model of ZFC, let $\mathbb{P} \in M$ be a partial order, and let $G \subset \mathbb{P}$ be $\mathbb{P}$-generic over M. $M[G] \cap OR = M \cap OR$.*

Proof By Lemma 6.14, we only need to see that $M[G] \cap OR \subset M$.

A straightforward induction yields that the $\in$-rank $\mathrm{rk}_\in(\tau^G)$ of τ^G is at most the $\in$-rank of τ, for every $\tau \in M^{\mathbb{P}}$. Now let $\xi \in M[G] \cap OR$, say $\xi = \tau^G$, where $\tau \in M^{\mathbb{P}}$. Then $\xi = \mathrm{rk}_\in(\xi) = \mathrm{rk}_\in(\tau^G) \leq rk(\tau) < M \cap \mathrm{OR}$, as desired. □

Because $M[G]$ is transitive and $\omega \in M[G]$, we know by Lemma 5.7 (1) through (3) that $M[G]$ is a model of the axioms (Ext), (Fund), and (Inf).

Lemma 6.16 *Let M be a transitive model of* ZFC, *let $\mathbb{P} \in M$ be a partial order, and let $G \subset \mathbb{P}$ be $\mathbb{P}$-generic over M. $M[G] \models$* (Pair).

Proof Let $x, y \in M[G]$, say $x = \tau^G$, $y = \sigma^G$, where $\tau, \sigma \in M^{\mathbb{P}}$. Let

$$\rho = \{(\tau, 1_{\mathbb{P}}), (\sigma, 1_{\mathbb{P}})\}.$$

Of course, $\rho \in M^{\mathbb{P}}$. But it is easy to see that $\rho^G = \{\tau^G, \sigma^G\} = \{x, y\}$, so that $\{x, y\} \in M[G]$. The result then follows via Lemma 5.7 (4). □

Lemma 6.17 *Let M be a transitive model of* ZFC, *let $\mathbb{P} \in M$ be a partial order, and let $G \subset \mathbb{P}$ be $\mathbb{P}$-generic over M. $M[G] \models$ (Union).*

Proof Let $x \in M[G]$, say $x = \tau^G$, where $\tau \in M^{\mathbb{P}}$. Let

$$\sigma = \{(\pi, p) \colon \exists \rho \exists q \exists q' \, (p \leq q \wedge p \leq q' \wedge (\pi, q) \in \rho \wedge (\rho, q') \in \tau)\}.$$

Of course, $\sigma \in M^{\mathbb{P}}$, and it is straightforward to verify that $\sigma^G = \bigcup x$. The result then follows via Lemma 5.7 (5). □

In order to verify $M[G]$ to satisfy (Aus), (Rep), and (Pow), we need the "forcing language."

Definition 6.18 Let M be a transitive model of ZFC, and let $\mathbb{P} \in M$ be a partial order. Let $p \in \mathbb{P}$, let $\varphi(v_1, \ldots, v_n)$ be a formula of the language of set theory, and let $\tau_1, \ldots, \tau_n \in M^{\mathbb{P}}$. We say that p forces $\varphi(\tau_1, \ldots, \tau_n)$ (over M), abbreviated by

$$p \Vdash_M^{\mathbb{P}} \varphi(\tau_1, \ldots, \tau_n),$$

iff for all G which are $\mathbb{P}$-generic over M and such that $p \in G$ we have that $M[G] \models \varphi(\tau_1^G, \ldots, \tau_n^G)$.

We also write $\Vdash^{\mathbb{P}}$ or just $\Vdash$ instead of $\Vdash_M^{\mathbb{P}}$. Notice that for a fixed φ,

$$\{(p, \tau_1, \ldots, \tau_n) : p \Vdash_M^{\mathbb{P}} \varphi(\tau_1, \ldots, \tau_n)\} \subset \mathbb{P} \times (M^{\mathbb{P}})^n \subset M.$$

We shall verify that this relation is in fact definable over M (from the parameter $\mathbb{P}$). In order to do that, we now define a relation $\vdash$ by working in M, and then prove that $\vdash$ and $\Vdash$ have the same extension.[1]

[1] "$\vdash$" will be used as a symbol for this purpose only temporarily, until p. 97.

Definition 6.19 Let M be a transitive model of ZFC, and let $\mathbb{P} \in M$ be a partial order. Let $p \in \mathbb{P}$.

(1) Let $\tau_1, \tau_2 \in M^{\mathbb{P}}$. We define $p \Vdash_M^{\mathbb{P}} \tau_1 = \tau_2$ to hold iff: for all $(\pi_1, s_1) \in \tau_1$,

$$\{q \leq p : q \leq s_1 \rightarrow \exists (\pi_2, s_2) \in \tau_2\ (q \leq s_2 \wedge q \Vdash_M^{\mathbb{P}} \pi_1 = \pi_2)\}$$

is dense below p and for all $(\pi_2, s_2) \in \tau_2$,

$$\{q \leq p : q \leq s_2 \rightarrow \exists (\pi_1, s_1) \in \tau_1\ (q \leq s_1 \wedge q \Vdash_M^{\mathbb{P}} \pi_1 = \pi_2)\}$$

is dense below p.

(2) Again let $\tau_1, \tau_2 \in M^{\mathbb{P}}$. We define $p \Vdash_M^{\mathbb{P}} \tau_1 \in \tau_2$ to hold iff

$$\{q \leq p : \exists (\pi, s) \in \tau_2\ (q \leq s \wedge q \Vdash_M^{\mathbb{P}} \pi = \tau_1)\}$$

is dense below p.

(3) Let $\varphi(v_1, \ldots, v_n)$, $\psi(v'_1, \ldots, v'_m)$ be formulae, and let $\tau_1, \ldots, \tau_n, \tau'_1, \ldots, \tau'_m \in M^{\mathbb{P}}$. We define $p \Vdash_M^{\mathbb{P}} \varphi(\tau_1, \ldots, \tau_n) \wedge \psi(\tau'_1, \ldots, \tau'_m)$ to hold iff both $p \Vdash_M^{\mathbb{P}} \varphi(\tau_1, \ldots, \tau_n)$ as well as $p \Vdash_M^{\mathbb{P}} \psi(\tau'_1, \ldots, \tau'_m)$ hold. We define $p \Vdash_M^{\mathbb{P}} \neg\varphi(\tau_1, \ldots, \tau_n)$ to hold iff for no $q \leq p$, $q \Vdash_M^{\mathbb{P}} \varphi(\tau_1, \ldots, \tau_n)$ holds.

(4) Let $\exists x \varphi(x, v_1, \ldots, v_n)$ be a formula, and let $\tau_1, \ldots, \tau_n \in M^{\mathbb{P}}$. We define $p \Vdash_M^{\mathbb{P}} \exists x \varphi(x, \tau_1, \ldots, \tau_n)$ to hold iff

$$\{q \leq p : \exists \sigma \in M^{\mathbb{P}} q \Vdash_M^{\mathbb{P}} \varphi(\sigma, \tau_1, \ldots, \tau_n)\}$$

is dense below p.

In what follows, we shall often write $\Vdash$ rather than $\Vdash_M^{\mathbb{P}}$.

The definition of $p \Vdash \tau_1 = \tau_2$ is by recursion on $(\mathrm{rk}_\in(\tau_1), \mathrm{rk}_\in(\tau_2))$, ordered lexicographically. $p \Vdash \tau_1 \in \tau_2$ is then defined with the help of $p \Vdash \pi = \tau_1$, where $(\pi, s) \in \tau_2$ for some s. Moreover, the definition of $p \Vdash \varphi(\tau_1, \ldots, \tau_n)$ for nonatomic φ is by recursion on the complexity of φ. The relation $\Vdash$ is thus well-defined. We obviously have:

Lemma 6.20 *Let M be a transitive model of* ZFC, *and let $\mathbb{P} \in M$ be a partial order. Let φ be a formula. Then $\{(p, \tau_1, \ldots, \tau_n) : p \Vdash \varphi(\tau_1, \ldots, \tau_n)\}$ is definable over M (from the parameter $\mathbb{P}$).*

We say that p *decides* $\varphi(\tau_1, \ldots, \tau_k)$ iff $p \Vdash \varphi(\tau_1, \ldots, \tau_k)$ or $p \Vdash \neg\varphi(\tau_1, \ldots, \tau_k)$. Definition 6.19 (3) trivially yields that for a given $\varphi(\tau_1, \ldots, \tau_k)$, there are densely many p which decide $\varphi(\tau_1, \ldots, \tau_k)$. The following is also straightforward to verify, cf. Problem 6.2.

Lemma 6.21 *Let M be a transitive model of* ZFC, *and let $\mathbb{P} \in M$ be a partial order. Let $p \in \mathbb{P}$, let φ be a formula, and let $\tau_1, \ldots, \tau_n \in M^{\mathbb{P}}$. Equivalent are:*

(1) $p \Vdash \varphi(\tau_1, \dots, \tau_n)$
(2) $\forall q \leq p \; q \Vdash \varphi(\tau_1, \dots, \tau_n)$
(3) $\{q \leq p : q \Vdash \varphi(\tau_1, \dots, \tau_n)\}$ *is dense below* p.

Theorem 6.22 (Forcing Theorem, part 1) *Let M be a transitive model of* ZFC, *let $\mathbb{P} \in M$ be a partial order, and let $G \subset \mathbb{P}$ be $\mathbb{P}$-generic over M. Let $\varphi(v_1, \dots, v_n)$ be a formula, and let $\tau_1, \dots, \tau_n \in M^{\mathbb{P}}$.*

(1) *If $p \in G$ and $p \Vdash_M^{\mathbb{P}} \varphi(\tau_1, \dots, \tau_n)$, then $M[G] \models \varphi(\tau_1^G, \dots, \tau_n^G)$.*
(2) *If $M[G] \models \varphi(\tau_1^G, \dots, \tau_n^G)$, then there is some $p \in G$ such that $p \Vdash_M^{\mathbb{P}} \varphi(\tau_1, \dots, \tau_n)$.*

Proof We prove (1) and (2) simultaneously. We first prove (1) and (2) for $\varphi \equiv v_1 = v_2$ by induction on $(\mathrm{rk}_{\in}(\tau_1), \mathrm{rk}_{\in}(\tau_2))$, ordered lexicographically.

(1): Suppose that $p \in G$ and $p \Vdash \tau_1 = \tau_2$. Let us verify that $\tau_1^G \subset \tau_2^G$. By symmetry, this will also show that $\tau_2^G \subset \tau_1^G$, and therefore $\tau_1^G = \tau_2^G$.

Let $x \in \tau_1^G$, say $x = \pi_1^G$, where $(\pi_1, s_1) \in \tau_1$ for some $s_1 \in G$. We need to see that $x \in \tau_2^G$. Pick $r \in G$ such that $r \leq p, r \leq s_1$. We still have $r \Vdash \tau_1 = \tau_2$ by Lemma 6.21, so that there is some $q \in G, q \leq r$, such that

$$q \leq s_1 \rightarrow \exists (\pi_2, s_2) \in \tau_2 \; (q \leq s_2 \wedge q \Vdash \pi_1 = \pi_2).$$

As $r \leq s_1$, we have that $q \leq s_1$. Hence we may pick some $(\pi_2, s_2) \in \tau_2$ with $q \leq s_2 \wedge q \Vdash \pi_1 = \pi_2$. As $q \in G$, we also have that $s_2 \in G$, and moreover we have that $\pi_1^G = \pi_2^G$ by our inductive hypothesis. But then $x = \pi_1^G = \pi_2^G \in \tau_2^G$, as $(\pi_2, s_2) \in \tau_2$ and $s_2 \in G$.

(2): Now suppose that $\tau_1^G = \tau_2^G$. Consider the following statement about a condition r:

$$\begin{aligned} \psi_1(r): \;\; \exists (\pi_1, s_1) \in \tau_1 \; (r \leq s_1 \wedge \forall (\pi_2, s_2) \in \tau_2 \\ \forall q \; (q \leq s_2 \wedge q \Vdash \pi_1 = \pi_2 \rightarrow q \perp r)). \end{aligned}$$

Assume we had $\psi_1(r)$ for some $r \in G$, and let $(\pi_1, s_1) \in \tau_1$ be a witness. As $r \leq s_1$, we also have $s_1 \in G$, so that $\pi_1^G \in \tau_1^G = \tau_2^G$. Pick $(\pi_2, s_2) \in \tau_2$ such that $s_2 \in G$ and $\pi_1^G = \pi_2^G$. By our inductive hypothesis, there is some $q_0 \in G$ with $q_0 \Vdash \pi_1 = \pi_2$. Pick $q \in G$ such that $q \leq q_0$ and $q \leq s_2$. Still $q \Vdash \pi_1 = \pi_2$ by Lemma 6.21. By $\psi_1(r)$, we must then have $q \perp r$. However, $q \in G$ as well as $r \in G$. Contradiction!

Therefore, we cannot have $\psi_1(r)$ for $r \in G$. The same argument shows that we cannot have $\psi_2(r)$ for $r \in G$, where

$$\begin{aligned} \psi_2(r): \;\; \exists (\pi_2, s_2) \in \tau_2 \; (r \leq s_2 \wedge \forall (\pi_1, s_1) \in \tau_1 \\ \forall q \; (q \leq s_1 \wedge q \Vdash \pi_1 = \pi_2 \rightarrow q \perp r)). \end{aligned}$$

Now let us consider

$$D = \{r : \psi_1(r) \vee \psi_2(r) \vee r \Vdash \tau_1 = \tau_2\}.$$

We claim that D is dense. To this end, let r be given. Suppose that $r \Vdash \tau_1 = \tau_2$ does not hold true. By the definition of $\Vdash$, there is then some $(\pi_1, s_1) \in \tau_1$ such that

$$\{q \leq r : q \leq s_1 \to \exists(\pi_2, s_2) \in \tau_2\ (q \leq s_2 \wedge q \Vdash \pi_1 = \pi_2)\} \tag{6.1}$$

is not dense below r, or there is some $(\pi_2, s_2) \in \tau_2$ such that

$$\{q \leq r : q \leq s_2 \to \exists(\pi_1, s_1) \in \tau_1\ (q \leq s_1 \wedge q \Vdash \pi_1 = \pi_2)\} \tag{6.2}$$

is not dense below r. Let us assume (6.1) to be true. We'll then show that there is some $p \leq r$ such that $\psi_1(p)$ holds true. (By symmetry, if (6.2) holds true, then $\psi_2(p)$ holds true for some $p \leq r$.) Let $(\pi_1, s_1) \in \tau_1$ witness that (6.1) holds true. There is some $p \leq r$ such that

$$\forall q \leq p(q \leq s_1 \wedge \forall(\pi_2, s_2) \in \tau_2\ \neg(q \leq s_2 \wedge q \Vdash \pi_1 = \pi_2)). \tag{6.3}$$

In particular, $p \leq s_1$. Also, (6.3) gives that if $(\pi_2, s_2) \in \tau_2, q \leq s_2, q \Vdash \pi_1 = \pi_2$, then $q \perp p$. That is, $p \leq r$ and $\psi_1(p)$ holds true. We have shown that D is dense.

But now there must be some $p \in G \cap D$. As $p \in G$, we have seen that $\psi_1(p)$ and $\psi_2(p)$ must both fail, so that $p \Vdash \tau_1 = \tau_2$ holds true, as desired.

We now prove (1) and (2) for $\varphi \equiv v_1 \in v_2$, exploiting the fact that (1) and (2) hold true for $\varphi \equiv v_1 = v_2$.

(1): Suppose that $p \in G$ and $p \Vdash \tau_1 \in \tau_2$. By definition,

$$D = \{q \leq p\colon \exists(\pi, s) \in \tau_2\ (q \leq s \wedge q \Vdash \pi = \tau_1)\}$$

is then dense below p. Pick $q \in D \cap G$. Let $(\pi, s) \in \tau_2$ be such that $q \leq s$ and $q \Vdash \pi = \tau_1$. As $q \in G, \pi^G = \tau_1^G$. But $s \in G$, too, and hence $\tau_1^G = \pi^G \in \tau_2^G$.

(2): Suppose that $\tau_1^G \in \tau_2^G$. There is then some $(\pi, s) \in \tau_2$ such that $s \in G$ and $\tau_1^G = \pi^G$. We therefore have some $r \in G$ with $r \Vdash \tau_1 = \pi$. Let $p \in G$ be such that $p \leq s, r$. Then $\forall q \leq p(q \leq s \wedge q \Vdash \tau_1 = \pi)$. Hence $p \Vdash \tau_1 \in \tau_2$.

Let us finally prove (1) and (2) for nonatomic formulae.

Let $\varphi(v_1, \ldots, v_n)$, $\psi(v'_1, \ldots, v'_m)$ be formulae, and let $\tau_1, \ldots, \tau_n, \tau'_1, \ldots, \tau'_m \in M^{\mathbb{P}}$. Suppose that (1) and (2) hold for $\varphi(\tau_1, \ldots, \tau_n)$ and for $\psi(\tau_1, \ldots, \tau_n)$. It is then trivial that (1) and (2) also hold for $\varphi(\tau_1, \ldots, \tau_n) \wedge \psi(\tau'_1, \ldots, \tau'_m)$. Let us show that (1) and (2) hold for $\neg\varphi(\tau_1, \ldots, \tau_n)$.

(1): Let $p \in G, p \Vdash \neg\varphi(\tau_1, \ldots, \tau_n)$. Suppose that $M[G] \models \varphi(\tau_1^G, \ldots, \tau_n^G)$. There is then some $q \in G$ such that $q \Vdash \varphi(\tau_1, \ldots, \tau_n)$. Pick $r \in G, r \leq p, q$. Then $r \leq p$ and $r \Vdash \varphi(\tau_1, \ldots, \tau_n)$. Contradiction! Hence $M[G] \models \neg\varphi(\tau_1^G, \ldots, \tau_n^G)$.

(2): Let $M[G] \models \neg\varphi(\tau_1^G, \ldots, \tau_n^G)$. It is easy to see that

$$D = \{q : q \Vdash \varphi(\tau_1, \ldots, \tau_n) \vee q \Vdash \neg\varphi(\tau_1, \ldots, \tau_n)\}$$

is dense. But if $q \in D \cap G$, then $q \vdash \neg\varphi(\tau_1, \dots, \tau_n)$. This is because otherwise $q \vdash \varphi(\tau_1, \dots, \tau_n)$ and then $M[G] \models \varphi(\tau_1^G, \dots, \tau_n^G)$. Contradiction!

Finally, let $\exists x\varphi(x, v_1, \dots, v_n)$ be a formula, and let $\tau_1, \dots, \tau_n \in M^{\mathbb{P}}$. Suppose that (1) and (2) hold for $\varphi(\sigma, \tau_1, \dots, \tau_n)$ whenever $\sigma \in M^{\mathbb{P}}$. We aim to show (1) and (2) for $\exists x\varphi(x, \tau_1, \dots, \tau_n)$.

(1): Suppose that $p \in G$, $p \vdash \exists x\varphi(x, \tau_1, \dots, \tau_n)$. By definition,

$$D = \{r \leq p : \exists \sigma \in M^{\mathbb{P}}\ r \vdash \varphi(\sigma, \tau_1, \dots, \tau_n)\}$$

is then dense below p. Pick $r \in D \cap G$. Then there is some $\sigma \in M^{\mathbb{P}}$ such that, using the inductive hypothesis, $M[G] \models \varphi(\sigma^G, \tau_1^G, \dots, \tau_n^G)$. But we then have that $M[G] \models \exists x\varphi(x, \tau_1^G, \dots, \tau_n^G)$.

(2): Let $M[G] \models \exists x\varphi(x, \tau_1^G, \dots, \tau_n^G)$. Pick $\sigma \in M^{\mathbb{P}}$ witnessing this, i.e., such that $M[G] \models \varphi(\sigma^G, \tau_1^G, \dots, \tau_n^G)$. By our inductive hypothesis, there is some $p \in G$ with $p \vdash \varphi(\sigma, \tau_1, \dots, \tau_n)$. But then $r \vdash \varphi(\sigma, \tau_1, \dots, \tau_n)$ for all $r \leq p$, which trivially implies that $p \vdash \exists x\varphi(x, \tau_1, \dots, \tau_n)$ by definition. □

One can in fact show that if $p \vdash \exists x\varphi(x, \tau_1, \dots, \tau_n)$, then there is some $\tau \in M^{\mathbb{P}}$ such that $p \vdash \varphi(\tau, \tau_1, \dots, \tau_n)$. This property is called *fullness*, cf. Problem 6.5.

Theorem 6.23 (Forcing Theorem, part 2) *Let M be a transitive model of* ZFC, *and let $\mathbb{P} \in M$ be a partial order such that for every $p \in \mathbb{P}$ there is some G with $p \in G$ such that G is $\mathbb{P}$-generic over M. Let $\varphi(v_1, \dots, v_n)$ be a formula, and let $\tau_1, \dots, \tau_n \in M^{\mathbb{P}}$.*

(1) *For all $p \in \mathbb{P}$,*

$$p \Vdash_M^{\mathbb{P}} \varphi(\tau_1, \dots, \tau_n) \iff p \vdash_M^{\mathbb{P}} \varphi(\tau_1, \dots, \tau_n).$$

(2) *Let $G \subset \mathbb{P}$ be $\mathbb{P}$-generic over M. Then*

$$M[G] \models \varphi(\tau_1^G, \dots, \tau_n^G) \iff \exists p \in G\ p \Vdash_M^{\mathbb{P}} \varphi(\tau_1, \dots, \tau_n).$$

Proof Let us first show (1):

"$\Longleftarrow$": Let $p \vdash \varphi(\tau_1, \dots, \tau_n)$. Let G be any $\mathbb{P}$-generic filter over M such that $p \in G$. Then $M[G] \models \varphi(\tau_1^G, \dots, \tau_n^G)$ by Theorem 6.22 (1). Therefore, $p \Vdash_M^{\mathbb{P}} \varphi(\tau_1, \dots, \tau_n)$.

"$\Longrightarrow$": Suppose that $p \Vdash_M^{\mathbb{P}} \varphi(\tau_1, \dots, \tau_n)$. We need to see that $\{q \leq p : q \vdash \varphi(\tau_1, \dots, \tau_n)\}$ is dense below p. If not, then there is some $q \leq p$ such that for all $r \leq q$, $r \vdash \varphi(\tau_1, \dots, \tau_n)$ does not hold true, i.e., $q \vdash \neg\varphi(\tau_1, \dots, \tau_n)$. But then $q \Vdash_M^{\mathbb{P}} \neg\varphi(\tau_1, \dots, \tau_n)$ by "$\Longleftarrow$". This contradicts $p \Vdash_M^{\mathbb{P}} \varphi(\tau_1, \dots, \tau_n)$, as $q \leq p$.

Let us show (2). Well, "$\Longrightarrow$" follows from (1) above plus Theorem 6.22 (2). "$\Longleftarrow$" is just by definition. □

By Theorem 6.23 (1), we shall not have any use for the notation "$\vdash$" any more.

Theorem 6.24 *Let M be a transitive model of* ZFC, *let $\mathbb{P} \in M$ be a partial order, and let $G \subset \mathbb{P}$ be $\mathbb{P}$-generic over M. $M[G] \models$* ZFC.

Proof We are left with having to verify that $M[G] \models$ (Aus), (Rep), (Pow), and (AC). (Cf. also the remark before Lemma 6.17.)

Let us begin with (Aus), i.e., separation. Let $\varphi(v_0, v_1, \ldots, v_n)$ be a formula, and let $\tau^G, \tau_1^G, \ldots, \tau_n^G \in M[G]$. We aim to see that

$$\{x \in \tau^G : M[G] \models \varphi(x, \tau_1^G, \ldots, \tau_n^G)\} \in M[G].$$

Well, consider

$$\pi = \{(\rho, p) : \exists q (p \leq q \wedge (\rho, q) \in \tau \wedge p \Vdash_M^{\mathbb{P}} \varphi(\rho, \tau_1, \ldots, \tau_n))\}.$$

Notice that $\pi \in M^{\mathbb{P}}$ by 6.23 (1).

Let $x \in M[G]$, say $x = \rho^G$. Then $\rho^G \in \pi^G$ iff $(\rho, p) \in \pi$ for some $p \in G$ iff there is some $q \geq p \in G$ with $(\rho, q) \in \tau$ and $p \Vdash_M^{\mathbb{P}} \varphi(\rho, \tau_1, \ldots, \tau_n)$ iff $\rho^G \in \tau^G$ and $M[G] \models \varphi(\rho, \tau_1^G, \ldots, \tau_n^G)$. Hence

$$\{x \in \tau^G : M[G] \models \varphi(x, \tau_1^G, \ldots, \tau_n^G)\} = \pi^G \in M[G].$$

A similar argument is used to show (Rep) in $M[G]$. Let $\varphi(v_0, v_1, v_2, \ldots, v_n)$ be a formula, let $\tau^G, \tau_2^G, \ldots, \tau_n^G \in M[G]$, and suppose that

$$M[G] \models \forall x \in \tau^G \, \exists y \, \varphi(x, y, \tau_2^G, \ldots, \tau_n^G).$$

We aim to see that there is some $a \in M[G]$ such that

$$M[G] \models \forall x \in \tau^G \, \exists y \in a \, \varphi(x, y, \tau_2^G, \ldots, \tau_n^G).$$

Consider

$$\begin{aligned} \pi = \{(\rho, p) : \exists (\overline{\rho}, \overline{p}) \in \tau \, (& p \leq \bar{p} \wedge \\ & p \Vdash_M^{\mathbb{P}} \varphi(\overline{\rho}, \rho, \tau_2, \ldots, \tau_n) \wedge \\ & \forall \rho' (p \Vdash_M^{\mathbb{P}} \varphi(\overline{\rho}, \rho', \tau_2, \ldots, \tau_n) \longrightarrow \mathrm{rk}_{\in}(\rho') \geq \mathrm{rk}_{\in}(\rho)))\}. \end{aligned}$$

Notice that $\pi \in M^{\mathbb{P}}$, again by Theorem 6.23 (1). (Without the clause $\forall \rho' [\ldots]$ in the bottom line of this definition of π, π would have ended up being a proper class in M rather than a set in M.)

Suppose that $x \in \tau^G$, say $x = \overline{\rho}^G$, where $(\overline{\rho}, \overline{p}) \in \tau$ for some $\overline{p} \in G$. Let ρ be such that $M[G] \models \varphi(\overline{\rho}^G, \rho^G, \tau_2^G, \ldots, \tau_n^G)$, and let $p \in G$ be such that $p \Vdash_M^{\mathbb{P}} \varphi(\overline{\rho}, \rho, \tau_2, \ldots, \tau_n)$. We may as well assume that for all ρ', if $p \Vdash_M^{\mathbb{P}} \varphi(\overline{\rho}, \rho', \tau_2, \ldots, \tau_n)$, then $\mathrm{rk}_{\in}(\rho') \geq \mathrm{rk}_{\in}(\rho)$, and $p \leq \bar{p}$. Then $\rho^G \in \pi^G$, and $a = \pi^G \in M[G]$ is as desired.

In order to verify (Pow) in $M[G]$, let $\tau^G \in M[G]$. As we already verified (Aus) in $M[G]$, it suffices to see that there is some $b \in M[G]$ such that $\{x \in M[G] : x \subset \tau^G\} \subset b$.

Set $N = \{\rho \colon \exists p\, (\rho, p) \in \tau\}$, and let $\pi = \{(\sigma, 1_{\mathbb{P}}) : \sigma \in M^{\mathbb{P}} \wedge \sigma \subset N \times \mathbb{P}\} \in M^{\mathbb{P}}$. Let $\sigma^G \subset \tau^G$. We want to see that $\sigma^G \in \pi^G$. Let $\sigma' = \{(\rho, p) : \rho \in N \wedge p \Vdash^{\mathbb{P}}_M \rho \in \sigma\}$. By Theorem 6.23 (1), $\sigma' \in M^{\mathbb{P}}$. We have $(\sigma', 1_{\mathbb{P}}) \in \pi$, so $\sigma'^G \in \pi^G$, and hence it suffices to verify that $\sigma'^G = \sigma^G$.

If $\rho^G \in \sigma'^G$, then $(\rho, p) \in \sigma'$ for some $p \in G$, which implies that $\rho \in N$ and $p \Vdash^{\mathbb{P}}_M \rho \in \sigma$, where $p \in G$. But then $\rho^G \in \sigma^G$. On the other hand, let $\rho^G \in \sigma^G$. There is then some $p \in G$ with $(\rho, p) \in \sigma$, which implies that $\rho \in N$ and $p \Vdash^{\mathbb{P}}_M \rho \in \sigma$. But then $(\rho, p) \in \sigma'$, where $p \in G$, and hence $\rho^G \in \sigma'^G$.

We have shown that $b = \pi^G$ is as desired.

Let us finally verify (AC) in $M[G]$. Let $x \in M[G]$, say $x = \tau^G$. Let $f \in M$, $f : \alpha \to \tau$ bijective (for some $\alpha < M \cap OR$). We aim to see that $M[G]$ has a surjection $g : \alpha \to x$.

In order to define a name for g, we shall use the following notation Let $y, z \in M[G]$, say $y = \rho^G, z = \sigma^G$. Write

$$\ulcorner \rho, \sigma \urcorner = \{(\{(\rho, 1), (\sigma, 1)\}, 1), (\{(\rho, 1)\}, 1)\}.$$

it is easy to see that $\ulcorner \rho, \sigma \urcorner \in M^{\mathbb{P}}$ and in fact $\ulcorner \rho, \sigma \urcorner^G = (y, z)$.

Now set

$$\pi = \{(\ulcorner \check{\xi}, \sigma \urcorner, 1) : \xi < \alpha \wedge \exists p\; f(\xi) = (\sigma, p)\}.$$

Obviously, $\pi \in M^{\mathbb{P}}$. Moreover, π^G is easily seen to be a function with domain α. Let us verify that $x \subset \operatorname{ran}(\pi^G)$.

Let $y \in x$, say $y = \sigma^G$, where $(\sigma, p) \in \tau$ for some $p \in G$. There is then some $\xi < \alpha$ with $f(\xi) = (\sigma, p)$, and hence $(\ulcorner \check{\xi}, \sigma \urcorner, 1) \in \pi$. But then $(\xi, \sigma^G) \in \pi^G$, i.e., $\pi^G(\xi) = \sigma^G = y$, so that $y \in \operatorname{ran}(\pi^G)$. □

6.2 Applications of Forcing

Let us now turn towards applications of forcing. By Corollary 5.32, "$V = L$" is consistent with **ZFC**. We now show that forcing may be used to show that "$V \neq L$," the negation of "$V = L$," is also consistent with **ZFC**.

Our proof will make use of the concept of a Π^0_1 statement: φ is Π^0_1 iff it can be written in the form $\forall n \in \omega\, \psi$, where ψ is recursive (cf. e.g. [11]). Π^0_1 statements are downward absolute between models of **ZFC** (compare Lemma 5.3); therefore, as every model of **ZFC** has an isomorphic copy of the standard natural numbers ω, we have that if φ is Π^0_1 and there is some **ZFC**-model which thinks that φ is true, then φ is really true (in V). In particular, this holds for $\varphi \equiv$ "T is consistent" for any recursively enumerable theory T, because "T does not prove $0 = 1$" may be written in a Π^0_1 fashion.

Theorem 6.25 *If* ZFC *is consistent, then so is* ZFC + *"$V \neq L$."*

Proof Let us first prove the following.

Claim 6.26 *If M is a transitive model of* ZFC *and if G is $\mathbb{C}$-generic over M, where $\mathbb{C}$ is* COHEN *forcing (cf. Definition 6.5), then $M[G]$ is a model of* ZFC + *"$V \neq L$."*

Proof By Theorem 6.24, we just need to verify $M[G] \models$ "$V \neq L$."

Let us suppose that $M[G] \models$ "$V = L$." By Lemma 5.28, we must then have $M[G] = J_\alpha$, where $\alpha = M[G] \cap OR$. However, $M[G] \cap OR = M \cap OR$ by Lemma 6.15, so that $M[G] = J_\alpha = L^M \subset M$ by Lemma 5.28 again. But $G \in M[G] \setminus M$ by Lemma 6.11, because $\mathbb{C}$ is certainly atomless. Contradiction! □

An inspection of the proofs of Theorem 6.24 and of Claim 6.26 shows that we may define a function $\Sigma \mapsto \Gamma(\Sigma)$ which maps finite subsets Σ of ZFC $\cup$ {"$V \neq L$"} to finite subsets $\Gamma = \Gamma(\Sigma)$ of ZFC such that the following holds true (provably in ZFC):

$$\begin{cases} \text{If } M \text{ is a transitive model of } \Gamma(\Sigma)\text{, and if } G \text{ is } \mathbb{C}\text{-generic over } M, \\ \text{then } M[G] \text{ is a transitive model of } \Sigma. \end{cases} \tag{6.4}$$

Let us now just assume that ZFC is consistent. This means that there is a (not necessarily well-founded) model $(N; E)$ of ZFC. We need the following statement about $(N; E)$:

There is a function $\Gamma \mapsto M(\Gamma)$ which maps finite subsets Γ of ZFC to elements $M = M(\Gamma)$ of N such that for all Γ,

$$(N; E) \models \text{"}M = M(\Gamma) \text{ is a countable transitive model of } \Gamma\text{."} \tag{6.5}$$

$M(\Gamma)$ may be obtained from Γ as follows. Given Γ, by Problem 5.14 there is an $(N; E)$—least α which is an ordinal from the point of view of $(N; E)$ such that

$$(N; E) \models \text{"}V_\alpha \models \Gamma\text{."}$$

We may then apply the LÖWENHEIM- - SKOLEM Theorem and Theorem 3.21 inside $(N; E)$ to find an M as desired. Set $M(\Gamma) = M$. Of course, in general the function $\Gamma \mapsto M(\Gamma)$ will not be *in* N, but we will only need that $M(\Gamma) \in N$ for every individual Γ.

Let us now show that ZFC does not prove "$V = L$." Let $\overline{\Sigma} \subset$ ZFC be finite, and set $\Sigma = \overline{\Sigma} \cup$ {"$V \neq L$"}. Setting $\Gamma = \Gamma(\Sigma)$ and $M = M(\Gamma)$, (6.5) gives that

$$(N; E) \models \text{"}M \text{ is a countable transitive model of } \Gamma\text{."}$$

By Lemma 6.4 there is then some $G \in N$ such that $(N; E) \models$ "G is $\mathbb{C}$-generic over M," and because (6.4) holds inside $(N; E)$ we get that[2]

$$(N; E) \models \text{"}M[G] \text{ is a transitive model of } \Sigma.\text{"}$$

This means that

$$(N; E) \models \text{"}\Sigma \text{ is consistent"}. \tag{6.6}$$

However, the consistency of Σ is a Π^0_1 statement, so that (6.6) yields that Σ is really consistent. But then $\overline{\Sigma}$ does not prove that "$V = L$."

Assuming that ZFC is consistent we have shown that ZFC does not prove that "$V = L$." □

All relative consistency results which use forcing may be produced in this fashion. In proving them, we thus may and shall always pretend to have a countable transitive model of ZFC at hand.

In order to prove the consistency of ZFC + ¬CH (relative to the consistency of ZFC), we need finite support products of COHEN forcing.

Definition 6.27 Let $\mathbb{C}$ be COHEN forcing, cf. Definition 6.5. Let α be an ordinal. For $p \in {}^{\alpha}\mathbb{C}$ let $\text{supp}(p)$ (the support of p) be the set of all $\xi < \alpha$ with $p(\xi) \neq \emptyset$. Let

$$\mathbb{C}(\alpha) = \{p \in {}^{\alpha}\mathbb{C} : \text{Card}(\text{supp}(p)) < \aleph_0\}.$$

For $p, q \in \mathbb{C}(\alpha)$, let us write $p \leq q$ iff for all $\xi < \alpha$, $p(\xi) \leq q(\xi)$ in the sense of COHEN forcing, i.e., $\exists n \; p(\xi) \restriction n = q(\xi)$.

$\mathbb{C}(\alpha)$ is often referred to as the finite support product of α COHEN forcings.

Definition 6.28 Let $\mathbb{P} = (\mathbb{P}; \leq)$ be a partial order, and let κ be an uncountable cardinal. $\mathbb{P}$ is said to be κ-KNASTER iff for all $A \subset \mathbb{P}$ of size κ there is some $B \subset A$ of size κ such that if $p, q \in B$, then $p \parallel q$. $\mathbb{P}$ is said to have the κ-chain condition (κ-c.c., for short) iff $\overline{\overline{A}} < \kappa$ whenever $A \subset \mathbb{P}$ is an antichain. $\mathbb{P}$ is said to have the *countable chain condition* (*c.c.c.*, for short) iff $\mathbb{P}$ has the $\aleph_1$-c.c.

Trivially, if $\mathbb{P}$ is κ-KNASTER, then $\mathbb{P}$ has the κ-c.c. Also, if $\text{Card}(\mathbb{P}) = \kappa$, then trivially $\mathbb{P}$ has the κ^+-c.c.

Lemma 6.29 *Let α be an ordinal. Then $\mathbb{C}(\alpha)$ is $\aleph_1$-KNASTER.*

Proof If α is at most countable, then $\overline{\overline{\mathbb{C}(\alpha)}} = \aleph_0$, so that $\mathbb{C}(\alpha)$ is trivially $\aleph_1$-KNASTER in this case.

Now let α be uncountable. Let $A \subset \mathbb{C}(\alpha)$ have size $\aleph_1$. We shall verify that there is some $B \subset A$ of size $\aleph_1$ such that any two conditions in B are compatible.

Let $X = \bigcup\{\text{supp}(p) : p \in A\}$. We must have that $X \subset \alpha$ has size $\aleph_1$. Let us pick some bijection $\pi : \omega_1 \to X$. This naturally induces $\sigma : A \to \mathbb{C}(\omega_1)$ as follows.

[2] Here, $M[G]$ is "$M[G]$ as computed inside $(N; E)$."

For $p \in A$, let $\text{supp}(\sigma(p)) = \{\xi < \omega_1 : \pi(\xi) \in \text{supp}(p)\}$, and if $\xi \in \text{supp}(\sigma(p))$, then let $\sigma(p)(\xi) = p(\pi(\xi))$. It now suffices to verify that there is some $\overline{B} \subset \sigma''A$ of size $\aleph_1$ such that any two conditions in $\overline{B}$ are compatible, because then any two conditions in $B = \sigma^{-1}{''}\overline{B} \subset A$ are compatible as well.

Let us write $D = \sigma''A$. By the Pigeonhole Principle, there is some $n < \omega$ such that $\{p \in D : \text{Card}(\text{supp}(p)) = n\}$ has size $\aleph_1$. Let $D_0 \subset D$ be $\{p \in D : \text{Card}(\text{supp}(p)) = n\}$ for the least such n. For $p \in D_0$, let us write $\text{supp}(p) = \{\xi_1^p, \ldots, \xi_n^p\}$, where $\xi_1^p < \ldots < \xi_n^p$.

A simple application of the Pigeonhole Principle yields the following.

Claim 6.30 *Let $1 \leq k \leq n$, and suppose $\{\xi_k^p : p \in D_0\}$ to be bounded in ω_1. There is then some $D_1 \subset D_0$ of size $\aleph_1$, some set $\{\xi_1, \ldots, \xi_k\}$ and some set $\{s_1, \ldots, s_k\} \subset \mathbb{C}$ such that for all $p \in D_1$, $\xi_1^p = \xi_1, \ldots, \xi_k^p = \xi_k$, $p(\xi_1) = s_1, \ldots, p(\xi_k) = s_k$.*

There is therefore some $k \leq n$, $k \geq 1$, such that $\{\xi_k^p : p \in D_0\}$ is *un*bounded in ω_1. Let k_0 be the least such.

Set $\gamma = \sup\{\xi_k^p : p \in D_0 \wedge k < k_0\} < \omega_1$. (If $k_0 = 1$, then set $\gamma = 0$. If $k_0 > 1$, then actually $\gamma = \sup\{\xi_{k_0-1}^p : p \in D_0\}$.)

If $k_0 > 1$, then we may apply Claim 6.30 to D_0 to get some $D_1 \subset D_0$ of size $\aleph_1$, some set $\{\xi_1, \ldots, \xi_{k_0-1}\}$, and some set $\{s_1, \ldots, s_{k_0-1}\} \subset \mathbb{C}$ such that for all $p \in D_1$, $\xi_1^p = \xi_1, \ldots, \xi_{k_0-1}^p = \xi_{k_0-1}$, $p(\xi_1) = s_1, \ldots, p(\xi_{k_0}) = s_{k_0}$. If $k_0 = 1$, then we just set $D_1 = D_0$.

Let us now recursively define $(\delta_i : i < \omega_1)$ and $(p_i : i < \omega_1)$ as follows. Let $\eta < \omega_1$ and suppose $(\delta_i : i < \eta)$ and $(p_i : i < \eta)$ have already been defined. Set $\delta_\eta = \sup\{\xi_n^{p_i} : i < \eta\}$. (If $\eta = 0$, then set $\delta_\eta = \delta_0 = \gamma$.) By the choice of k_0, there is then some $p \in D_1$ such that $\xi_{k_0}^p > \delta_\eta$; let p_η be some such p. Now write $\overline{B} = \{p_i : i < \omega_1\}$. We have that if $p_i \in \overline{B}$, then $\delta_i < \xi_{k_0}^{p_i} < \ldots < \xi_n^{p_i} \leq \delta_{i+1}$. We therefore have that any two conditions in $\overline{B}$ are compatible: if $p, q \in \overline{B}$, then we may define $r \in \mathbb{C}(\omega_1)$ by: $\text{supp}(r) = \text{supp}(p) \cup \text{supp}(q) = \{\xi_1, \ldots, \xi_{k_0-1}, \xi_{k_0}^p, \ldots, \xi_n^p, \xi_{k_0}^q, \ldots, \xi_n^q\}$, $r(\xi_k) = p(\xi_k) = q(\xi_k)$ for $1 \leq k < k_0$, $r(\xi_k^p) = p(\xi_k^p)$ for $k_0 \leq k \leq n$, and $r(\xi_k^q) = q(\xi_k^q)$ for $k_0 \leq k \leq n$. $\overline{B}$ is thus as desired. □

The combinatorial heart of this latter argument leads to the following lemma which is very useful for the analysis of many forcings. The proof is pretty much the same as the proof of the previous lemma. (Cf. Problem 6.6.)

Lemma 6.31 (Δ-Lemma) *Let κ be an uncountable regular cardinal, and let $\mu < \kappa$ be an infinite cardinal such that $\rho^\gamma < \kappa$ for all $\rho < \kappa$ and $\gamma < \mu$. Let $A \subset [\kappa]^{<\mu}$ with $\overline{\overline{A}} = \kappa$. There is then some $B \subset A$ with $\overline{\overline{B}} = \kappa$ which forms a Δ-*system*, which means that there is some $r \in [\kappa]^{<\mu}$ (the "root" of B) such that for all $x, y \in B$ with $x \neq y$, $x \cap y = r$.*

Notice that if M is a transitive model of **ZFC** and G is $\mathbb{P}$-generic over M for some $\mathbb{P} \in M$ and if ρ is a cardinal of $M[G]$, then ρ is also a cardinal of M. The following lemma provides a covering fact and gives a criterion for when cardinals of M will not be collapsed in $M[G]$. We will later study forcings which do collapse cardinals, cf. Definitions 6.41 and 6.43.

Lemma 6.32 *Let M be a transitive model of* ZFC, *and let κ be a cardinal of M. Let $\mathbb{P} = (\mathbb{P}; \leq) \in M$ be a partial order such that $M \models$ "$\mathbb{P}$ has the κ-c.c." Let G be $\mathbb{P}$-generic over M. Let $X \subset M$, $X \in M[G]$, and write $\mu = \mathrm{Card}^{M[G]}(X)$. There is then some $Y \in M$ such that $Y \supset X$ and*

$$M \models \mathrm{Card}(Y) \begin{cases} \leq \mu, \text{ if } \kappa \leq \mu^+, \\ < \kappa, \text{ if } \kappa \geq \mu^+ \text{ and } \mu < \mathrm{cf}(\kappa), \text{ and} \\ \leq \kappa, \text{ if } \kappa \geq \mu \text{ and } \mu \geq \mathrm{cf}(\kappa). \end{cases} \tag{6.7}$$

In particular, if $\rho \geq \kappa$ is a cardinal in M such that $\rho = \kappa$ and ρ is regular in M or else $\rho > \kappa$, then ρ remains a cardinal in $M[G]$.

Proof Let $X \in M[G]$ be given, $X \subset M$. Pick $f \in M[G]$, $f : \mu \to X$ bijective. Let $f = \tau^G$. Pick $p \in G$ such that

$$p \Vdash_M^{\mathbb{P}} \text{"}\tau \text{ is a function with domain } \check{\mu}.\text{"}$$

For each $\xi < \mu$, let $B_\xi = \{\eta : \exists q \leq p \; q \Vdash_M^{\mathbb{P}} \tau(\check{\xi}) = \check{\eta}\}$. Working in M, for each $\eta \in B_\xi$ we may pick $q_\xi^\eta \leq p$ such that $q_\xi^\eta \Vdash_M^{\mathbb{P}} \tau(\check{\xi}) = \check{\eta}$. If $\eta \neq \eta'$, then $q_\xi^\eta \perp q_\xi^{\eta'}$, so that $M \models \mathrm{Card}(B_\xi) < \kappa$, as $M \models$ "$\mathbb{P}$ has the κ-c.c."

Now set $Y = \bigcup\{B_\xi : \xi < \mu\}$. Of course, $Y \in M$ and $Y \supset X$. If κ_ξ is the cardinality of B_ξ inside M, then

$$M \models \mathrm{Card}(Y) \leq \sum_{\xi<\mu} \kappa_\xi. \tag{6.8}$$

It is straightforward to verify that (6.8) yields (6.7).

Now suppose that $\rho \geq \kappa$ is a cardinal in M such that if $\rho = \kappa$, then ρ is regular. Suppose that $\mu = \mathrm{Card}^{M[G]}(\rho) < \rho$. We may then cover the set $X = \rho$ by a set $Y \in M$ such that (6.7) holds true. Let us now argue in M. If $\kappa \leq \mu^+$, or $\kappa \geq \mu^+$ and $\mu < \mathrm{cf}(\kappa)$, or $\kappa < \rho$, then $M \models \mathrm{Card}(Y) < \rho$, which is nonsense. Otherwise $\kappa \geq \mu \geq \mathrm{cf}(\kappa)$ and $\rho = \kappa$, so that ρ is regular by hypothesis and hence in fact $\rho = \kappa = \mu$, which contradicts $\mu < \rho$. □

In the light of Theorem 5.31, the following result shows that CH cannot be decided on the basis of ZFC.

Theorem 6.33 (P. Cohen) *If* ZFC *is consistent, then so is* ZFC + ¬CH. *In fact, if* ZFC *is consistent, then so is* ZFC + $2^{\aleph_0} = \aleph_2$.

Proof In the light of the discussion above (cf. the proof of Theorem 6.25), we may argue under the hypothesis that there be a transitive model of ZFC.

We first prove the first part of the theorem. Let M be a countable transitive model of ZFC. Let $\alpha \in M$, where $\alpha \geq \omega_2^M$. Then $\mathbb{C}(\alpha) \in M$. Let G be $\mathbb{C}(\alpha)$-generic over M. Inside $M[G]$ we may define $F : \alpha \to {}^\omega\omega$ by setting $F(\xi) = \bigcup\{p(\xi) : p \in G\}$ for $\xi < \alpha$. For all $n < \omega$, $\xi < \alpha$

$$D_{n,\xi} = \{p \in \mathbb{C}(\alpha) : n \in \text{dom}(p(\xi))\} \in M$$

is dense in $\mathbb{C}(\alpha)$, so that F is well-defined. For all $\xi, \xi' < \alpha$ with $\xi \neq \xi'$,

$$D^{\xi,\xi'} = \{p \in \mathbb{C}(\alpha) : \exists n \in \text{dom}(p(\xi)) \cap \text{dom}(p(\xi'))\ p(\xi)(n) \neq p(\xi')(n)\} \in M$$

is dense in $\mathbb{C}(\alpha)$, so that $F(\xi) \neq F(\xi')$ for $\xi \neq \xi'$. $F \in M[G]$ is therefore an injection from α into ${}^\omega\omega$.

In order to verify $M[G] \models \neg CH$, we need to verify $\alpha \geq \omega_2^{M[G]}$. For this it will be enough to show that M, $M[G]$ have the same cardinals. However, as $M \models$ "$\mathbb{C}(\alpha)$ has the c.c.c." by Lemma 6.29, this immediately follows from Lemma 6.32.

We now prove the second part of the theorem. By our hypothesis that there be a transitive model of ZFC, we have as in the proof of Theorem 6.25 that there is some $\gamma < \omega_1$ such that J_γ is a (countable transitive) model of ZFC. Let $\alpha = \omega_2^{J_\gamma}$. Then $\mathbb{C}(\alpha) \in J_\gamma$, and we may pick some G which is $\mathbb{C}(\alpha)$-generic over J_γ. By the argument for the first part of the theorem, $2^{\aleph_0} \geq \aleph_2$ in $J_\gamma[G]$. We are hence left with having to verify that $2^{\aleph_0} \leq \aleph_2$ in $J_\gamma[G]$.

Let $x \in {}^\omega\omega \cap J_\gamma[G]$, say $x = \tau^G$. For $n < \omega$ let

$$E_n = \{p \in \mathbb{C}(\alpha) : \exists m < \omega\ p \Vdash_M^{\mathbb{C}(\alpha)} \tau(\check{n}) = \check{m}\}.$$

Each E_n is dense, and we may pick some maximal antichain $A_n \subset E_n$ for $n < \omega$.

Let

$$\sigma = \sigma(\tau) = \{(\ulcorner n, m \urcorner, p) : p \in A_n,\ p \Vdash_M^{\mathbb{C}(\alpha)} \tau(\check{n}) = \check{m}\}.$$

We claim that $\sigma^G = \tau^G$.

First let $(n, m) \in \sigma^G$. There is then some $p \in G$ such that $(\ulcorner \check{n}, \check{m} \urcorner, p) \in \sigma$, which implies that $p \Vdash_M^{\mathbb{C}(\alpha)} \tau(\check{n}) = \check{m}$, and hence $(n, m) \in \tau^G$. Now let $(n, m) \in \tau^G$. There is then some $p \in G$ such that $p \Vdash_M^{\mathbb{C}(\alpha)} \tau(\check{n}) = \check{m}$. As A_n is a maximal antichain, there is also some $q \in G \cap A_n$, which implies that $q \Vdash_M^{\mathbb{C}(\alpha)} \tau(\check{n}) = \check{s}$ for some $s < \omega$. But as p and q are both in G, p and q are compatible, so that we must have $s = m$. Therefore $(\ulcorner n, m \urcorner, q) \in \sigma$, where $q \in G$, i.e., $(n, m) \in \sigma^G$. $\sigma = \sigma(\tau)$ is often referred to as a "nice name" for x (or, for τ).

We have shown that for every $x \in {}^\omega\omega \cap J_\gamma[G]$ there is some "nice name" $\sigma \in J_\gamma^{\mathbb{C}(\alpha)}$ such that $x = \sigma^G$, each element of σ is of the form $(\ulcorner \check{n}, \check{m} \urcorner, p)$, and for all n there are at most countably many m, p such that $(\ulcorner \check{n}, \check{m} \urcorner, p) \in \sigma$. By the HAUSDORFF Formula 4.19 we may compute inside J_γ that there are $\alpha^{\aleph_0} = \alpha$ such names. In $J_\gamma[G]$ we may hence define an injection from ${}^\omega\omega \cap J_\gamma[G]$ into α, so that $2^{\aleph_0} \leq \alpha = \aleph_2$ in $J_\gamma[G]$, as desired. □

The proof of Theorem 6.33 shows that if we assume that $M \models$ "$\mathsf{ZFC} + \alpha^{\aleph_0} = \alpha$, then $2^{\aleph_0} = \alpha$ holds true in $M[G]$ whenever G is $\mathbb{C}(\alpha)$-generic over M. Hence by Lemma 4.22 we get that $2^{\aleph_0}$ may be any cardinal κ with $\mathrm{cf}(\kappa) > \omega$ (cf. Problem 6.9).

We now consider variants of COHEN forcing for cardinals above ω.

Definition 6.34 Let $\kappa \geq \omega$ be a cardinal. Let $\mathbb{C}_\kappa = {}^{<\kappa}\kappa$, i.e., the set of all f such that there is some $\gamma < \kappa$ with $f : \gamma \to \kappa$. For $p, q \in \mathbb{C}_\kappa$, let $p \leq q$ iff $p \supset q$ (iff $\exists \gamma\ p \upharpoonright \gamma = q$). The partial order $(\mathbb{C}_\kappa, \leq)$ is called COHEN *forcing at* κ.

Of course, $\mathbb{C}_\omega = \mathbb{C}$. If $\kappa^{<\kappa} = \kappa$ (which is true for $\kappa = \omega$ and only possible for regular κ), then $\mathrm{Card}(\mathbb{C}_\kappa) = \kappa$, so that in this case forcing with $\mathbb{C}_\kappa$ preserves all cardinals above κ^+ by Lemma 6.32 (though cf. Problem 6.8). We now develop a technique for showing that forcing with $\mathbb{C}_\kappa$ never collapses cardinals below κ.

Definition 6.35 Let $\mathbb{P} = (\mathbb{P}, \leq)$ be a partial order, and let κ be an infinite regular cardinal. $\mathbb{P}$ is called $<\kappa$*-closed* iff for all $\gamma < \kappa$ and for all sequences $(p_\xi : \xi < \gamma)$ of conditions in $\mathbb{P}$ such that $p_{\xi'} \leq p_\xi$ for all $\xi' \geq \xi$ there is some condition $q \in \mathbb{P}$ with $q \leq p_\xi$ for all $\xi < \gamma$. $\mathbb{P}$ is called $<\kappa$*-distributive* iff for every $\gamma < \kappa$ and for every collection $(D_\xi : \xi < \gamma)$ of open dense subsets of $\mathbb{P}$, $\bigcap_{\xi<\gamma} D_\xi$ is open dense.

The proofs of the following two lemmas are trivial.

Lemma 6.36 *Let κ be an infinite regular cardinal. Then $\mathbb{C}_\kappa$ is $<\kappa$-closed.*

Lemma 6.37 *Let $\mathbb{P} = (\mathbb{P}, \leq)$ be a partial order, and let κ be an infinite regular cardinal. If $\mathbb{P}$ is $<\kappa$-closed, then $\mathbb{P}$ is $<\kappa$-distributive.*

Not every $<\kappa$-distributive forcing is $<\kappa$-closed, cf. Problem 6.16.

Lemma 6.38 *Let M be a transitive model of* ZFC, *let $\mathbb{P} = (\mathbb{P}; \leq)$ be a partial order in M, and let κ be a regular cardinal of M such that $M \models$ "$\mathbb{P}$ is $<\kappa$-distributive." Let G be $\mathbb{P}$-generic over M. Then*[3]

$$^{<\kappa}M \cap M[G] = {}^{<\kappa}M \cap M.$$

Proof Let $f \in M[G]$, $f : \gamma \to M$ for some $\gamma < \kappa$. We may pick some $x \in M$ such that $\mathrm{ran}(f) \subset x$, and by the Forcing Theorem we may pick some $\tau \in M^{\mathbb{P}}$ and some $p \in G$ such that

$$p \Vdash_M^{\mathbb{P}} \tau : \check{\gamma} \to \check{x}.$$

For each $\xi < \gamma$,

$$D_\xi = \{q \in \mathbb{P} : q \perp p \vee \exists y \in x\ q \Vdash_M^{\mathbb{P}} \tau(\check{\xi}) = \check{y}\}$$

[3] If α is an ordinal and X is any set or class, then we write $^{<\alpha}X$ for $\bigcup_{\xi<\alpha} {}^{\xi}X$.

is easily seen to be open dense. As $\bigcap_{\xi<\gamma} D_\xi$ is open dense, we may pick $q \in G \cap \bigcap_{\xi<\gamma} D_\xi$. As $q \| p$, this means that for every $\xi < \gamma$, there is a (unique) $y \in x$ with $q \Vdash_M^{\mathbb{P}} \tau(\check{\xi}) = \check{y}$. Setting

$$g = \{(\xi, y) \in \gamma \times x : q \Vdash_M^{\mathbb{P}} \tau(\check{\xi}) = \check{y}\},$$

we then get $q \Vdash_M^{\mathbb{P}} \tau = \check{g}$. Therefore, $f = \tau^G = g \in M$. □

Problem 6.10 shows that the converse to Lemma 6.38 is true also. The following just generalizes Definition 6.27.

Definition 6.39 Let α be an ordinal, and let κ be an infinite regular cardinal. For $p \in {}^\alpha(\mathbb{C}_\kappa)$ let supp(p) (the support of p) be the set of all $\xi < \alpha$ with $p(\xi) \neq \emptyset$. Let

$$\mathbb{C}_\kappa(\alpha) = \{p \in {}^\alpha(\mathbb{C}_\kappa) : \text{Card}(\text{supp}(p)) < \kappa\}.$$

For $p, q \in \mathbb{C}_\kappa(\alpha)$, let us write $p \leq q$ iff for all $\xi < \alpha$, $p(\xi) \leq q(\xi)$ in the sense of $\mathbb{C}_\kappa$, i.e., $\exists \gamma\ p(\xi) \restriction \gamma = q(\xi)$.

Lemma 6.40 *Let M be a transitive model of* ZFC*, let κ be a regular cardinal of M, let α be an ordinal in M, and let $(\mathbb{C}_\kappa(\alpha))^M$ be M's version of $\mathbb{C}_\kappa(\alpha)$. Let G be $(\mathbb{C}_\kappa(\alpha))^M$-generic over M. Then exactly the cardinals of M which are not in the half-open interval $(\kappa, (\kappa^{<\kappa})^M]$ remain cardinals of $M[G]$.*

Proof Of course, $\mathbb{C}_\kappa(\alpha)$ is $< \kappa$-closed. The Δ-Lemma 6.31 implies that $M \models$ "$(\mathbb{C}_\kappa(\alpha))^M$ has the $((\kappa^{<\kappa})^+)^M$-c.c.," so that no M-cardinals outside the half-open interval $(\kappa, (\kappa^{<\kappa})^M]$ will get collapsed. On the other hand, Problem 6.8 shows that all the M-cardinals inside the half-open interval $(\kappa, (\kappa^{<\kappa})^M]$ will get collapsed to κ. □

If $\kappa^{<\kappa} = \kappa$ in M, $(\mathbb{C}_\kappa(\alpha))^M$ will therefore not collapse any M-cardinals. Moreover, we may have $2^\kappa = \alpha$ in a forcing extension (cf. Problem 6.9).

We shall now study forcings which collapse cardinals.

Definition 6.41 Let μ be a regular cardinal, and let $\kappa \geq \mu$. We let $\text{Col}(\mu, \kappa) = {}^{<\mu}\kappa$, i.e., the set of all functions f such that there is some $\gamma < \mu$ with $f : \gamma \to \kappa$. For $p, q \in \text{Col}(\mu, \kappa)$, let $p \leq q$ iff $p \supset q$ (iff $\exists \gamma\ p \restriction \gamma = q$). The partial order $(\text{Col}(\mu, \kappa); \leq)$ is called the *collapse of κ to μ*.

Notice that $\text{Col}(\mu, \mu) = \mathbb{C}_\mu$.

Lemma 6.42 *Let M be a transitive model of* ZFC*, let μ be a regular cardinal of M, let $\kappa \geq \mu$ be a cardinal of M, and let $\mathbb{P} = (\text{Col}(\mu, \kappa))^M$ be M's version of* Col(μ, κ)*. Let G be $\mathbb{P}$-generic over M. Then every M-cardinal $\leq \mu$ is still a cardinal in $M[G]$, and in $M[G]$,* Card$(\kappa) = \mu = \cdot 2^{<\mu}$*. Moreover, cardinals above $((\kappa^{<\mu})^+)^M$ remain cardinals in $M[G]$.*

Proof $\mathbb{P}$ is certainly $< \mu$-closed inside M, so that every M-cardinal $\leq \mu$ is still a cardinal in $M[G]$. For $\xi < \mu$ let

$$D_\xi = \{p \in \mathbb{P} : \xi \in \text{dom}(p)\},$$

and for $\eta < \kappa$ let

$$D^\eta = \{p \in \mathbb{P} : \eta \in \text{ran}(p)\}.$$

For all $\xi < \mu$ and $\eta < \kappa$, $D_\xi \in M$ and $D^\eta \in M$ are both dense in $\mathbb{P}$. Therefore, $f = \bigcup G$ is a surjective function from μ onto κ, so that $\text{Card}(\kappa) = \mu$ in $M[G]$.

To show that $2^{<\mu} = \mu$ in $M[G]$, let us fix $\alpha < \mu$. If $X \in \mathscr{P}(\alpha) \cap M = \mathscr{P}(\alpha) \cap M[G]$, then[4]

$$D_X = \{p \in \mathbb{P} : \exists \gamma < \mu\, (\gamma \cdot (\alpha+1) \subset \text{dom}(p) \wedge \{\xi < \alpha : p(\gamma \cdot \alpha + \xi) = 1\} = X)\}$$

is in M and is dense in $\mathbb{P}$, so that we may define $F \in M[G]$, $F \colon \mathscr{P}(\alpha) \cap M[G] \to \mu$, by setting

$$F(X) = \text{ the least } \gamma \text{ with } \{\xi < \alpha : (\bigcup G)(\gamma \cdot \alpha + \xi) = 1\} = X.$$

F is certainly injective, so that $\text{Card}(2^\alpha) \leq \mu$ in $M[G]$. This shows that $2^{<\mu} = \mu$ in $M[G]$.

$\mathbb{P}$ has size $\kappa^{<\mu}$ in M, so that cardinals above $((\kappa^{<\mu})^+)^M$ remain cardinals in $M[G]$. □

If M is a transitive model of ZFC, and if G is $(\text{Col}(\omega_1, 2^{\aleph_0}))^M$-generic over M, then CH holds is $M[G]$ by Lemmas 6.38 and 6.42. More generally, if κ is regular in M and H is $(\text{Col}(\kappa^+, 2^\kappa))^M$-generic over M, then $2^\kappa = \kappa^+$ holds $M[H]$. Theorem 6.46 will produce a stronger result.

Definition 6.43 Let μ be a regular cardinal, and let X be a set of ordinals which are all of size $\geq \mu$. We let

$$\text{Col}^*(\mu, X) = \{p : p \text{ is a function with domain } X \text{ and } \forall \xi \in X\ p(\xi) \in \text{Col}(\mu, \xi)\}.$$

For $p \in \text{Col}^*(\mu, X)$, let $\text{supp}(p) = \{\xi \in X \colon p(\xi) \neq \emptyset\}$. We let

$$\text{Col}(\mu, X) = \{p \in \text{Col}^*(\mu, X) : \text{Card}(\text{supp}(p)) < \mu\}.$$

For $p, q \in \text{Col}(\mu, X)$ we write $p \leq q$ iff for all $\xi \in X$ we have that $p(\xi) \leq q(\xi)$ in the sense of $\text{Col}(\mu, \xi)$. If $\kappa > \mu$, then we also write $\text{Col}(\mu, < \kappa)$ for $\text{Col}(\mu, [\mu, \kappa))$. The partial order $(\text{Col}(\mu, < \kappa); \leq)$ is called the LEVY *collapse of* κ *to* μ.

[4] In what follows, we use ordinal arithmetic.

Lemma 6.44 *Let μ be a regular cardinal, and let $\kappa > \mu$ be a regular cardinal such that $\rho^\gamma < \kappa$ for all $\rho < \kappa$ and $\gamma < \mu$. Then $\mathrm{Col}(\mu, < \kappa)$ has the κ-c.c.*

Proof This immediately follows from the Δ-Lemma 6.31. □

As $\mathrm{Col}(\mu, < \kappa)$ is certainly $< \mu$-closed, this immediately implies the following.

Lemma 6.45 *Let M be a transitive model of ZFC, and let $\mu < \kappa$ be regular cardinals of M such that inside M, $\rho^\gamma < \kappa$ for all $\rho < \kappa$ and $\gamma < \mu$. Let $\mathbb{P} = (\mathrm{Col}(\mu, < \kappa))^M$ be M's version of the LEVY collapse of κ to μ. Then all M-cardinals strictly between μ and κ will have size μ in $M[G]$, and all M-cardinals outside of the open interval (μ, κ) will remain cardinals in $M[G]$. In particular, $\kappa = \mu^+$ in $M[G]$.*

The LEVY collapse $\mathrm{Col}(\omega, < \kappa)$ will play a crucial role in Chap. 8.

We may force $\diamondsuit_\kappa$ which was shown to be true in L, cf. Definition 5.34 and Theorem 5.35.

Theorem 6.46 *Let M be a transitive model of ZFC, let κ be an uncountable regular cardinal in M, and let $\mathbb{P} \in M$ be defined inside M as follows. $\mathbb{P} = \{(c_\alpha : \alpha \leq \beta) : \beta < \kappa \wedge \forall \alpha \leq \beta \; c_\alpha \subset \alpha\}$, ordered by end-extension. Let G be $\mathbb{P}$-generic over M. Let $S \subset \kappa$, $S \in M$, be stationary in M. Then $\diamondsuit_\kappa(S)$ holds true in $M[G]$.*

Proof An easy density argument shows that there are $C_\alpha \subset \alpha$, $\alpha < \kappa$, such that $\bigcup G = (C_\alpha : \alpha < \kappa)$. We claim that $\bigcup G \upharpoonright S = (C_\alpha : \alpha \in S)$ witnesses that $\diamondsuit_\kappa(S)$ holds true in $M[G]$.

Let $\tau, \rho \in M^{\mathbb{P}}$ and $p \in G$ be such that

$$p \Vdash \tau \subset \check{\kappa} \text{ is club in } \check{\kappa}, \text{ and} \rho \subset \kappa.$$

Let $p_0 \leq p$ be arbitrary. It suffices to show that there is some $q \leq p_0$ such that if $q = (c_\alpha : \alpha \leq \beta)$, then

$$q \Vdash \check{\beta} \in \tau \cap \check{S} \wedge \rho \cap \check{\beta} = \check{c_\beta}. \tag{6.9}$$

Let us work inside M. Notice that $\mathbb{P}$ is $< \kappa$-closed. We may thus easily construct a sequence $(p_i : 1 \leq i < \kappa)$ of conditions in $\mathbb{P}$ such that there are $F = (c_\alpha : \alpha < \kappa)$ and $(\beta_i : i < \kappa)$, such that for all $0 \leq i < j < \kappa$, $p_i = (c_\alpha : \alpha \leq \beta_i)$, $\sup_{k<j} \beta_k < \beta_j$ (in particular, $p_j < p_i$), and there is some $\xi \in \beta_j \setminus \sup_{k<j} \beta_k$ and some $a_j \in M$ such that, writing $\tilde{\beta}_j$ for $\sup_{k<j} \beta_k$,

$$p_j \Vdash \check{\xi} \in \tau \wedge \rho \cap \check{\tilde{\beta}}_j = \check{a}_j.$$

As S is stationary and $\{\tilde{\beta}_i : i < \kappa\}$ is club in κ, we may pick some limit ordinal i_0 such that $\tilde{\beta}_{i_0} \in S$. Set

$$A = \bigcup_{i<i_0} a_i \text{ and } q = (\bigcup_{i<i_0} p_i) \cup \{(\tilde{\beta}_{i_0}, A)\}.$$

Write $\beta = \tilde{\beta}_{i_0}$. We have that $q \in \mathbb{P}$, $q < p_0$, and $q \Vdash \check{\beta} \in \tau$ by the choice of $(p_i : i < i_0)$. Also, $q \Vdash (\rho \cap \check{\beta}) = \check{A}$. As $\beta \in S$, (6.9) is shown. □

The forcing $\mathbb{P}$ used in the previous proof is forcing eqivalent to $\text{Col}(\kappa, \kappa)$ in the sense of Lemma 6.48 below.

We now work towards showing that ω_1 may be singular in **ZF**, cf. Theorem 6.69.

Definition 6.47 Let $\mathbb{P} = (\mathbb{P}; \leq_{\mathbb{P}})$, $\mathbb{Q} = (\mathbb{Q}; \leq_{\mathbb{Q}})$ be partial orders. We call a map $\pi : \mathbb{P} \to \mathbb{Q}$ a *homomorphism* iff for all $p, q \in \mathbb{P}$,

(a) $p \leq_{\mathbb{P}} q \Longrightarrow \pi(p) \leq_{\mathbb{Q}} \pi(q)$ and
(b) $p \perp_{\mathbb{P}} q \Longrightarrow \pi(p) \perp_{\mathbb{Q}} \pi(q)$.

A homomorphism $\pi : \mathbb{P} \to \mathbb{Q}$ is called *dense* iff $\text{ran}(\pi)$ is dense in $\mathbb{Q}$, i.e., for every $q \in \mathbb{Q}$ there is some $p \in \mathbb{P}$ such that $\pi(p) \leq_{\mathbb{Q}} q$.

If $\pi : \mathbb{P} \to \mathbb{Q}$ is a homomorphism, then (a) implies that $p \parallel_{\mathbb{P}} q \Longrightarrow \pi(p) \parallel_{\mathbb{Q}} \pi(q)$, so that (b) gives $p \parallel_{\mathbb{P}} q \Longleftrightarrow \pi(p) \parallel_{\mathbb{Q}} \pi(q)$ for all $p, q \in \mathbb{P}$.

Lemma 6.48 *Let M be a transitive model of* **ZFC**, *let $\mathbb{P} = (\mathbb{P}; \leq_{\mathbb{P}})$, $\mathbb{Q} = (\mathbb{Q}; \leq_{\mathbb{Q}}) \in M$ be partial orders and let $\pi : \mathbb{P} \to \mathbb{Q}$ be a dense homomorphism, where $\pi \in M$. If $G \subset \mathbb{P}$ is $\mathbb{P}$-generic over M, then $H = \{p \in \mathbb{Q} : \exists q \in G\, \pi(q) \leq_{\mathbb{Q}} p\}$ is $\mathbb{Q}$-generic over M, $G = \{p \in \mathbb{P} : \pi(p) \in H\}$, and $M[G] = M[H]$. Also, if $H \subset \mathbb{Q}$ is $\mathbb{Q}$-generic over M, then $G = \{p \in \mathbb{P} : \pi(p) \in H\}$ is $\mathbb{P}$-generic over M and $M[H] = M[G]$.*

Proof First let $G \subset \mathbb{P}$ be $\mathbb{P}$-generic over M, and set $H = \{p \in \mathbb{Q} : \exists q \in G\, \pi(q) \leq_{\mathbb{Q}} p\}$. To see that H is a filter, let $p, p' \in H$. Then there are $q, q' \in G$ with $\pi(q) \leq_{\mathbb{Q}} p$ and $\pi(q') \leq_{\mathbb{Q}} p'$. If $r \leq_{\mathbb{P}} q, q'$, then $\pi(r) \leq_{Q} p, p'$. Now let $D \in M$ be dense in $\mathbb{Q}$. We need to see $D \cap H \neq \emptyset$. Let $D^* = \{s \in \mathbb{P} : \exists r \in D\, \pi(s) \leq_{\mathbb{Q}} r\} \in M$. D^* is dense in $\mathbb{P}$: given $p \in \mathbb{P}$, there is some $r \in D$ with $r \leq_{\mathbb{Q}} \pi(p)$, and because π is dense there is some $s \in \mathbb{P}$ with $\pi(s) \leq_{\mathbb{Q}} r$; in particular, $\pi(s) \parallel_{\mathbb{Q}} \pi(p)$, so that $s \parallel_{\mathbb{P}} p$, and if $q \leq_{\mathbb{P}} s, p$, then $\pi(q) \leq_{\mathbb{Q}} \pi(s) \leq_{\mathbb{Q}} r$; i.e., $q \in D^*$ and $q \leq_{\mathbb{P}} p$. Now let $p \in D^* \cap G$. Then $\pi(p) \leq_{\mathbb{Q}} r$ for some $r \in D$, where $p \in G$, so that $r \in D \cap H$.

Let us now show that $G = \{p \in \mathbb{P} : \pi(p) \in H\}$. If $\pi(p) \in H$, then there is some $q \in G$ with $\pi(q) \leq_{\mathbb{P}} \pi(p)$. As $D = \{r \in \mathbb{P} : r \leq_{\mathbb{P}} p \vee r \perp_{\mathbb{P}} p\}$ is dense in $\mathbb{P}$, we may pick $r \in D \cap G$. There is some $s \in G$ with $s \leq_{\mathbb{P}} r, q$; then $\pi(s), \pi(p) \in H$, hence $\pi(s) \| _{\mathbb{Q}} \pi(p)$, and hence $s \| _{\mathbb{P}} p$, so that $r \| _{\mathbb{P}} p$. But then $r \leq_{\mathbb{P}} p$, as $r \in D$, and so $p \in G$.

Conversely, let $H \subset \mathbb{Q}$ be $\mathbb{Q}$-generic over M, and set $G = \{p \in \mathbb{P} : \pi(p) \in H\}$. It is again easy to see that G is a filter. Now let $D \in M$ be dense in $\mathbb{P}$. Let $D' = \{\pi(p) : p \in D\}$. D' is dense in $\mathbb{Q}$: given $p \in \mathbb{Q}$, there is some $q \in \mathbb{P}$ with $\pi(q) \leq_{\mathbb{Q}} p$, as π is dense, and there is some $r \in D$ such that $r \leq_{\mathbb{P}} q$, as D is dense; but then $\pi(r) \in D'$ and $\pi(r) \leq_{\mathbb{Q}} p$. Now let $p \in D' \cap H$. Then $p = \pi(q)$ for some $q \in D \cap G$. □

Lemma 6.49 *Let κ be an infinite cardinal, and let $\mathbb{P}$ be an atomless partial order such that*

$$1_{\mathbb{P}} \Vdash \mathrm{Card}(\check{\kappa}) = \check{\aleph}_0.$$

Then for every $p \in \mathbb{P}$ there is an antichain $A \subset \{q \in \mathbb{P} \colon q \leq_{\mathbb{P}} p\}$ of size κ.

Proof Let us fix $p \in \mathbb{P}$.

Let us first assume that $\kappa = \omega$. Let us construct a sequences $(p_n \colon n < \omega)$ and $(q_n \colon n < \omega)$ of conditions at follows. Set $p_0 = p$. Given p_n, let q_n and p_{n+1} two incompatible extensions of p_n. We then have that $\{q_n \colon n < \omega\}$ is an antichain of size $\aleph_0$.

Let us now assume that $\mathrm{cf}(\kappa) = \omega < \kappa$. Let $(\kappa_n \colon n < \omega)$ be a sequence of uncountable regular cardinals which is cofinal in κ. We construct a sequence $(p_n \colon n < \omega)$ of conditions and a sequence $(A_n \colon n < \omega)$ of antichains in $\mathbb{P}$ as follows. Set $p_0 = p$. Given p_n, notice that

$$p_n \Vdash \mathrm{Card}(\check{\kappa}) = \check{\aleph}_0,$$

so that by Lemma 6.32 there must be an antichain $A \subset \{q \in \mathbb{P} \colon q \leq p_n\}$ of size κ_n. Let A_n be some such antichain, and let $p_{n+1} \in A_n$ be arbitrary. It is now easy to see that

$$\bigcup \{A_n \setminus \{p_{n+1}\} \colon n < \omega\} \subset \{q \in \mathbb{P} \colon q \leq p\}$$

is an antichain of size $\sum_{n<\omega} \kappa_n = \kappa$.

Finally, let us assume that $\mathrm{cf}(\kappa) > \omega$. Let $r \leq p$ and $\tau \in V^{\mathbb{P}}$ be such that

$$r \Vdash \tau \colon \check{\omega} \to \check{\kappa} \text{ is surjective.} \tag{6.10}$$

Let us suppose that every antichain $A \subset \{q \in \mathbb{P} \colon q \leq r\}$ is smaller than κ. For every $n < \omega$, let A_n be a maximal antichain in

$$\{q \in \mathbb{P} \colon q \leq r \wedge \exists \xi \; q \Vdash \tau(\check{n}) = \check{\xi}\}.$$

By our hypothesis, $\mathrm{Card}(A_n) < \kappa$ for every $n < \omega$, so that by $\mathrm{cf}(\kappa) > \omega$ there cannot be a surjective function

$$f \colon (\omega \times \bigcup \{A_n \colon n < \omega\}) \to \kappa. \tag{6.11}$$

However, we may define a function f as in (6.11) by setting, for $n < \omega$ and $p \in \bigcup \{A_n \colon n < \omega\}$,

$$f((n, p)) = \begin{cases} \xi, \text{ if } p \Vdash \tau(\check{n}) = \check{\xi}, \\ 0, \text{ if there is no } \xi < \kappa \text{ such that } p \Vdash \tau(\check{n}) = \check{\xi}. \end{cases} \tag{6.12}$$

If $\xi < \kappa$, then by (6.10) there is some $r' \leq r$ and some $n < \omega$ such that $r' \Vdash \tau(\check{n}) = \check{\xi}$. But then $r' \| q$ for some $q \in A_n$ by the choice of A_n, so that $q \Vdash \tau(\check{n}) = \check{\xi}$. Therefore, f is surjective. Contradiction! □

Definition 6.50 Let $\mathbb{P} = (\mathbb{P}, \leq)$ be a partial order. $\mathbb{P}$ is called *separative* iff whenever p is not stronger than q then there is some $r \leq p$ such that r and q are incompatible.

Every separative partial order $\mathbb{P}$ such that for every $p \in \mathbb{P}$ there is some $q \in \mathbb{P}$ with $q <_{\mathbb{P}} p$ is easily seen to be atomless.

Lemma 6.51 *Let μ be an infinite cardinal, and let $\mathbb{P}$ be a separative partial order such that* $\text{Card}(\mathbb{P}) = \mu$ *and*

$$1_{\mathbb{P}} \Vdash \check{\mu} \text{ is countable.}$$

Then there is a dense homomorphism $\pi : \text{Col}(\omega, \mu) \to \mathbb{P}$.

Proof Let τ be a name such that

$$1_{\mathbb{P}} \Vdash \tau : \check{\omega} \to \dot{G} \text{ is onto.}$$

Let us construct $\pi(p)$ by recursion on $\text{lh}(p)$, where $p \in \text{Col}(\omega, \mu)$. Set $\pi(\emptyset) = 1_{\mathbb{P}}$. Let us now suppose that $p \in \text{Col}(\omega, \mu)$ and $\pi(p)$ has been defined, where $n = \text{lh}(p)$. As $\text{Card}(\mathbb{P}) = \mu$, by Lemma 6.49, we may let $A \subset \mathbb{P}$ be a maximal antichain of size μ consisting of $q \in \mathbb{P}$ such that $q \leq_{\mathbb{P}} \pi(p)$ and q decides $\tau(\check{n})$, i.e., there is some $\xi < \mu$ such that

$$q \Vdash \tau(\check{n}) = \check{\xi}.$$

We may write $A = \{q_i : i < \mu\}$, where q_i is different from (and thus incomaptible with) q_j for $i \neq j$. We may then set $\pi(p \cup \{(n, i)\}) = q_i$.

It is easy to see that π is a homomorphism. Also, an easy induction shows that for each $n < \omega$,

$$A_n = \{\pi(p) \colon p \in \text{Col}(\omega, \mu) \wedge \text{lh}(p) = n\}$$

is a maximal antichain of $q \in \mathbb{P}$ such that q decides $\tau(\check{0}), \ldots, \tau((n-1)\check{})$.

Let us show that π is dense. Pick $r \in \mathbb{P}$. As $r \Vdash \check{r} \in \dot{G}$, there is some $s \leq_{\mathbb{P}} r$ and some $n < \omega$ such that $s \Vdash \tau(\check{n}) = \check{r}$. Let $t \leq_{\mathbb{P}} s$ be such that t decides $\tau(\check{0}), \ldots, \tau(\check{n})$. There is then some $p \in \text{Col}(\omega, \mu)$ such that $\pi(p) \in A_{n+1}$ and $\pi(p) \| t$. We must then have that $\pi(p) \Vdash \tau(\check{n}) = \check{r}$, which implies that $\pi(p) \Vdash \check{r} \in \dot{G}$. As $\mathbb{P}$ is separative, this gives that $\pi(p) \leq_{\mathbb{P}} r$. □

Definition 6.52 Let $\mathbb{P} = (\mathbb{P}; \leq)$ be a partial order. $\mathbb{P}$ is called homogenous iff for all $p, q \in \mathbb{P}$ there is some dense endomorphism[5] $\pi : \mathbb{P} \to \mathbb{P}$ such that $\pi(p) \parallel q$.

Lemma 6.53 $\mathbb{C}$ *is homogeneous. If α is an ordinal, then $\mathbb{C}(\alpha)$ is homogeneous.*

---

[5] i.e., a homomorphism to itself.

Proof Let us first show that $\mathbb{C}$ is homogenous. Let us fix $p, q \in \mathbb{C}$. Let us then define $\pi : \mathbb{C} \to \mathbb{C}$ as follows. If $r \in \mathbb{C}$, then $\mathrm{dom}(\pi(r)) = \mathrm{dom}(r)$, and if $n \in \mathrm{dom}(r)$, then

$$\pi(r)(n) = \begin{cases} q(n), \text{ if } n \in \mathrm{dom}(p) \cap \mathrm{dom}(q) \text{ and } r(n) = p(n), \\ p(n), \text{ if } n \in \mathrm{dom}(p) \cap \mathrm{dom}(q) \text{ and } r(n) = q(n), \text{ and} \\ r(n) \text{ otherwise.} \end{cases} \tag{6.13}$$

Then if $n \in \mathrm{dom}(p) \cap \mathrm{dom}(q)$, $\pi(p)(n) = q(n)$, so that $\pi(p) \parallel q$. It is easy to see that π is a dense endomorphism.

Now if α is an ordinal, and if $p, q \in \mathbb{C}(\alpha)$, then for each $\xi \in \mathrm{supp}(p) \cap \mathrm{supp}(q)$ there is a dense endomorphism $\pi_\xi : \mathbb{C} \to \mathbb{C}$ such that $\pi_\xi(p(\xi)) \parallel q(\xi)$ in the sense of COHEN forcing. These endomorphisms then easily induce an endomorphism $\pi : \mathbb{C}(\alpha) \to \mathbb{C}(\alpha)$ such that $\pi(p) \parallel q$ in the sense of $\mathbb{C}(\alpha)$. Again, π will be dense. □

The endomorphism constructed in the previous proof is actually an automorphism, i.e. bijective.

In much the same way as Lemma 6.53 we may prove the following.

Lemma 6.54 *Let μ be a regular cardinal, and let X be a set of ordinals which are all of size $\geq \mu$. Then* $\mathrm{Col}(\mu, X)$ *is homogeneous.*

Proof Let $p, q \in \mathrm{Col}(\mu, X)$ be given. We may then define $\pi : \mathrm{Col}(\mu, X) \to \mathrm{Col}(\mu, X)$ as follows. Given $r \in \mathrm{Col}(\mu, X)$, let $\mathrm{supp}(\pi(r)) = \mathrm{supp}(r)$ and $\mathrm{dom}(\pi(r)(\eta)) = \mathrm{dom}(r(\eta))$ for all $\eta \in X$, and if $\eta \in X$ and $\xi \in \mathrm{dom}(r(\eta))$, then let

$$\pi(r)(\eta)(\xi) \begin{cases} q(\eta)(\xi), \text{ if } \xi \in \mathrm{dom}(p(\eta)) \cap \mathrm{dom}(q(\eta)) \text{ and } r(\eta)(\xi) = p(\eta)(\xi), \\ p(\eta)(\xi), \text{ if } \xi \in \mathrm{dom}(p(\eta)) \cap \mathrm{dom}(q(\eta)) \text{ and } r(\eta)(\xi) = q(\eta)(\xi), \text{ and} \\ r(\eta)(\xi) \text{ otherwise.} \end{cases} \tag{6.14}$$

It is easy to see that π is a dense automorphism of $\mathrm{Col}(\mu, X)$ such that $\pi(p) || q$. □

Definition 6.55 Let M be a transitive model of ZFC, and let $\mathbb{P} = (\mathbb{P}; \leq) \in M$ be a partial order. Let $\pi : \mathbb{P} \to \mathbb{P}$ be a dense endomorphism, $\pi \in M$. The π induces a map

$$\tilde{\pi} : M^{\mathbb{P}} \to M^{\mathbb{P}}$$

as follows:

$$\tilde{\pi}(\tau) = \{(\tilde{\pi}(\sigma), \pi(p)) : (\sigma, p) \in \tau\}.$$

Lemma 6.56 *Let M be a transitive model of* ZFC, *let $\mathbb{P} = (\mathbb{P}; \leq)$ be a partial order, and let $\pi : \mathbb{P} \to \mathbb{P}$ be a dense endomorphism with $\pi \in M$. Let $p \in \mathbb{P}$, let $\varphi(v_1, \ldots, v_n)$ be a formula, and let $\tau_1, \ldots, \tau_n \in M^{\mathbb{P}}$ Then*

$$p \Vdash^{\mathbb{P}}_{M} \varphi(\tau_1, \ldots, \tau_n) \Longleftrightarrow \pi(p) \Vdash^{\mathbb{P}}_{M} \varphi(\tilde{\pi}(\tau_1), \ldots, \tilde{\pi}(\tau_n)).$$

Proof We first show:

Claim 6.57 *Let $G \subset \mathbb{P}$ be $\mathbb{P}$-generic over M, and let $H = \{p : \exists q \in G\, \pi(q) \leq p\}$. Then for all $\tau \in M^{\mathbb{P}}$, $\tau^G = \tilde{\pi}(\tau)^H$.*

The proof is an easy induction on the rank of τ. Notice that $\sigma^G \in \tau^G$ iff $(\sigma, p) \in \tau$ for some $p \in G$ iff $(\tilde{\pi}(\sigma), \pi(p)) \in \tilde{\pi}(\tau)$ for some $p \in G$ (i.e., $\pi(p) \in H$) iff $\tilde{\pi}(\sigma)^H \in \tilde{\pi}(\tau)^H$.

The same argument shows:

Claim 6.58 *Let $G \subset \mathbb{P}$ be $\mathbb{P}$-generic over M, and let $H = \{p : \pi(p) \in G\}$. Then for all $\tau \in M^{\mathbb{P}}$, $\tau^H = \tilde{\pi}(\tau)^G$.*

Now suppose that $p \Vdash^{\mathbb{P}}_{M} \varphi(\tau_1, \ldots, \tau_n)$. Let $G \subset \mathbb{P}$ be $\mathbb{P}$-generic over M such that $\pi(p) \in G$.[6] Setting $H = \{p : \pi(p) \in G\}$, H is $\mathbb{P}$-generic over M by Lemma 6.48, and $p \in H$. By $p \Vdash^{\mathbb{P}}_{M} \varphi(\tau_1, \ldots, \tau_n)$, $M[H] \models \varphi(\tau_1^H, \ldots, \tau_n^H)$. But $M[H] = M[G]$ by Lemma 6.48 and $\tau_1^H = \tilde{\pi}(\tau_1)^G, \ldots, \tau_n^H = \tilde{\pi}(\tau_1)^G$ by Claim 6.58, so that $M[G] \models \varphi(\tilde{\pi}(\tau_1)^G, \ldots, \tilde{\pi}(\tau_n)^H)$. We have shown that $\pi(p) \Vdash^{\mathbb{P}}_{M} \varphi(\tilde{\pi}(\tau_1), \ldots, \tilde{\pi}(\tau_n))$.

Conversely suppose that $\pi(p) \Vdash^{\mathbb{P}}_{M} \varphi(\tilde{\pi}(\tau_1), \ldots, \tilde{\pi}(\tau_n))$. Let $G \subset \mathbb{P}$ be $\mathbb{P}$-generic over M by Lemma 6.48, and $p \in G$. Setting $H = \{p : \exists q \in G\, \pi(q) \leq p\}$, H is $\mathbb{P}$-generic over M such that $\pi(p) \in H$. By $\pi(p) \Vdash^{\mathbb{P}}_{M} (\tilde{\pi}(\tau_1), \ldots, \tilde{\pi}(\tau_n))$, $M[H] \models (\tilde{\pi}(\tau_1)^H, \ldots, \tilde{\pi}(\tau_n)^H)$. But $M[G] = M[H]$ by Lemma 6.48 and $\tau_1^G = \tilde{\pi}(\tau_1)^G, \ldots, \tau_n^G = \tilde{\pi}(\tau_n)^H$ by Claim 6.57, so that $M[G] \models \varphi(\tau_1^G, \ldots, \tau_n^G)$. We have shown that $p \Vdash^{\mathbb{P}}_{M} \varphi(\tau_1, \ldots, \tau_n)$. □

Definition 6.59 Let M be a transitive model of ZFC, and let $\mathbb{P} = (\mathbb{P}; \leq)$ be a partial order. Then $\tau \in M^{\mathbb{P}}$ is called homogenous iff for all dense endomorphisms $\pi : \mathbb{P} \to \mathbb{P}$ with $\pi \in M$, $\tilde{\pi}(\tau) = \tau$. If $\tau_1, \ldots, \tau_n \in M^{\mathbb{P}}$, then $\mathbb{P}$ is called homogenous with respect to $\tau_1, \ldots, \tau_n$ iff for all $p, q \in \mathbb{P}$ there is some dense endomorphism $\pi : \mathbb{P} \to \mathbb{P}$ such that $\pi(p) \parallel q$, and $\tilde{\pi}(\tau_1) = \tau_1, \ldots, \tilde{\pi}(\tau_n) = \tau_n$.

Hence $\mathbb{P}$ is homogenous iff $\mathbb{P}$ is homogenous with respect to the empty sequence $\emptyset$ of names. Moreover, if $\mathbb{P}$ is homogenous with respect to $\sigma_1, \ldots, \sigma_m$ and $\tau_1, \ldots, \tau_n \in M^{\mathbb{P}}$ are homogenous, then $\mathbb{P}$ is homogenous with respect to $\sigma_1, \ldots, \sigma_n, \tau_1, \ldots, \tau_m$.

Lemma 6.60 *Let M be a transitive model of* ZFC, *and let $\mathbb{P} = (\mathbb{P}, \leq) \in M$ be a separative partial order. For every $x \in M$, $\check{x}$ is homogenous.*

Proof We must have $\pi(1) = 1$ for every dense homomorphism $\pi : \mathbb{P} \to \mathbb{P}$. This is because if $\pi(1) < 1$, then there is some $r \leq 1$ such that $\pi(1)$, r are incompatible. By density, there would be some s such that $\pi(s) \leq r$. Then $\pi(s)$ and $\pi(1)$ are incompatible, which is nonsense. □

---

[6] We may assume without loss of generality that such a G exists, as otherwise we might work with the transitive collapse of a countable (sufficiently) elementary substructure of M.

Lemma 6.61 *Let M be a transitive model of* ZFC*, and let $\mathbb{P} = (\mathbb{P}, \leq) \in M$ be a partial order. Let $\varphi(v_1, \ldots, v_n)$ be a formula, and let $\tau_1, \ldots, \tau_n \in M^{\mathbb{P}}$ be such that $\mathbb{P}$ is homogenous with respect to $\tau_1, \ldots, \tau_n$. Then either $1 \Vdash_M^{\mathbb{P}} \varphi(\tau_1, \ldots, \tau_n)$ or else $1 \Vdash_M^{\mathbb{P}} \neg\varphi(\tau_1, \ldots, \tau_n)$.*

Proof Otherwise there are $p, q \in \mathbb{P}$ such that $p \Vdash_M^{\mathbb{P}} \varphi(\tau_1, \ldots, \tau_n)$ and $q \Vdash_M^{\mathbb{P}} \neg\varphi(\tau_1, \ldots, \tau_n)$. Pick a dense endomorphism $\pi : \mathbb{P} \to \mathbb{P}$ such that $\pi(p) \parallel q$ and $\tilde{\pi}(\tau_1) = \tau_1, \ldots, \tilde{\pi}(\tau_n) = \tau_n$. By Lemma 6.56, we then have $\pi(p) \Vdash_M^{\mathbb{P}} \varphi(\tilde{\pi}(\tau_1), \ldots, \tilde{\pi}(\tau_n))$, i.e., $\pi(p) \Vdash_M^{\mathbb{P}} \varphi(\tau_1, \ldots, \tau_n)$, and $q \Vdash_M^{\mathbb{P}} \neg\varphi(\tau_1, \ldots, \tau_n)$, so that $\pi(p), q$ cannot be compatible. Contradiction! □

Corollary 6.62 *Let M be a transitive model of* ZFC*, and let $\mathbb{P} = (\mathbb{P}; \leq)$ be a partial order. Let $G \subset \mathbb{P}$ be $\mathbb{P}$-generic over M. Let $x \in M[G]$, where $x \subset M$. Suppose also that*

$$M[G] \models \forall y(y \in x \longleftrightarrow \varphi(y, \tau_1^G, \ldots, \tau_n^G))$$

for some formula φ and $\tau_1, \ldots, \tau_n \in M^{\mathbb{P}}$ such that $\mathbb{P}$ is homogenous with respect to $\tau_1, \ldots, \tau_n$. Then $x \in M$.

In particular, if $\mathbb{P}$ is homogenous, then every $x \in M[G] \cap \mathsf{OD}_M^{M[G]}$ such that $x \subset M$ is an element of M. In particular, if $\mathbb{P}$ is homogenous, then $\mathsf{HOD}_M^{M[G]} \subset M$.

Proof Let $y \in M$. Then $y \in x$ iff $\exists p \in G\ p \Vdash_M^{\mathbb{P}} \varphi(\check{y}, \tau_1, \ldots, \tau_n)$. But because $\mathbb{P}$ is homogenous with respect to $\check{y}, \tau_1, \ldots, \tau_n$, $p \Vdash_M^{\mathbb{P}} \varphi(\check{y}, \tau_1, \ldots, \tau_n)$ is equivalent to $1 \Vdash_M^{\mathbb{P}} \varphi(\check{y}, \tau_1, \ldots, \tau_n)$. We may therefore compute x inside M as $\{y : 1 \Vdash_M^{\mathbb{P}} \varphi(\check{y}, \tau_1, \ldots, \tau_n)\}$. □

As an example, we get that a COHEN real is not definable in the generic extension:

Corollary 6.63 *Let M be a transitive model of* ZFC*, and let G be $\mathbb{C}$-generic over M. Then neither G nor $\bigcup G$ is definable in $M[G]$ from parameters in M.*

Definition 6.64 Let $\mathbb{P} = (\mathbb{P}; \leq_{\mathbb{P}})$, $\mathbb{Q} = (\mathbb{Q}; \leq_{\mathbb{Q}})$ be partial orders. The product $\mathbb{P} \times \mathbb{Q}$ of $\mathbb{P}$, $\mathbb{Q}$ is defined to be $\mathbb{P} \times \mathbb{Q} = (\mathbb{P} \times \mathbb{Q}; \leq_{\mathbb{P}\times\mathbb{Q}})$, where for $(p, q), (p', q') \in \mathbb{P} \times \mathbb{Q}$ we set $(p, q) \leq_{\mathbb{P}\times\mathbb{Q}} (p', q')$ iff $p \leq_{\mathbb{P}} p'$ and $q \leq_{\mathbb{Q}} q'$.

Lemma 6.65 (Product Lemma) *Let M be a transitive model of* ZFC*, and let $\mathbb{P} = (\mathbb{P}; \leq_{\mathbb{P}})$ and $\mathbb{Q} = (\mathbb{Q}; \leq_{\mathbb{Q}})$ be partial orders in M. If G is $\mathbb{P}$-generic over M and H is $\mathbb{Q}$-generic over $M[G]$, then $G \times H$ is $\mathbb{P} \times \mathbb{Q}$-generic over M. On the other hand, if $K \subset \mathbb{P} \times \mathbb{Q}$ is $\mathbb{P} \times \mathbb{Q}$-generic over M, then, setting*

$$G = \{p \in \mathbb{P} : \exists q \in \mathbb{Q}\ (p, q) \in K\}, \text{ and}$$
$$H = \{q \in \mathbb{Q} : \exists p \in \mathbb{P}\ (p, q) \in K\}$$

G is $\mathbb{P}$-generic over M and H is $\mathbb{Q}$-generic over $M[G]$.

Proof First let G be $\mathbb{P}$-generic over M and H be $\mathbb{Q}$-generic over $M[G]$. It is clear that $G \times H$ is a filter. Let us show that $G \times H$ is $\mathbb{P} \times \mathbb{Q}$-generic over M. Let $D \subset \mathbb{P} \times \mathbb{Q}$ be dense. We need to see that $D \cap (G \times H) \neq \emptyset$.

Let $D^* = \{q \in \mathbb{Q} : \exists p \in G\ (p, q) \in D\}$. D^* is dense in $\mathbb{Q}$: Given $q \in \mathbb{Q}$, let $D' = \{p \in \mathbb{P} : \exists q' \leq_{\mathbb{Q}} q\ (p, q') \in D\}$. $D' \in M$ and D' is clearly dense in $\mathbb{P}$, so that there is some $p \in D' \cap G$. But then there is some $q' \leq_{\mathbb{Q}} q$ with $(p, q') \in D$, i.e., $q' \in D^*$ and $q' \leq_{\mathbb{Q}} q$.

Now $D^* \in M[G]$, and thus there is some $q \in D^* \cap H$. This means that there is some $p \in G$ with $(p, q) \in D$ and $(p, q) \in G \times H$.

Now let $K \subset \mathbb{P} \times \mathbb{Q}$ be $\mathbb{P} \times \mathbb{Q}$-generic over M, and set $G = \{p \in \mathbb{P} : \exists q \in \mathbb{Q}\ (p, q) \in K\}$ and $H = \{q \in \mathbb{Q} : \exists p \in \mathbb{P}\ (p, q) \in K\}$. Let $D \subset \mathbb{P}$ be dense in $\mathbb{P}$, where $D \in M$. Then $D' = \{(p, q) \in \mathbb{P} \times \mathbb{Q} \colon p \in D\}$ is clearly dense in $\mathbb{P} \times \mathbb{Q}$ and $D' \in M$, so that there is some $(p, q) \in D' \cap K$, i.e., $p \in D \cap G$. This shows that G is $\mathbb{P}$-generic over M.

Now let $D \subset \mathbb{Q}$ be dense in $\mathbb{Q}$, where $D \in M[G]$. Let $D = \tau^G$, and $p^* \Vdash_M^{\mathbb{P}}$ "τ is dense in $\check{\mathbb{Q}}$", where $p^* \in G$. Let

$$D^* = \{(p, q) \in \mathbb{P} \times \mathbb{Q} : p \leq p^* \wedge p \Vdash_M^{\mathbb{P}} \check{q} \in \tau\}.$$

D^* is dense below $(p^*, 1_{\mathbb{Q}})$: Given $p \leq_{\mathbb{P}} p^*, q \in \mathbb{Q}$, we have $p \Vdash_M^{\mathbb{P}}$ "τ is dense in $\check{\mathbb{Q}}$." There is then some $p' \leq_{\mathbb{P}} p$ and some $q' \leq_{\mathbb{Q}} q$ with $p' \Vdash_M^{\mathbb{P}}$ "$\check{q}' \in \tau$." Then $(p', q') \in D^*$ and $(p', q') \leq_{\mathbb{P} \times \mathbb{Q}} (p^*, 1_{\mathbb{Q}})$. Now let $(p, q) \in D^* \cap K$. Then $p \in G$ and $p \Vdash_M^{\mathbb{P}} \check{q} \in \tau$, so that $q \in \tau^G = D$. Therefore $q \in H \cap D$. □

In the situation of Lemma 6.65, G and H are called *mutually generic*.

Lemma 6.66 *Let M be a transitive model of* ZFC*, and let $\alpha \in M$. Let G be $\mathbb{C}(\alpha)$-generic over M, and let $x \in {}^\omega\omega \cap M[G]$. Then x is $\mathbb{C}$-generic over M in the following sense: there is some $\mathbb{C}$-generic $H \in M[G]$ over M such that $x \in M[H]$. In addition, if $\alpha \geq \omega_1^M$, then there is also some $\mathbb{C}(\alpha)$-generic $K \in M[G]$ over $M[H]$ such that $M[G] = M[H][K]$.*

Proof Fix $x \in {}^\omega\omega \cap M[G]$, say $x = \tau^G$. Because $M \models$ "$\mathbb{C}(\alpha)$ has the c.c.c", there is some $\sigma \in M^{\mathbb{C}(\alpha)}$ and there is a sequence $(A_n : n < \omega) \in M$ of countable antichains in $\mathbb{C}(\alpha)$ such that $\sigma^G = \tau^G$ and

$$(\rho, p) \in \sigma \Longrightarrow \rho = \ulcorner \check{n}, \check{m} \urcorner \text{ and} p \in A_n \text{ for some } n, m < \omega,$$

cf. the proof of Theorem 6.33. In particular, $X = \bigcup\{\text{supp}(p) : \exists \rho\ (\rho, p) \in \sigma\}$ is at most countable. Obviously,

$$\mathbb{C}(\alpha) \cong \{p \in {}^X\mathbb{C} : \text{Card}(\text{supp}(p)) < \aleph_0\} \times \{p \in^{\alpha \setminus X} \mathbb{C} : \text{Card}(\text{supp}(p)) < \aleph_0\}.$$

It is easy to verify that

$$\{p \in {}^X\mathbb{C} : \text{Card}(\text{supp}(p)) < \aleph_0\} \cong \mathbb{C}$$

and if $\alpha \geq \omega_1^M$, then

$$\{p \in {}^{\alpha \setminus X}\mathbb{C} : \mathrm{Card}(\mathrm{supp}(p)) < \aleph_0\} \cong \mathbb{C}(\alpha).$$

The rest is then immediate by the Product Lemma 6.65. □

Corollary 6.67 *Let M be a transitive model of* ZFC*, and let $\alpha \in M$, $\alpha \geq \omega_1^M$. Let G be $\mathbb{C}(\alpha)$-generic over M. Then in $M[G]$ there is no* $\mathsf{OD}_{\mathbb{R}}$*-wellordering of the reals.*

Proof Suppose that there is a formula $\varphi(v_0, v_1, v_2, \ldots, v_n, v_{n+1}, \ldots, v_{n+m})$ and there are $\gamma_2, \ldots, \gamma_n \in M \cap \mathrm{OR}$ and $x_{n+1}, \ldots, x_{n+m} \in {}^\omega\omega \cap M[G]$ such that

$$M[G] \models \text{“}\{(u, v) : \varphi(u, v, \gamma_2, \ldots, \gamma_n, x_{n+1}, \ldots, x_{n+m})\}\text{ is a wellordering of } {}^\omega\omega.\text{”}$$

Let $H \in M[G]$ be $\mathbb{C}$-generic over M such that $x_{n+1}, \ldots, x_{n+m} \in M[H]$, and let $K \in M[G]$ be $\mathbb{C}(\alpha)$-generic over $M[H]$ such that $M[G] = M[H][K]$. The choice of H and K is possible by Lemma 6.66.

Then every $x \in {}^\omega\omega \cap M[G] = {}^\omega\omega \cap M[H][K]$ is $\mathsf{OD}^{M[H][K]}_{\{x_{n+1},\ldots,x_{n+m}\}}$, so that by the homogeneity of $\mathbb{C}(\alpha)$ in $M[H]$ (cf. Lemma 6.53) every such x is in $M[H]$ (cf. Corollary 6.62). But this is nonsense! □

Theorem 6.68 (P. Cohen) *If* ZFC *is consistent, then so is* $\mathsf{ZF} + \neg\mathsf{AC}$.

Proof Let M be a countable transitive model of ZFC, and let G be $\mathbb{C}(\omega_1^M)$-generic over M. Let

$$N = \mathsf{HOD}^{M[G]}_{{}^\omega\omega \cap M[G]}.$$

We have that $N \models \mathsf{ZF}$ by Theorem 5.44. However, by Corollary 6.67 there is no wellorder of the reals in N. □

The following is a strengthening of Theorem 6.68. (Cf. also Problem 11.11.)

Theorem 6.69 (Feferman–Levy) *If* ZFC *is consistent, then so is* $\mathsf{ZF} + \mathrm{cf}(\omega_1) = \omega$.

Proof Let J_α be a countable model of ZFC. We aim to find a “symmetric extension” of J_α in which $\mathsf{ZF} + \mathrm{cf}(\omega_1) = \omega$ holds true.

Let $\kappa = \aleph_\omega^{J_\alpha}$. Let G be $Col(\omega, < \kappa)$-generic over J_α, and write $N = J_\alpha[G]$. With the help of Lemma 6.65, it is straightforward to see that $\omega_1^N = \kappa^{+J_\alpha}$. Now let

$$({}^\omega\kappa)^* = \bigcup \{{}^\omega\kappa \cap J_\alpha[G \restriction \xi] : \xi < \kappa\},$$

and set

$$M = \mathsf{HOD}^N_{({}^\omega\kappa)^* \cup \{({}^\omega\kappa)^*\}}.$$

By Theorem 5.44, $M \models \mathsf{ZF}$. We aim to verify that

$$\kappa = \omega_1^M \text{ and } M \models \text{cf}(\omega_1) = \omega. \tag{6.15}$$

Notice that for every $\xi < \kappa$, $J_\alpha[G \upharpoonright \xi + 1] \models$ "ξ is countable," so that there is some bijective $f : \omega \to \xi$ with $f \in ({}^\omega\kappa)^*$. In particular, every $\xi < \kappa$ is countable in M, i.e., $\kappa \leq \omega_1^M$. Moreover, because $(\aleph_n^{J_\alpha} : n < \omega) \in J_\alpha \cap {}^\omega\kappa \subset ({}^\omega\kappa)^*$, we have that $M \models \text{cf}(\kappa) = \omega$. In order to verify (6.15) it thus suffices to show that κ is a cardinal in M.

If κ were not a cardinal in M, then $M \models$ "κ is countable," and there would then be some bijection $f : \omega \to \kappa$ with $f \in M$. Such a bijection cannot be an element of $({}^\omega\kappa)^*$. This is because if $f \in J_\alpha[G \upharpoonright \xi]$, say, where $\xi < \kappa$, then $G \upharpoonright \xi$ is $\text{Col}(\omega, <\xi)$-generic over J_α by the Product Lemma 6.65 and every J_α-cardinal above ξ (in particular, κ) will remain a cardinal in $J_\alpha[G \upharpoonright \xi]$ by Lemma 6.32.

In order to show (6.15), it thus suffices to verify that

$$^\omega\kappa \cap M = ({}^\omega\kappa)^*. \tag{6.16}$$

To this end, let $f \in {}^\omega\kappa \cap M$. There is then a formula $\varphi(v_0, v_1, v_2, \ldots, v_n, v_1', \ldots, v_m', v)$, there are ordinals $\gamma_2, \ldots, \gamma_n < \alpha$, and there are $f_1, \ldots, f_m \in ({}^\omega\kappa)^*$ such that for all $(n, \eta) \in \omega \times \kappa$,

$$f(n) = \eta \iff N \models \varphi(n, \eta, \gamma_2, \ldots, \gamma_n, f_1, \ldots, f_m, ({}^\omega\kappa)^*).$$

Let $\xi < \kappa$ be such that $f_1, \ldots, f_m \in J_\alpha[G \upharpoonright \xi]$. By the Product Lemma 6.65, $G \upharpoonright \xi$ is $\text{Col}(\omega, <\xi)$-generic over J_α, and $G \upharpoonright [\xi, \kappa)$ is $\text{Col}(\omega, [\xi, \kappa))$-generic over $J_\alpha[G \upharpoonright \xi]$.

Claim 6.70 *There is some $\tau \in J_\alpha[G \upharpoonright \xi]$ such that $\tau^{G \upharpoonright [\xi,\kappa)} = ({}^\omega\kappa)^*$ and τ is homogenous for* $\text{Col}(\omega, [\xi, \kappa))$.

Proof The proof for $\xi > \omega$ is only notationally different from the proof for $\xi = \omega$, so let us assume that $\xi = \omega$. I.e., we assume that $f_1, \ldots, f_m \in J_\alpha$.

Let $\lambda = \kappa^{+J_\alpha} = (\aleph_{\omega+1})^{J_\alpha}$, and let

$$\tau = \{(\sigma, p) : \sigma \in J_\lambda \wedge \exists \delta < \kappa \; p \Vdash_{J_\alpha}^{\text{Col}(\omega,<\kappa)} \sigma \in {}^\omega\kappa \cap J_\alpha[\dot{G} \cap \text{Col}(\omega, <\check{\delta})]\}.$$

Let us verify that $\tau^G = ({}^\omega\kappa)^*$. First let $f \in ({}^\omega\kappa)^*$, say $f = \sigma^G$. We may assume that $\sigma \in J_\lambda$, cf. the proof of Theorem 6.33. As $f \in ({}^\omega\kappa)^*$, there is some $\delta < \kappa$ such that $f \in {}^\omega\kappa \cap J_\alpha[G \cap \text{Col}(\omega, <\delta)]$, and there is then some $p \in G$ such that $p \Vdash_{J_\alpha}^{\text{Col}(\omega,<\kappa)} \sigma \in {}^\omega\kappa \cap J_\alpha[\dot{G} \cap \text{Col}(\omega, <\delta)]$. But then $(\sigma, p) \in \tau$, so that $f = \sigma^G \in \tau^G$. Now let $f \in \tau^G$, say $f = \sigma^G$, where $(\sigma, p) \in \tau$ for some $p \in G$. There is then some $\delta < \mu$ with $p \Vdash_{J_\alpha}^{\text{Col}(\omega,<\kappa)} \sigma \in {}^\omega\kappa \cap J_\alpha[\dot{G} \cap \text{Col}(\omega, <\check{\delta})]$, and hence $\sigma^G \in {}^\omega\kappa \cap J_\alpha[G \cap \text{Col}(\omega, <\delta)]$, i.e., $f = \sigma^G \in ({}^\omega\kappa)^*$.

Let us now verify that $\mathrm{Col}(\omega, <\kappa)$ is homogenous with respect to τ. Let $p, q \in \mathrm{Col}(\omega, <\kappa)$ be given. We may then define $\pi : \mathrm{Col}(\omega, <\kappa) \to \mathrm{Col}(\omega, <\kappa)$ as in (6.14) in the proof of Lemma 6.54 (where $\mu = \omega$ and $X = [\omega, \kappa)$).

We have that $\tilde{\pi}(\dot{G}) = \{(\tilde{\pi}(\check{p}), \pi(p)) : p \in \mathrm{Col}(\omega, <\kappa)\} = \{(\check{p}, \pi(p)) : p \in \mathrm{Col}(\omega, <\kappa)\}$, as π is an automorphism, so that $\tilde{\pi}(\dot{G})^G = \{p : \pi(p) \in G\} = \pi^{-1}"G$, where $(\pi^{-1}"G) \cap \mathrm{Col}(\omega, <\delta)$ is $\mathrm{Col}(\omega, <\delta)$ generic over J_α for every $\delta < \kappa$ by Lemma 6.48 (and the definition of π). Also, $J_\alpha[(\pi^{-1}"G) \cap \mathrm{Col}(\omega, <\delta)] = J_\alpha[G \cap \mathrm{Col}(\omega, <\delta)]$ for every $\delta < \kappa$ by Lemma 6.48, which is certainly true independently from the particular choice of G, so that in fact

$$1_{\mathrm{Col}(\omega,<\kappa)} \Vdash_{J_\alpha}^{\mathrm{Col}(\omega,<\kappa)} J_\alpha[\tilde{\pi}(\dot{G}) \cap \mathrm{Col}(\omega, <\delta)] = J_\alpha[\dot{G} \cap \mathrm{Col}(\omega, <\delta)]. \tag{6.17}$$

But we may now show $\tilde{\pi}(\tau) = \tau$ as follows. With the help of (6.17), we have

$$\begin{aligned}
(\sigma, p) \in \tau &\Longleftrightarrow \sigma \in J_\lambda \wedge \exists\delta < \kappa \; p \Vdash_{J_\alpha}^{\mathrm{Col}(\omega,<\kappa)} \sigma \in J_\alpha[\dot{G} \cap \mathrm{Col}(\omega, <\delta)] \\
&\Longleftrightarrow \tilde{\pi}(\sigma) \in J_\lambda \wedge \exists\delta < \kappa \; \pi(p) \Vdash_{J_\alpha}^{\mathrm{Col}(\omega,<\kappa)} \tilde{\pi}(\sigma) \in J_\alpha[\tilde{\pi}(\dot{G}) \cap \mathrm{Col}(\omega, <\delta)] \\
&\Longleftrightarrow \tilde{\pi}(\sigma) \in J_\lambda \wedge \exists\delta < \kappa \; \pi(p) \Vdash_{J_\alpha}^{\mathrm{Col}(\omega,<\kappa)} \tilde{\pi}(\sigma) \in J_\alpha[\dot{G} \cap \mathrm{Col}(\omega, <\delta)] \\
&\Longleftrightarrow (\tilde{\pi}(\sigma), \pi(p)) \in \tau.
\end{aligned}$$

Therefore $\tau = \{(\sigma, p) : (\sigma, p) \in \tau\} = \{(\tilde{\pi}(\sigma), \pi(p)) : (\sigma, p) \in \tau\} = \tilde{\pi}(\tau)$, as desired. This shows Claim 6.70. □

By Claim 6.70 and Corollary 6.62 we get that in fact $f \in J_\alpha[G \restriction \xi]$. Hence $f \in ({}^\omega\kappa)^*$. We verified (6.16). □

Elaborate forcings are studied e.g. in [9, 24, 37] and [44].

6.3 Problems

6.1. Let $(\mathbb{P}; \leq)$ be a partial order, and let D be dense in $\mathbb{P}$. Use the HAUSDORFF Maximality Principle 2.11 to construct an antichain $A \subset D$ such that $\forall p \in D \exists q \in A \; q \parallel p$. Conclude that A is a maximal antichain in $\mathbb{P}$.

6.2. Prove Lemma 6.21!
In what follows, we shall always assume that M is a (countable, if convenient) transitive model of ZFC, $\mathbb{P} = (\mathbb{P}, \leq) \in M$ is a partial order, and G is $\mathbb{P}$-generic over M.

6.3. Let $\kappa > \mathrm{Card}(\mathbb{P})$ be a regular cardinal in M (and hence in $M[G]$). Show that $(M, M[G])$ has the *κ-approximation property* which means that if $A \subset \kappa$, $A \in M[G]$, is such that $A \cap \xi \in M$ for all $\xi < \kappa$, then $A \in M$.

6.4. Show that if N is a transitive model of ZFC^- and if H is $\mathbb{Q}$-generic over N, where $\mathbb{Q} \in N$ is a partial order, then $N[H] \models \mathsf{ZFC}^-$.

6.5. Suppose that $p \Vdash^{\mathbb{P}}_{M} \exists x \varphi(x, \tau_1, \ldots, \tau_n)$. Show that there is some $\tau \in M^{\mathbb{P}}$ such that $p \Vdash \varphi(\tau, \tau_1, \ldots, \tau_n)$. (This is called "*fullness*.") Let $X \prec (H_\theta)^M$, where θ is regular in M and $\mathbb{P} \in X$. Let $X[G] = \{\tau^G : \tau \in M^{\mathbb{P}} \cap X\}$ and $(H_\theta)^M[G] = \{\tau^G : \tau \in M^{\mathbb{P}} \cap (H_\theta)^M\}$. Show that $(H_\theta)^M[G] = (H_\theta)^{M[G]}$ and (using fullness)

$$X[G] \prec (H_\theta)^{M[G]}.$$

6.6. Prove Lemma 6.31!

6.7. Let H be $\mathbb{C}$-generic over M. Let $s \in {}^{<\omega}\omega$, and let $H_s = \{(p \upharpoonright [\mathrm{dom}(s), \mathrm{dom}(p))) \cup (s \upharpoonright \mathrm{dom}(p)) : p \in H\}$. Show that H_s is $\mathbb{C}$-generic over M.

6.8. Let κ be an infinite cardinal of M, and write $\lambda = (\kappa^{<\kappa})^M$. Let H be $(\mathbb{C}_\kappa)^M$-generic over M. Show that in $M[H]$, there is a surjection $f : \kappa \to \lambda$. (Cf. the proof of Lemma 6.42.) Conclude that $(\mathbb{C}_\kappa)^M$ collapses exactly the M-cardinals in the half-open interval $(\kappa, \lambda]$.

6.9 Assume M to satisfy GCH. Let $\kappa \in M$ be an M-cardinal such that $M \models \mathrm{cf}(\kappa) > \omega$. Show that if H is $\mathbb{C}(\kappa)$-generic over M, then $M[H] \models 2^{\aleph_0} = \kappa$. More generally, show that if μ is an infinite regular cardinal in M, $\kappa \in M$ is an M-cardinal with $M \models \mathrm{cf}(\kappa) > \mu$, and if H is $\mathbb{C}_\mu(\kappa)$-generic over M then M and $M[H]$ have the same cardinals and $M[H] \models 2^\mu = \kappa$.

6.10. Show that the converse to Lemma 6.38 is also true, i.e., if $\mathbb{P}$ is separative and

$${}^{<\kappa}M \cap M[G] = {}^{<\kappa}M \cap M,$$

then $\mathbb{P}$ is $< \kappa$-distributive in M.

6.11. Let M be a transitive model of ZFC such that if $\alpha = M \cap \mathrm{OR}$, then $\mathrm{Card}(\alpha) = \aleph_1$. Show that there is a transitive model M' of ZFC with $M' \cap \mathrm{OR} = \alpha$ and $M' \neq M$.

6.12 **(Solovay)** Let us assume G and K to be mutually $\mathbb{P}$-generic over M. Show that $M[G] \cap M[K] = M$. [Hint. Let $\tau^G = \sigma^K$, where τ and σ are $\mathbb{P}$-names. We may also construe τ and σ as $(\mathbb{P} \times \mathbb{P})$-names, and we may pick $(p, q) \in \mathbb{P} \times \mathbb{P}$ such that $(p, q) \Vdash \tau = \sigma$. Show that for every $y \in M$, p decides "$\check{y} \in \tau$," i.e., $p \Vdash_{\mathbb{P}} \check{y} \in \tau$ or $p \Vdash_{\mathbb{P}} \check{y} \notin \tau$.]

6.13. Let κ be inaccessible in M. Show that if H is $\mathbb{C}(\kappa)$-generic over M, then κ is weakly inaccessible in $M[H]$.

6.14. **(R. Solovay)** Suppose that $M \models \omega_1^L = \omega_1$ and $A \subset \omega_1^M$, $A \in M$. Show that there is some poset $\mathbb{R}$ which has the c.c.c. such that if H is $\mathbb{R}$-generic over M, then in $M[H]$ there is some $x \subset \omega$ with $M[H] \models A \in L[x]$. [Hint. First work in M. Let $\{x_i : i < \omega_1\} \in L$ be an almost disjoint collection of subsets of ω, cf. Problem 1.3. Let

$$\mathbb{R} = \{(s, t) : s \in {}^{<\omega}2, t \in [\omega_1]^{<\omega}\},$$

ordered by $(s', t') \leq (s, t)$ iff $s' \supset s$, $t' \supset t$, and if $i \in t \cap A$, then

$$\{n \in \text{dom}(s') \setminus \text{dom}(s) \colon s'(n) = 1\} \cap x_i = \emptyset.$$

Show that $\mathbb{R}$ has the c.c.c. Now stepping out of M, if H is $\mathbb{R}$-generic over M and if $x \subset \omega$ is such that its characteristic function is $\bigcup\{s \colon \exists t\, (s, t) \in G\}$, then $i \in A$ iff x and x_i are almost disjoint.]

6.15. (a) Suppose that $M \models$ "$S \subset \omega_1$ is stationary." Let $M \models$ "$\mathbb{P}$ has the c.c.c. or is ω-closed. Show that $M[G] \models$ "S is stationary."

(b) Let $M \models$ "$S \subset \kappa$ is stationary, where κ is uncoutable and regular, and $\text{cf}(\alpha) = \omega$ for all $\alpha \in S$." Suppose also that $M \models$ "$\mathbb{P}$ is ω-closed." Then $M[G] \models$ "S is stationary." [Hint. Fix $p \in \mathbb{P}$ such that $p \Vdash^{\mathbb{P}}_M \tau$ is club in κ. In M pick some nice $X \prec H_\theta$ with $p \in X$ and $\sup(X \cap \kappa) \in S$. Pick $(\alpha_n \colon n < \omega)$ cofinal in $\alpha = \sup(X \cap \kappa)$. Construct $(p_n \colon n < \omega)$, a decreasing sequence of conditions in X, such that $p_n \Vdash \tau \setminus \check{\alpha}_n \neq \emptyset$. Let q be stronger than all p_n. Then $q \Vdash \check{\alpha} \in \tau \cap \check{S}$.]

6.16. Assume $M \models$ "$S \subset \omega_1$ is stationary." Show that there is some ω-distributive forcing $\mathbb{Q} \in M$ such that if H is $\mathbb{Q}$-generic over M, then S contains a club in $M[H]$. [Hint. In M, let

$$\mathbb{Q} = \{p \colon \exists \alpha < \omega_1\, (p \text{ is a closed subset of } S, \text{otp}(p) = \alpha + 1)\},$$

ordered by end-extension.] Show that in fact if $T \subset S$ is stationary in M, then T is still stationary in $M[H]$.

More generally, let κ be an infinite regular cardinal in M, and let $M \models$ "$S \subset \kappa^+$ is stationary and $< \kappa$-closed." Show that there is a $< \kappa$-closed κ-distributive forcing $\mathbb{Q} \in M$ such that if H is $\mathbb{Q}$-generic over M, then S contains a club in $M[H]$.

6.17. **(J. Silver)** Let H, H^* be transitive models of a sufficiently large fragment of ZFC, let $\mathbb{P} \in H$ be a partial order, let $\sigma \colon H \to H^*$ be an elementary embedding, let G be $\mathbb{P}$-generic over H, and let K be $\sigma(\mathbb{P})$-generic over H^* such that $\sigma''G \subset H$. There is then an elementary embedding $\tilde{\sigma} \colon H[G] \to H^*[K]$ such that $\tilde{\sigma} \supset \sigma$.

6.18. Let X be a large cardinal concept, e.g., $X =$ inaccessible, measurable, etc. We say that "κ is an X-cardinal" is preserved by small forcing iff the following holds true. Let κ be an X-cardinal in M, and assume $\mathbb{P} \in (V_\kappa)^M$ to be a poset. Then κ is still an X-cardinal in $M[G]$. Show that the following statements are preserved by small forcing. "κ is inaccessible," "κ is Mahlo," "κ is weakly compact," and "κ is measurable." [Hint: To prove that "κ is measurable" is preserved by small forcing, let U be any measure on κ in M. Show that

$$U^* = \{Y \subset \kappa \colon \exists X \in U\; Y \supset X\},$$

as defined in $M[G]$, witnesses that κ is still measurable in $M[G]$.]

6.19. In M, let κ be regular, and let F be a non-trivial filter on κ. For $X, Y \in F^+$ (in the sense of Definition 4.30) let us write $X \leq_{F^+} Y$ iff

$$\kappa \setminus (X \setminus Y) = (\kappa \setminus X) \cup Y \in F.$$

Let G be $(F^+, \leq_{F^+})$-generic over M. Show that in $M[G]$, G is a non-trivial M-ultrafilter which extends F in the following sense.

(a) $F \subset G$,
(b) if $X, Y \in G$, then $X \cap Y \in G$,
(c) if $X \in G$ and $Y \supset X$, $Y \in \mathscr{P}(\kappa) \cap M$, then $Y \in G$,
(d) $\emptyset \notin G$, and
(e) if $X \in \mathscr{P}(\kappa) \cap M$, then either $X \in G$ or $\kappa \setminus X \in G$.

We may then make sense of $\text{Ult}(M; G) \subset M[G]$.
If F is θ-closed in M ($\theta < \kappa$), then G is M-θ-closed in the sense that if $(X_i : i < \theta) \in M$, $X_i \in G$ for all $i < \theta$, then $\bigcap_{i<\theta} X_i \in G$. If F is normal in M, then G is M-normal in the sense that if $(X_i : i < \kappa) \in M$, $X_i \in G$ for all $i < \kappa$, then $\Delta_{i<\kappa} X_i \in G$.

6.20. Let $\kappa \geq \aleph_0$. Recall that F_{κ^+} is the club filter on κ^+, and write $F = F_{\kappa^+}$. *Assume* that $(F^+, \leq_{F^+})$ has the κ^{++}-c.c. Let

$$\pi : H \cong X \prec H_{(2^{\kappa^+})^+},$$

where $\kappa + 1 \subset X$, $\text{Card}(X) = \kappa$, $(F^+, \leq_{F^+}) \in X$, and H is transitive. Let $\pi(\delta) = \kappa^+$ and $\pi(\mathbb{Q}) = (F^+, \leq_{F^+})$. Write $g = \{X \subset \kappa^+ : \delta \in \pi(X)\}$. Show that g is $\mathbb{Q}$-generic over H. [Hint. Show that if $\{A_i : i < \kappa^+\} \in \text{ran}(\pi)$ is a maximal antichain in $(F^+, \leq_{F^+})$, then

$$\{\xi < \kappa^+ : \xi \in \bigcup_{i<\xi} A_i\}$$

contains a club C in $\text{ran}(\pi)$, so that $\delta \in C$.] Conclude that if $M \models$ "$(F^+, \leq_{F^+})^M$ has the κ^{++}-c.c.," and if H is $(F^+, \leq_{F^+})^M$-generic over M, then $\text{ult}(M; H)$ is well-founded.

6.21. (**Petr Vopěnka**) Show that for every α there is some partial order $\mathbb{V} = \mathbb{V}(\alpha) \in \mathsf{HOD}$ such that for every $A \subset \alpha$, $A \in V$, there is some $G \in V$ such that G is $\mathbb{V}$-generic over HOD and $A \in \mathsf{HOD}[G]$.
[Hint. Let $D = \{Y \subset \mathscr{P}(\alpha) : Y \text{ is } \mathsf{OD}\}$, and let f be OD such that $f : \theta \to D$ is bijective. Let $\leq$ be such that $(\theta, \leq) \overset{\pi}{\cong} (D, \subset)$, so that $\mathbb{V} = (\theta, \leq) \in \mathsf{HOD}$. Show that

$$G = \{\xi < \theta : A \in \pi(\xi)\}$$

is $\mathbb{V}$-generic over HOD and $A \in \mathsf{HOD}[G]$.]

6.22. Let g be a group of automorphisms of $\mathbb{P}$. Let F be a collection of subgroups of g which satisfies the following.

(a) $g \in F$,
(b) if $H \in F$, and H' is a subgroup of G with $H \subset H'$, then $H' \in F$,
(c) if $H, H' \in F$, then $H \cap H' \in F$, and
(d) if $H \in F$ and $\pi \in g$, then $\pi \circ H \circ \pi^{-1} = \{\pi \circ \sigma \circ \pi^{-1} : \sigma \in H\} \in F$.

(Such an F is called a *filter* on g.) For $\tau \in V^{\mathbb{P}}$ let us write

$$\mathrm{sym}_g(\tau) = \{\pi \in g : \pi(\tau) = \tau\}.$$

A name $\tau \in V^{\mathbb{P}}$ is called *symmetric* iff $\mathrm{sym}_g(\tau) \in F$, and τ is called *hereditarily symmetric* iff for every (finite) sequence $((\tau_i, p_i) : 0 \leq i \leq n)$ such that $\tau_0 = \tau$ and $(\tau_{i+1}, p_{i+1}) \in \tau_i$ for $0 \leq i < n$ we have that all τ_i, $0 \leq i \leq n$, are symmetric.

Now let

$$N = \{\tau^G : M \models \tau \in M^{\mathbb{P}} \text{ is hereditarily symmetric}\}.$$

Show that N is a model of ZF.

Chapter 7
Descriptive Set Theory

7.1 Definable Sets of Reals

Descriptive set theory is the study of definable sets of real numbers. However, rather than working with $\mathbb{R}$, descriptive set theorists often work with a space which can be shown to be homeomorphic to the space of all irrational numbers.

Let $X \neq \emptyset$ be an arbitrary set. If $s \in {}^{<\omega}X$ then we declare $U_s = \{x \in {}^{\omega}X : s \subset x\}$ to be a *basic open* set. $A \subset {}^{\omega}X$ is declared to be *open* iff A is the union of basic open sets. (As $\emptyset = \bigcup \emptyset$, $\emptyset$ is also open.) Complements of open sets are called *closed*. Notice that each U_s is also closed, because

$${}^{\omega}X \setminus U_s = \bigcup \{U_t : lh(t) = lh(s) \wedge t \neq s\}.$$

Here, $lh(t) = dom(t) = \overline{\overline{t}}$ is the *length of* t. If $X = \omega$, then the space ${}^{\omega}\omega$, together with the topology just defined, is called the BAIRE *space*. We shall often refer to the elements of ${}^{\omega}\omega$ as "reals." If $X = \{0, 1\}$, then the space ${}^{\omega}2$, together with the topology just defined, is called the CANTOR *space*. In this chapter, we shall focus our attention on the BAIRE space, but most statements carry over, *mutatis mutandis*, to the CANTOR space.

If $x, y \in {}^{\omega}\omega$, $x \neq y$, then their *distance* $d(x, y)$ is defined to be $\frac{1}{2^n}$, where n is least such that $x(n) \neq y(n)$. It is easy to see now that the topology we defined on ${}^{\omega}\omega$ is exactly the one which is induced by the distance function d, so that ${}^{\omega}\omega$ is a *Polish space*, i.e. a complete seperable metric space. (Cf. Problem 7.1.)

A *tree* T *on* X is a subset of ${}^{<\omega}X$ which is closed under initial segments, i.e., if $s \in T$ and $n \leq lh(s)$, then $s \upharpoonright n \in T$. Then $(T, \subset\upharpoonright T)$ is a tree in the sense of Definition 4.43. If T is a tree on X, then we write $[T]$ for the set of all $x \in {}^{\omega}X$ such that $x \upharpoonright n \in T$ for all $n < \omega$. A tree T on X is called *perfect* iff $T \neq \emptyset$ and whenever $s \in T$, then s has $\overline{\overline{X}}$ pairwise incompatible extensions t in T, i.e., there is $(t_i : i < \overline{\overline{X}})$ such that for all $i, j < \overline{\overline{X}}$, $s \subset t_i$, $t_i \in T$, $t_i \not\subset t_j$, and $t_j \not\subset t_i$. Perfect trees admit a CANTOR–BENDIXSON analysis (cf. p. 5), cf. Problem 7.5.

R. Schindler, *Set Theory*, Universitext, DOI: 10.1007/978-3-319-06725-4_7,

If $T \neq \emptyset$ is a tree on ω, then T is perfect iff whenever $s \in T$, then there are t, $t' \in T$ with $s \subset t$, $s \subset t'$, $t \not\subset t'$, and $t' \not\subset t$. Recall that a set of reals A is called *perfect* iff $A \neq \emptyset$, A is closed, and every element of A is an accumulation point of A, cf. Definition 1.7.

Lemma 7.1 *If $A \subset {}^{\omega}\omega$ is closed, then $A = [T]$ for some tree T on ω. On the other hand, if T is a tree on ω then $[T]$ is closed.*

Moreover, $A \subset {}^{\omega}\omega$ is perfect iff $A = [T]$ for some tree $T \neq \emptyset$ on ω which is perfect.

Proof We show the first part of the lemma. Let $A \subset {}^{\omega}\omega$ be closed. Let $T = \{s \in {}^{<\omega}\omega \colon \exists x \in A\, s \subset x\}$. It is easy to see that $A \subset [T]$. Let $x \in [T]$. For each $n < \omega$ there is some $x_n \in A$ with $x \upharpoonright n \subset x_n$. But then $x = \lim_{n\to\infty} x_n \in A$, because A is closed. Therefore, $A = [T]$.

It is easy to verify that $[T]$ is closed whenever T is a tree on ω.

The second part of the lemma is easy to check. □

If $A \subset {}^{\omega}\omega$, then

$$\bigcap\{B \subset {}^{\omega}\omega \colon B \supset A \wedge B \text{ is closed}\}$$

is the smallest closed set in which A is contained, called the *closure of* A.

A *σ-algebra on* a set Y is a collection $S \subset \mathscr{P}(Y)$ which is closed under relative complements as well as countable unions and intersections.

Definition 7.2 A set $A \subset {}^{\omega}\omega$ is called Borel iff A is in the smallest σ-algebra containing all closed (open) subsets *of* ${}^{\omega}\omega$.

The simplest Borel sets are the open and closed sets. Countable intersections of open sets are often called G_δ- and countable unions of closed sets F_σ-sets. The Borel sets form a natural hierarchy, cf. Problem 7.3.

Let $\alpha \geq \omega$. A *tree T on $\omega \times \alpha$* is a set of pairs (s, t) with $s \in {}^{<\omega}\omega$, $t \in {}^{<\omega}\alpha$, and $lh(s) = lh(t)$, such that T is closed under initial segments, i.e. if $(s, t) \in T$ and $n \leq lh(s)$ then $(s \upharpoonright n, t \upharpoonright n) \in T$. If T is a tree on $\omega \times \alpha$, then we write $[T]$ for the set of all $(x, y) \in {}^{\omega}\omega \times {}^{\omega}\alpha$ such that $(x \upharpoonright n, y \upharpoonright n) \in T$ for all $n < \omega$. If T is a tree on $\omega \times \alpha$ then $p[T]$, the *projection of* T, is the set of all $x \in {}^{\omega}\omega$ such that there is some $f \in {}^{\omega}\alpha$ so that for all n, $(x \upharpoonright n, f \upharpoonright n) \in T$. If $x \in {}^{\omega}\omega$ then we let T_x denote the set of all $t \in {}^{<\omega}\alpha$ such that $(x \upharpoonright lh(t), t) \in T$. Obviously,

$$x \in p[T] \Longleftrightarrow \exists y\, (x, y) \in [T] \Longleftrightarrow (T_x, \supset) \text{ is ill-founded.} \tag{7.1}$$

Definition 7.3 A set $A \subset {}^{\omega}\omega$ is called *analytic* iff there is a tree T on $\omega \times \omega$ with $A = p[T]$. $A \subset {}^{\omega}\omega$ is called *coanalytic* iff ${}^{\omega}\omega \setminus A$ is analytic.

We will show below, cf. Lemma 7.11, that there are analytic sets which are not Borel. A classical result of Souslin says that a set of reals is Borel if and only if it is analytic as well as coanalytic, cf. Theorem 7.5.

Lemma 7.4 *Let $A \subset {}^\omega\omega$ be a* BOREL*-set. Then A is analytic (and hence also coanalytic).*

Proof Let us first show the following two statements.

1. If every A_n, $n < \omega$, is analytic, then so is $\bigcup_{n<\omega} A_n$.
2. If every A_n, $n < \omega$, is analytic, then so is $\bigcap_{n<\omega} A_n$.

(1): Let $A_n = p[T_n]$, where T_n is a tree on $\omega \times \omega$. Let T on $\omega \times \omega$ be defined by[1]

$$(s,t) \in T \text{ iff } s = t = \emptyset \vee \exists n < \omega \, \exists t' \, (t = n^\frown t' \wedge (s \upharpoonright (\text{lh}(s) - 1), t') \in T_n).$$

It is straightforward to verify that $p[T] = \bigcup_{n<\omega} A_n$.

(2): Again let $A_n = p[T_n]$, where T_n is a tree on $\omega \times \omega$, and let $e \colon \omega \times \omega \to \omega$ be bijective such that $e(n,k) \leq e(n,l)$ whenever $k \leq l$. If $t \in {}^{<\omega}\omega$, say $t = (m_0, \ldots, m_{i-1})$, and $n < \omega$, then we write t^n for $(m_{e(n,0)}, \ldots, m_{e(n,k-1)})$, where k is least with $e(n,k) \geq i$. (If $e(n,0) \geq i$, then $t^n = \emptyset$.) We now let T on $\omega \times \omega$ be defined by

$$(s,t) \in T \text{ iff } \text{lh}(s) = \text{lh}(t) \wedge \forall n < \omega (s \upharpoonright \text{lh}(t^n), t^n) \in T_n.$$

It is straightforward to verify that $p[T] = \bigcap_{n<\omega} A_n$.

Now Lemma 7.1 quite trivially yields that every closed set is analytic. In particular, every basic open set U_s is analytic, and hence by (1) every open set is analytic. (1) and (2) then imply that every BOREL set is analytic. □

Theorem 7.5 (Souslin) *Let $A \subset {}^\omega\omega$. Then A is* BOREL *if and only if A is analytic as well as coanalytic.*

This theorem readily follows from Lemma 7.4 and the following one, Lemma 7.6. If A and B are disjoint sets, then we say that C *separates* A and B iff $C \supset A$ and $C \cap B = \emptyset$.

Lemma 7.6 *Let $A, B \subset {}^\omega\omega$ disjoint analytic sets. Then A, B are separable by a* BOREL *set C.*

Proof Let $A = p[T]$, $B = p[U]$, where T, U are trees on ω^2. For $s, t \in {}^{<\omega}\omega$, let

$$A_t^s = \{x \in A \colon s \subset x \wedge \exists y (t \subset y \wedge (x,y) \in [T])\},$$

and let

$$B_t^s = \{x \in B \colon s \subset x \wedge \exists y (t \subset y \wedge (x,y) \in [U])\}.$$

We have that $A_\emptyset^\emptyset = A$ and $B_\emptyset^\emptyset = B$, and we always have

[1] Here, $n^\frown t'$ is that sequence which starts with n, followed by $t'(0), \ldots, t'(\text{lh}(t') - 1)$. This notation as well as self-explaining variants thereof will frequently be used in what follows.

$$(s,t) \in T \Longrightarrow A^s_t = \bigcup\{A^{s^\frown n}_{t^\frown m} : (s^\frown n, t^\frown m) \in T\}$$

and

$$(s,t) \in U \Longrightarrow B^s_t = \bigcup\{B^{s^\frown n}_{t^\frown m} : (s^\frown n, t^\frown m) \in U\}.$$

Let us assume that A, B are not separable. We aim to derive a contradiction.

Let us define four reals x, y, u, v. We shall define $x(n)$, $y(n)$, $u(n)$, $v(n)$ recursively. We shall inductively maintain that $(x \restriction n, y \restriction n) \in T$, $(u \restriction n, v \restriction n) \in U$, and $A^{x \restriction n}_{y \restriction n}$, $B^{u \restriction n}_{v \restriction n}$ are not separable, which is true for $n = 0$.

Now suppose that $x \restriction n, y \restriction n, u \restriction n, v \restriction n$ have been chosen in such a way that $(x \restriction n, y \restriction n) \in T$, $(u \restriction n, v \restriction n) \in U$, and $A^{x \restriction n}_{y \restriction n}$, $B^{u \restriction n}_{v \restriction n}$ are not separable.

Assume that for all $i, j, k, l < \omega$ such that $(x \restriction n^\frown i, y \restriction n^\frown j) \in T$ and $(u \restriction n^\frown k, v \restriction n^\frown l) \in U$ there is a BOREL set $C^{i,k}_{j,l}$ separating

$$A^{x \restriction n^\frown i}_{y \restriction n^\frown j} \text{ and } B^{u \restriction n^\frown k}_{v \restriction n^\frown l},$$

i.e.,

$$A^{x \restriction n^\frown i}_{y \restriction n^\frown j} \subset C^{i,k}_{j,l} \subset {}^\omega\omega \setminus B^{u \restriction n^\frown k}_{v \restriction n^\frown l}.$$

It is then easy to see that the BOREL set

$$\bigcup_{i,j} \bigcap_{k,l} C^{i,k}_{j,l} \text{ separates } A^{x \restriction n}_{y \restriction n}, B^{u \restriction n}_{v \restriction n},$$

i.e.,

$$A^{x \restriction n}_{y \restriction n} \subset \bigcup_{i,j} \bigcap_{k,l} C^{i,k}_{j,l} \subset {}^\omega\omega \setminus B^{u \restriction n}_{v \restriction n}.$$

There must hence be $i, j, k, l < \omega$ such that

$$(x \restriction n^\frown i, y \restriction n^\frown j) \in T, (u \restriction n^\frown k, v \restriction n^\frown l) \in U,$$

and

$$A^{x \restriction n^\frown i}_{y \restriction n^\frown j} \text{ and } B^{u \restriction n^\frown k}_{v \restriction n^\frown l}$$

cannot be separated by a BOREL set. Set $x(n) = i$, $y(n) = j$, $u(n) = k$, $v(n) = l$.

Now of course $\bigcap_n A^{x \restriction n}_{y \restriction n} = \{x\}$ and $\bigcap_n B^{u \restriction n}_{v \restriction n} = \{u\}$. We have that $x \in A$, as witnessed by $(x, y) \in [T]$, and we have that $u \in B$ as witnessed by $(u, v) \in [U]$. As A, B are disjoint, $x \neq u$, and we may pick two disjoint open sets F, G such that $x \in F$ and $y \in G$. Because F is open, $A^{x \restriction n}_{y \restriction n} \subset F$ for all but finitely many n. For the same reason, $B^{u \restriction n}_{v \restriction n} \subset G$ for all but finitely many n. In particular, there is some $n < \omega$ such that

$$A^{x \restriction n}_{y \restriction n} \subset F \subset {}^{\omega}\omega \setminus G \subset {}^{\omega}\omega \setminus B^{u \restriction n}_{v \restriction n}.$$

So $A^{x \restriction n}_{y \restriction n}$ and $B^{u \restriction n}_{v \restriction n}$ can be separated by a BOREL (in fact, open) set after all. Contradiction! □

Definition 7.7 Let $A \subset {}^{\omega}\omega$, and let $\alpha \geq \omega$ be an ordinal. If $A = p[T]$, where T is a tree on $\omega \times \alpha$, then A is called *α-Souslin.*

The $\aleph_0$-Souslin sets are hence exactly the analytic sets.

Lemma 7.8 *If $A \subset {}^{\omega}\omega$ is coanalytic, then A is $\aleph_1$-Souslin.*

Proof Let ${}^{\omega}\omega \setminus A = p[T]$, where T is a tree on $\omega \times \omega$. Therefore, $x \in A$ iff $(T_x, \supset)$ is well-founded. If $T_x = (T_x, \supset)$ is well-founded, then, as $\overline{\overline{T_x}} = \aleph_0$, it can be ranked by some function $f \colon T_x \to \omega_1$ such that

$$s \supsetneq t \Longrightarrow f(s) < f(t) \tag{7.2}$$

and vice versa, cf. Lemma 3.17. Therefore, $x \in A$ iff there is some $f \colon T_x \to \omega_1$ with (7.2).

Now we construct S to be a tree "searching for" some such ranking. Let $e \colon \omega \to {}^{<\omega}\omega$ be a bijection such that if $n < lh(s)$, then $e^{-1}(s \restriction n) < e^{-1}(s)$. We let $(s, h) \in S$ iff $s \in {}^{<\omega}\omega$ and, setting

$$T_s = \{t \in {}^{<\omega}\omega \colon lh(t) \leq lh(s) \wedge (s \restriction lh(t), t) \in T\},$$

$h \colon lh(s) \longrightarrow \omega_1$ is such that

$$\forall k < lh(s) \forall l < lh(s)\ (e(k) \in T_s \wedge e(l) \in T_s \wedge e(k) \supsetneq e(l) \Longrightarrow h(k) < h(l)).$$

For $(s', h'), (s, h) \in S$ we write $(s', h') \leq_S (s, h)$ iff $s' \supset s$ and $h' \supset h$.

It is easy to verify now that $x \in A$ iff $(S, \supset)$ is ill-founded. □

The tree S constructed in the previous proof is called "the" SHOENFIELD *tree* for A.

Corollary 7.9 *Every coanalytic set $A \subset {}^{\omega}\omega$ is the union of $\aleph_1$ many* BOREL *sets.*

Proof Let $A \subset {}^{\omega}\omega$ be coanalytic. Let S be "the" SHOENFIELD tree on $\omega \times \omega_1$ for A, as being constructed in the proof of Lemma 7.8, so that $x \in A$ iff $x \in p[S]$. If $\xi < \omega_1$, let us write $S \restriction \xi$ for the set of all $(s, t) \in S$ with $\text{ran}(t) \subset \xi$. Obviously,

$$p[S] = \bigcup_{\xi < \omega_1} p[S \restriction \xi]. \tag{7.3}$$

Let $\xi < \omega_1$. Using any bijection of ξ with ω, we may construe $S \restriction \xi$ as a tree on $\omega \times \omega$. The sets $p[S \restriction \xi]$ and ${}^{\omega}\omega \setminus A$ are then disjoint analytic sets and may hence by Lemma 7.6 be separated by a BOREL set B_ξ.

We then have $A = \bigcup_{\xi<\omega_1} p[S \upharpoonright \xi] \subset \bigcup_{\xi<\omega_1} B_\xi \subset A$, so that in fact

$$A = \bigcup_{\xi<\omega_1} B_\xi$$

as desired. □

Let us now consider the spaces $({}^\omega\omega)^k$, where $1 \leq k < \omega$, equipped with the product topology. It is not hard to see that $({}^\omega\omega)^k$ is actually homeomorphic to the Baire space ${}^\omega\omega$.

Let $\alpha \geq \omega$. A tree T on $\omega^k \times \alpha$ is a set of $(k+1)$-tuples $(s_0, s_1, \dots, s_k)$ with $s_0, s_1, \dots, s_{k-1} \in {}^{<\omega}\omega$, $s_k \in {}^{<\omega}\alpha$, and $lh(s_0) = \dots = lh(s_k)$, such that T is closed under initial segments, i.e., if $(s_0, \dots, s_k) \in T$ and $n \leq lh(s_0)$ then $(s_0 \upharpoonright n, \dots, s_k \upharpoonright n) \in T$. We shall write $[T]$ for the set of all $(x_0, x_1, \dots, x_{k-1}, f)$ such that for all n, $(x_0 \upharpoonright n, x_1 \upharpoonright n, \dots, x_{k-1} \upharpoonright n, f \upharpoonright n) \in T$.

A set $A \subset ({}^\omega\omega)^k$ can easily be verified to be closed iff there is a tree T on $\omega^{k-1} \times \omega$ with $A = [T]$. $A \subset ({}^\omega\omega)^k$ is perfect iff there is a tree $T \neq \emptyset$ on $\omega^{k-1} \times \omega$ with $A = [T]$ and T is perfect, i.e., whenever $(s_0, \dots, s_k) \in T$ then there are extensions $(t_0, \dots, t_k), (t'_0, \dots, t'_k) \in T$ of $(s_0, \dots, s_k)$ with $lh(t_0) = lh(t'_0)$ and $(t_0, \dots, t_k) \neq (t'_0, \dots, t'_k)$ (Cf. Lemma 7.1).

We may now define the projective hierarchy. Let T be a tree on $\omega^k \times \alpha$. The projection of T onto the first $l \leq k$ many coordinates, written $p_l[T]$ or just $p[T]$, is the set of all $(x_0, \dots, x_{l-1})$ such that there are $(x_l, \dots, x_{k-1}, f)$ with

$$(x_0, \dots, x_{l-1}, x_l, \dots, x_{k-1}, f) \in [T].$$

A set $A \subset ({}^\omega\omega)^k$ is *analytic* iff there is a tree T on $\omega^k \times \omega$ with $A = p_k[T]$. (Hence A is analytic iff

$$A = \{(x_0, \dots, x_{k-1}) \colon \exists x_k (x_0, \dots, x_{k-1}, x_k) \in B\}$$

for some closed set $B \subset ({}^\omega\omega)^{k+1}$.)

$A \subset ({}^\omega\omega)^k$ is *coanalytic* iff A is the complement of an analytic set, $A = ({}^\omega\omega)^k \setminus B$, where B is analytic. The analytic sets are also called $\utilde{\Sigma}^1_1$, the coanalytic sets $\utilde{\Pi}^1_1$.

$A \subset ({}^\omega\omega)^k$ is $\utilde{\Sigma}^1_{n+1}$ iff

$$A = \{(x_1, \dots, x_{k-1}) \colon \exists x_k (x_1, \dots, x_{k-1}, x_k) \in B\}$$

for some $\utilde{\Pi}^1_n$ set $B \subset ({}^\omega\omega)^{k+1}$. $A \subset ({}^\omega\omega)^k$ is $\utilde{\Pi}^1_{n+1}$ iff A is the complement of a $\utilde{\Sigma}^1_{n+1}$ set.

Definition 7.10 A set $A \subset ({}^\omega\omega)^k$, some $k < \omega$, is called *projective* iff there is some $n < \omega$ such that A is $\utilde{\Sigma}^1_n$.

Of course there are exactly $2^{\aleph_0}$ projective sets.

We now aim to verify that $\underset{\sim}{\Sigma}^1_1 \neq \underset{\sim}{\Pi}^1_1$ by showing that there is a universal $\underset{\sim}{\Sigma}^1_1$-set which cannot be $\underset{\sim}{\Pi}^1_1$.

Let us pick a bijection

$$e\colon \{(s,t)\colon s \in {}^{<\omega}\omega, t \in {}^{<\omega}\omega, lh(s) = lh(t)\} \to \omega$$

such that for all s, t and $i < \omega$, $e(s \restriction i, t \restriction i) \leq e(s,t)$. Let us say that $u \in {}^{<\omega}\omega$ codes a finite tree iff $T_u = \{(s,t)\colon s \in {}^{<\omega}\omega, t \in {}^{<\omega}\omega, lh(s) = lh(t), u(e(s,t)) = 1\}$ is a (finite) tree, i.e., is closed under initial segments.

We may now define a tree U on $\omega \times \omega \times \omega$ as follows. We set $(s,t,u) \in U$ iff $s, t, u \in {}^{<\omega}\omega$, $lh(s) = lh(t) = lh(u)$, u codes a finite tree, and if $i < \omega$ is such that $e(s \restriction i, t \restriction i) \in \text{dom}(u)$, then $u(e(s \restriction i, t \restriction i)) = 1$.

We claim that for all $A \subset {}^{\omega}\omega$, A is $\underset{\sim}{\Sigma}^1_1$ iff there is some $z \in {}^{\omega}\omega$ such that

$$A = p[U_z], \text{ where } U_z = \{(s,t)\colon \exists u\, ((s,t,u) \in U \wedge u = z \restriction \text{lh}(u))\}. \tag{7.4}$$

Well, let $A \subset {}^{\omega}\omega$ be $\underset{\sim}{\Sigma}^1_1$, and let $A = p[T]$, where T is on $\omega \times \omega$. Let $z \in {}^{\omega}\omega$, where $z(n) = 1$ iff there is $(s,t) \in T$ such that $e(s,t) = n$. Then for all $x, y \in {}^{\omega}\omega$, $(x,y,z) \in [U]$ iff $(x,y) \in [U_z] = [T]$, so that $A = p[U_z]$. On the other hand, if $z \in {}^{\omega}\omega$ and $A = p[U_z]$, then A is clearly $\underset{\sim}{\Sigma}^1_1$.

The set $B = \{(x,z)\colon \exists y\, (x,y,z) \in [U]\}$ is easily be seen to be $\underset{\sim}{\Sigma}^1_1$, and $A \subset {}^{\omega}\omega$ is $\underset{\sim}{\Sigma}^1_1$ iff there is some $z \in {}^{\omega}\omega$ with

$$A = \{x\colon (x,z) \in B\}. \tag{7.5}$$

(This uses (7.4).) A set B with these properties is called a *universal* $\underset{\sim}{\Sigma}^1_1$-set.

Let $B \subset ({}^{\omega}\omega)^2$ be a universal $\underset{\sim}{\Sigma}^1_1$-set. We claim that B cannot be also a $\underset{\sim}{\Pi}^1_1$-set. Otherwise

$$A = \{x \in {}^{\omega}\omega\colon (x,x) \notin B\}$$

would be a $\underset{\sim}{\Sigma}^1_1$-subset of ${}^{\omega}\omega$, and there would thus be some $z \in {}^{\omega}\omega$ such (7.5) holds. In particular, $(z,z) \notin B$ iff $z \in A$ iff $(z,z) \in B$.

We have shown:

Lemma 7.11 *There is an analytic set of reals which is not coanalytic.*

We may also think of the universal $\underset{\sim}{\Sigma}^1_1$-set B constructed above, in fact of *any* $B \subset ({}^{\omega}\omega)^2$, as a subset of ${}^{\omega}\omega$, in the following way. If $x, y \in {}^{\omega}\omega$, then let $x \oplus y$ denote that $z \in {}^{\omega}\omega$ such that $z(2n) = x(n)$ and $z(2n+1) = y(n)$ for all $n < \omega$. (Clearly, $(x,y) \mapsto x \oplus y$ is a continuous, in fact LIPSCHITZ, bijection between $({}^{\omega}\omega)^2$

and ${}^{\omega}\omega$.) If $B \subset ({}^{\omega}\omega)^2$, then $B^{\oplus} = \{x \oplus y \colon (x, y) \in B\}$ "codes" B in the sense that B may be easily read off from $B^{\oplus}$.

Let us define two important sets which are in $\underset{\sim}{\Pi}^1_1 \setminus \underset{\sim}{\Sigma}^1_1$, namely WF and WO.

Let $n, m \mapsto \langle n, m \rangle$ be the GÖDEL pairing function, cf. p. 35. Every real $x \in {}^{\omega}\omega$ induces a binary relation R_x on ω as follows:

$$(n, m) \in R_x \iff x(\langle n, m \rangle) = 1. \tag{7.6}$$

We let

$$\mathsf{WF} = \{x \in {}^{\omega}\omega \colon R_x \text{ is well-founded } \}, \text{ and}$$
$$\mathsf{WO} = \{x \in {}^{\omega}\omega \colon R_x \text{ is a well-ordering}\}.$$

The sets WF and WO are coanalytic, cf. Problem 7.6.

WF and WO are in fact *complete* coanalytic sets in the sense that if $B \subset {}^{\omega}\omega$ is coanalytic, then there are continuous (in fact, LIPSCHITZ) functions $f \colon {}^{\omega}\omega \to {}^{\omega}\omega$ and $g \colon {}^{\omega}\omega \to {}^{\omega}\omega$ such that for all $x \in {}^{\omega}\omega$,

$$x \in B \iff f(x) \in \mathsf{WF} \iff g(x) \in \mathsf{WO}.$$

We may construct such a function g for WO as follows. (Then $f = g$ will also work for WF.) Let $B \subset {}^{\omega}\omega$ be coanalytic. There is then a tree T on $\omega \times \omega$ with $x \in B$ iff $(T_x; \supset)$ is well-founded. With the help of some bijection $e \colon \omega \to {}^{<\omega}\omega$ which is such that for all $s \in {}^{<\omega}\omega$ and $i < \omega$ we have that $e^{-1}(s \restriction i) \leq e^{-1}(s)$, T_x induces an order R^x on ω as follows[2]:

$$\begin{aligned} n R^x m \iff & [(e(n) \in T_x \wedge e(m) \in T_x \wedge e(n) \supsetneq e(m)) \vee \\ & (e(n) \in T_x \wedge e(m) \in T_x \wedge e(n) \perp e(m) \wedge e(n) <_{\text{lex}} e(m)) \vee \\ & (e(n) \notin T_x \wedge (e(m) \in T_x \vee (e(m) \notin T_x \wedge n < m))]. \end{aligned}$$

Let us then define $g(x)$ to be such that $R_{g(x)} = R^x$, i.e., $g(x)$ is that real $y \in {}^{\omega}\omega$ such that

$$y(\langle n, m \rangle) = \begin{cases} 1 \text{ iff } (n, m) \in R^x, \text{ and} \\ 0 \text{ otherwise.} \end{cases}$$

This defines $g \colon {}^{\omega}\omega \to {}^{\omega}\omega$. Notice that g is continuous. In fact, if $x \restriction n = y \restriction n$, then T_x and T_y agree upon the first n levels, so that $R^x \restriction n = R^y \restriction n$ and hence $g(x) \restriction n = g(y) \restriction n$.

It is easy to see that if $g(x) \in \mathsf{WO}$, then $(T_x; \supset)$ must be well-founded, so that $x \in B$. On the other hand, suppose that $g(x) \notin \mathsf{WO}$, and let $(n_i : i < \omega)$ be such that $(n_{i+1}, n_i) \in R^x$ for all $i < \omega$. Clearly, $e(n_i) \in T_x$ for all $i < \omega$. Moreover,

---

[2] We here write $s \perp t$ iff s and t are incomparable, i.e., $s \restriction (\text{lh}(s) \cap \text{lh}(t)) \neq t \restriction (\text{lh}(s) \cap \text{lh}(t))$. Also, $<_{\text{lex}}$ is the lexicographic ordering.

$$\forall k\, \exists i(k)\, \forall i \geq i(k)\, (\text{lh}(e(n_i)) \geq k \wedge e(n_i) \restriction k = e(n_{i(k)}) \restriction k).$$

This gives that

$$\bigcup_{k<\omega} e(n_{i(k)}) \restriction k \in [T_x],$$

so that $x \in B$. We have verified that $x \in B$ iff $g(x) \in \mathsf{WO}$.

If $x \in \mathsf{WO}$ and R_x is defined as in (7.6), then we write $||x||$ for the order type of R_x, i.e., for the (countable) ordinal α such that $(\alpha; \in) \simeq (\omega; R_x)$, cf. Definition 3.22.

Lemma 7.12 (Boundedness Lemma) *Let $A \subset \mathsf{WO}$ be analytic. Then $\{||x||: x \in A\}$ is bounded below ω_1.*

Proof Suppose not, i.e., let $A \subset \mathsf{WO}$ be analytic such that $\{||x||: x \in A\}$ is unbounded in ω_1. Let $B \subset {}^\omega\omega$ be an arbitrary coanalytic set. There is a tree T on $\omega \times \omega$ such that $x \in B$ iff $(T_x; \supset)$ is well-founded. We may thus write $x \in B$ iff there is a ranking $f: T_x \to \alpha$, where $\alpha = ||z||$ for some $z \in A$. It is straightforward to verify that B is then $\underset{\sim}{\Sigma}^1_1$. Therefore, every coanalytic set would be analytic. Contradiction! □

Similar to our proof of Lemma 7.11 one may show that for all $n < \omega$, $\underset{\sim}{\Sigma}^1_n$ is different from $\underset{\sim}{\Pi}^1_n$, in fact $\underset{\sim}{\Sigma}^1_n \setminus \underset{\sim}{\Pi}^1_n \neq \emptyset$ and $\underset{\sim}{\Pi}^1_n \setminus \underset{\sim}{\Sigma}^1_n \neq \emptyset$. (Cf. Problem 7.6.)

One also defines

$$\underset{\sim}{\Delta}^1_n = \underset{\sim}{\Sigma}^1_n \cap \underset{\sim}{\Pi}^1_n$$

for $n < \omega$. By SOUSLIN's Theorem 7.5, $\underset{\sim}{\Delta}^1_1$ is the family of all BOREL sets.

Let us consider a $\underset{\sim}{\Sigma}^1_2$ set $A \subset {}^\omega\omega$. As every coanalytic set $B \subset ({}^\omega\omega)^2$ is of the form $p_2[T]$, where T is on $\omega^2 \times \omega_1$, by the proof of Lemma 7.8, we get that $A = p_1[T]$. Via some bijection $g: \omega_1 \to \omega \times \omega_1$, we thus see that $A = p[S]$, where S is on $\omega \times \omega_1$. This tree S is also called "the" SHOENFIELD *tree* for A.

This argument shows:

Lemma 7.13 *If $A \subset {}^\omega\omega$ is $\underset{\sim}{\Sigma}^1_2$ then A is $\aleph_1$-Souslin.*

Lemma 7.14 *Let $A \subset {}^\omega\omega$ be $\aleph_n$-Souslin, where $n < \omega$. Then $A = \bigcup_{i<\omega_n} A_i$, where each A_i is analytic.*

Proof By induction on n: There is nothing to prove for $n = 0$. Now let $n > 0$ and suppose the statement to be true for $n - 1$. Let $A = p[T]$, where T is on $\omega \times \omega_n$. For $\alpha < \omega_n$, let $T \restriction \alpha$ be the set of all $(s, t) \in T$ with $\text{ran}(t) \subset \alpha$. Because $\text{cf}(\omega_n) > \omega$, $p[T] = \bigcup_{\alpha<\omega_n} p[T \restriction \alpha]$.

By the inductive hypothesis, for each $\alpha < \omega_n$, $p[T \restriction \alpha] = \bigcup_{i<\omega_{n-1}} A_i^\alpha$, where each A_i^α is analytic. Therefore,

$$A = p[T] = \bigcup_{\alpha<\omega_n} \bigcup_{i<\omega_{n-1}} A_i^\alpha,$$

a representation as desired. □

If U is a tree on $\omega^k \times \alpha$ and $(s_0, s_1, \dots, s_k) \in U$, then we write $U_{(s_0,s_1,\dots,s_k)}$ for the tree

$$\{(t_0, t_1, \dots, t_k) \in U : (s_0 \subset t_0 \wedge s_1 \subset t_1 \wedge \dots \wedge s_k \subset t_k) \vee (s_0 \supset t_0 \wedge s_1 \supset t_1 \wedge \dots \wedge s_k \supset t_k)\}.$$

Theorem 7.15 (Souslin, Mansfield) *Let $A \subset {}^\omega\omega$ be κ-Souslin. Then either A has at most κ elements or else A contains a perfect subset.*

Proof This is shown by a "CANTOR BENDIXSON analysis" of A, cf. the proof of Theorem 1.9.

If U is a tree on $\omega \times \kappa$ then we set

$$U' = \{(s,t) \in U : \mathrm{Card}(p[U_{(s,t)}]) > 1\}. \tag{7.7}$$

Let us now fix a tree T on $\omega \times \kappa$ such that $A = p[T]$. Let us inductively define trees T_i, $i \in \mathrm{OR}$, as follows.

$T_0 = T$,
$T_{i+1} = (T_i)'$, and
$T_\lambda = \bigcap_{i<\lambda} T_i$ for limit ordinals λ.

Notice that, inductively, each T_i is in fact a tree on $\omega \times \kappa$. Moreover, $T_i \supset T_j$ whenever $i \leq j$.

As $\mathrm{Card}(T) \leq \kappa$, there must be some $\theta < \kappa^+$ such that $T_{\theta+1} = T_\theta$. Let us write T^∞ for this tree.

The argument now splits into two cases.

Case 1: $T^\infty = \emptyset$.

Let $x \in A = p[T]$. Pick $g \in {}^\omega\kappa$ such that $(x \restriction n, g \restriction n) \in T$ for all n. As $T^\infty = \emptyset$, there must be a largest i such that $(x \restriction n, g \restriction n) \in T_i$ for all n. Let n be maximal such that $(x \restriction n, g \restriction n) \in T_{i+1}$. Then $p[(T_i)_{(x \restriction n+1, g \restriction n+1)}]$ has exactly one element, namely x, as

$$(x \restriction n+1, g \restriction n+1) \in T_i \setminus T_{i+1}.$$

We have seen that

$$A = \bigcup \{p[(T_i)_{(s,t)}] : (s,t) \in T_i \setminus T_{i+1}\},$$

where $p[(T_i)_{(s,t)}]$ has exactly one element in case $(s,t) \in T_i \setminus T_{i+1}$. Because $\mathrm{Card}(T) \leq \kappa$, this shows that $Card(A) \leq \kappa$.

Case 2: $T^\infty \neq \emptyset$.
Then $\mathrm{Card}(p[T^\infty_{(s,t)}]) > 1$ for all $(s,t) \in T^\infty$.

Let us recursively construct $(s_u, t_u) \in T^\infty$, where $u \in {}^{<\omega}2$. Set $(s_\emptyset, t_\emptyset) = (\emptyset, \emptyset)$. Suppose that (s_u, t_u) has been chosen. As $p[T^\infty_{(s_u,t_u)}]$ has at least two elements, we may pick $(s_{u^\frown 0}, t_{u^\frown 0}) \in T^\infty$, $(s_{u^\frown 1}, t_{u^\frown 1}) \in T^\infty$ such that

$$\mathrm{lh}(s_{u^\frown 0}) = \mathrm{lh}(s_{u^\frown 1}) > lh(s_u) \text{ and } s_{u^\frown 0} \neq s_{u^\frown 1}.$$

For $z \in {}^\omega 2$, let $x_z = \bigcup\{s_{z\restriction n} : n < \omega\}$. We have $x_z \in A = p[T]$, as being witnessed by $(t_{z\restriction n} : n < \omega)$, for each $z \in {}^\omega 2$. Moreover, $\{x_z : z \in {}^\omega 2\}$ is a perfect set. □

Corollary 7.16 *Every uncountable analytic set of reals has a perfect subset.*

We now need to define the effective projective hierarchy.

Let $x \in {}^\omega\omega$. A set $A \subset ({}^\omega\omega)^k$ is called $\Sigma^1_1(x)$ iff $A = p[T]$, where T is a tree on ω^{k+1} which is definable over the structure $(V_\omega; \in, x)$. A is called $\Pi^1_n(x)$ iff A is the complement of a $\Sigma^1_n(x)$ set, and A is called $\Sigma^1_{n+1}(x)$ if A is the projection of a $\Pi^1_n(x)$ set. We also set $\Delta^1_n(x) = \Sigma^1_n(x) \cap \Pi^1_n(x)$. Notice that

$$\underset{\sim}{\Sigma}{}^1_1 = \bigcup_{x \in {}^\omega\omega} \Sigma^1_1(x),$$

and therefore analogous facts hold for the other projective pointclasses $\underset{\sim}{\Pi}{}^1_n$ and $\underset{\sim}{\Sigma}{}^1_{n+1}$ as well. We write Σ^1_n instead of $\Sigma^1_n(0)$, Π^1_n instead of $\Pi^1_n(0)$, and Δ^1_n instead of $\Delta^1_n(0)$.

The following is often very useful.

Lemma 7.17 *Let $x \in {}^\omega\omega$. A set $A \subset ({}^\omega\omega)^k$ is $\Sigma^1_n(x)$ iff there is a formula φ such that*[3] *for all $y \in {}^\omega\omega$,*

$$y \in A \iff \exists z_1 \in {}^\omega\omega \forall z_2 \in {}^\omega\omega \ldots Qz_n \in {}^\omega\omega \\ (V_\omega; \in, x, y, z_1, \ldots, z_n) \models \varphi(x, y, z_1, \ldots, z_n). \tag{7.8}$$

Proof By induction on n. The only non-trivial step of this induction is the base, $n = 1$. We first verify

Claim 7.18 *$\Sigma^1_1(x)$ is closed under $\forall^\omega$ and $\exists^\omega$, i.e., if $B \subset ({}^\omega\omega)^k \times \omega$ is $\Sigma^1_1(x)$,*[4] *then so are*

$$\{\mathbf{y} : \forall n \in \omega\, (\mathbf{y}, n) \in B\} \text{ and} \\ \{\mathbf{y} : \exists n \in \omega\, (\mathbf{y}, n) \in B\}.$$

Proof "$\forall^\omega$": Let us assume that $k = 1$. Let $(y, n) \in B$ iff $(y, n) \in p[T]$, where T is on ω^3. We may then define U on ω^2 by setting $(s, t) \in U$ iff for all $n < lh(s)$,

---

[3] In what follows, Q is $\exists$ or $\forall$ depending on whether n is odd or even.

[4] By identifying $n < \omega$ with the constant function $c_n : \omega \to \omega$ with value n, we may construe $({}^\omega\omega)^k \times \omega$ as a subset of $({}^\omega\omega)^{k+1}$.

$(s, n, t) \in T$. It is then easy to see that $y \in p[U]$ iff for all $n < \omega$, $(y, n) \in p[T] = B$. The proof for "$\exists^\omega$" is easy. □

Let us now prove Lemma 7.17 for $n = 1$. First let $A \in \Sigma^1_1(x)$, say $A = p[T]$, where $(s, t) \in T$ iff $(V_\omega; \in, x) \models \psi(s, t)$ for some formula ψ. Then

$$y \in A \Longleftrightarrow \exists z \in {}^\omega\omega\, (V_\omega; \in, x, y, z) \models \forall n < \omega\, \psi((y \restriction n, z \restriction n)).$$

Now we prove by induction on the complexity of φ that if A be as in (7.8) with $n = 1$, then A is $\Sigma^1_1(x)$. We shall use Problem 5.2. Let in what follows $(\omega; E_A) \overset{\pi}{\cong} (\mathrm{HF}; \in)$ be as in Problem 5.2. The relation "$\pi(n) = u$" is $\Delta_1^{\mathsf{ZFC}^{-\infty}}$ by Problem 5.2.

First let φ be Π_1. We let $(s, t) \in T$ iff $s, t \in {}^{<\omega}\omega$, $\mathrm{lh}(s) = \mathrm{lh}(t)$, and $(V_{\mathrm{lh}(s)}; \in, x \cap V_{\mathrm{lh}(s)}, s, t) \models \varphi(x \cap V_{\mathrm{lh}(s)}, s, t)$. It is easy to see that $A = p[T]$.

If $\varphi \equiv \exists u\, \bar{\varphi}$, where $\bar{\varphi}$ is Π_n, then

$$y \in A \Longleftrightarrow \exists z \in {}^\omega\omega \exists n\, (V_\omega; \in, x, y, z) \models \forall u (u = \pi(n) \rightarrow \bar{\varphi}(x, y, z, u)),$$

so that the result follows from the inductive hypothesis and Claim 7.18.

Finally, let $\varphi \equiv \forall u\, \bar{\varphi}$, where $\bar{\varphi}$ is Π_n. Then

$$y \in A \Longleftrightarrow \exists z \in {}^\omega\omega \forall n\, (V_\omega; \in, x, y, z) \models \exists u (u = \pi(n) \wedge \bar{\varphi}(x, y, z, u)),$$

so that the result also follows from the inductive hypothesis and Claim 7.18. □

Lemma 7.19 *Let $z \in {}^\omega\omega$. Then ${}^\omega\omega \cap L[z]$ as well as $<_{L[z]} \restriction ({}^\omega\omega \cap L[z])$ are $\Sigma^1_2(z)$.*

Proof The proof in the general case is only notationally different from the proof in the case $z = \emptyset$, so let us assume that $z = \emptyset$. We have (using Theorem 5.31) that $x \in {}^\omega\omega \cap L$ iff $x \in {}^\omega\omega \cap J_\alpha$ for some $\alpha < \omega_1$ iff (using Lemma 5.28) $x \in {}^\omega\omega \cap M$ for some countable transitive model of "$V = L$," which is true if and only if there is some z such that, setting $(n, m) \in E \Longleftrightarrow z(n, m) = 1$, E is well-founded (i.e., there is no $y \in {}^\omega\omega$ such that for all $n < \omega$, $(y)_{n+1} E (y)_n$),

$$(\omega; E) \models \text{"}V = L,\text{"}$$

and x is a real number in the transitive collapse of $(\omega; E)$. It is easy to verify that this can be written in a Σ^1_2 fashion. (Cf. Problem 7.10.)

This shows that ${}^\omega\omega \cap L$ is Σ^1_2.

An entirely analoguous argument shows that $<_L \restriction ({}^\omega\omega \cap L)$ is Σ^1_2. □

In general, the complexity of ${}^\omega\omega \cap L$ given by Lemma 7.19 is optimal, as we aim to show now.

Lemma 7.20 *Let $\kappa \geq \omega$, and let $A \subset {}^\omega\omega$ be κ-SOUSLIN, say $A = p[T]$, where T is a tree on $\omega \times \kappa$. If $A \neq \emptyset$, then $A \cap L[T] \neq \emptyset$. Moreover, if A does not contain a perfect subset, e.g., if $\mathrm{Card}(A) < 2^{\aleph_0}$, then $A \in L[T]$.*

Proof Suppose that $A \neq \emptyset$, i.e., $[T] \neq \emptyset$. This means that the relation

$$(T, \supset) \in L[T] \tag{7.9}$$

is ill-founded in V. By the absoluteness of well-foundedness, cf. Lemma 5.6, the relation (7.9) is then ill-founded in $L[T]$, which implies that $[T] \cap L[T] \neq \emptyset$, i.e., $A \cap L[T] = p[T] \cap L[T] \neq \emptyset$.

Now let $(T_i : i \leq \theta)$, where $\theta < \kappa^+$, be the "CANTOR–BENDIXSON analysis" of A as in the proof of Theorem 7.15. If $U \in L[T]$ is a tree on $\omega \times \kappa$, and if U' is defined as in (7.7), then U' computed in V is the same as U' computed in $L[T]$; this is because "$\text{Card}(p[U_{(s,t)}]) > 1$" is absolute between V and $L[T]$ by Lemma 5.6. We therefore in fact get that the construction producing $(T_i : i \leq \theta)$ is absolute between V and $L[T]$, so that

$$(T_i : i \leq \theta) \in L[T]. \tag{7.10}$$

Now let us suppose that A does not contain a perfect subset, so that $T_\theta = \emptyset$. We aim to show that $A \in L[T]$. Let $x \in A$, say $x \in p[T_i] \backslash p[T_{i+1}]$. By construction, there is then some $n < \omega$ and $t \in {}^{<\omega}\omega$ with $\text{lh}(t) = n$ such that

$$\text{Card}(p[(T_i)_{(x \restriction n,t)}]) = 1. \tag{7.11}$$

By Lemma 5.6, there must be some $(x', y') \in [(T_i)_{(x \restriction n,t)}] \cap L[T]$. However, by (7.11), any such x' must be equal to x, and thus x is easily definable from T_i and $(x \restriction n, t)$, so that $x \in L[T]$. We have shown that $A \subset L[T]$ and in fact $A \in L[T]$. □

Corollary 7.21 (Shoenfield absoluteness) *Let $x \in {}^\omega\omega$, and let $A \subset {}^\omega\omega$ be $\Sigma^1_2(x)$. If $A \neq \emptyset$, then $A \cap L[x] \neq \emptyset$. Moreover, if A does not contain a perfect subset, then $A \in L[x]$*

Proof Let S be "the" SHOENFIELD tree for A, cf. p. 135. An inspection of the construction of S, cf. the proof of Lemma 7.8, shows that $S \in L[x]$ follows from the assumption that A be $\Sigma^1_2(x)$. The conclusion then follows from Lemma 7.20. □

This implies that ${}^\omega\omega \cap L[x]$ can in general not be better than the complexity given by Lemma 7.19, namely $\Sigma^1_2(x)$, unless ${}^\omega\omega \subset L[x]$. This is because if ${}^\omega\omega \cap L[x]$ were $\Pi^1_2(x)$, then ${}^\omega\omega \backslash L[x]$ would be $\Sigma^1_2(x)$, hence if ${}^\omega\omega \backslash L[x] \neq \emptyset$, then $({}^\omega\omega \backslash L[x]) \cap L[x] \neq \emptyset$ by Corollary 7.21, which is nonsense.

Also, if $(2^{\aleph_0})^{L[x]} = \omega_1^{L[x]} < 2^{\aleph_0}$, then by Lemma 7.19 there is a *largest* $\Sigma^1_2(x)$-set of reals, namely ${}^\omega\omega \cap L[x]$.

Definition 7.22 Let $A \subset ({}^\omega\omega)^2$. We say that a partial function $F : {}^\omega\omega \to {}^\omega\omega$ *uniformizes* A iff for all $x \in {}^\omega\omega$, if there is some $y \in {}^\omega\omega$ such that $(x, y) \in A$, then $x \in \text{dom}(F)$ and $(x, F(y)) \in A$.

Theorem 7.23 (Kondŏ, Addison) *Let $A \subset {}^\omega\omega \times {}^\omega\omega$ be Π^1_1. Then* A *can be uniformized by a function whose graph is Π^1_1.*

Proof We shall prove that each nonempty $\Pi^1_1(z)$ set $A \subset {}^\omega\omega$ has a member x such that $\{x\}$ is $\Pi^1_1(z)$. As the definition of $\{x\}$ will be uniform in the parameter z, this proof will readily imply the theorem. For notational convenience, we shall assume that $z = 0$. Hence let $A \subset {}^\omega\omega$ be given such that A is Π^1_1 and $A \neq \emptyset$. Let T be a tree on $\omega \times \omega$ such that

$$x \in A \Longleftrightarrow T_x \text{ is well-founded.}$$

Let us fix an enumeration $(s_n : n < \omega)$ of ${}^{<\omega}\omega$ such that if $s_n \subsetneq s_m$, then $n < m$. x

We first define maps $\varphi_n : A \to \omega_1$ by setting

$$\varphi_n(x) = \begin{cases} ||s_n||_{T_x} & \text{, if } s_n \in T_x \\ 0 & \text{, else.} \end{cases}$$

Here, $||s||_{T_x}$ is the rank of s in $(T_x, \supset)$ in the sense of Definition 3.18, which is well-defined for $x \in A$.

Claim 7.24 *If* $\lim_{k \to \omega} x_k = x$, *where each* x_k *is in* A, *and for all* n, $\varphi_n(x_k)$ *is eventually constant, i.e.*

$$\exists \alpha_n \exists k_n \forall k \geq k_n \, \varphi_n(x_k) = \alpha_n,$$

then x *is in* A *and* $\varphi_n(x) \leq \alpha_n$.

Proof Suppose that $(x_k : k < \omega)$ is as described, but $x = \lim_{k \to \omega} x_k \notin A$. Then T_x is ill-founded, and we may pick some $y \in [T_x]$. Let $y \upharpoonright i = s_{n_i}$ for $i < \omega$. If $k \geq k_{n_i}$, $k_{n_{i+1}}$ is large enough, then

$$\alpha_{n_{i+1}} = ||s_{n_{i+1}}||_{T_{x_k}} < ||s_{n_i}||_{T_{x_k}} = \alpha_{n_i}.$$

Hence $(\alpha_{n_i} : i < \omega)$ is a descending sequence of ordinals. This contradiction shows that $x \in A$ after all. It is easy to see that $\varphi_n(x) \leq \alpha_n$. □

We shall now pick some $x \in A$. Let $x \in B$ iff $x \in A$ and for all y and for all n,

$$[x \upharpoonright n = y \upharpoonright n \wedge \forall m < n (y \in A \wedge \varphi_m(x) = \varphi_m(y))]$$
$$\longrightarrow [x(n) < y(n) \vee (x(n) = y(n) \wedge (y \notin A \vee (y \in A \wedge \varphi_n(x) \leq \varphi_n(y))))].$$

A moment of reflection shows that $B = \{x\} \subset A$ for some x.

It remains to be shown that B is Π^1_1. Well, $\forall m < n(y \in A \wedge \varphi_m(x) = \varphi_m(y))$ says that for all $m < n$ there are order-preserving embeddings $f : (T_x)_{s_m} \to (T_y)_{s_m}$ and $g : (T_y)_{s_m} \to (T_x)_{s_m}$, and is hence Σ^1_1 by Lemmas 7.17 and 7.18. Similarily,

$$y \notin A \vee (y \in A \wedge \varphi_n(x) \leq \varphi_n(y))$$

says that there is no order-preserving embedding $f\colon (T_y)_{s_n} \to (T_x)_t$ for some $t \supsetneq s_n$, and is hence Π^1_1 by Lemmas 7.17 and 7.18. We may thus rewrite "$x \in B$" in a Π^1_1 fashion. □

Let $A, B \subset {}^\omega\omega$. We say that $\bar{A}$ and $\bar{B}$ *reduce* A and B iff $\bar{A} \subset A$, $\bar{B} \subset B$, $\bar{A} \cup \bar{B} = A \cup B$, and $\bar{A} \cap \bar{B} = \emptyset$. If $\Gamma \subset \mathscr{P}({}^\omega\omega)$, then we say that Γ has the *reduction property* iff for all $A, B \in \Gamma$ there are $\bar{A}, \bar{B} \in \Gamma$ such that $\bar{A}$ and $\bar{B}$ reduce A and B.

Recall that if $A, B \subset {}^\omega\omega$ are disjoint, then we say that C *separates* A and B iff $A \subset C$ and $C \cap B = \emptyset$. If $\Gamma \subset \mathscr{P}({}^\omega\omega)$, then we say that Γ has the *separation property* iff for all $A, B \in \Gamma$ there is some $C \in \Gamma$ such that also ${}^\omega\omega \setminus C \in \Gamma$ and C separates A and B.

Lemma 7.25 *The following hold true.*

(a) $\utilde{\Pi}^1_1$ *has the reduction property.*
(b) $\utilde{\Sigma}^1_1$ *has the separation property.*
(c) $\utilde{\Sigma}^1_1$ *does not have the reduction property.*

Proof (a) Let $A, B \in \utilde{\Pi}^1_1$, $A, B \subset {}^\omega\omega$. Let $C = (A \times \{0\}) \cup (B \times \{1\}) \in \utilde{\Pi}^1_1$, and let $F\colon {}^\omega\omega \to {}^\omega\omega$, $F \in \utilde{\Pi}^1_1$ uniformize C. Then $\bar{A} = \{x \in A\colon F(x) = 0\} \in \utilde{\Pi}^1_1$ and $\bar{B} = \{x \in B\colon F(x) = 1\} \in \utilde{\Pi}^1_1$ reduce A and B.

(b) This easily follows from (a).

(c) Let us assume that $\utilde{\Sigma}^1_1$ has the separation property. Let $U \subset ({}^\omega\omega)^2$ be a universal $\utilde{\Sigma}^1_1$-set, and define $A = \{x \in {}^\omega\omega\colon ((x)_0, x) \in U\}$ and $B = \{x \in {}^\omega\omega\colon ((x)_1, x) \in U\}$.[5] As $A, B \in \utilde{\Sigma}^1_1$, we may pick $\bar{A}, \bar{B} \in \utilde{\Sigma}^1_1$ such that $\bar{A}$ and $\bar{B}$ reduce A and B, and we may then pick a Borel set C such that C separates $\bar{A}$ and $\bar{B}$. Let $a, b \in {}^\omega\omega$ be such that $C = \{x \in {}^\omega\omega\colon (a, x) \in U\}$ and ${}^\omega\omega \setminus C = \{x \in {}^\omega\omega\colon (b, x) \in U\}$. It is easy to verify that then $b \oplus a \in C$ iff $b \oplus a \in {}^\omega\omega \setminus C$. Contradiction! □

It follows from Lemma 7.25 (c) and the proof of Lemma 7.25 (a) that Theorem 7.23 is false with Π^1_1 replaced by $\utilde{\Sigma}^1_1$.

7.2 Descriptive Set Theory and Constructibility

Definition 7.26 We say that ω_1 is *inaccessible to the reals* iff $\omega_1^{L[x]} < \omega_1$ for every $x \in {}^\omega\omega$.

Lemma 7.27 ω_1 *is inaccessible to the reals iff* ω_1^V *is an inaccessible cardinal in* $L[x]$ *for every* $x \in {}^\omega\omega$.

[5] Here, $(x)_0$ and $(x)_1$ are defined to be the unique reals such that $(x)_0 \oplus (x)_1 = x$.

Proof Suppose that ω_1 is inaccessible to the reals, and let $x \in {}^\omega\omega$. We have to show that ω_1^V is not a successor cardinal in $L[x]$. Suppose that $\omega_1^V = \delta^{+L[x]}$. Let $f : \omega \to \delta$, $f \in V$ be a bijection, and define $z \in {}^\omega\omega$ by

$$z(n) = \begin{cases} x(\frac{n}{2}) & \text{if } n \text{ is even} \\ 1 & \text{if } n = 3^k \cdot 5^l \text{ and } f(k) < f(l) \\ 0 & \text{otherwise.} \end{cases}$$

Then $x, f \in L[z]$, and thus $\omega_1^V = \omega_1^{L[z]}$. Contradiction! □

We shall later see a model in which ω_1 is inaccessible to the reals, cf. Theorem 8.20.

By Theorem 7.15, every uncountable $\boldsymbol{\Sigma}^1_1$-set of reals has a perfect subset. The following statement gives a characterization of when every $\boldsymbol{\Pi}^1_1$-set of reals has a perfect subset in terms of "inner model theory," cf. Corollary 7.29.

Theorem 7.28 *Let $x \in {}^\omega\omega$. The following statements are equivalent.*

(1) *Every uncountable $\Sigma^1_2(x)$-set of reals has a perfect subset.*
(2) *Every uncountable $\Pi^1_1(x)$-set of reals has a perfect subset.*
(3) $\omega_1^{L[x]} < \omega_1$.

Let $\Gamma \subset \mathscr{P}({}^\omega\omega)$. We say that Γ has the perfect subset property iff every uncountable $A \in \Gamma$ has a perfect subset.

Corollary 7.29 *The class of coanalytic sets has the perfect subset property if and only if ω_1 is inaccessible to the reals.*

Proof of Theorem 7.28. Let us suppose that $x = 0$. The proof relativizes to any real different from 0.

(1) $\Longrightarrow$ (2) is trivial. Let us prove (2) $\Rightarrow$ (3). Suppose that $\omega_1^L = \omega_1$. Let $x \in A$ iff

$$x \in {}^\omega\omega \cap L \wedge x \in \mathsf{WO} \wedge \forall y \in ({}^\omega\omega \cap L)\,(y <_L x \to (y \notin \mathsf{WO} \vee ||y|| \neq ||x||)).$$

A is Σ^1_2 by Lemma 7.19 and Problem 7.6. By the Boundedness Lemma 7.12, if $B \subset A$ is analytic, $\{||x|| : x \in B\}$ is bounded below ω_1, and hence B is countable. (In particular, A does not contain a perfect subset.)

As A is Σ^1_2, there is a coanalytic set $B \subset ({}^\omega\omega)^2$ such that

$$A = \{x \in {}^\omega\omega : \exists y \in {}^\omega\omega\,(x, y) \in B\}.$$

By the Uniformization Theorem 7.23, let $F \subset B$ be a uniformizing function whose graph is Π^1_1.

We have

$$A = \{x \in {}^\omega\omega \colon \exists y \in {}^\omega\omega \; y = F(x)\}.$$

As A is uncountable, (the graph of) F is an uncountable Π^1_1 subset of $({}^\omega\omega)^2$.

Suppose that F has a perfect subset, say $P \subset F$, where P is perfect. Write

$$Q = \{x \in {}^\omega\omega \colon \exists y \in {}^\omega\omega \, (x, y) \in P\}.$$

As F is a function, $\text{Card}(Q) = \text{Card}(P) = 2^{\aleph_0}$, so that Q is an uncountable analytic subset of A. Contradiction!

Hence F is an uncountable coanalytic set without a perfect subset.

Finally, (3) $\Longrightarrow$ (1) is given by Corollary 7.21. Let $A \subset {}^\omega\omega$ be an uncountable Σ^1_2 set. If A does not have a perfect subset, then $A \in L$ by Corollary 7.21. However, $\omega_1^L < \omega_1$ implies that A is then countable. Contradiction! □

There is a more cumbersome argument of proving (3) $\Longrightarrow$ (1) of Theorem 7.28, using forcing, cf. the proof of Lemma 8.18.

Excellent textbooks on "classical" descriptive set theory are [27] and [20]. Modern variants of descriptive set theory are dealt with e.g. in [4, 12, 14, 19, 21].

7.3 Problems

7.1. Show that the topology we defined on the Baire space ${}^\omega\omega$ is exactly the one which is induced by the distance function d. Conclude that ${}^\omega\omega$ is a Polish space.

7.2. Show that there is a continuous bijection $f \colon {}^\omega\omega \to {}^\omega 2$.

7.3. (BOREL *hierarchy*) Let $\utilde{\Sigma}^0_1$ denote the set of all open $A \subset {}^\omega\omega$ and $\utilde{\Pi}^0_1$ the set of all closed $A \subset {}^\omega\omega$. Having defined $\utilde{\Sigma}^0_\beta$ and $\utilde{\Pi}^0_\beta$ for all $\beta < \alpha$, let $\utilde{\Sigma}^0_\alpha$ be the set of all $\bigcup_{n<\omega} A_n$, where $\{A_n \colon n < \omega\} \subset \bigcup_{\beta<\alpha} \utilde{\Pi}^0_\beta$, and let $\utilde{\Pi}^0_\alpha$ be the set of all ${}^\omega\omega \setminus A$, where $A \in \utilde{\Sigma}^0_\alpha$. Show that $\utilde{\Sigma}^0_{\omega_1+1} = \utilde{\Sigma}^0_{\omega_1}$, and that $\utilde{\Sigma}^0_{\omega_1}$ is the set of all BOREL subsets of ${}^\omega\omega$.

7.4. Let $(\bar{M}; \bar{A}_0, \ldots, \bar{A}_k, \bar{B}_0, \ldots, \bar{B}_l)$ and $(M; A_0, \ldots, A_k, B_0, \ldots, B_l)$ be models of the same type, let N be a transitive model of ZFC, and assume $(\bar{M}; \bar{A}_0, \ldots, \bar{A}_k) \in N$, $(M; A_0, \ldots, A_k, B_0, \ldots, B_l) \in N$, $\bar{M}$ is countable in N, and in V there is some elementary embedding

$$\pi \colon (\bar{M}; \bar{A}_0, \ldots, \bar{A}_k, \bar{B}_0, \ldots, \bar{B}_l) \to (M; A_0, \ldots, A_k, B_0, \ldots, B_l).$$

Show that in N there are $\bar{B}_0{}', \ldots, \bar{B}_l{}'$ and an elementary embedding

$$\pi' \colon (\bar{M}; \bar{A}_0, \ldots, \bar{A}_k, \bar{B}_0', \ldots, \bar{B}_l') \to (M; A_0, \ldots, A_k, B_0, \ldots, B_l).$$

[Hint. Construct a tree of height ω "seaching for" some such $\bar{B}_0', \ldots, \bar{B}_l'$, and π'.] For the conclusion to hold it actually suffices that N is a transitive model which contains an admissible set $\bar{N}$ which in turn contains $(\bar{M}; \bar{A}_0, \ldots, \bar{A}_k)$ and $(M; A_0, \ldots, A_k, B_0, \ldots, B_l)$ such that $\bar{M}$ is countable in $\bar{N}$, cf. Problem 5.28.

7.5. Let κ be a regular infinite cardinal, and let $T \neq \emptyset$ be a tree on κ. Let

$$T' = \{s \in T : \exists \{t_i : i < \kappa\} \subset T (\forall i \forall j\, (s \subset t_i \wedge (i \neq j \to t_i \not\subset t_j)))\},$$

and define $T^0 = T$, $T^{\alpha+1} = (T^\alpha)'$, and $T^\lambda = \bigcap \{T^\alpha : \alpha < \lambda\}$ for limit ordinals λ. Show that there is some $\theta < \kappa^+$ with $T^{\theta+1} = T^\theta$, call it T^∞. If $s \in T^\alpha \setminus T^{\alpha+1}$, then we say that α is the CANTOR–BENDIXSON *rank of* s, and if $s \in T^\infty$, then we say that ∞ is the CANTOR–BENDIXSON *rank of* s. Show that if $T^\infty \neq \emptyset$, then T^∞ is perfect.

7.6. Show that the sets WF and WO are both coanalytic and in fact Π^1_1. Show also that for every $n \geq 1$, $\underset{\sim}{\Sigma}^1_n \setminus \underset{\sim}{\Pi}^1_n \neq \emptyset$ and $\underset{\sim}{\Pi}^1_n \setminus \underset{\sim}{\Sigma}^1_n \neq \emptyset$.

7.7. Show that $A \subset {}^\omega\omega$ is coanalytic iff there is some map $s \mapsto <_s$, where $s \in {}^{<\omega}\omega$, such that for all $s, t \in {}^{<\omega}\omega$ with $s \subset t$, $<_t$ is an order on $\mathrm{lh}(t)$ which extends $<_s$, and for all $x \in {}^\omega\omega$, $x \in A$ iff $<_x = \bigcup_{s \subset x} <_s$ is a well-ordering. (Hint: Proof of Lemma 7.8.)

Let $(M; E)$, (M', E') be models of the language $\mathscr{L}_\in$ of set theory. We say that $(M'; E')$ is an *end-extension* of $(M; E)$ iff $M \subset M'$, $E \subset E'$, and if $x \in M$, $y \in M'$, and $yE'x$, then $y \in M$.

7.8. Show that for all countable transitive M there is a end-extension $(M'; E')$ of $(M; \in\restriction M)$ such that $(M'; E') \models V = L$. [Hint. This holds in L. Then use Corollary 7.21.] If $M \setminus L \neq \emptyset$, then $(M'; E')$ cannot be well-founded.

7.9. Show that if there is a transitive model of ZFC + "there is a supercompact cardinal", then some such model exists in L. [Hint. Corollary 7.21.]

7.10. Fill in the details in the proof of Lemma 7.19! Show that "there is a real x which is not an element of L" may be written in a Σ^1_3 fashion. Conclude that it is consistent to have a non-empty Π^1_2 set $A \subset {}^\omega\omega$ such that $A \cap L = \emptyset$. (Compare Corollary 7.21.)

7.11. Let $n < \omega$. Show that every Σ^1_{n+1}-formula is equivalent to a Σ^{HC}_n-formula in the following sense. Let $A \subset {}^\omega\omega$ be $\Sigma^1_{n+1}(z)$, where $z \in {}^\omega\omega$. There is then a Σ_n-formula $\varphi(v, w)$ such that for all $x \in {}^\omega\omega$, $x \in A \iff \mathrm{HC} \models \varphi(x, z)$. Conclude that if $z \in {}^\omega\omega$, $\varphi(v)$ is Σ_1, and $V \models \varphi(z)$, then $L[z] \models \varphi(z)$. [Hint. Corollary 7.21.] Show also that it is consistent to have some $a \in \mathrm{HC}$ and Σ_1-formula $\varphi(v)$ such that $V \models \varphi(a)$, but $L[a] \models \neg\varphi(a)$. [Hint. $a = \omega_1^L$, $\varphi(v) \equiv$ "v is countable", and V is $\mathrm{Col}(\omega, \omega_1^L)$-generic over L.]

Let $A \subset {}^\omega\omega$. A function $\varphi\colon A \to \mathrm{OR}$ is called a *norm on A*.

7.12. Let $A \subset {}^\omega\omega$ be Π^1_1. Show that there is a norm $\varphi\colon A \to \mathrm{OR}$ on A such that there are $R, S \subset ({}^\omega\omega)^2$, $R \in \Pi^1_1$ and $S \in \Sigma^1_1$, such that for all $y \in A$,

$$(x, y) \in R \Longleftrightarrow (x \in A \wedge \varphi(x) \leq \varphi(y)) \text{ and}$$
$$(x, y) \in S \Longleftrightarrow (x \in A \wedge \varphi(x) \leq \varphi(y)).$$

[Hint. Proof of Theorem 7.23.] Use this to show that Π^1_1 has the reduction property.

Let $A \subset {}^\omega\omega$. A sequence $(\varphi_n : n < \omega)$ of norms on A is called a *scale on A* iff the following holds true. Let $\{x_k : k < \omega\} \subset A$ be such that $x = \lim_{k\to\infty} x_k$ and such that for every $n < \omega$ there are $k(n) < \omega$ and λ_n such that $\varphi_n(x_k) = \lambda_n$ for every $k \geq k(n)$. Then $x \in A$ and $\varphi_n(x) \leq \lambda_n$ for all $n < \omega$.

7.13. Let $A \subset {}^\omega\omega$, and let $(\varphi_n : n < \omega)$ be a scale on A. Let $(s, f) \in T$ iff $s \in {}^{<\omega}\omega$, f is a finite sequence of ordinals of the same length as s, and there is some $x \in A$ such that $s = x \restriction \mathrm{lh}(s)$ and $f(n) = \varphi_n(x)$ for all $n < \mathrm{lh}(s)$. Show that $A = p[T]$.

7.14. Let $A \subset {}^\omega\omega$ be Π^1_1. Show that there is a scale $(\varphi_n : n < \omega)$ on A such that there are $R, S \subset \omega \times ({}^\omega\omega)^2$, $R \in \Pi^1_1$ and $S \in \Sigma^1_1$, such that for all $y \in A$,

$$(n, x, y) \in R \Longleftrightarrow (x \in A \wedge \varphi_n(x) \leq \varphi_n(y)) \text{ and}$$
$$(n, x, y) \in S \Longleftrightarrow (x \in A \wedge \varphi_n(x) \leq \varphi_n(y)).$$

[Hint. Proof of Theorem 7.23.] Use this to derive the conclusion of Theorem 7.23.

Let $x, z \in {}^\omega\omega$. Then x is called a $\Delta^1_1(z)$ *singleton* iff $\{x\}$ is a $\Sigma^1_1(z)$, and hence $\Delta^1_1(z)$, set.

7.15. Let $x, z \in {}^\omega\omega$. Show that x is a $\Delta^1_1(z)$ singleton iff $x \in J_{\omega_1^z}$. [Hint. "$\Longrightarrow$": Let $\{x\} = p[T]$ where T on $\omega \times \omega$ is in $J_{\omega_1^z}[z]$. If $s \in {}^{<\omega}\omega$, $s \neq x \restriction \mathrm{lh}(s)$, then $T_s = \{(s', t') \in T : s' \subset s \vee s' \supset s\}$ is well-founded and hence has a ranking in $J_{\omega_1^z}[z]$ by Problem 5.28. Then use Problem 5.25 (c).]

7.16. Show that for every $z \in {}^\omega\omega$ there is a $\Sigma^1_1(z)$-set A such that A does not contain any $\Delta^1_1(z)$ singleton. Show that there is actually such an A which is $\Pi^0_1(z)$, i.e., A is closed and $A = p[T]$ for some tree T on ω which is definable over $(V_\omega; \in, z)$.

Conclude that Theorem 7.23 is false for Σ^1_1 and that in fact there is a closed $R \subset ({}^\omega\omega)^2$ which cannot be uniformized by an analytic function. [Hint. Suppose that every $\Pi^0_1(z)$ set contains a $\Delta^1_1(z)$ singleton. Say $A \subset {}^\omega\omega$ is $\Sigma^1_1(z)$. Then

$$x \in A \Longleftrightarrow \exists y \in \Delta^1_1(x \oplus z)\, (x, y) \in B, \tag{7.12}$$

where B is $\Pi_1^0(z)$. Using Problem 5.27 , show that the right hand side of (7.12) can be written in a $\Pi_1^1(z)$ fashion, so that every $\Sigma_1^1(z)$ set would also be $\Pi_1^1(z)$.]

Chapter 8
Solovay's Model

In this chapter we shall construct a model of ZF in which every set of reals is LEBESGUE measurable and has the property of BAIRE.

8.1 Lebesgue Measurability and the Property of Baire

Definition 8.1 Let $s \in {}^{<\omega}\omega$, and let $U_s = \{x \in {}^{\omega}\omega \colon s \subset x\}$. We recursively define the *measure* $\mu(U_s)$ of U_s, for $s \in {}^{<\omega}\omega$ as follows. Set $\mu(U_\emptyset) = \mu({}^{\omega}\omega) = 1$. Having defined $\mu(U_s)$, we let $\mu(U_{s^\frown n}) = \frac{1}{2^{n+1}} \cdot \mu(U_s)$. Now let $A \subset {}^{\omega}\omega$ be open, say $A = \bigcup\{U_s \colon s \in X\}$, where $U_s \cap U_t = \emptyset$ for all $s \neq t, s, t \in X$. We define the *measure* $\mu(A)$ of A to be

$$\sum_{s \in X} \mu(U_s).$$

If $B \subset {}^{\omega}\omega$ is arbitrary, then we define the *outer measure* $\mu^*(B)$ of B to be

$$\inf\{\mu(A) \colon B \subset A \wedge A \text{ is open}\}.$$

A set $B \subset {}^{\omega}\omega$ is called a *null set*, or just *null*, iff $\mu^*(B) = 0$.

It is easy to verify that if every B_n, $n < \omega$, is null, then so is $\bigcup_{n<\omega} B_n$.

Usually, a set $B \subset {}^{\omega}\omega$ is called LEBESGUE *measurable* iff for all $X \subset {}^{\omega}\omega$,

$$\mu^*(X) = \mu^*(X \cap B) + \mu^*(X \setminus B). \tag{8.1}$$

If $B \subset {}^{\omega}\omega$ is LEBESGUE measurable, then one also writes $\mu(B)$ for $\mu^*(B)$ and calls it the LEBESGUE *measure* of B. It is not hard to verify that the family of sets which are LEBESGUE measurable forms a σ-algebra containing all the open sets, so that in particular all BOREL sets are LEBESGUE measurable. (Cf. Problem 8.2.) For our purpose, we'll define LEBESGUE measurability as follows.

R. Schindler, *Set Theory*, Universitext, DOI: 10.1007/978-3-319-06725-4_8,

Definition 8.2 Let $A \subset {}^\omega\omega$. We say that A is LEBESGUE *measurable* iff there is a BOREL set $B \subset {}^\omega\omega$ such that

$$A \Delta B = (A \setminus B) \cup (B \setminus A) \text{ is null.}$$

The definitions given carry over, *mutatis mutandis*, to the CANTOR space ${}^\omega 2$, with the difference that if $s \in {}^{<\omega}2$, then for $U_s = \{x \in {}^\omega 2\colon\ s \subset x\}$, $\mu(U_s) = \frac{1}{2^{\mathrm{lh}(s)}}$.

Let $\mathscr{B}$ be the σ-algebra of all BOREL sets $B \subset {}^\omega\omega$. For $A, B \in \mathscr{B}$, let us write $A \leq B$ iff $A \subset B$ modulo a null set, i.e., iff $A \setminus B$ is null. Write $A \sim B$ iff $A \leq B$ and $B \leq A$ (i.e., iff $A \Delta B$ is null), and let $[A]$ denote the equivalence class of A with respect to $\sim$, i.e. $[A] = \{B\colon B \sim A\}$. The order $\leq$ on $\mathscr{B}$ induces an order, which we shall also denote by $\leq$, on the set of of all equivalence classes by setting $[A] \leq [B]$ iff $A \leq B$. (Notice that if $A' \in [A]$ and $B' \in [B]$, then $A \leq B$ iff $A' \leq B'$.)

We shall write $\mathscr{B}/\text{null}$ for the set $\{[A]\colon A \in \mathscr{B} \wedge \mu(A) > 0\}$, equipped with the order $\leq$. The partial order $\mathscr{B}/\text{null}$ is called the *measure algebra*, or, *random algebra*. We'll see later, cf. Lemma 8.8, that forcing with $\mathscr{B}/\text{null}$ amounts to adding a single real, a "random real."

Recall Definition 1.11 for $\mathbb{R}$ which, *mutatis mutandis*, carries over to ${}^\omega\omega$. Let $A \subset {}^\omega\omega$. Then A is *nowhere dense* iff for all nonempty open $B \subset {}^\omega\omega$ there is some nonempty open $B' \subset B$ such that $B' \cap A = \emptyset$. A set $B \subset {}^\omega\omega$ is called *meager* (or *of first category*) iff B is the countable union of nowhere dense sets.

Usually, a set $B \subset {}^\omega\omega$ is said to have the BAIRE *property* iff there is some open set $A \subset {}^\omega\omega$ such that $B \Delta A$ is meager. It is not hard to verify that the family of sets which have the BAIRE property forms a σ-algebra containing all the open sets, so that in particular all BOREL sets have the BAIRE property. (Cf. Problem 8.2.)

For our purposes, we may then define the BAIRE property as follows.

Definition 8.3 Let $A \subset {}^\omega\omega$. We say that A has the BAIRE *property* iff there is a BOREL set $B \subset {}^\omega\omega$ such that

$$A \Delta B = (A \setminus B) \cup (B \setminus A) \text{ is meager.}$$

Again the definitions given carry over, *mutatis mutandis*, to the CANTOR space ${}^\omega 2$.

We may now define a partial order $\mathscr{B}/\text{meager}$ in exactly the same way as we defined the measure algebra $\mathscr{B}/\text{null}$, except that we start with declaring $A \leq B$ iff $A \subset B$ modulo a meager set, i.e., iff $A \setminus B$ is meager, for $A, B \in \mathscr{B}$. It turns out, though, that forcing with $\mathscr{B}/\text{meager}$ is tantamount to forcing with COHEN forcing, which is why we refer to $\mathscr{B}/\text{meager}$ as the COHEN *algebra*.

Lemma 8.4 *There is a dense homomorphism*

$$i\colon \mathbb{C} \to \mathscr{B}/\mathit{meager}.$$

Proof Let $[B] \in \mathscr{B}/\text{meager}$, so that B is a nonmeager BOREL set. There is a nonempty open set $A \subset {}^\omega\omega$ such that $A \setminus B$ is meager, and there is hence a nonempty basic open set U_s, $s \in {}^{<\omega}\omega$, with $U_s \setminus B$ being meager, i.e., $[U_s] \leq [B]$. But this means that $i\colon \mathbb{C} \to \mathscr{B}/\text{meager}$ defined by $i(s) = [U_s]$ is dense. □

Lemma 8.5 *Both $\mathscr{B}/null$ as well as $\mathscr{B}/meager$ have the c.c.c.*

Proof For $\mathscr{B}$/meager this immediately follows from the preceding lemma. Now suppose $\{[B_i]\colon i < \omega_1\}$ to be an antichain in $\mathscr{B}$/null. Set

$$B_i' = B_i \setminus \bigcup_{j<i} B_j$$

for $i < \omega_1$. Notice that $B_i \cap B_j$ is null whenever $i \neq j$, as $\{[B_i]\colon i < \omega_1\}$ is an antichain, so that $\bigcup_{j<i}(B_i \cap B_j)$ is null whenever $i < \omega_1$. But then $B_i' = B_i \setminus \bigcup_{j<i}(B_i \cap B_j) \sim B_i$, i.e., $[B_i'] = [B_i]$ for all $i < \omega_1$, $B_i' \cap B_j' = \emptyset$ whenever $i \neq j$, and $\mu(B_i') > 0$ for all $i < \omega_1$. By the Pigeonhole Principle, there will be an $n < \omega$ such that $\mu(B_i') > \frac{1}{n}$ for $\aleph_1$ many $i < \omega_1$. This gives a contradiction! □

We now want to verify that all $\underset{\sim}{\Sigma}{}^1_1$ as well as all $\underset{\sim}{\Pi}{}^1_1$-sets are LEBESGUE measurable and have the BAIRE property.

Let $\mathbb{P}$ be an atomless partial order. By Lemma 6.11, there is then no $\mathbb{P}$-generic filter over V. As in the following definition, it is often very convenient, though, to pretend that there is and say things like "pick G which is $\mathbb{P}$-generic over V." To make such talk rigorous, we should instead talk about filters which are generic over collapses of countable elementary substructures of rank initial segments of V, or have the letter "V" not denote the true universe V of all sets but rather e.g. a countable transitive model of ZFC.

Definition 8.6 Let $A \subset {}^\omega\omega$, and let κ be an uncountable cardinal. We say that A is *κ-universally* BAIRE iff $A = p[T]$, where T is a tree on $\omega \times \alpha$ for some ordinal α, and there is some tree U on $\omega \times \beta$ for some ordinal β such that $p[U] \cap p[T] = \emptyset$ and for all posets $\mathbb{P} \in H_\kappa$,

$$1_{\mathbb{P}} \Vdash p[\check{U}] \cup p[\check{T}] = {}^\omega\omega.$$

A is called *universally* BAIRE iff A is κ-universally BAIRE for all uncountable cardinals κ.

Notice that if $p[U] \cap p[T] = \emptyset$ in V, then $p[U] \cap p[T] = \emptyset$ in $V[G]$ for all generic extensions[1] $V[G]$ of V: if $p[U] \cap p[T] \neq \emptyset$ in $V[G]$, then, setting $U \oplus T = \{(s, f, g)\colon (s, f) \in U \wedge (s, g) \in T\}$, $p[U \oplus T] \neq \emptyset$ in $V[G]$ and hence $p[U \oplus T] \neq \emptyset$ in V by absoluteness of wellfoundedness, cf. Lemma 5.6, and thus $p[U] \cap p[T] \neq \emptyset$ in V. Therefore, if T, U witness that A is κ-universally BAIRE, then $p[T]$ and $p[U]$ project to complements of each other in all $V[G]$, where G is $\mathbb{P}$-generic over V for some $\mathbb{P} \in H_\kappa$. Also, if T, U as well as T', U' both witness that A is κ-universally

[1] cf. the remark before Definition 8.6.

BAIRE, then $p[T] = p[T']$ in all $V[G]$, where G is $\mathbb{P}$-generic over V for some $\mathbb{P} \in H_\kappa$: if, say $x \in (p[T] \cap V[G]) \setminus p[T']$, then $x \in p[U']$, so that by the argument just given $p[T] \cap p[U'] \neq \emptyset$ in V; but $p[T] = A$ and $p[U'] = {}^\omega\omega \setminus A$ in V.

Therefore, if $A \subset {}^\omega\omega$ is κ-universally BAIRE and if G is $\mathbb{P}$-generic over V, where $\mathbb{P} \in H_\kappa$, then we may unambiguously define the new version A^G of A in $V[G]$ as $p[T] \cap V[G]$, where T, U witness A is κ-universally BAIRE. We often just write A^* rather then A^G, provided that G is clear from the context.

Let us consider a closely related situation. Let M and N be inner models such that $M \subset N$. (We allow M and N to exists in $V[G]$, a generic extension of V.) Let A be a BOREL set in M, say $M \models$ "$A = p[T] \wedge {}^\omega\omega \setminus A = p[U]$," where T and U are trees on ω^2, $T, U \in M$. By the absoluteness of wellfoundedness, cf. Lemma 5.6, we must have that $N \models p[T] \cap p[U] = \emptyset$. Let us define a simple variant of the SHOENFIELD tree S on $\omega \times \omega_1^N$ by $(s, h) \in S$ iff $s \in {}^{<\omega}\omega$ and, setting

$$T_s = \{t \in {}^{<\omega}\omega\colon \mathrm{lh}(t) \leq lh(s) \wedge (s \restriction lh(t), t) \in T\} \text{ and}$$
$$U_s = \{t \in {}^{<\omega}\omega\colon \mathrm{lh}(t) \leq lh(s) \wedge (s \restriction lh(t), t) \in U\},$$

$h\colon lh(s) \to \omega_1^N$ is such that

$$\forall k < \frac{\mathrm{lh}(s)-1}{2} \forall l < \frac{\mathrm{lh}(s)-1}{2} [(e(k) \in T_s \wedge e(l) \in T_s \wedge e(k) \supsetneq e(l) \Rightarrow h(2k) < h(2l)) \wedge$$
$$(e(k) \in U_s \wedge e(l) \in U_s \wedge e(k) \supsetneq e(l) \Rightarrow h(2k+1) < h(2l+1))].$$

(Here, $e\colon \omega \to {}^{<\omega}\omega$ is a bijection such that if $n < \mathrm{lh}(s)$, then $e^{-1}(s \restriction n) < e^{-1}(s)$.) It is straightforward to verify that $S \in L[T, U] \subset M$, and that *both in M and N*,

$$p[T] \cup p[U] \neq {}^\omega\omega \Longleftrightarrow [S] \neq \emptyset.$$

By the absoluteness of wellfoundedness, cf. Lemma 5.6, we must then have that $N \models p[T] \cup p[U] = {}^\omega\omega$. (In particular, every BOREL set is universally Baire, as being witnessed by a pair of trees on ω^2, cf. Lemma 8.7.)

As above, we may now also show that if $M \models$ "$A = p[T'] \wedge {}^\omega\omega \setminus A = p[U']$," where T' and U' are trees on ω^2, $T', U' \in M$, then $N \models p[T'] = p[T]$. We may therefore now unambiguously write A^N for $p[T]$, as computed in N. If M, N are clear from the context, we often just write A^* rather than A^N. Of course if $M = V$ and $N = V[G]$ is a generic extension of V, then $A^N = A^G$, so that "A^*" has an unambiguous meaning.

Still let M and N be inner models such that $M \subset N$ (possibly in $V[G]$ rather than V, e.g. $M = V$, $N = V[G]$). Suppose that A is a BOREL set in M, or A is universally BAIRE in M and N is a generic extension of M. The following facts are easy to verify, cf. Problem 8.4:

1. For $s \in {}^{<\omega}\omega$, $((U_s)^M)^* = (U_s)^N$.
2. $({}^\omega\omega \cap M \setminus A)^* = ({}^\omega\omega \cap N) \setminus A^*$.
3. $(\bigcap_{n<\omega} A_n)^* = \bigcap_{n<\omega} (A_n)^*$.

Lemma 8.7 *Every analytic (and hence also every coanalytic) set is universally* BAIRE.

Proof Let $A \subset {}^\omega\omega$ be analytic, say $A = p[T]$, where T is on $\omega \times \omega$. Let κ be any uncountable cardinal. Let S on $\omega \times \kappa$ be the "κ version" of the SHOENFIELD tree (cf. the proof of Lemmas 7.8 and 7.13): $(s, h) \in S$ iff $s \in {}^{<\omega}\omega$ and, setting

$$T_s = \{t \in {}^{<\omega}\omega\colon \mathrm{lh}(t) \leq lh(s) \wedge (s \upharpoonright lh(t), t) \in T\},$$

$h\colon lh(s) \to \kappa$ is such that

$$\forall k < lh(s) \forall l < lh(s)(e(k) \in T_s \wedge e(l) \in T_s \wedge e(k) \supsetneq e(l) \Rightarrow h(k) < h(l)).$$

(Again, $e\colon \omega \to {}^{<\omega}\omega$ is a bijection such that if $n < lh(s)$, then $e^{-1}(s \upharpoonright n) < e^{-1}(s)$.) It is straightforward to see that T, S witness that A is κ-universally BAIRE. □

The previous proof shows that in fact if $A \subset {}^\omega\omega$ is coanalytic, then A is universally BAIRE "in a strong sense": Say A is $\Sigma^1_1(x)$ with $x \in {}^\omega\omega$, then the trees witnessing that A is universally BAIRE may be taken *as elements of* $L[x]$.

By Lemma 8.7, if A is analytic or coanalytic (for instance, if A is just BOREL) and if G is $\mathbb{P}$-generic over V, where $\mathbb{P} \in V$, then A^G (i.e., A^*) is well-defined.

Lemma 8.8 (1) *Let G be $\mathscr{B}$/null-generic over V. Then there is a unique $x_G \in {}^\omega\omega \cap V[G]$ such that for all $B \in \mathscr{B}$,*

$$x_G \in B^* \iff [B] \in G.$$

(2) *Let G be $\mathscr{B}$/meager-generic over V. Then there is a unique $x_G \in {}^\omega\omega \cap V[G]$ such that for all $B \in \mathscr{B}$,*

$$x_G \in B^* \iff [B] \in G.$$

Proof The same proof works for (1) and (2). Let us first show uniqueness. Let $x, y \in {}^\omega\omega \cap V[G]$ be such that $x \in B^* \Leftrightarrow [B] \in G \Leftrightarrow y \in B^*$ for all $B \in \mathscr{B}$. In particular, $x \in U^*_s \Leftrightarrow [U^*_s] \in G \Leftrightarrow y \in U^*_s$ for all $s \in {}^{<\omega}\omega$. If $x \neq y$, then there is some $s \in {}^{<\omega}\omega$ with $\mathrm{lh}(s) > 0$ and $x \in U^*_s$ but $y \notin U^*_s$; this gives a contradiction!

Let us now show existence. Working in $V[G]$, let us recursively construct $\{s_n\colon n < \omega\} \subset {}^{<\omega}\omega$ with $lh(s_n) = n$ and $[U_{s_n}] \in G$ for all $n < \omega$ as follows. Set $s_0 = \emptyset$. Of course, $[U_\emptyset] = [{}^\omega\omega] \in G$. Given s_n with $\mathrm{lh}(s_n) = n$ and $[U_{s_n}] \in G$, we pick s_{n+1} as follows. Let $[B] \leq [U_{s_n}]$, where $[B] \in \mathscr{B}$/null (or $[B] \in \mathscr{B}$/meager), i.e., B is not null (or not meager). Therefore, one of $B \cap U_{s_n{}^\frown k}$, $k < \omega$, is not null (or not meager). This argument shows that

$$D = \{[B] \in \mathscr{B}/\text{null (or meager)}\colon \exists k\, [B] \leq [U_{s_n{}^\frown k}]\}$$

is dense below $[U_{s_n}]$. There is hence some $k < \omega$ such that $[U_{s_n{}^\frown k}] \in G$, and we may set $s_{n+1} = s_n{}^\frown k$ for this k. Let us also set $x = x_G = \bigcup_{n<\omega} s_n$.

We claim that

$$\text{for all } B \in \mathscr{B}, x \in B^* \iff [B] \in G. \tag{8.2}$$

Well, (8.2) is true for each basic open set by the construction of x. An easy density argument shows that for each $B \in \mathscr{B}$, exactly one of $[B]$, $[{}^\omega\omega \backslash B]$ has to be in G. Therefore, if (8.2) is true for $B \in \mathscr{B}$, then it is also true for ${}^\omega\omega \backslash B$. Now let $B_n \in \mathscr{B}$, $n < \omega$, such that for each $n < \omega$, $x \in B_n^* \Leftrightarrow [B_n] \in G$. Another easy density arguement (similar to the one above) yields that at least one element of $\{[\bigcap_{n<\omega} B_n]\} \cup \{[{}^\omega\omega \backslash B_n]\colon\ n < \omega\}$ must be in G. We therefore must have that $x \in (\bigcap_{n<\omega} B_n)^* = \bigcap_{n<\omega} B_n^*$ iff $x \in B_n^*$ for all $n < \omega$ iff $[B_n] \in G$ for all $n < \omega$ iff $[\bigcap B_n] \in G$. We have shown (8.2). □

If M is an inner model, if G is $(\mathscr{B}/\text{null})^M$-generic over M and if x_G is unique with $x_G \in B^* \iff [B] \in G$ for all BOREL sets B of M, then x_G is called a *random real* over M. (Here, B^* is computed in V.)

If, on the other hand, G is $(\mathscr{B}/\text{meager})^M$-generic over M, and if x_G is unique with $x_G \in B^* \iff [B] \in G$ for all BOREL sets B of M, then Lemma 8.4 above shows that x is a COHEN *real* over M. (Again, B^* is computed in V.)

Lemma 8.9 *Let M be a transitive model of* ZFC.

(1) *$x \in {}^\omega\omega$ is a random real over M iff $x \notin B^*$ for all $B \in \mathscr{B}^M$ which are null sets in M.*
(2) *$x \in {}^\omega\omega$ is a* COHEN *real over M iff $x \notin B^*$ for all $B \in \mathscr{B}^M$ which are meager sets in M.*

Proof (1) First let $x = x_G \in {}^\omega\omega$ be random over M. Let $B \in \mathscr{B}^M$ be a null set in M. As $[({}^\omega\omega \cap M)\backslash B] = [{}^\omega\omega \cap M] \in G$, we have that $x \in (({}^\omega\omega \cap M)\backslash B)^* = ({}^\omega\omega \cap M[G])\backslash B^*$, i.e., $x \notin B^*$.

Now suppose $x \in {}^\omega\omega$ to be such that $x \notin B^*$ for all $B \in \mathscr{B}^M$ which are null sets in M. Let

$$G = \left\{[B]\colon B \in \mathscr{B}^M \wedge x \in B^*\right\}.$$

It suffices to verify that G is a filter which is generic over M.

Well, G is easily seen to be a filter. Now let $A \in M$ be a maximal antichain. As $M \models$ "$\mathscr{B}$/null has the c.c.c." by Lemma 8.5, A is countable in M, say $A = \{[B_n]\colon n < \omega\}$, where $(B_n\colon n < \omega) \in M$. Also $({}^\omega\omega \cap M)\backslash \bigcup_{n<\omega} B_n$ must be a null set in M, as A is a maximal antichain. But then $x \notin (({}^\omega\omega \cap M)\backslash \bigcup_{n<\omega} B_n)^*$, and hence $x \in (\bigcup_{n<\omega} B_n)^* = \bigcup_{n<\omega} B_n^*$, i.e., $x \in B_n^*$ for some $n < \omega$, so that $[B_n] \in G$ for some $n < \omega$, as desired.

The proof of (2) is entirely analogous. □

Lemma 8.10 *Let M be a transitive model of* ZFC.

(1) *If $(2^{\aleph_0})^M$ is countable, then $A = \{x \in {}^\omega\omega\colon x$ is not random over $M\}$ is a null set.*

(2) *If* $(2^{\aleph_0})^M$ *is countable, then* $B = \{x \in {}^\omega\omega\colon x$ *is not* COHEN *over* $M\}$ *is a meager set.*

Proof There are (provably in ZFC) $2^{\aleph_0}$ BOREL sets. Therefore, as $(2^{\aleph_0})^M$ is countable, there are only countably many $B \in M$ such that B is a BOREL set from the point of view of M.

(1) By Lemma 8.9 (1), $x \in A$ iff

$$x \in \bigcup \{B^*\colon B \in \mathscr{B}^M \wedge B \text{ is null in } M\}.$$

As $(2^{\aleph_0})^M$ is countable, A is thus a countable union of null sets, and hence A is null.

(2) By Lemma 8.9 (2), $x \in B$ iff

$$x \in \bigcup \{B^*\colon B \in \mathscr{B}^M \wedge B \text{ is meager in } M\}.$$

As $(2^{\aleph_0})^M$ is countable, B is thus a countable union of meager sets, and hence B is meager. □

Definition 8.11 Let M be a transitive model of ZFC. We say that $x \in {}^\omega\omega \cap V$ is *generic over M* if there is a poset $\mathbb{P} \in M$ and there is a $\mathbb{P}$-generic filter G over M such that $M[G]$ is the $\subset$-least transitive model N of ZFC with $M \cup \{x\} \subset N$. In this situation, we also write $M[x]$ instead of $M[G]$.

Now let $A \subset {}^\omega\omega$. We say that A is SOLOVAY *over M* iff there is a formula φ and there are parameters $a_1, \ldots, a_k \in M$ such that for all $x \in {}^\omega\omega$, if x is generic over M, then

$$x \in A \Longleftrightarrow M[x] \models \varphi(x, a_1, \ldots, a_k).$$

Lemma 8.12 *Let M be a transitive model of ZFC, and let* $A \subset {}^\omega\omega$ *be* SOLOVAY *over M.*

(1) *There is a* BOREL *set* $B \subset {}^\omega\omega$ *such that for every* $x \in {}^\omega\omega$ *which is random over* M, $x \in A \Leftrightarrow x \in B$.
(2) *There is a* BOREL *set* $C \subset {}^\omega\omega$ *such that for every* $x \in {}^\omega\omega$ *which is* COHEN *over* M, $x \in A \Leftrightarrow x \in C$.

Proof (1) Let φ be a formula, and let $a_1, \ldots, a_k \in M$ be such that for all $x \in {}^\omega\omega$ which are generic over M, $x \in A \Leftrightarrow M[x] \models \varphi(x, a_1, \ldots, a_k)$. Let $\tau \in M^{(\mathscr{B}/\text{null})^M}$ be a (canonical) name for the random real which is added by forcing with $(\mathscr{B}/\text{null})^M$ over M, i.e., if G is $(\mathscr{B}/\text{null})^M$-generic over M, then $\tau^G = x_G$. Let $E \subset (\mathscr{B}/\text{null})^M$ be a maximal antichain of $[D] \in (\mathscr{B}/\text{null})^M$ such that

$$[D] \Vdash_M^{(\mathscr{B}/\text{null})^M} \varphi(\tau, \check{a}_1, \ldots, \check{a}_k).$$

As $(\mathscr{B}/\text{null})$ has the c.c.c. by Lemma 8.5, E is at most countable in M, say

$$E = \{[D_n]: n < \omega\},$$

where $D_n \in \mathscr{B}$ for every $n < \omega$. Set

$$B = \bigcup_{n<\omega} D_n^* = \left(\bigcup_{n<\omega} D_n\right)^*.$$

B is a countable union of BOREL sets, and hence BOREL. We claim that for all $x \in {}^\omega\omega$ which are random over M, $x \in A \Leftrightarrow x \in B$.

Let $x = x_G$ be random over M. Then $x \in A$ iff $M[x] = M[G] \models \varphi(x, a_1, \ldots, a_k)$ iff there is some $[D] \in E \cap G$ such that

$$[D] \Vdash_M^{(\mathscr{B}/\text{ null})^M} \varphi(\tau, \check{a}_1, \ldots, \check{a}_k)$$

iff there is some $n < \omega$ with $[D_n] \in G$ iff $[\bigcup_{n<\omega} D_n] \in G$ iff $x = x_G \in (\bigcup_{n<\omega} D_n)^* = B$, as desired.

The proof of (2) is entirely analogous. □

Lemmas 8.10 and 8.12 now immediately give the following

Corollary 8.13 *Let M be a transitive model of ZFC, such that $(2^{\aleph_0})^M$ is countable. Let $A \subset {}^\omega\omega$ be* SOLOVAY *over M. Then A is* LEBESGUE *measurable and has the* BAIRE *property.*

Theorem 8.14 (Feng, Magidor, Woodin) *Let $A \subset {}^\omega\omega$ be $(2^{\aleph_0})^+$-universally* BAIRE. *Then A is* LEBESGUE *measurable and has the* BAIRE *property.*

Proof Let T, U witness that A is $(2^{\aleph_0})^+$-universally BAIRE. Let θ be a regular cardinal such that $\theta > (2^{\aleph_0})^+$ and $T, U \in H_\theta$. Let

$$\pi: M \rightarrow H_\theta$$

be an elementary embedding where M is countable and transitive and $T, U \in ran(\pi)$, say $\pi(\bar{T}) = T$ and $\pi(\bar{U}) = U$. In order to prove the theorem, by the proof of Corollary 8.13 it suffices to verify that for every $x \in {}^\omega\omega$ which is either random over M or else COHEN over M,

$$x \in A \Longleftrightarrow M[x] \models x \in p[\bar{T}].$$

Well, let $x \in {}^\omega\omega$ be either random over M or else COHEN over M. As

$$M \models \text{“}\bar{T}, \bar{U} \text{ witness that } p[\bar{T}] \text{ is } (2^{\aleph_0})^+\text{-universally BAIRE”},$$

we must have that either $M[x] \models x \in p[\bar{T}]$ or else $M[x] \models x \in p[\bar{U}]$.

Suppose that $x \in p[\bar{T}]$, say $(x \restriction n, f \restriction n) \in \bar{T}$ for all $n < \omega$. Then $(x \restriction n, \pi(f \restriction n)) \in T$ for every $n < \omega$, and hence $x \in p[T]$. (We don't need $f \in M$ here, just

$f \upharpoonright n \in M$ for each $n < \omega$, which is trivial.) In the same way, $x \in p[\bar{U}]$ implies that $x \in p[U]$.

This shows that $x \in A \Leftrightarrow x \in p[\bar{T}]$, as desired. □

Corollary 8.15 *Every analytic as well as every coanalytic set is* LEBESGUE *measurable and has the* BAIRE *property.*

Definition 8.16 Let $A \subset {}^{\omega}\omega$. A is said to have the BERNSTEIN *property* iff for every perfect set $P \subset {}^{\omega}\omega$, $P \cap A$ or $P \setminus A$ contains a perfect subset.

Lemma 8.17 *Let M be a transitive model of* ZFC *such that $(2^{\aleph_0})^M$ is countable. Let $A \subset {}^{\omega}\omega$ be* SOLOVAY *over M. Let $P \subset {}^{\omega}\omega$ be a perfect set such that $P = [T]$ for some perfect tree on ω with $T \in M$. Then $P \cap A$ or $P \setminus A$ contains a perfect subset.*

Proof Let φ be a formula, and let $a_1, \ldots, a_k \in M$ be such that for all $x \in {}^{\omega}\omega$ which are generic over M, $x \in A \Leftrightarrow M[x] \models \varphi(x, a_1, \ldots, a_k)$. As $(2^{\aleph_0})^M$ is countable, there is some $G \in V$ which is $\mathbb{C}$-generic over M. As T is perfect, we may pick some $(t_s : s \in {}^{<\omega}2)$ such that for all $s \in {}^{<\omega}2$, $t_s \in T$ and $t_{s^\frown 0}$ and $t_{s^\frown 1}$ are two incompatible extensions of t_s in T of the same length. Let $x_G = \bigcup G$ be the COHEN real over M given by G. Let $y \in {}^{\omega}2$ be defined by

$$y(n) = \begin{cases} 0\,, \text{ if } x(n) \text{ is even, and} \\ 1\,, \text{ if } x(n) \text{ is odd.} \end{cases}$$

It is easy to verify that then

$$\bigcup_{n<\omega} t_{y \upharpoonright n} \in [T] \setminus M.$$

Write $z = \bigcup_{n<\omega} t_{y \upharpoonright n}$.

Let us suppose that $z \in A$. Let $\tau \in M^{\mathbb{C}}$ be a canonical name for $\bigcup_{n<\omega} t_{y \upharpoonright n}$, and let $p \in \mathbb{C}$ be such that

$$p \Vdash^{\mathbb{C}}_{M} \tau \in [T] \setminus ({}^{\omega}\omega \cap M)^{\vee} \wedge \varphi(\tau, a_1, \ldots, a_k). \tag{8.3}$$

Let $(D_n : n < \omega) \in V$ enumerate the sets in M which are dense in $\mathbb{C}$. By recursion on $\mathrm{lh}(s)$, where $s \in {}^{<\omega}2$, we now construct conditions $p_s \leq p$ and sequences $t_s \in {}^{<\omega}\omega$ such that $p_s \Vdash \tau \upharpoonright \mathrm{lh}(\check{t_s}) = \check{t_s}$ as follows.

Put $p_\emptyset = p$ and $t_\emptyset = \emptyset$. Now suppose p_s and t_s have been constructed. There must be extensions $p^0, p^1 \leq p_s$, $m \geq lh(t_s)$ and $n^0 \neq n^1 \in \omega$ such that

$$p^0 \Vdash \tau(\check{m}) = \check{n^0} \text{ and } p^1 \Vdash \tau(\check{m}) = \check{n^1},$$

because otherwise $p_s \Vdash \tau \in ({}^{\omega}\omega \cap M)^{\vee}$ by the homogeneity of $\mathbb{C}$, cf. Lemma 6.53. We may thus pick $p_{s^\frown 0} \neq p_{s^\frown 1}$, both in $D_{lh(s)}$, and $t_{s^\frown 0} \neq t_{s^\frown 1}$ with $lh(t_{s^\frown 0}) = lh(t_{s^\frown 1})$ such that

$$p_{s\frown 0} \Vdash^{\mathbb{C}}_{M} \tau \restriction lh((t_{s\frown 0})^{\check{}}) = (t_{s\frown 0})^{\check{}} \text{and } p_{s\frown 1} \Vdash^{\mathbb{C}}_{M} \tau \restriction lh((t_{s\frown 1})^{\check{}}) = (t_{s\frown 1})^{\check{}}.$$

In the end, for each $x \in {}^{\omega}2$, $\{p_{x\restriction n} : n < \omega\}$ generates a $\mathbb{C}$-generic filter $g_x \in V$ over M such that by (8.1), $\tau^{g_x} \notin M$, $\tau^{g_x} \in [T]$, and $\tau^{g_x} \in A$. By construction, the set $\{t^{g_x} : x \in {}^{\omega}2\}$ is thus a perfect subset of $P \cap A$.

If $z \notin A$, then a symmetric argument yields that $P \setminus A$ contains a perfect subset. □

Theorem 8.18 *Let $A \subset {}^{\omega}\omega$ be $\aleph_1$-universally* BAIRE. *Then A has the* BERNSTEIN *property.*

Proof We amalgamate the arguments for Theorem 8.14 and Lemma 8.17. Let $P \subset {}^{\omega}\omega$ be perfect. Let T, U witness that A is $\aleph_1$-universally BAIRE, and let S be a perfect tree on ω such that $P = [S]$, cf. Lemma 7.1. Let θ be a regular cardinal such that $\theta > (2^{\aleph_0})^+$ and $T, U \in H_\theta$. Let

$$\pi : M \to H_\theta$$

be an elementary embedding where M is countable and transitive and $S, T, U \in ran(\pi)$, say $\pi(\bar{T}) = T$ and $\pi(\bar{U}) = U$. Notice that $\pi(S) = S$.

Let $g \in V$ be $\mathbb{C}$-generic over M. As in the proof of Lemma 8.17,

$$(P \setminus M) \cap M[g] \neq \emptyset. \tag{8.4}$$

As in the proof of Theorem 8.14,

$$p[\bar{T}]^{M[g]} = A \cap M[g] \text{ and } p[\bar{U}]^{M[g]} = ({}^{\omega}\omega \setminus A) \cap M[g].$$

We may then finish off the argument exactly as in the proof of Lemma 8.17, with "$x \in p[\bar{T}]$" playing the role of "$\varphi(x, a_1, \ldots, a_k)$." □

Corollary 8.19 *Every analytic as well as every coanalytic set of reals has the* BERNSTEIN *property.*

We now want to start producing a model of ZF + "every set of reals is LEBESGUE measurable and has the BAIRE property". Modulo of what has been done so far, the remaining issue will be an analysis of forcing.

8.2 Solovay's Theorem

We shall now analyze the situation after LEVY collapsing an inaccessible cardinal, cf. Definitions 4.41 and 6.43. We shall also need the concept of $\mathsf{OD}_{^{\omega}\omega}$, cf. Definition 5.42. We may construe Theorem 8.23 as an ultimate extension of Corollaries 8.15 and 7.16.

Theorem 8.20 *Let κ be an inaccessible cardinal, and let G be* $\mathrm{Col}(\omega, {<}\kappa)$*-generic over V. Then in $V[G]$, ω_1 is inaccessible to the reals.*

Proof Recall that $\mathrm{Col}(\omega, {<}\kappa)$ has the κ-c.c., cf. Lemma 6.44. Therefore, $V[G]$ and V have the same cardinals $\geq\kappa$ and $\kappa = \omega_1^{V[G]}$, cf. Lemma 6.32.

If $f \in {}^{\omega}\mathrm{OR} \cap V[G]$, then there is some $\tau \in V^{\mathrm{Col}(\omega, <\kappa)}$ such that $f = \tau^G$ and

$$\tau = \{((n, \alpha)^{\vee}, p) \colon n < \omega \wedge \alpha \in \mathrm{OR} \wedge p \in A_n \wedge p \Vdash \tau(\check{n}) = \check{\alpha}\},$$

where each $A_n, n < \omega$, is a maximal antichain in V of $p \in \mathrm{Col}(\omega, {<}\kappa)$ with $\exists\alpha p \Vdash \tau(\check{n}) = \check{\alpha}$. As $\mathrm{Col}(\omega, {<}\kappa)$ has the κ-c.c., $(A_n \colon n < \omega) \in V_\kappa = H_\kappa$, say $(A_n \colon n < \omega) \in H_\lambda$, where $\lambda < \kappa$. Let us write

$$G \restriction \lambda = \{p \restriction \lambda \colon p \in G\} \quad \text{and}$$
$$G \restriction [\lambda, \kappa) = \{p \restriction [\lambda, \kappa) \colon p \in G\}.$$

By the Product Lemma 6.65, $G \restriction \lambda$ is $\mathrm{Col}(\omega, {<}\lambda)$-generic over V and $G \restriction [\lambda, \kappa)$ is $\mathrm{Col}(\omega, [\lambda, \kappa))$-generic over $V[G \restriction \lambda]$. We now have $f(n) = \alpha$ iff $\exists p \in A_n \cap G\ p \Vdash \tau(\check{n}) = \check{\alpha}$ iff $\exists p \in A_n \cap G \restriction \lambda\ p \Vdash \tau(\check{n}) = \check{\alpha}$, so that in fact $f \in V[G \restriction \lambda]$.

We thus have shown the following.

Claim 8.21 *For each $f \in {}^{\omega}\omega \cap V[G]$ there is some $\lambda < \kappa$ such that $f \in V[G \restriction \lambda]$.*

It is not hard to verify that Claim 8.21 implies that ω_1 is inaccessible to the reals, cf. Problem 8.5. □

Theorem 7.28 and Lemma 7.27 now immediately yields:

Corollary 8.22 (Specker) *Let κ be an inaccessible cardinal, and let G be* Col $(\omega, {<}\kappa)$*-generic over V. Then in $V[G]$, every uncountable $\underset{\sim}{\Sigma}^1_2$-set of reals has a perfect subset. On the other hand, if every uncountable coanalytic set of reals has a perfect subset, then ω_1^V is inaccessible in L.*

Theorem 8.23 (Solovay) *Let κ be an inaccessible cardinal, and let G be* Col $(\omega, {<}\kappa)$*-generic over V. Then in $V[G]$, every set of reals which is $\mathsf{OD}_{{}^{\omega}\omega}$ is* Lebesgue *measurable and has the* Baire *property, and every uncountable set of reals which is $\mathsf{OD}_{{}^{\omega}\omega}$ has a perfect subset.*

Proof We continue from where we left off the proof of Theorem 8.20. The following is the key technical fact.

Claim 8.24 *Let $\lambda < \kappa$. Let $\mathbb{P} \in H_\kappa^{V[G \restriction \lambda]}$ be a partial order, and let $s \in V[G]$ be $\mathbb{P}$-generic over $V[G \restriction \lambda]$. There is then some $H \in V[G]$ which is* $\mathrm{Col}(\omega, {<}\kappa)$*-generic over $V[G \restriction \lambda][s]$ such that*

$$V[G] = V[G \restriction \lambda][s][H].$$

With the help of Claim 8.24, the proof of SOLOVAY's Theorem 8.23 may be finished as follows.

Let us fix $A \subset {}^\omega\omega \cap V[G]$ which is $\mathsf{OD}_{\omega_\omega}$ in $V[G]$. Let φ be a formula, let $\alpha_1, \dots, \alpha_k$ be ordinals, and let $x_1, \dots, x_l \in {}^\omega\omega \cap V[G]$ such that for all $x \in {}^\omega\omega \cap V[G]$,

$$x \in A \Longleftrightarrow V[G] \models \varphi(x, \alpha_1, \dots, \alpha_k, x_1, \dots, x_l).$$

By Claim 8.21, we may pick some $\lambda < \kappa$ such that $G \upharpoonright \lambda$ is $\mathrm{Col}(\omega, {<}\lambda)$-generic over V and $x_1, \dots, x_k \in V[G \upharpoonright \lambda]$. Let $x \in {}^\omega\omega \cap V[G]$ be generic over $V[G \upharpoonright \lambda]$.[2] By Claim 8.24, there is some $H \in V[G]$ which is $\mathrm{Col}(\omega, {<}\kappa)$-generic over $V[G \upharpoonright \lambda][x]$ such that

$$V[G] = V[G \upharpoonright \lambda][x][H].$$

We then have

$$\begin{aligned} x \in A &\Longleftrightarrow V[G] \models \varphi(x, \alpha_1, \dots, \alpha_k, x_1, \dots, x_l) \\ &\Longleftrightarrow V[G \upharpoonright \lambda][x][H] \models \varphi(x, \alpha_1, \dots, \alpha_k, x_1, \dots, x_l) \\ &\Longleftrightarrow \exists p \in H \Vdash^{\mathrm{Col}(\omega, <\kappa)}_{V[G \upharpoonright \lambda][x]} \varphi(\check{x}, \check{\alpha}_1, \dots, \check{\alpha}_k, \check{x}_1, \dots, \check{x}_l). \end{aligned}$$

However, $\mathrm{Col}(\omega, {<}\kappa)$ is homogeneous by Lemma 6.54, and therefore by Lemma 6.61 the last line is equivalent to

$$1_{\mathrm{Col}(\omega, <\kappa)} \Vdash^{\mathrm{Col}(\omega, <\kappa)}_{V[G \upharpoonright \lambda][x]} \varphi(\check{x}, \check{\alpha}_1, \dots, \check{\alpha}_k, \check{x}_1, \dots, \check{x}_l).$$

By the definability of "$\Vdash$" over $V[G \upharpoonright \lambda][x]$, cf. Theorem 6.23 (1), A is thus in fact SOLOVAY over $V[G \upharpoonright \lambda]$. However, $(2^{\aleph_0})^{V[G \upharpoonright \lambda]}$ is certainly countable from the point of view of $V[G]$, so that in $V[G]$, A is LEBESGUE measurable and has the BAIRE property by Corollary 8.13.

Now let us assume that A is uncountable in $V[G]$. We use the proof of Lemma 8.17 to show that A contains a perfect subset in $V[G]$. As A is uncountable in $V[G]$, $A \backslash V[G \upharpoonright \lambda] \neq \emptyset$. By Claim 8.21, there is hence some α with $\lambda < \alpha < \kappa$ such that

$$(A \cap V[G \upharpoonright \alpha]) \backslash V[G \upharpoonright \lambda] \neq \emptyset.$$

By the Product Lemma 6.65, $G \upharpoonright [\lambda, \alpha)$ is $\mathrm{Col}(\omega, [\lambda, \alpha))$-generic over $V[G \upharpoonright \lambda]$. Setting $P = {}^\omega\omega \cap V[G]$ and replacing $\mathbb{C}$ with $\mathrm{Col}(\omega, [\lambda, \alpha))$, the argument for Lemma 8.17 now proves that A contains a perfect subset.

In order to finish the proof of SOLOVAY's theorem, it therefore remains to show Claim 8.24.

As the case $\lambda > 0$ is only notationally different frome the case $\lambda = 0$, let us assume that $\lambda = 0$.

---

[2] It can be shown that *every* real in $V[G]$ is generic over $V[G \upharpoonright \lambda]$ in the sense of Definition 8.11, but we won't need that.

So let us fix $\mathbb{P} \in H_\kappa$, a partial order, and let $s \in V[G]$ be $\mathbb{P}$-generic over V. We aim to construct some $H \in V[G]$ which is $\mathrm{Col}(\omega, {<}\kappa)$-generic over $V[s]$ such that

$$V[G] = V[s][H].$$

Well, we have $s \in V[G \upharpoonright \mu+1]$ for some $\mu < \kappa$ by Claim 8.21. As $\mathrm{Col}(\omega, {<}\mu+1)$ has the same cardinality as μ and

$$1 \Vdash^{\mathrm{Col}(\omega, {<}\mu+1)} \check{\mu} \text{ is countable,}$$

by Lemma 6.51 there is a dense homomorphism

$$i \colon \mathrm{Col}(\omega, \mu) \to \mathrm{Col}(\omega, {<}\mu + 1).$$

By Lemma 6.48,

$$G_0 = \{p \in \mathrm{Col}(\omega, \mu) \colon i(p) \in G \upharpoonright (\mu + 1)\}$$

is a $\mathrm{Col}(\omega, \mu)$-generic filter over V with $V[G_0] = V[G \upharpoonright \mu + 1]$.

Also, $\mathrm{Col}(\omega, \mu) \cong \mathrm{Col}(\omega, \{\mu + 1\})$, so that if

$$j \colon \mathrm{Col}(\omega, \mu) \to \mathrm{Col}(\omega, \{\mu + 1\})$$

is an isomorphism, then

$$G_1 = \{p \in \mathrm{Col}(\omega, \mu) \colon \exists q \in G \; j(p) = q(\mu + 1)\}$$

is a $\mathrm{Col}(\omega, \mu)$-generic filter over $V[G_0] = V[G \upharpoonright \mu+1]$ with $V[G_0][G_1] = V[G \upharpoonright (\mu + 2)]$. Recall that $s \in V[G_0] = V[G \upharpoonright \mu + 1]$.

Claim 8.25 *There is some* $\mathrm{Col}(\omega, \mu)$*-generic filter* H^* *over* $V[s]$ *with*

$$V[s][H^*] = V[G_0][G_1]. \tag{8.5}$$

Suppose Claim 8.25 to be true. We then have that

$$\begin{aligned} V[G] &= V[G \upharpoonright (\mu + 2)][G \upharpoonright [\mu + 2, \kappa)] \\ &= V[G_0][G_1][G \upharpoonright [\mu + 2, \kappa)] \\ &= V[s][H^*][G \upharpoonright [\mu + 2, \kappa)]. \end{aligned}$$

Let $i' \colon \mathrm{Col}(\omega, \mu) \to \mathrm{Col}(\omega, {<}\mu + 2)$ be a dense homomorphism, and let us set

$$H_0 = \{p \in \mathrm{Col}(\omega, {<}\, \mu + 2) \colon \exists q \in H^* \; i'(q) \leq p\}.$$

Then H_0 is $\mathrm{Col}(\omega < \mu + 2)$-generic over $V[s]$ by Lemma 6.48 and $V[s][H^*] = V[s][H_0]$. If we finally set

$$H = \{p \in \mathrm{Col}(\omega, {<}\kappa)\colon p \restriction \mu + 2 \in H_0 \wedge p \restriction [\mu + 2, \kappa) \in G \restriction [\mu + 2, \kappa)\},$$

then H is $\mathrm{Col}(\omega, {<}\kappa)$-generic over $V[s]$ and

$$\begin{aligned} V[s][H] &= V[s][H_0][G \restriction [\mu + 2, \kappa)] \\ &= V[s][H^*][G \restriction [\mu + 2, \kappa)] \\ &= V[G], \end{aligned}$$

as desired. We have shown Claim 8.24, modulo Claim 8.25. □

It thus remains to verify Claim 8.25. We aim to produce some H^* which is $\mathrm{Col}(\omega, \mu)$-generic over $V[s]$ such that (8.5) holds true.

As $s \in V[G_0]$, we may pick some $\tau \in V^{\mathrm{Col}(\omega,\mu)}$ such that

$$s = \tau^{G_0}. \tag{8.6}$$

Let us recursively define inside $V[s]$ a sequence $(\mathbb{Q}_\alpha\colon\ \alpha \in \mathrm{OR})$ of subsets of $\mathrm{Col}(\omega, \mu)$ as follows. Set $p \in \mathbb{Q}_0$ iff for all $r \in \mathbb{P}$,

$$\begin{cases} p \Vdash \check{r} \in \tau \Longrightarrow r \in s \wedge \\ p \Vdash \check{r} \notin \tau \Longrightarrow r \notin s. \end{cases} \tag{8.7}$$

Having defined $\mathbb{Q}_\alpha$, set $p \in \mathbb{Q}_{\alpha+1}$ iff for all open dense sets $D \subset \mathrm{Col}(\omega, \mu)$, $D \in V$,

$$\exists p' \leq p\, (p' \in D \cap \mathbb{Q}_\alpha).$$

If λ is a limit ordinal, and $\mathbb{Q}_\alpha$ is defined for every $\alpha < \lambda$, then we set $\mathbb{Q}_\lambda = \bigcap_{\alpha<\lambda} \mathbb{Q}_\alpha$.

For each α, if $p \in \mathbb{Q}_\alpha$ and $p \leq p'$, $p' \in \mathrm{Col}(\omega, \mu)$, then $p' \in \mathbb{Q}_\alpha$. This gives that if $\alpha \leq \beta$, then $\mathbb{Q}_\beta \subset \mathbb{Q}_\alpha$. Let δ be least such that $\mathbb{Q}_{\delta+1} = \mathbb{Q}_\delta$. Set

$$\bar{\mathbb{Q}} = \mathbb{Q}_\delta \text{ and } \mathbb{Q} = \bar{\mathbb{Q}} \times \mathrm{Col}(\omega, \mu).$$

We construe $\bar{\mathbb{Q}}$ and $\mathbb{Q}$ as partial orders, with the order relation given by the restriction of the order relation of $\mathrm{Col}(\omega, \mu)$ and $\mathrm{Col}(\omega, \mu) \times \mathrm{Col}(\omega, \mu)$ to $\bar{\mathbb{Q}}$ and $\mathbb{Q}$, respectively. If $p \in \bar{\mathbb{Q}}$, $p \leq q$, and $q \in \mathrm{Col}(\omega, \mu)$, then $q \in \bar{\mathbb{Q}}$.

For the record, notice that $\mathbb{Q}$ was defined inside $V[s]$, and the parameters we need for this are μ, $\mathbb{P}$, τ, and s. Let us write $\Psi(v_0, v_1, v_2, v_3, v_4)$ for the defining formula, i.e.,

$$V[s] \models \forall \mathbb{Q}'\, (\mathbb{Q}' = \mathbb{Q} \longleftrightarrow \Psi(\mathbb{Q}', \mu, \mathbb{P}, \tau, s)). \tag{8.8}$$

Subclaim 8.26 $G_0 \subset \bar{\mathbb{Q}}$.

Proof Suppose that $p \in G_0 \backslash \mathbb{Q}_\beta$, where β is minimal such that $G_0 \backslash \mathbb{Q}_\beta \neq \emptyset$. We cannot have $\beta = 0$, by (8.6) and the definition of $\mathbb{Q}_0$. Also, β cannot be a limit ordinal, so that $\beta = \alpha + 1$ for some α. We may pick some open dense set $D \subset \mathrm{Col}(\omega, \mu)$, $D \in V$, such that

$$\forall p' \leq p\, (p' \in D \rightarrow p' \notin \mathbb{Q}_\alpha). \tag{8.9}$$

Let $p^* \in D \cap G_0$. If $p' \leq p^*, p$ with $p' \in G_0$, then $p' \in D$, as D is open, and hence $p' \notin \mathbb{Q}_\alpha$ by (8.9). But then $p' \in G_0 \backslash \mathbb{Q}_\alpha$, hence $G_0 \backslash \mathbb{Q}_\alpha \neq \emptyset$, which contradicts the choice of β. □

Subclaim 8.26 trivially implies that $\mathbb{Q} \neq \emptyset$, so that:

Subclaim 8.27 *$\mathbb{Q}$ is separative and has the same cardinality as μ inside $V[s]$.*

Subclaim 8.28 *Let $p \in \bar{\mathbb{Q}}$. In $V[G]$, there is then some G_0' which is $\mathrm{Col}(\omega, \mu)$-generic over V such that $p \in G_0'$ and*

$$s = \tau^{G_0'}.$$

Proof Let $\bar{p} \in \bar{\mathbb{Q}}$. Let $D \subset \mathrm{Col}(\omega, \mu)$, $D \in V$, be open dense in $\mathrm{Col}(\omega, \mu)$. By $\bar{p} \in \bar{\mathbb{Q}} = \mathbb{Q}_{\delta+1}$ there is some $p' \leq \bar{p}$ with $p' \in D \cap \mathbb{Q}_\delta = \bar{\mathbb{Q}}$. In $V[G]$, there are only countably many dense subsets of $\mathrm{Col}(\omega, \mu)$ which are in V, so that given $p \in \bar{\mathbb{Q}}$ we may work in $V[G]$ and produce some G_0' which is $\mathrm{Col}(\omega, \mu)$-generic over V such that $p \in G_0'$ and $G_0' \subset \bar{\mathbb{Q}}$.

Let $r \in \mathbb{P}$, and suppose that $p' \in G_0'$ decides $\check{r} \in \tau$. As $p' \in G_0' \subset \bar{\mathbb{Q}} \subset \mathbb{Q}_0$, we must have (8.7). Therefore, G_0' is as desired. □

With Subclaim 8.26, $G_0 \times G_1 \subset \mathbb{Q}$ is a filter. We now show:

Subclaim 8.29 *$G_0 \times G_1$ is $\mathbb{Q}$-generic over $V[s]$.*

Proof Let $D \in V[s]$ be dense in $\mathbb{Q}$. We need to see that $D \cap (G_0 \times G_1) \neq \emptyset$. Suppose that $D \cap (G_0 \times G_1) = \emptyset$.

Let $\rho \in V^{\mathbb{P}}$ be such that $\rho^s = D$. Recall (8.8), and let $\Phi(v_0, v_1, v_2, v_3, v_4)$ be a formula such that whenever $G_0'' \times G_1''$ is $\mathrm{Col}(\omega, \mu) \times \mathrm{Col}(\omega, \mu)$-generic over V, then

$$V[G_0'' \times G_1''] \models \Phi(G_0'', G_1'', \mu, \mathbb{P}, \tau)$$

iff the following holds true.

If $s' = \tau^{G_0''}$, then s' is $\mathbb{P}$-generic over V, and if

$$D' = \{(p, q) \in \mathrm{Col}(\omega, \mu) \times \mathrm{Col}(\omega, \mu) \colon \exists r \in s'\, r \Vdash_V^{\mathbb{P}} (p, q)^\vee \in \rho\}$$

and inside $V[s']$, $\Psi(\mathbb{Q}', \mu, \mathbb{P}, \tau, s')$ holds true for exactly one $\mathbb{Q}'$, then D' is dense in $\mathbb{Q}'$ and

$$D' \cap (G_0'' \times G_1'') = \emptyset.$$

By hypothesis, $V[G_0 \times G_1] \models \Phi(G_0, G_1, \mu, \mathbb{P}, \tau)$. Let $(p, q) \in (G_0 \times G_1)$ be such that

$$(p, q) \Vdash_V^{\mathrm{Col}(\omega,\mu)\times\mathrm{Col}(\omega,\mu)} \Phi(\dot{G}_0, \dot{G}_1, \check{\mu}, \check{\mathbb{P}}, \check{\tau}), \tag{8.10}$$

where $\dot{G}_h \in V^{\mathrm{Col}(\omega,\mu)\times\mathrm{Col}(\omega,\mu)}$ is the canonical name for G_h, $h \in \{0, 1\}$. By Subclaim 8.26, $(p, q) \in \mathbb{Q}$. As D is dense in $\mathbb{Q}$, there is some $(p', q') \leq (p, q)$ such that $(p', q') \in D$.

By Subclaim 8.28, there is some $G_0' \times G_1'$ inside $V[G]$ which is $\mathrm{Col}(\omega, \mu) \times \mathrm{Col}(\omega, \mu)$-generic over V such that $(p', q') \in G_0' \times G_1'$ and

$$s = \tau^{G_0'}. \tag{8.11}$$

By (8.10), $V[G_0' \times G_1'] \models \Phi(G_0', G_1', \mu, \mathbb{P}, \tau)$. By (8.11), the s' which $\Phi(G_0', G_1', \mu, \mathbb{P}, \tau)$ describes in $V[G_0' \times G_1']$ is equal to s, which then also gives that the D' which $\Phi(G_0', G_1', \mu, \mathbb{P}, \tau)$ describes in $V[G_0' \times G_1']$ must be equal to D and that the $\mathbb{Q}'$ which $\Psi(\mathbb{Q}', \mu, \mathbb{P}, \tau, s)$ describes in $V[G_0' \times G_1']$ as part of $\Phi(G_0', G_1' \mu, \mathbb{P}, \tau)$ must be equal to $\mathbb{Q}$. Therefore, $V[G_0' \times G_1'] \models \Phi(G_0', G_1', \mu, \mathbb{P}, \tau)$ yields that $D \cap (G_0' \times G_1') = \emptyset$.

However, $(p', q') \in (G_0' \times G_1') \cap D$. Contradiction! □

Now by Subclaim 8.27, inside $V[s]$, there is thus a dense homomorphism

$$k\colon \mathrm{Col}(\omega, \mu) \to \mathbb{Q}.$$

By Subclaim 8.29, if we set

$$H^* = \{p \in \mathrm{Col}(\omega, \mu)\colon k(p) \in G_0 \times G_1\},$$

then H^* is $\mathrm{Col}(\omega, \mu)$-generic over $V[s]$ and $V[s][H^*] = V[s][G_0 \times G_1] = V[G_0][G_1]$. Therefore, H^* is as desired. □

This proof has the following corollary.

Theorem 8.30 (Solovay) *Let κ be an inaccessible cardinal, and let G be* Col *$(\omega, {<}\kappa)$-generic over V. Set*

$$N = \mathsf{HOD}^{V[G]}_{({}^\omega\omega \cap V[G])}.$$

Then in N, ZF + DC *holds and every set of reals is* Lebesgue *measurable and has the* Baire *property and every uncountable set of reals has a perfect subset.*

Proof In the light of Theorem 8.23, we are left with having to prove that DC holds true in N. If $f \in {}^\omega N \cap V[G]$, then $f \subset N \cap V[G \restriction \lambda]$ for some $\lambda < \kappa$ by the proof of Theorem 8.20. It is easy to see that this implies $f \in N$. □

Definition 8.31 Let $A \subset [\omega]^\omega$ be uncountable. We say that A is *Ramsey* iff there is some $x \in [\omega]^\omega$ such that $[x]^\omega \subset A$ or $[x]^\omega \cap A = \emptyset$.

In the presence of (AC), it is not hard to construct an $A \subset [\omega]^\omega$ with $\text{Card}(A) = 2^{\aleph_0}$ which is not RAMSEY, cf. Problem 8.11.

With the help of MATHIAS *forcing*, cf. p. 181, the arguments developed in this Chapter may be used to show that every uncountable $A \subset [\omega]^\omega$ is RAMSEY in the model of Theorem 8.30. Cf. Problem 12.13. Cf. also [13].

8.3 Problems

8.1. Let $A \subset {}^\omega\omega$ be open. Show that there is some $X \subset {}^{<\omega}\omega$ such that $A = \bigcup_{s \in X} U_s$ and $U_s \cap U_{s'} = \emptyset$ for all $s \neq s'$. Show also that $\sum_{s \in X} \mu(U_s)$ is independent of this representation of A, so that $\mu(A)$ is well-defined according to Definition 8.1.

8.2. Let $\mathscr{L}$ be the set of all $B \subset {}^\omega\omega$ such that for all $X \subset {}^\omega\omega$ (8.1) holds true. Let $\mathscr{C}$ be be the set of all $B \subset {}^\omega\omega$ such that there is some open set $A \subset {}^\omega\omega$ with $B \Delta A$ being meager. Show that both $\mathscr{L}$ and $\mathscr{C}$ form a σ-algebra containing all the open sets.

8.3. A set $A \subset {}^\omega 2$ is called a *flip set* iff for all $x, x' \in {}^\omega 2$ such that $\text{Card}(\{n < \omega \colon x(n) \neq x'(n)\}) = 1$, $x \in A \iff x' \notin A$. Show that if $A \subset {}^\omega 2$ is a flip set, then A is not LEBESGUE measurable and A does not have the property of BAIRE. Show in ZF + "there is a uniform ultrafilter on ω" that there is a flip set.

8.4. Verify the statements (1) through (3) from p. 150.

8.5. Show that Claim 8.21 implies that ω_1 is inaccessible to the reals. [Hint. Use Problem 6.18.]

8.6. Let κ be weakly compact, let G be $\text{Col}(\omega, {<}\kappa)$-generic over V, and let H be $\mathbb{Q}$-generic over $V[G]$, where $\mathbb{Q} \in V[G]$ and $V[G] \models$ "$\mathbb{Q}$ has the c.c.c." Show that if $x \in {}^\omega\omega \cap V[G][H]$, then there is some $\bar{\mathbb{Q}} \in V_\kappa$ and some g which is $\bar{\mathbb{Q}}$-generic over V such that $x \in V[g]$.

8.7. A set $A \subset \omega_1$ is called *reshaped* iff for all $\xi < \omega_1$,

$$L[A \cap \xi] \models \xi \text{ is countable.}$$

Show that if ω_1^V is not MAHLO in L, then there is a reshaped $A \subset \omega_1$. Show also that if $A \subset \omega_1$ is reshaped, then there is some poset $\mathbb{P}$ which has the c.c.c. such that if G is $\mathbb{P}$-generic over V, then in $V[G]$ there is a real x with $A \in L[x]$. Conclude that if ω_1^V is inaccessible in L, then ω_1 need not be inaccessible to the reals. [Hint: There is an almost disjoint collection $\{x_i \colon i < \omega_1\}$ of subsets of ω such that for each i, x_i is uniformly definable from $(x_j : j < i)$ and $A \cap i$ inside $L[A \cap i]$. Then use Problem 6.14.]

8.8. (**R. Jensen**) Show that if $V = L[B]$, where $B \subset \omega_1$, then there is an ω–distributive $\mathbb{P}$ such that if G is $\mathbb{P}$-generic over V, then in $V[G]$ there is a reshaped $A \subset \omega_1$. [Hint. Let $p \in \mathbb{P}$ iff $p\colon \alpha \to 2$, where $\alpha < \omega_1$ and for all $\xi \leq \alpha$, $L[B \cap \xi, p \upharpoonright \xi] \models$ "ξ is countable," ordered by end–extension.]

8.9. Let $E \subset \mathrm{OR}$ be "universally BAIRE in the codes" in the following sense. There are trees T, U witnessing that $p[T]$ is a universally BAIRE set of reals, and for all ordinals ξ, $\xi \in E$ iff

$$\Vdash_V^{\mathrm{Col}(\omega,\xi)} \exists x \in \mathbb{R}\, (x \in p[T] \wedge \xi = ||x||).$$

Show that E satisfies full condensation in the sense of Definition 5.30.

8.10. Show that if κ is a strong cardinal and $A \subset {}^{\omega}\omega$ is κ-universally BAIRE, then A is universally BAIRE.

8.11. Show in ZFC that there is some $A \subset [\omega]^{\omega}$ with $\mathrm{Card}(A) = 2^{\aleph_0}$ which is not RAMSEY.

Chapter 9
The Raisonnier Filter

By Corollary 8.22, it is impossible to construct just from a model of ZFC a model in which the statements from the conclusions of Solovay's Theorems 8.23 and 8.30 hold true. We now aim to consider LEBESGUE measurability and prove a theorem of SAHARON SHELAH, Theorem 9.1.

9.1 Rapid Filters on ω

Theorem 9.1 (Shelah) *Suppose that every* $\underset{\sim}{\Sigma}^1_3$*-set of reals is* LEBESGUE *measurable. Then* ω_1^V *is inaccessible to the reals.*

Our proof will make use of FUBINI's Theorem as well as the 0–1–Law of HEWITT–SAVAGE; we refer the reader to any standard textbook on Measure theory, e.g. [38]. In order to prove this theorem, we need the concept of a rapid filter.

Definition 9.2 Let $F \subset \mathscr{P}(\omega)$ be a filter on ω. We say that F is rapid iff F is non-trivial, F extends the FRÉCHET filter, and for every monotone f: $\omega \to \omega$ there is some $b \in F$ such that

$$\forall n < \omega \ \overline{\overline{b \cap f(n)}} \leq n. \tag{9.1}$$

We first want to construct, assuming that ω_1^V is not inaccessible to the reals, an interesting rapid filter.

Notice that by identifying any $a \in \mathscr{P}(\omega)$ with its characteristic function, we may identify $\mathscr{P}(\omega)$ with the CANTOR space ${}^{\omega}2$. We construe ${}^{\omega}2 \cong \mathscr{P}(\omega)$ as being equipped with the natural topology, cf. 123. We shall verify that no rapid filter is LEBESGUE measurable, cf. Theorem 9.16.

Theorem 9.3 *Assume that* ω_1^V *is not inaccessible to the reals, but every* $\underset{\sim}{\Sigma}^1_2$*-set of reals is* LEBESGUE *measurable. There is then a rapid filter* F *on* ω *such that* F *is* $\underset{\sim}{\Sigma}^1_3$.

R. Schindler, *Set Theory*, Universitext, DOI: 10.1007/978-3-319-06725-4_9,

Proof Let us fix $a \in {}^{\omega}\omega$ such that $\omega_1^V = \omega_1^{L[a]}$, cf. the proof of Lemma 7.27. We have $\overline{\overline{{}^{\omega}\omega \cap L[a]}} = \overline{\overline{{}^{\omega}2 \cap L[a]}} = \aleph_1$. Let us write $X = {}^{\omega}2 \cap L[a]$.

If $x, y \in {}^{\omega}2$, $x \neq y$, let us write $h(x, y)$ for the "distance" of x and y, i.e., $h(x, y)$ is the least $n < \omega$ such that $x \restriction n \neq y \restriction n$ (Hence $h(x, y) > 0$). For $Y \subset {}^{\omega}2$, let us write $H(Y)$ for

$$\{h(x, y) : x, y \in Y \wedge x \neq y\};$$

$H(Y)$ is thus a set of positive integers.

Definition 9.4 We define $F_X \subset \mathscr{P}(\omega)$ by setting $a \in F_X$ iff there is a covering $(X_n : n < \omega)$ of X, i.e., $X \subset \bigcup_{n<\omega} X_n$, where $X_n \subset {}^{\omega}2$ for each $n < \omega$, such that

$$\bigcup_{n<\omega} H(X_n) \subset a.$$

F_X is called the RAISONNIER filter.

Claim 9.5 *F_X is a non-trivial filter extending the* FRÉCHET *filter.*

Proof Trivially, if $a \in F_X$ and $b \supset a$, where $b \subset \omega$, then $b \in F_X$. Also, $F_X \neq \emptyset$, because $\omega \in F_X$. Let us suppose that $a \in F_X$ and $b \in F_X$, witnessed by $(X_n^a : n < \omega)$ and $(X_n^b : n < \omega)$ respectively. Let $\gamma : \omega \times \omega \to \omega$ be bijective. Set, for $n, m < \omega$, $X_{\gamma(n,m)} = X_n^a \cap X_m^b$. We have that

$$X = X \cap \bigcup_{n<\omega} X_n^a = X \cap \bigcup_{n<\omega} \bigcup_{m<\omega} (X_n^a \cap X_m^b) = X \cap \bigcup_{p<\omega} X_p,$$

so that $(X_p : p < \omega)$ is a covering of X. Let $q \in H(X_{\gamma(n,m)})$, say $q = h(x, y)$, where $x, y \in X_{\gamma(n,m)} = X_n^a \cap X_m^b$, $x \neq y$. Then $q = h(x, y) \in a$, as $H(X_n^a) \subset a$, and $q \in h(x, y) \in b$, as $H(X_m^b) \subset b$. We have shown that

$$\bigcup_{p<\omega} H(X_p) \subset a \cap b,$$

so that $a \cap b \in F_X$.

Also, F_X is non-trivial: if $X \subset \bigcup_{n<\omega} X_n$, then at least one X_n has two (in fact uncountably many) elements, because X is uncountable; therefore $H(X_n) \neq \emptyset$, and hence $\emptyset \notin F_X$.

To show that F_X extends the FRÉCHET filter, let $(X_n : n < 2^m)$ be an enumeration of all

$$U_s = \{x \in {}^{\omega}2 : x \supset s\},$$

where $s \in {}^{m}2$. We have that $\bigcup_{n<2^m} H(X_n) = \omega \setminus m \in F_X$. □

Claim 9.6 *F_X is $\underset{\sim}{\Sigma}^1_3$.*

Proof Let us first verify that $a \in F_X$ iff there is a covering $(Y_n : n < \omega)$ of X with $\bigcup_{n<\omega} H(Y_n) \subset a$, where Y_n is a *closed* subset of ${}^\omega 2$ for each $n < \omega$. Namely, let $a \in F_X$, as being witnessed by $(X_n : n < \omega)$. For $n < \omega$, let Y_n be the closure of X_n. Trivially, $X_n \subset Y_n$, and therefore $H(X_n) \subset H(Y_n)$. We claim that in fact $H(X_n) = H(Y_n)$. Well, if $n = h(x, y) \in H(Y_n)$, we may pick $x', y' \in X_n$ with $x' \upharpoonright n = x \upharpoonright n$ and $y' \upharpoonright n = y \upharpoonright n$. But then $n = h(x, y) = h(x', y')$, so that $n \in H(X_n)$. We now have that $\bigcup_{n<\omega} H(Y_n) = \bigcup_{n<\omega} H(X_n) \subset a$.

This now gives the following characterization of F_X. $a \in F_X$ iff $\exists (T_n : n < \omega)$ such that T_n is a tree on ${}^{<\omega}2$ for each $n < \omega$,

$$\begin{aligned}
&\forall x\ (x \in X \to \exists n \forall m\ x \upharpoonright m \in T_n) \wedge \\
&\forall x\ \forall y \forall n (x \neq y \wedge \forall m\ x \upharpoonright m \in T_n \wedge \forall m\ y \upharpoonright m \in T_n \\
&\quad \to \exists m \in a \setminus \{0\} (x \upharpoonright m - 1 = y \upharpoonright m - 1 \wedge x \upharpoonright m \neq y \upharpoonright m)).
\end{aligned}$$

Because $X = {}^\omega 2 \cap L[a]$ is $\underset{\sim}{\Sigma}^1_2(a)$, cf. Lemma 7.19, this easily gives that F_X is $\underset{\sim}{\Sigma}^1_3(a)$ by Lemma 7.17. $\square$

We have verified that F_X is a nontrivial filter which is $\underset{\sim}{\Sigma}^1_3$. In order to finish the proof of Theorem 9.3, we now need to see that F_X is rapid. Let $f : \omega \to \omega$ be monotone. We need to find some $b \in F_X$ such that (9.1) hold true.

Claim 9.7 *For every $f : \omega \to \omega$, ${}^\omega\omega \cap L[a, f]$ is a null set.*

Proof Because ${}^\omega\omega \cap L[a, f]$ is $\underset{\sim}{\Sigma}^1_2$, cf. Lemma 7.19, it is LEBESGUE measurable by hypothesis (Recall that we assume all $\underset{\sim}{\Sigma}^1_2$ sets of reals to be LEBESGUE measurable). Set

$$A = \{(x, y) \in ({}^\omega\omega)^2 \cap L[a, f] : x <_{L[a,f]} y\}.$$

For each $y \in {}^\omega\omega \cap L[a, f]$,

$$\{x : (x, y) \in A\}$$

is countable, and hence null. By FUBINI's Theorem, we therefore first get A to be null and then also

$$\{x : \{y : (x, y) \in A\} \text{ is not null}\}$$

to be null. If ${}^\omega\omega \cap L[a, f]$ is not null, there is then some x_0 such that

$$\{y : (x_0, y) \in A\} \text{ is null;}$$

but then

$$\begin{aligned}
{}^\omega\omega \cap L[a, f] = \{&x \in {}^\omega\omega \cap L[a, f] : x \leq_{L[a,f]} x_0\} \cup \\
&\{y \in {}^\omega\omega \cap L[a, f] : (x_0, y) \in A\}
\end{aligned}$$

is the union of two null sets and hence null.

We have shown that ${}^{\omega}\omega \cap L[a, f]$ is a null set. □

Claim 9.8 *Let $(n_k : k < \omega)$ be a sequence of positive integers. There is a family $(G_k : k < \omega)$ of open subsets of ${}^{\omega}2$ such that*

$$\mu(G_k) = \frac{1}{2^{n_k}}$$

for all $k < \omega$ and such that $(G_k : k < \omega)$ is independent *in that if $N \subset \omega$ is finite,*

$$\mu\left(\bigcap \{G_k : k \in N\}\right) = \prod_{k \in N} \mu(G_k).$$

Proof Write $r_{-1} = 0$. For $k \in \omega$, set $r_k = \Sigma_{l=0}^{k} n_l$, and put

$$G_k = \{x \in {}^{\omega}2 : \forall n (r_{k-1} \leq n < r_k \rightarrow x(n) = 1)\}.$$

It is easy to see that $\mu(G_k) = 2^{-n_k}$ and if $N \subset \omega$ is finite, then

$$\mu\left(\bigcap \{G_k : k \in N\}\right) = 2^{-\sum_{k \in N} n_k} = \prod_{k \in N} 2^{-n_k} = \prod_{k \in N} \mu(G_k),$$

so that $(G_k : k < \omega)$ is as desired. □

Claim 9.9 *Let $Y \subset {}^{\omega}2$ be null. There is then some closed set $C \subset {}^{\omega}2$ such that $Y \cap C = \emptyset$, $\mu(C) > 0$, and in fact for all $s \in {}^{\omega}2$, if $U_s \cap C \neq \emptyset$, then*

$$\mu(U_s \cap C) \geq \frac{1}{2^{3lh(s)+1}}.$$

Proof Let $C_0 \subset {}^{\omega}2$ be closed such that $\mu(C_0) \geq \frac{2}{3}$ and $Y \cap C_0 = \emptyset$. Let $C_0 = [T_0]$, where $T_0 \subset {}^{<\omega}2$ is a tree. Let us recursively define trees $T_k \subset {}^{<\omega}2$, $k > 0$, as follows.

$$T_k = \{t \in T_{k-1} : \exists s \in {}^{k}2\, \exists x \in [T_{k-1}] \cap U_s \\ \mu([T_{k-1}] \cap U_s) \geq \frac{1}{8^k} \wedge t = x \upharpoonright lh(t)\}.$$

Set $C_k = [T_k]$ for $k < \omega$. Also set $T_\infty = \bigcap_{k<\omega} T_k$ and $C = [T_\infty] = \bigcap_{k<\omega} C_k$. We claim that C is as desired.

As $T_\infty \subset T_0$, $Y \cap C = \emptyset$ is trivial.

In the step from C_{k-1} to C_k we consider 2^k many U_s, $s \in {}^{k}2$, and throw out those sets $C_{k-1} \cap U_s$ such that $\mu(C_{k-1} \cap U_s) < \frac{1}{8^k}$. Therefore, $\mu(C_k) \geq \mu(C_{k-1}) - 2^k \cdot \frac{1}{8^k} = \mu(C_{k-1}) - \frac{1}{4^k}$. This means that $\mu(C) \geq \mu(C_0) - \sum_{k=1}^{\infty} (\mu(C_{k-1}) - \mu(C_k)) \geq \mu(C_0) - \sum_{k=1}^{\infty} \frac{1}{4^k} \geq \frac{2}{3} - \frac{1}{3} = \frac{1}{3} > 0$.

Now let $s \in {}^{<\omega}2$ be such that $U_s \cap C \neq \emptyset$. Then, setting $k = \mathrm{lh}(s)$, $\mu(C_{k-1} \cap U_s) \geq \frac{1}{8^k}$ and $C_k \cap U_s = C_{k-1} \cap U_s$, so that $\mu(C_k \cap U_s) \geq \frac{1}{8^k}$. In the step from $C_{l-1} \cap U_s$ to $C_l \cap U_s$, $l > k$, we consider 2^{l-k} many U_t, $t \in {}^{l}2$, $t \supset s$, and throw out those sets $C_{l-1} \cap U_t$ such that $\mu(C_{l-1} \cap U_t) < \frac{1}{8^l}$; therefore, $\mu(C \cap U_s) \geq \mu(C_k \cap U_s) - \sum_{l=k+1}^{\infty}(\mu(C_l \cap U_s) - \mu(C_{l-1} \cap U_s)) \geq \frac{1}{8^k} - \sum_{l=k+1}^{\infty} 2^{l-k} \cdot \frac{1}{8^l} = \frac{1}{8^k} - \frac{1}{8^k}\sum_{p=1}^{\infty}\frac{1}{4^p} = \frac{1}{8^k} \cdot (1 - \frac{1}{3}) > \frac{1}{8^k} \cdot \frac{1}{2} = \frac{1}{2^{3k+1}}$. □

Claim 9.10 *For every monotone $f: \omega \to \omega$, there is some $b \in F_X$ such that*

$$\forall n < \omega\, \overline{\overline{b \cap f(n)}} \leq n \cdot (3n+1)^2 \cdot 2^{4n}. \tag{9.2}$$

Proof Using Claim 9.8, we may pick a sequence

$$(G_{s,m,n}: s \in {}^{<\omega}2 \wedge m, n < \omega)$$

of open subsets of ${}^{\omega}2$ such that $\mu(G_{s,m,n}) = \frac{1}{2^{m+n}}$ for all s, m, n, and such that the sequence is independent in that if $N \subset {}^{<\omega}2 \times \omega \times \omega$ is finite,

$$\mu\left(\bigcap\{G_{s,m,n}: s, m, n) \in N\}\right) = \prod_{(s,m,n)\in N} \mu(G_{s,m,n}).$$

Fix $f: \omega \to \omega$ monotone. Let

$$G = \bigcap_{n<\omega} \bigcup_{n' \geq n} \bigcup_{m \geq n'} \left\{(x, y) \in ({}^{\omega}2)^2: y \in G_{x \restriction f(m), m, n'}\right\}.$$

Obviously, G is a G_δ subset of $({}^{\omega}2)^2$. Let $x \in {}^{\omega}2$. Setting $G^x = \{y \in {}^{\omega}2: (x, y) \in G\}$, we have that $G^x \subset \bigcup_{n' \geq n} \bigcup_{m \geq n'} G_{x \restriction f(m), m, n'}$ for every $n < \omega$. However, $\mu(G_{x \restriction f(m), m, n'}) = \frac{1}{2^{m+n'}}$, and for each $\varepsilon > 0$ there is some $n < \omega$ such that $\sum_{n' \geq n} \sum_{m \geq n'} \frac{1}{2^{m+n'}} < \varepsilon$. Thus, G^x is null for every $x \in {}^{\omega}2$.

Let us define

$$G^* = \bigcup\{G^x: x \in X\}.$$

We aim to see that G^* is null. Well, for each $y \in G^*$ we may let $x(y)$ be the $<_{L[a]}$-least $x \in X$ such that $y \in G^x$, and we may set

$$A = \{(y, z) \in (G^*)^2: x(y) <_{L[a]} x(z)\}.$$

A is $\underset{\sim}{\Sigma}^1_2$ by Lemma 7.19 and hence LEBESGUE measurable by our hypothesis. For each $z \in G^*$, $\{x(y): (y, z) \in A\}$ is at most countable, and for any $x \in X$, $\{y \in G^*: x(y) = x\} \subset G^x$ is null. Therefore, $\{y \in G^*: (y, z) \in A\}$ is null for every $z \in G^*$, so that A is null by FUBINI's Theorem. Hence

$$\{y \in G^*: \{z \in G^*: (y, z) \in A\} \text{ is not null }\}$$

is a null set, again by FUBINI's Theorem. If G^* were not null, then we could pick some $y_0 \in G^*$ such that $\{z \in G^*: (y_0, z) \in A\}$ is null. But then

$$\begin{aligned} G^* = \{z \in G^*: (z, y_0) \in A\} \cup \\ \{z \in G^*: x(z) = x(y_0)\} \\ \{z \in G^*: (y_0, z) \in A\} \end{aligned}$$

would be null after all.

We have shown that G^* is a null set, so that by Claim 9.9, we may pick some closed set $C \subset {}^\omega 2$ such that $G^* \cap C = \emptyset$, $\mu(C) > 0$, and in fact for all $s \in {}^{<\omega}2$, if $U_s \cap C \neq \emptyset$, then $\mu(U_s \cap C) \geq \frac{1}{2^{3lh(s)+1}}$.

For $x \in X$ and $n < \omega$, let

$$O_n^x = \bigcup_{n' \geq n} \bigcup_{m \geq n'} G_{x \restriction f(m), m, n'}.$$

Each O_n^x, $n < \omega$, is open. Let $x \in X$. Suppose that for every $n < \omega$ and every $s \in {}^{<\omega}2$, if $C \cap U_s \neq \emptyset$, then $C \cap U_s \cap O_n^x = \emptyset$. We may then define $(z_n : n < \omega) \in {}^\omega 2$ and a monotone $(k_n : n < \omega) \in {}^\omega \omega$ with $C \cap U_{z_n \restriction k_n} \neq \emptyset$ and $z_{n+1} \restriction k_n = z_n \restriction k_n$ for all $n < \omega$ as follows. Let $z_0 \in C$ and $k_0 = 0$. If z_n and k_n have been defined such that $z_n \in C$, then $C \cap U_{z_n \restriction k_n} \cap O_n^x \neq \emptyset$, so that as O_n^x is open we may pick z_{n+1} and $n_{k+1} > n_k$ such that $z_{n+1} \restriction k_n = z_n \restriction k_n$, $U_{z_{n+1} \restriction k_{n+1}} \subset O_n^x$, and $z_{n+1} \in C$. Then

$$\bigcup_{n<\omega} z_n \restriction k_n \in \bigcap_{n<\omega} O_n^x \cap C = G^x \cap C \subset G^* \cap C = \emptyset.$$

Contradiction! There is thus for each $x \in X$, a pair $(n(x), s(x))$ such that $n(x) < \omega$, $s(x) \in {}^{<\omega}2$, and

$$C \cap U_{s(x)} \neq \emptyset, \text{ yet } C \cap U_{s(x)} \cap O_{n(x)}^x = \emptyset. \tag{9.3}$$

Let $e: \omega \times {}^{<\omega}2 \to \omega$ be bijective such that $e(n, s) \geq n$ and $e(n, s) \geq \text{lh}(s)$ for all $n < \omega$ and $s \in {}^{<\omega}2$, and let $(X_m : m < \omega)$ be an enumeration of the set of all

$$\{x \in X : n(x) = n \wedge s(x) = s \wedge x \restriction f(e(n, s)) = t\},$$

where $n < \omega$, $s \in {}^{<\omega}2$, and $t \in {}^{<\omega}2$. We may write $X = \bigcup_{m<\omega} X_m$, so that, setting

$$b = \bigcup_{m<\omega} H(X_m),$$

we have that $b \in F_X$. We aim to verify that (9.2) holds true.

So let us fix $n < \omega$. We have that

$$\begin{aligned} b \cap f(n) = \{h(x, y) < f(n) \colon x, y \in X, x \neq y, n(x) = n(y), \\ s(x) = s(y), \text{ and} \\ x \restriction f(e(n(x), s(x))) = y \restriction f(e(n(x), s(x)))\}. \end{aligned}$$

Obviously, if $x, y \in X, x \neq y$, witness that $h(x, y) \in b \cap f(n)$, then, setting $m = n(x) = n(y)$ and $s = s(x) = s(y)$,

$$f(e(m, s)) < h(x, y) < f(n),$$

so that

$$e(m, s) < n \tag{9.4}$$

by the monotonicity of f.

Let $m < \omega$ and $s \in {}^{<\omega}2$ be such that $e(m, s) < n$. Let us write

$$\begin{aligned} b_{m,s} = \{h(x, y) < f(n) \colon x, y \in X, x \neq y, n(x) = n(y) = m, \\ s(x) = s(y) = s, \text{ and} \\ x \restriction f(e(m, s)) = y \restriction f(e(m, s))\}. \end{aligned}$$

As $b \cap f(n) = \bigcup_{e(m,s)<n} b_{m,s}$, in order to show that (9.2) holds true it suffices to show that

$$\overline{\overline{b_{m,s}}} \leq (3n + 1)^2 \cdot 2^{4n} = ((3n + 1) \cdot 2^{2n})^2. \tag{9.5}$$

Let again $m < \omega$ and $s \in {}^{<\omega}2$ be such that $e(m, s) < n$. Notice that if $x, x', y \in X$, $x \restriction f(n) = x' \restriction f(n)$, and $h(x, y) < f(n)$, then $h(x', y) = h(x, y)$. This implies that

$$(\mathrm{Card}(\{t \in {}^{f(n)}2 \colon \exists x \in X(t \subset x \wedge n(x) = m \wedge s(x) = s)\}))^2 \leq \overline{\overline{b_{m,s}}}. \tag{9.6}$$

But we have that

$$\begin{aligned} \{t \in {}^{f(n)}2 \colon \exists x \in X(t \subset x \wedge n(x) = m \wedge s(x) = s)\} \subset \\ \{t \in {}^{f(n)}2 \colon C \cap U_s \neq \emptyset \wedge C \cap U_s \cap G_{t,n,m} = \emptyset\}. \end{aligned}$$

This is because if $x \in X, t \subset x, n(x) = m$, and $s(x) = s$, then by (9.3), $C \cap U_s \neq \emptyset$ and

$$\emptyset = C \cap U_s \cap O^x_m = C \cap U_s \cap \left(\bigcup_{n' \geq m} \bigcup_{p \geq n'} G_{x|f(p),p,n'} \right).$$

But $e(m, s) < n$, so that $m < n$ by the property of e, and thus

$$G_{t,n,m} \subset \left(\bigcup_{n' \geq m} \bigcup_{p \geq n'} G_{x|f(p),p,n'} \right).$$

By (9.6), in order to verify (9.5) it thus suffices to show that if $C \cap U_s \neq \emptyset$, then

$$\text{Card}(\{t \in {}^{f(n)}2 \colon C \cap U_s \cap G_{t,n,m} = \emptyset\}) \leq (3n+1) \cdot 2^{2n}. \tag{9.7}$$

Suppose that $C \cap U_s \neq \emptyset$, where $s \in {}^{<\omega}2$. Let us write q for the cardinality of

$$\left\{ t \in {}^{f(n)}2 \colon C \cap U_s \cap G_{t,n,m} = \emptyset \right\}.$$

We have that

$$C \cap U_s \subset \bigcap \left\{ {}^{\omega}2 \setminus G_{t,n,m} \colon t \in {}^{f(n)}2,\, C \cap U_s \cap G_{t,n,m} = \emptyset \right\}.$$

As $\mu(G_{t,n,m}) = \frac{1}{2^{n+m}}$, and because the $G_{t,n,m}$'s are independent in the sense of Claim 9.8,

$$\mu(C \cap U_s) \leq \sum_{k=0}^{q} \binom{q}{k} \left(-\frac{1}{2^{n+m}} \right)^k = \left(1 - \frac{1}{2^{n+m}} \right)^q.$$

By the choice of C, $\mu(C \cap U_s) \geq \frac{1}{2^{3\text{lh}(s)+1}}$, so that $\frac{1}{2^{3\text{lh}(s)+1}} \leq (1 - \frac{1}{2^{n+m}})^q$. By (9.4) and the properties of e, we have that $m < n$ and $\text{lh}(s) < n$ and therefore

$$\frac{1}{2^{3n+1}} \leq \frac{1}{2^{3\text{lh}(s)+1}} \leq \left(1 - \frac{1}{2^{2n}} \right)^q.$$

We always have $\log_{2^{2n}}(2) \leq 1$, which gives us $2 \leq \left(\frac{2^{2n}}{2^{2n}-1} \right)^{2^{2n}}$, thus $1 \leq \log_2 \left(\frac{2^{2n}}{2^{2n}-1} \right) \cdot 2^{2n}$, and hence

$$q \leq (3n+1) \cdot \left(\log_2 \left(\frac{2^{2n}}{2^{2n}-1} \right) \right)^{-1} \leq (3n+1) \cdot 2^{2n},$$

as we had wished. □

Using Claim 9.10 it is now easy to prove that F_X is rapid. Let $g \colon \omega \to \omega$ be monotone, and let $f \colon \omega \to \omega$ be defined by

$$f(n) = g((n+1) \cdot (3n+4)^2 \cdot 2^{4n+4}).$$

By Claim 9.10, there is some $b \in F_X$ such that for all $n < \omega$,

$$\overline{\overline{b \cap f(n)}} \leq n \cdot (3n+1)^2 \cdot 2^{4n}.$$

Let $n < \omega$, and let $n' \leq n$ be largest such that $n' \cdot (3n'+1)^2 \cdot 2^{4n'} \leq n$. Then

$$\begin{aligned}\overline{\overline{b \cap g(n)}} &\leq \overline{\overline{b \cap g((n'+1) \cdot (3n'+4)^2 \cdot 2^{4n'+4})}} \\ &= \overline{\overline{b \cap f(n')}} \leq n' \cdot (3n'+1)^2 \cdot 2^{4n'} \leq n,\end{aligned}$$

as desired.

This finishes the proof of Theorem 9.3. □

9.2 Mokobodzki's Theorem

Lemma 9.11 (Sierpiński) *Let $F \subset {}^{\omega}2$ be a non-trivial filter which extends the* FRÉCHET *filter and is* LEBESGUE *measurable. Then F is null.*

Proof For $s \in {}^{n}2$, where $n < \omega$, we may define a homomorphism $\varphi_s: {}^{\omega}2 \to {}^{\omega}2$ by

$$\varphi_s(x)(k) = \begin{cases} 1 - x(k) & \text{if } k < n \wedge s(k) = 1 \\ x(k) & \text{otherwise} . \end{cases}$$

Because F is assumed to extend the FRÉCHET filter, we have $\{\varphi_s(x): x \in F\} = F$ for every $s \in {}^{n}2$. The 0–1–Law of HEWITT–SAVAGE then implies that either $\mu(F) = 0$ or $\mu(F) = 1$.

Suppose that $\mu(F) = 1$. Let us define a homeomorphism $\varphi: {}^{\omega}2 \to {}^{\omega}2$ by

$$\varphi(x)(k) = 1 - x(k)$$

for $k < \omega$. It is easy to see that φ respects μ, i.e., $\{\varphi(x): x \in X\}$ is LEBESGUE measurable and $\mu(\{\varphi(x): x \in X\}) = \mu(X)$ for all LEBESGUE measurable $X \subset {}^{\omega}2$. $\mu(F) = 1$ yields that $\mu(\{\varphi(x): x \in F\}) = 1$. We may then pick some $x_0 \in F \cap \{x: \varphi(x) \in F\}$. As F is a filter, the characteristic function of the intersection of the two sets for which x_0 and $\varphi(x_0)$ are the respective characteristic functions is then in F again, i.e.,

$$\emptyset = \{k < \omega: x_0(k) = 1 \wedge \varphi(x_0)(k) = 1\} \in F.$$

This contradicts the fact that F is assumed to be non-trivial.

We have shown that $\mu(F) = 0$. □

Definition 9.12 Let $\mathbf{I} = \{I_n : n < \omega\}$ be a partition of ω into intervals, and let $\mathbf{J} = (J_n : n < \omega)$ be such that $J_n \subset {}^{I_n}2$ for $n < \omega$. We write

$$(\mathbf{I}, \mathbf{J}) = \{x \in {}^{\omega}2 : x \restriction I_n \in J_n \text{ for infinitely many } n < \omega\}.$$

A set $N \subset {}^{\omega}2$ is called small iff for every sequence $(\varepsilon_n : n < \omega)$ of positive reals there is a partition $\mathbf{I} = \{I_n : n < \omega\}$ of ω into intervals and there is a sequence $\mathbf{J} = (J_n : n < \omega)$ with $J_n \subset {}^{I_n}2$ for $n < \omega$ such that

(1) $N \subset (\mathbf{I}, \mathbf{J})$, and
(2) $\mu(\{x \in {}^{\omega}2 : x \restriction I_k \in J_k\}) < \varepsilon_k$ for every $k < \omega$.

If $N \subset {}^{\omega}2$ is small and if $(\mathbf{I}, \mathbf{J})$ is as in (1) and (2) of Definition 9.12, then for every $n_0 < \omega$,

$$(\mathbf{I}, \mathbf{J}) = \{x \in {}^{\omega}2 : x \restriction I_n \in J_n \text{ for infinitely many } n \geq n_0\}.$$

Hence $\mu((\mathbf{I}, \mathbf{J})) \leq \sum_{n=n_0}^{\infty} \varepsilon_n$ for every $n_0 < \omega$. We may choose $(\varepsilon_n : n < \omega)$ in such a way that $\sum_{n=0}^{\infty} \varepsilon_n < \infty$, so that we get that $(\mathbf{I}, \mathbf{J})$ is a null set. Hence N is a null set. We show that every null set can be covered by two small sets:

Lemma 9.13 *If $A \subset {}^{\omega}2$ is null, then there are small sets $N_0, N_1 \subset {}^{\omega}2$ with $A \subset N_0 \cup N_1$.*

Proof Let $A \subset {}^{\omega}2$ be null. For each $n < \omega$, we may pick an open set $O_n \subset {}^{\omega}2$ such that $A \subset O_n$ and $\mu(O_n) < \frac{1}{2^n}$. Let

$$O_n = \bigcup_{m<\omega} U_{s_m^n},$$

where $s_m^n \in {}^{<\omega}2$ and $U_{s_m^n} \cap U_{s_{m'}^n} = \emptyset$ for $s_m^n \neq s_{m'}^n$. Notice that by $\mu(O_n) < \frac{1}{2^n}$,

$$\min\{\mathrm{lh}(s_m^n) : m < \omega\} > n. \tag{9.8}$$

Set

$$F_n = \{s \in {}^{n}2 : \exists k \exists m \; s = s_m^k\}.$$

With the help of (9.8),

$$A \subset \{x \in {}^{\omega}2 : x \restriction n \in F_n \text{ for infinitely many } n < \omega\}. \tag{9.9}$$

Also

$$\inf_{m<\omega}\left(\left\{\sum_{n=m}^{\infty} \mu(\{x \in {}^{\omega}2 : x \restriction n \in F_n\right\}\right) = 0.$$

Let $(\varepsilon_n : n < \omega)$ be a sequence of positive reals. In the light of what we are supposed to prove, we may assume without loss of generality that $\sum_{n=0}^{\infty} \varepsilon_n < \infty$. Let us recursively construct $(n_k : k < \omega)$ and $(m_k : k < \omega)$ as follows.

Let $n_0 = 0$, $m_0 = 0$,

$$m_{k+1} = \min \left\{ m > n_k : 2^{n_k} \cdot \sum_{n=m}^{\infty} \mu \left(\{ x \in {}^{\omega}2 : x \upharpoonright n \in F_n \} \right) < \varepsilon_k \right\}$$

and

$$n_{k+1} = \min \left\{ m > m_{k+1} : 2^{m_{k+1}} \cdot \sum_{n=m}^{\infty} \mu \left(\{ x \in {}^{\omega}2 : x \upharpoonright n \in F_n \} \right) < \varepsilon_k \right\}.$$

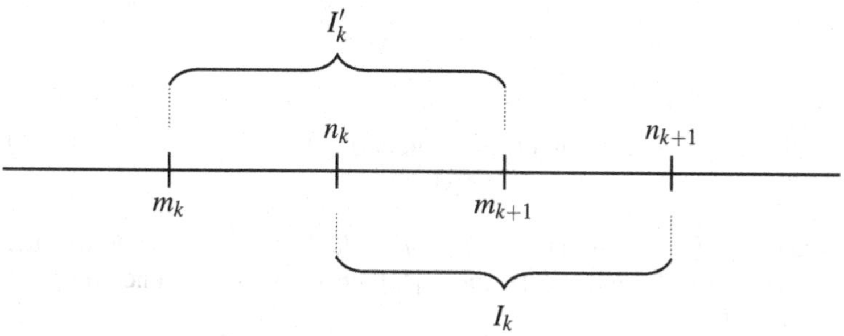

Let, for $k < \omega$,

$$I_k = [n_k, n_{k+1}) \text{ and } I'_k = [m_k, m_{k+1}),$$

and set

$$J_k = \{ s \in {}^{I_k}2 : \exists i \in [m_{k+1}, n_{k+1}] \exists t \in F_i \; s \upharpoonright [n_k, i) = t \upharpoonright [n_k, i) \}$$

and

$$J'_k = \{ s \in {}^{I'_k}2 : \exists i \in [n_k, m_{k+1}] \exists t \in F_i \; s \upharpoonright [m_k, i) = t \upharpoonright [m_k, i) \}$$

We claim that both $(\mathbf{I}, \mathbf{J})$ and $(\mathbf{I}', \mathbf{J}')$ satisfy (1) and (2) of Definition 9.12.

As for (2),

$$\begin{aligned}
&\mu \left(\{ x \in {}^{\omega}2 : x \upharpoonright I_k \in J_k \} \right) \\
&\quad = \mu \left(\{ x \in {}^{\omega}2 : \exists i \in [m_{k+1}, n_{k+1}] \exists t \in F_i \; x \upharpoonright [n_k, i) = t \upharpoonright [n_k, i) \} \right) \\
&\quad \leq 2^{n_k} \cdot \sum_{i=m_{k+1}}^{n_{k+1}} \mu \left(\{ x \in {}^{\omega}2 : \exists t \in F_i \; x \upharpoonright i = t \} \right) < \varepsilon_k.
\end{aligned}$$

Symmetrically,

$$\begin{aligned}
&\mu\left(\left\{x \in {}^{\omega}2 : x \restriction I'_k \in J'_k\right\}\right) \\
&\quad = \mu\left(\left\{x \in {}^{\omega}2 : \exists i \in [n_k, m_{k+1}] \exists t \in F_i\ x \restriction [m_k, i) = t \restriction [m_k, i)\right\}\right) \\
&\quad \le 2^{m_k} \cdot \sum_{i=n_k}^{m_{k+1}} \mu\left(\left\{x \in {}^{\omega}2 : \exists t \in F_i\ x \restriction i = t\right\}\right) < \varepsilon_k.
\end{aligned}$$

To verify that $A \subset (\mathbf{I}, \mathbf{J}) \cup (\mathbf{I}', \mathbf{J}')$, let $x \in A$. By (9.9), there are infinitely many $n < \omega$ with

$$x \restriction n \in F_n \text{ and } n \in \bigcup_{k=0}^{\infty} [m_{k+1}, n_{k+1}] \tag{9.10}$$

or with

$$x \restriction n \in F_n \text{ and } n \in \bigcup_{k=0}^{\infty} [n_k, m_{k+1}]. \tag{9.11}$$

If $x \restriction n \in F_n$ and $n \in [m_{k+1}, n_{k+1}]$, then $x \restriction I_k \in J_k$; hence if (9.10) holds true, then $x \in (\mathbf{I}, \mathbf{J})$. If $x \restriction n \in F_n$ and $n \in [n_k, m_{k+1}]$, then $x \restriction I'_k \in J'_k$; hence if (9.11) holds true, then $x \in (\mathbf{I}', \mathbf{J}')$. □

Lemma 9.14 *Let F be a non-trivial filter on ω which extends the* FRÉCHET *filter and is* LEBESGUE *measurable. Then F is small.*

Proof By Lemma 9.11 we know that F is null. Let us fix a sequence $(\varepsilon'_k : k < \omega)$ of positive reals. We may assume without loss of generality that $\varepsilon'_k < 1$ for all $k < \omega$. Let $\delta_k = \min\{\frac{\varepsilon'_k}{2}, 2^{m_{k+1}-n_{k+1}}\} < \varepsilon'_k$ and $\varepsilon_k = \frac{(\varepsilon'_k)^2}{8} < \varepsilon'_k$. We may write

$$F \subset (\mathbf{I}, \mathbf{J}) \cup (\mathbf{I}', \mathbf{J}'), \tag{9.12}$$

where $(\mathbf{I}, \mathbf{J})$ and $(\mathbf{I}', \mathbf{J}')$ are exactly as constructed in the proof of Lemma 9.13. We are also going to use the notations $n_k, m_k, I_k, I'_k, J_k, J'_k$ for $k < \omega$ from the proof of Lemma 9.13; in particular, $\mu(\{x : x \restriction I_k \in J_k\}) < \varepsilon_k$ and $\mu(\{x : x \restriction I'_k \in J'_k\}) < \varepsilon_k$ for every $k < \omega$. We aim to find

$$(\mathbf{I}^*, \mathbf{J}^*) \supset F \tag{9.13}$$

such that for every $k < \omega$, $\mu\left(\left\{x : x \restriction I^*_k \in J^*_k\right\}\right) < \varepsilon'_k$.

For $k < \omega$, let

$$H_k = \left\{t \in {}^{[n_k, m_{k+1})}2 : \mu\left(\left\{x \in {}^{\omega}2 : t^\frown x \restriction [m_{k+1}, n_{k+1}) \in J_k\right\}\right) \ge \delta_k\right\}.$$

For all $k < \omega$,

$$\begin{aligned}\mu\left(\{x\colon x \restriction [n_k, m_{k+1}) \in H_k\}\right) &\leq 2^{n_{k+1}-m_{k+1}} \cdot \mu\left(\{x\colon x \restriction I_k \in J_k\}\right)\\ &\leq \frac{1}{\delta_k} \cdot \mu\left(\{x\colon x \restriction I_k \in J_k\}\right)\\ &< \frac{1}{\delta_k} \cdot \varepsilon_k = \frac{\varepsilon'_k}{4}.\end{aligned} \tag{9.14}$$

We also define, for $k < \omega$,

$$H'_k = \left\{t \in {}^{[n_k, m_{k+1})}2\colon \mu\left(\left\{x \in {}^{\omega}2\colon x \restriction [m_k, n_k)^\frown t \in J'_k\right\}\right) \geq \delta_k\right\},$$

so that in analogy with (9.14)

$$\mu(\{x\colon x \restriction [n_k, m_{k+1}) \in H'_k\}) < \frac{\varepsilon'_k}{4}. \tag{9.15}$$

Now let us write

$$\begin{aligned}N_0 &= (\mathbf{I}, \mathbf{J}),\\ N_1 &= (\mathbf{I}', \mathbf{J}'), \text{ and}\\ N_2 &= (\mathbf{I}', \mathbf{J}''),\end{aligned}$$

where

$$J''_k = \{s^\frown t\colon s \in {}^{[m_k, n_k)}2 \wedge t \in H_k \cup H'_k\}.$$

With the help of (9.14) and (9.15), we have that

$$\mu\left(\{x\colon x \restriction [m_k, m_{k+1}) \in J'_k \cup J''_k\}\right) < \varepsilon_k + 2 \cdot \frac{\varepsilon'_k}{4} < \varepsilon'_k.$$

Hence if $F \subset N_1 \cup N_2$, then we found a covering of F as in (9.13). Let us thus assume that $F \not\subset N_1 \cup N_2$.

Let $x_0 \in F$ be such that $x_0 \notin N_1 \cup N_2$. Then $x_0 \in N_0$ by (9.12), so that for infinitely many $k < \omega$,

$$x_0 \restriction I_k \in J_k.$$

For $k < \omega$, let $I^*_{2k} = [n_k, m_{k+1})$, $I^*_{2k+1} = [m_{k+1}, n_{k+1})$, $J^*_{2k} = \emptyset$, and also $J^*_{2k+1} = \emptyset$, unless $x_0 \restriction I_k \in J_k$ in which case

$$\begin{aligned}J^0_{2k+1} &= L_{2k+1} \cup L'_{2k+1}, \text{ where}\\ L_{2k+1} &= \{s \in {}^{[m_{k+1}, n_{k+1})}2\colon x_0 \restriction [n_k, m_{k+1})^\frown s \in J_k\} \text{ and}\\ L'_{2k+1} &= \{s \in {}^{[m_{k+1}, n_{k+1})}2\colon s^\frown x_0 \restriction [n_{k+1}, m_{k+2}) \in J'_{k+1}\}.\end{aligned}$$

As $x_0 \notin N_2$, we must have $x_0 \upharpoonright [m_k, m_{k+1}) \notin H_k$ and $x_0 \upharpoonright [m_k, m_{k+1}) \notin H'_k$ and hence

$$\mu(\{x : x \upharpoonright [m_{k+1}, n_{k+1}) \in J^0_{2k+1}\}) < 2\delta_k = \varepsilon'_k$$

for all but finitely many $k < \omega$. Therefore, the proof of the following Claim will provide a covering of F as in (9.13) and finish the proof of Lemma 9.14.

Claim 9.15 $F \subset (\mathbf{I}^*, \mathbf{J}^*)$.

Proof Suppose not, and pick $y \in F \backslash (\mathbf{I}^*, \mathbf{J}^*)$. Define $z \in {}^{\omega}2$ by

$$z(n) = \begin{cases} y(n) & \text{if } \exists k\ (x_0 \upharpoonright [n_k, n_{k+1}) \in J_k \text{ and } n \in [m_{k+1}, n_{k+1})) \\ x_0(n) & \text{otherwise.} \end{cases}$$

As F is a filter and $x_0(n) = y(n) = 1$ implies $z(n) = 1$, we have that $z \in F$. By (9.12), we must have $z \in N_0$ or $z \in N_1$.

Say $z \in N_0$. Consider $I_k = [n_k, n_{k+1}]$. If $x_0 \upharpoonright I_k \notin J_k$, then $z \upharpoonright I_k = x_0 \upharpoonright I_k \notin J_k$. But as $z \in N_0$, we must have $z \upharpoonright I_k \in J_k$ for infinitely many k. We must then have $x_0 \upharpoonright I_k \in J_k$ and then

$$z \upharpoonright I_k = x_0 \upharpoonright [n_k, m_{k+1})^\frown y \upharpoonright [m_{k+1}, n_{k+1}) \in J_k$$

for any such k, which implies that $y \upharpoonright [m_{k+1}, n_{k+1}) \in L_{2k+1} \subset J^0_{2k+1}$. However, $y \notin (\mathbf{I}^*, \mathbf{J}^*)$, so there can be only finitely many such k. Contradiction!

Say $z \in N_1$. Consider $I'_{k+1} = [m_{k+1}, m_{k+2})$. If $x_0 \upharpoonright [n_k, n_{k+1}) \notin J_k$, then $z \upharpoonright [m_{k+1}, m_{k+2}) = x_0 \upharpoonright [m_{k+1}, m_{k+2})$, which by $x_0 \notin N_1$ can only be in J'_{k+1} for finitely many k. But $z \in N_1$, so we must have $z \upharpoonright I'_{k+1} \in J'_{k+1}$ and hence $x_0 \upharpoonright [n_k, n_{k+1}) \in J_k$ for infinitely many k. For such k,

$$z \upharpoonright I'_{k+1} = y \upharpoonright [m_{k+1}, n_{k+1})^\frown x_0 \upharpoonright [n_{k+1}, m_{k+2}) \in J'_{k+1},$$

which implies that $y \upharpoonright [m_{k+1}, n_{k+1}) \in L'_{2k+1} \subset I^0_{2k+1}$. But again $y \notin (\mathbf{I}^*, \mathbf{J}^*)$, so there can be only finitely many such k. Contradiction!

We have shown that $F \subset (\mathbf{I}^*, \mathbf{J}^*)$. □

In the light of Theorem 9.3, Shelah's Theorem 9.1 is now an immediate consequence of the following.

Theorem 9.16 (Mokobodzki) *No rapid filter* $F \subset {}^{\omega}2$ *is* LEBESGUE *measurable.*

Proof Let $\epsilon_n = \frac{1}{2^{n+1}}$ for $n < \omega$. By Lemma 9.14 we may write

$$F \subset (\mathbf{I}, \mathbf{J}),$$

where for every $n < \omega$,

$$\mu(\{{}^{\omega}2 : x \upharpoonright I_n \in J_n\}) < \varepsilon_n.$$

For $n < \omega$, let

$$J_n^* = \{s \in J_n : \forall t \in {}^{I_n}2 (\forall k \in I_n \; s(k) \le t(k) \to t \in J_n)\}.$$

We claim that

$$F \subset (\mathbf{I}, \mathbf{J}^*). \tag{9.16}$$

Suppose that $x \in F \setminus (\mathbf{I}, \mathbf{J}^*)$. As $x \in (\mathbf{I}, \mathbf{J})$, $X = \{n < \omega : x \restriction I_n \in J_n\}$ is infinite. As $x \notin (\mathbf{I}, \mathbf{J}^*)$, there must be an $l < \omega$ such that for all $n \in X \setminus l$ we may pick some $t_n \in {}^{I_n}2$ such that for all $k \in I_n$, $x(k) \le t_n(k)$, but $t_n \notin J_n$. Define $y \in {}^{\omega}2$ by

$$y(k) = \begin{cases} t_n(k) \text{ if } n \in X \setminus l \text{ and } k \in I_n \\ x(k) \text{ otherwise.} \end{cases}$$

Obviously, $y \notin (\mathbf{I}, \mathbf{J})$. But F is a filter, so that $x \in F$ implies $y \in F \subset (\mathbf{I}, \mathbf{J})$. Contradiction! We have shown that (9.16) holds true.

As $J_n^* \subset J_n$ for every $n < \omega$, we still have that

$$\mu(\{{}^{\omega}2 : x \restriction I_n \in J_n^*\}) < \varepsilon_n = \frac{1}{2^{n+1}} \tag{9.17}$$

for all $n < \omega$. Let us write

$$\#(n) = \min\{\overline{\overline{\{k \in I_n : s(k) = 1\}}} : s \in J_n^*\},$$

and

$$J_n^{*,\min} = \{s \in J_n^* : \overline{\overline{\{k \in I_n : s(k) = 1\}}} = \#(n)\}.$$

We must have that

$$\#(n) \ge n + 1. \tag{9.18}$$

This is because if $s \in J_n^*$ is such that $\overline{\overline{\{k \in I_n : s(k) = 1\}}} \le n$, then $\mu(\{x \in {}^{\omega}2 : x \restriction I_n \in J_n^*\}) \ge \frac{1}{2^n}$, contradicting (9.17).

Let us now define $f : \omega \to \omega$ by

$$f(n) = \max\{\{\max(k) : s(k) = 1\} : s \in J_n^{*,\min}\}$$

for $n < \omega$. If F is rapid, then we may pick some $b \in F$ such that

$$\forall n < \omega \; \overline{\overline{\{k : b(k) = 1\} \cap f(n)}} \le n.$$

By (9.16), $b \in (\mathbf{I}, \mathbf{J}^*)$. If $b \restriction I_n \in J_n^*$, then (9.18) gives that $\{k \leq \max(I_n) : b(k) = 1\}$ is contained in I_n and has maximum $f(n)$. Hence there can be at most one such n. In particular, $b \notin (\mathbf{I}, \mathbf{J}^*)$. Contradiction!

We have shown that F cannot be rapid. □

The book [3] contains exciting material extending the topic of the current chapter. We also refer the reader to [5].

9.3 Problems

Let $F \subset \mathscr{P}(\omega)$ be a non-trivial filter on ω extending the FRÉCHET filter. We say that F is a *p-point* iff for every $f \in {}^\omega\omega$ there is some $X \in F$ such that $f \restriction X$ is constant or finite-to-one (by which we mean that $\{n : f(n) = m\}$ is finite for every $m < \omega$). We say that F is a *q-point* iff for every $f \in {}^\omega\omega$ which is finite-to-one there is some $X \in F$ such that $f \restriction X$ is injective. F is called *selective*, or *Ramsey*, iff F is both a p-point as well as a q-point. F is called *nowhere dense* iff for every $f : \omega \to \mathbb{R}$ there is some $X \in F$ such that $f''X$ is nowhere dense.

9.1. Let F be a p-point.
(a) Show that if $(X_n : n < \omega)$ is such that $X_n \in F$ for all $n < \omega$, then there is some $Y \in F$ such that $Y \setminus X_n$ is finite for all $n < \omega$.
(b) Show that if $\{X_n : n < \omega\}$ is such that $\bigcup_{n<\omega} X_n = \omega$, $X_n \notin F$ for all $n < \omega$, and $X_n \cap X_m = \emptyset$ for all $n \neq m$, then there is some $X \in F$ such that $X \cap X_n$ has finitely many elements for every $n < \omega$. If F is assumed to be selective, then we may in fact pick $X \in F$ in such a way that $X \cap X_n$ has exactly one element for every $n < \omega$.

9.2. (a) Show that if F is a p-point, then F is nowhere dense. In fact, if F is a p-point, then F is discrete (by which we mean that for every $f : \omega \to \mathbb{R}$ there is some $X \in F$ such that for every $x \in f''X$ there are $a < x < b$ such that $(a, b) \cap f''X = \{x\}$).
(b) Show that if F is a q-point, then F is rapid.

9.3. Let U be a selective ultrafilter on ω.
(a) Let $(X_n : n < \omega)$ be such that $X_n \in F$ for all $n < \omega$. Show that there is some $Y \in U$ such that for all $\{n, m\} \subset Y$ with $n < m$, $m \in X_n$. [Hint. First use Problem 9.1 (a) to get some $Z \in U$ and some $g : \omega \to \omega$ such that $Z \setminus g(n) \subset X_n$ for all $n < \omega$. Suppose w.l.o.g. that g is strictly inceasing, and write

$$f(n) = \underbrace{g \circ \ldots \circ g}_{n \text{ times}}(0).$$

By Problem 9.1 (b), let $Z' \in U$ be such that for every $n < \omega$, there is exactly one $m \in Z'$ with $g(n) \leq m < g(n + 1)$, call it m_n. One of $\{m_{2n} : n < \omega\}$, $\{m_{2n+1} : n < \omega\}$ is in U, call it Z^*. Verify that $Y = Z \cap Z^*$ is as desired.]

(b) Let $(X_s : s \in {}^{<\omega}\omega)$ be such that $X_s \in F$ for all $s \in {}^{<\omega}\omega$. Show that there is some $Y \in U$ such that for all strictly increasing $s \in {}^{<\omega}\omega$ with $\operatorname{ran}(s) \subset Y$, $s(n) \in X_{s\restriction n}$ for all $n \in \operatorname{lh}(s)$.

9.4. Show that if CH holds, then there is a selective ultrafilter. [Hint. Let $(\{X_n^\alpha : n < \omega\} : \alpha < \omega_1)$ enumerate all $\{X_n : n < \omega\}$ such that $\bigcup_{n<\omega} X_n = \omega$ and $X_n \cap X_m = \emptyset$ for all $n \neq m$. Recursively construct a sequence $(Y_\alpha : \alpha < \omega_1)$ of infinite subsets of ω such that if $\alpha < \beta$, then $Y_\beta \setminus Y_\alpha$ is finite and $Y_{\alpha+1} = Y_\alpha \cap X_n^\alpha$ for some n for which $Y_\alpha \cap X_n^\alpha$ is infinite, if such an n exists, and otherwise $\operatorname{Card}(Y_{\alpha+1} \cap X_n^\alpha) \leq 1$ for all n. Set $F = \{X \subset \omega : \exists \alpha\, X \setminus X_\alpha$ is finite $\}$.]

9.5. Show that if CH holds, then there is a q-point which is not selective. [Hint. Let $U, U_0, U_1, \ldots$ be non-isomorphic selective ultrafilters, and let $X \in U^*$ iff $\{m : \{n : \langle m, n\rangle \in X_m\} \in U\}$.]
Let $\mathbb{P} \in V$ be a partial order, an let G be $\mathbb{P}$-generic over V. Then $z \in {}^\omega\omega \cap V[G]$ is called *unbounded* iff for every $x \in {}^\omega\omega \cap V$, $\{n < \omega : x(n) < z(n)\}$ is infinite. $z \in {}^\omega\omega \cap V[G]$ is called *dominating* iff for every $x \in {}^\omega\omega \cap V$, $\{n < \omega : x(n) < z(n)\}$ is cofinite, i.e., there are only finitely many $n < \omega$ with $z(n) \leq x(n)$.

9.6. Let z be a COHEN real over V. Then z is unbounded.
Let α be any ordinal, and let G be $\mathbb{C}(\alpha)$-generic over V. Show that $V[G]$ does not contain a dominating real. [Hint. Use Lemma 6.29 and the proofs of Lemmas 6.53 and 6.61.]
Let

$$b = \min\{\operatorname{Card}(F) : \forall x \in {}^\omega\omega \exists z \in {}^\omega\omega \cap F\, \{n : x(n) < z(n)\} \text{ is infinite}\}, \text{ and}$$

$$d = \min\{\operatorname{Card}(F) : \forall x \in {}^\omega\omega \exists z \in {}^\omega\omega \cap F\, \{n : x(n) < z(n)\} \text{ is cofinite}\}.$$

9.7. $b \leq d$. Let $\alpha \geq \aleph_2$ be a cardinal, and let G be $\mathbb{C}(\alpha)$-generic over V. Suppose that $V \models$ CH. Show that in $V[G]$, $\aleph_1 = b < \alpha \leq d$.
Let $\mathbb{D}$ consist of all (x, n), where $x \in {}^\omega\omega$ and $n < \omega$, ordered by $(x', n') \leq (x, n)$ iff $n' \geq n$, $x' \restriction n = x \restriction n$, and $x'(k) \geq x(k)$ for all $k \geq n$.

9.8. If (x, n), $(x', n') \in \mathbb{D}$, where $n = n'$ and $x \restriction n = x' \restriction n$, then (x, n) is compatible with $(x'n')$. Conclude that $\mathbb{D}$ has the c.c.c. Show that if G is $\mathbb{D}$-generic over V, then $V[G]$ contains a dominating real.
Let $F \subset \mathscr{P}(\omega)$, and let

$$\mathbb{M}_F = \{(s, X) : s \in [\omega]^{<\omega} \wedge X \in F \wedge (s \neq \emptyset \rightarrow \min(X) > \max(s))\}, \quad (9.19)$$

ordered by $(s', X') \leq (s, X)$ iff $s' \supset s$, $X' \subset X$, and $s' \setminus s \subset X$. $\mathbb{M}_F$ is called MATHIAS *forcing* for F.

9.9. Let F be a filter on ω.
(a) Show that $\mathbb{M}_F$ has the c.c.c.

(b) Show that if G is $\mathbb{M}_F$-generic over V, then, setting $x_G = \bigcup\{s : \exists X\, (s, X) \in G\}$, $x_G \setminus X$ is finite for all $X \in F$.

9.10. Let F be a non-trivial filter on ω extending the FRÉCHET filter. Asuume that either (a) F is not an ultrafilter, or else (b) is an ultrafilter, but not selective. Show that if G is $\mathbb{M}_F$-generic over V, then there is a COHEN real over V in $V[G]$.

9.11. Let U be a non-trivial filter on ω extending the FRÉCHET filter such that U is not a p-point. Show that if G is $\mathbb{M}_U$-generic over V, then $V[G]$ contains a dominating real.

Chapter 10
Measurable Cardinals

Measurable cardinals (cf. Definition 4.54) and elementary embeddings induced by them (cf. Theorem 4.55) play a crucial role in contemporary set theory. We here develop the theory of iterated ultrapowers, of $0^\sharp$, and of short and long extenders.

10.1 Iterations of V

Theorem 10.3 and Lemma 10.4 of this section will be used in the proof of Theorem 13.3.

Definition 10.1 Let κ be a measurable cardinal, and let U be a measure on κ, i.e., a $< \kappa$-closed uniform ultrafilter on κ. Let γ be an ordinal, or $\gamma = \infty$. Then the system

$$\mathscr{I} = (M_\alpha, \pi_{\alpha\beta} : \alpha \leq \beta < \gamma)$$

is called the *(linear) putative iteration of V of length γ given by U* iff the following hold true.

(1) $M_0 = V$, and if $\alpha + 1 < \gamma$, then M_α is an inner model.
(2) If $\alpha \leq \beta \leq \delta < \gamma$, then $\pi_{\alpha\beta} : M_\alpha \to M_\beta$ is an elementary embedding, and $\pi_{\alpha\delta} = \pi_{\beta\delta} \circ \pi_{\alpha\beta}$.
(3) If $\alpha + 1 < \gamma$, then $M_{\alpha+1} = \text{ult}(M_\alpha; \pi_{0\alpha}(U))$ and $\pi_{\alpha\alpha+1}$ is the canonical ultrapower embedding.
(4) If $\lambda < \gamma$ is a limit ordinal, then $(M_\lambda, \pi_{\alpha\lambda} : \alpha < \lambda)$ is the direct limit of $(M_\alpha, \pi_{\alpha\beta} : \alpha \leq \beta < \lambda)$.

The system $\mathscr{I}$ is called the *(linear) iteration of V of length γ given by U* if either γ is a limit ordinal or else the last model $M_{\gamma-1}$ is well-founded (and may therefore be identified with an inner model).

Notice that by (2), $\pi_{\alpha\alpha} = \text{id}$ for all $\alpha < \gamma$. Also, if we write $\kappa_\alpha = \pi_{0\alpha}(\kappa)$ and $U_\alpha = \pi_{0\alpha}(U)$, then

R. Schindler, *Set Theory*, Universitext, DOI: 10.1007/978-3-319-06725-4_10,

$$M_\alpha \models \text{“}U_\alpha \text{ is a measure on } \kappa_\alpha.\text{”}$$

Therefore, (3) makes sense and is to be understood in the sense of Definition 4.57. The requirement that the M_α for $\alpha + 1 < \gamma$ be inner models is tantamount to requiring that they be transitive. For $\lambda < \gamma$ a limit ordinal, the requirement that $(M_\lambda, \pi_{\alpha\lambda}: \alpha < \lambda)$ is the direct limit of $(M_\alpha, \pi_{\alpha\beta}: \alpha \leq \beta < \lambda)$ means, by virtue of (2), that $M_\lambda = \bigcup\{\mathrm{ran}(\pi_{\alpha\lambda}): \alpha < \lambda\}$.

Definition 10.2 Let κ be a measurable cardinal, and let U be a measure on κ. Then V is called *iterable by U and its images* iff for every γ, if

$$\mathscr{I} = (M_\alpha, \pi_{\alpha\beta}: \alpha \leq \beta < \gamma + 1)$$

is the (linear) putative iteration of V of length $\gamma + 1$ given by U, then $\mathscr{I}$ is an iteration, i.e., M_γ is well-founded (and may therefore be identified with an inner model).

Theorem 10.3 *Let κ be a measurable cardinal, and let U be a measure on κ. Then V is iterable by U and its images.*

Proof Let γ be an ordinal, and let

$$(M_\alpha, \pi_{\alpha\beta}: \alpha \leq \beta < \gamma + 1) \tag{10.1}$$

be the (linear) putative iteration of V of length $\gamma + 1$ given by U. Let

$$\sigma: \overline{V} \cong X \prec_{\Sigma_{1002}} V,$$

where $\{\kappa, U, \gamma\} \subset X$, X is countable, and $\overline{V}$ is transitive. Let $\overline{\kappa} = \sigma^{-1}(\kappa)$, $\overline{U} = \sigma^{-1}(U)$, and $\overline{\gamma} = \sigma^{-1}(\gamma)$. We may also set, for $\alpha \in \mathrm{ran}(\sigma) \cap (\gamma + 1)$,

$$\overline{M}_{\sigma^{-1}(\alpha)} = \sigma^{-1}(M_\alpha),$$

and for $\alpha \leq \beta$, $\alpha, \beta \in \mathrm{ran}(\sigma) \cap (\gamma + 1)$,[1]

$$\overline{\pi}_{\sigma^{-1}(\alpha), \sigma^{-1}(\beta)} = \sigma^{-1}(\pi_{\alpha\beta}).$$

Then, from the point of view of $\overline{V}$,

$$(\overline{M}_{\alpha, \overline{\pi}_{\alpha\beta}}: \alpha \leq \beta < \overline{\gamma} + 1)$$

is the (linear) putative iteration of $\overline{V}$ of length $\overline{\gamma} + 1$ given by $\overline{U}$.[2]

---

[1] For a proper class X, we write $\sigma^{-1}(X)$ for $\bigcup\{\sigma^{-1}(X \cap V_\alpha): X \cap V_\alpha \in \mathrm{ran}(\sigma)\}$.

[2] We here use the fact that the ultrapower construction may also be applied with transitive models of a sufficiently large *fragment* of ZFC. We leave the straightforward details to the reader.

We shall now recursively, for $\alpha < \overline{\gamma} + 1$, construct embeddings

$$\sigma_\alpha : \overline{M}_\alpha \rightarrow_{\Sigma_{1000}} V$$

such that whenever $\alpha \leq \beta < \overline{\gamma} + 1$, then

$$\sigma_\beta \circ \overline{\pi}_{\alpha\beta} = \sigma_\alpha . \tag{10.2}$$

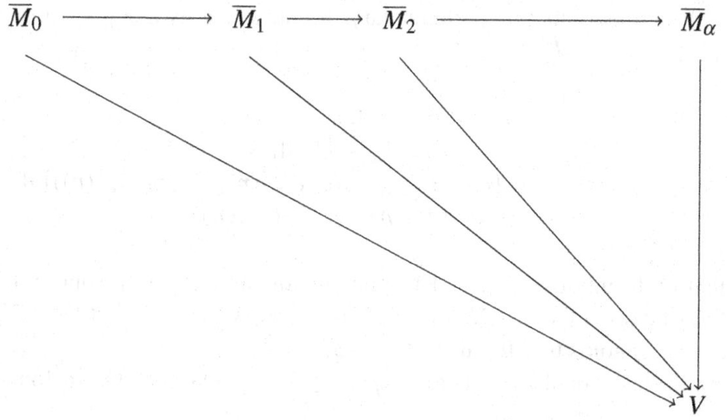

We set $\sigma_0 = \sigma$. Now let $\delta \leq \overline{\gamma}$, and suppose all σ_α, $\alpha < \delta$, are already construed such that (10.2) holds true for all $\alpha \leq \beta < \delta$.

Let us first suppose δ to be a limit ordinal, so that $(\bar{M}_\delta, (\bar{\pi}_{\alpha\delta} : \alpha < \delta))$ is the direct limt of $((\bar{M}_\alpha : \alpha < \delta), (\bar{\pi}_{\alpha\beta} : \alpha \leq \beta < \delta))$. We then define $\sigma_\delta : \overline{M}_\delta \to V$ by setting

$$\sigma_\delta(x) = \sigma_\alpha \circ \overline{\pi}_{\alpha\delta}^{-1}(x),$$

whenever $x \in \operatorname{ran}(\overline{\pi}_{\alpha\delta})$. For every $x \in \overline{M}_\delta$ there is some $\alpha < \delta$ with $x \in \operatorname{ran}(\overline{\pi}_{\alpha\delta})$, and if $x = \overline{\pi}_{\alpha\delta}(x') = \overline{\pi}_{\bar{\alpha}\delta}(x'')$ with $\bar{\alpha} \leq \alpha$, then, using (10.2),

$$\sigma_\alpha(x') = \sigma_\alpha(\overline{\pi}_{\alpha\delta}^{-1}(x)) = \sigma_\alpha \circ \overline{\pi}_{\bar{\alpha}\alpha} \circ \overline{\pi}_{\bar{\alpha}\delta}^{-1}(x) = \sigma_{\bar{\alpha}}(x'').$$

This means that σ_δ is well-defined, and it is easy to verify that σ_δ is Σ_{1000}-elementary and (10.2) holds true for all $\alpha \leq \beta \leq \delta$.

Now suppose δ to be a successor ordinal, say $\delta = \xi + 1$. Set $\overline{\kappa}_\xi = \overline{\pi}_{0\xi}(\overline{\kappa})$ and $\overline{U}_\xi = \overline{\pi}_{0\xi}(\overline{U})$. We have that $\bar{M}_\delta = \operatorname{ult}(\bar{M}_\xi ; \bar{U}_\xi)$, which is given by equivalence relations (mod $\bar{U}_\xi$) of functions $f \in {}^{\overline{\kappa}_\xi}\overline{M}_\xi \cap \overline{M}_\xi$.

If φ is a Σ_{1000}-formula, and $f_1, \ldots, f_k \in {}^{\overline{\kappa}_\xi}\overline{M}_\xi \cap \overline{M}_\xi$, then we write $X_{\varphi, f_1, \ldots, f_k}$ for

$$\{\eta < \overline{\kappa}_\xi : \overline{M}_\xi \models \varphi(f_1(\eta), \ldots, f_k(\eta))\}.$$

By Łoś' Theorem 4.56, $X_{\varphi, f_1, \ldots, f_k} \in \overline{U}_\xi$ iff

$$\overline{M}_\delta \models \varphi([f_1], \ldots, [f_k]).$$

Because $\bar{M}_\xi$ and hence $\bar{U}_\xi$ is countable, $\sigma_\xi{}''\overline{U}_\xi$ is a countable subset of U. As U is $< \aleph_1$-closed, we thus have that $\bigcap \sigma_\xi{}''\overline{U}_\xi \neq \emptyset$. Say $\rho \in \bigcap \sigma_\xi{}''\overline{U}_\xi$.

We may now define $\sigma_\delta \colon \overline{M}_\delta \to V$ by setting

$$\sigma_\delta([f]) = \sigma_\xi(f)(\rho).$$

This is well-defined and Σ_{1000}-elementary, because if φ is a Σ_{1000}-formula and $f_1, \ldots, f_k \in {}^{\kappa_\xi}\overline{M}_\xi \cap \overline{M}_\xi$, then

$$\begin{aligned} \overline{M}_{\xi+1} \models \varphi([f_1], \ldots [f_k]) \text{ iff} \\ X_{\varphi, f_1, \ldots, f_k} \in \overline{U}_\xi \text{ iff} \\ \rho \in \sigma_\xi(X_{\varphi, f_1, \ldots, f_k}) = \{\eta < \kappa \colon V \models \varphi(\sigma_\xi(f_1)(\eta), \ldots, \sigma_\xi(f_k)(\eta))\} \text{ iff} \\ V \models \varphi(\sigma_\xi(f_1)(\rho), \ldots, \sigma_\xi(f_k)(\rho)). \end{aligned}$$

We use that uniformly over $\bar{M}_\xi$ and V, bounded quantification in front of a Σ_{1000}-formula may be rewritten in a Σ_{1000} way. It is also easy to verify that $\sigma_\delta = \overline{\pi}_{\xi\delta} \circ \sigma_\xi$ and hence (10.2) holds true for all $\alpha \leq \beta \leq \delta$.

But now the last model $\bar{M}_{\bar{\gamma}}$ of $(\overline{M}_\alpha, \overline{\pi}_{\alpha\beta} \colon \alpha \leq \beta < \overline{\gamma}+1)$ cannot be ill-founded, as

$$\sigma_{\bar{\gamma}} \colon \bar{M}_{\bar{\gamma}} \to_{\Sigma_{1000}} V.$$

By the elementarity of σ, the last model M_γ of $(M_\alpha, \pi_{\alpha\beta} \colon \alpha \leq \beta < \gamma + 1)$ cannot be ill-founded either. This means that (10.1) is in fact a (linear) iteration of V of length $\gamma + 1$ given by U, as desired. □

Lemma 10.4 (Shift Lemma) *Let κ be a measurable cardinal, and let U be a normal measure on κ. Let*

$$(M_\alpha, \pi_{\alpha\beta} \colon \alpha \leq \beta \in \mathrm{OR})$$

be the (linear) iteration of $V = M_0$ which is given by U. For $\alpha \in \mathrm{OR}$, set $U_\alpha = \pi_{0\alpha}(U)$ and $\kappa_\alpha = \mathrm{crit}(U_\alpha) = \pi_{0\alpha}(\kappa)$. Let $\alpha \leq \beta$, and let $\varphi \colon \alpha \to \beta$ be order preserving. There is then a natural elementary embedding

$$\pi^\varphi_{\alpha\beta} \colon M_\alpha \to M_\beta,$$

called the shift map given by φ *such that $\pi^\varphi_{\alpha\beta}(\kappa_\alpha) = \kappa_\beta$, and for all $\overline{\alpha} < \alpha$, $\pi^\varphi_{\alpha\beta}(\kappa_{\overline{\alpha}}) = \kappa_{\varphi(\overline{\alpha})}$ and in fact*

$$\pi^\varphi_{\alpha\beta} \circ \pi_{\overline{\alpha}\alpha} = \pi_{\overline{\beta}\beta} \circ \pi^{\varphi \restriction \overline{\alpha}}_{\overline{\alpha}\overline{\beta}} \tag{10.3}$$

for all $\overline{\beta}$ with $\mathrm{ran}(\varphi \restriction \overline{\alpha}) \subset \overline{\beta} < \beta$.

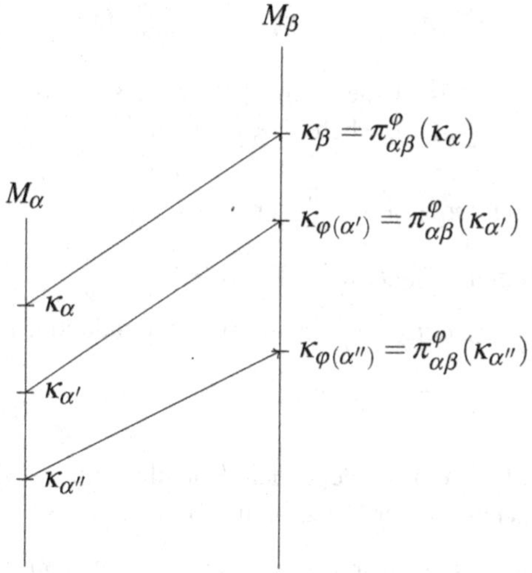

Proof by induction on β. The statement is trivial for $\beta = 0$, setting $\pi^{\varphi}_{\alpha\beta} = \text{id}$. Now let $\beta > 0$.

Let us first suppose β to be a successor ordinal, say $\beta = \overline{\beta} + 1$. If $\overline{\beta} \notin \text{ran}(\varphi)$, then we may construe φ as a map from α to $\overline{\beta}$ and simply set $\pi^{\varphi}_{\alpha\beta} = \pi_{\overline{\beta}\beta} \circ \pi^{\varphi}_{\alpha\overline{\beta}}$. Let us thus assume $\overline{\beta} \in \text{ran}(\varphi)$, which implies that α is a successor ordinal as well, say $\alpha = \overline{\alpha} + 1$, and $\varphi(\overline{\alpha}) = \overline{\beta}$. By (4.8), we have that

$$M_\alpha = \{\pi_{\overline{\alpha},\alpha}(f)(\kappa_{\overline{\alpha}}) \colon f : \kappa_{\overline{\alpha}} \to M_{\overline{\alpha}},\ f \in M_{\overline{\alpha}}\}.$$

We may thus define $\pi^{\varphi}_{\alpha\beta}$ by setting

$$\pi^{\varphi}_{\alpha\beta}(\pi_{\overline{\alpha}\alpha}(f)(\kappa_{\overline{\alpha}})) = \pi_{\overline{\beta}\beta} \circ \pi^{\varphi\restriction\overline{\alpha}}_{\overline{\alpha}\overline{\beta}}(f)(\kappa_{\overline{\beta}}),$$

where $f \in M_{\overline{\alpha}}$, $f \colon \kappa_{\overline{\alpha}} \to M_{\overline{\alpha}}$. This is well-defined because if ψ is a formula and if $f_1, \ldots, f_k \in {}^{\kappa_{\overline{\alpha}}}M_{\overline{\alpha}} \cap M_{\overline{\alpha}}$, then

$$\begin{gathered} M_\alpha \models \psi(\pi_{\overline{\alpha}\alpha}(f_1)(\kappa_{\overline{\alpha}}), \ldots, f_k(\kappa_{\overline{\alpha}})) \iff \\ \{\xi < \kappa_{\overline{\alpha}} \colon M_{\overline{\alpha}} \models \psi(f_1(\xi), \ldots, f_k(\xi))\} \in U_{\overline{\alpha}} \iff \\ \{\xi < \kappa_{\overline{\beta}} \colon M_{\overline{\beta}} \models \psi(\pi^{\varphi\restriction\overline{\alpha}}_{\overline{\alpha}\overline{\beta}}(f_1)(\xi), \ldots, \pi^{\varphi\restriction\overline{\alpha}}_{\overline{\alpha}\overline{\beta}}(f_k)(\xi))\} \in U_{\overline{\beta}}, \text{ by using } \pi^{\varphi\restriction\overline{\alpha}}_{\overline{\alpha}\overline{\beta}}, \iff \\ M_\beta \models \psi(\pi_{\overline{\beta}\beta} \circ \pi^{\varphi\restriction\overline{\alpha}}_{\overline{\alpha}\overline{\beta}}(f_1)(\kappa_{\overline{\beta}}), \ldots, \pi_{\overline{\beta}\beta} \circ \pi^{\varphi\restriction\overline{\alpha}}_{\overline{\alpha}\overline{\beta}}(f_k)(\kappa_{\overline{\beta}})). \end{gathered}$$

It is easy to verify, using the inductive hypotheses, that $\pi^{\varphi}_{\alpha\beta}$ is as desired.

Now suppose β to be a limit ordinal. If φ is not cofinal in β, say $\text{ran}(\varphi) \subset \overline{\beta} < \beta$, then we may construe φ as a map from α to $\overline{\beta}$ and simply set $\pi^{\varphi}_{\alpha\beta} = \pi_{\bar{\beta}\beta} \circ \pi^{\varphi}_{\alpha\bar{\beta}}$. Let us thus assume that φ is cofinal in β, which implies that α is a limit ordinal as well. We then define $\pi^{\varphi}_{\alpha\beta}$ by setting

$$\pi^{\varphi}_{\alpha\beta}(\pi_{\overline{\alpha}\alpha}(x)) = \pi_{\varphi(\overline{\alpha})\beta} \circ \pi^{\varphi\restriction\overline{\alpha}}_{\overline{\alpha}\varphi(\overline{\alpha})}(x).$$

Notice that each $y \in M_\alpha$ is of the form $\pi_{\overline{\alpha}\alpha}(x)$ for some $\overline{\alpha} < \alpha$ and $x \in M_{\overline{\alpha}}$. Moreover, if $\pi_{\overline{\alpha}\alpha}(x) = \pi_{\overline{\alpha}'\alpha}(x')$, where $\overline{\alpha} < \overline{\alpha}'$, then

$$\begin{aligned}\pi_{\overline{\alpha}\overline{\alpha}'}(x) = x' \text{ and } \pi_{\varphi(\overline{\alpha})\beta} &\circ \pi^{\varphi\restriction\overline{\alpha}}_{\overline{\alpha}\varphi(\overline{\alpha})}(x)\\ &= \pi_{\varphi(\overline{\alpha}')\beta} \circ \pi_{\varphi(\overline{\alpha})\varphi(\overline{\alpha}')} \circ \pi^{\varphi\restriction\overline{\alpha}}_{\overline{\alpha}\varphi(\overline{\alpha})}(x)\\ &= \pi_{\varphi(\overline{\alpha}')\beta} \circ \pi^{\varphi\restriction\overline{\alpha}'}_{\overline{\alpha}'\varphi(\overline{\alpha}')} \circ \pi_{\overline{\alpha}\overline{\alpha}'}(x) \text{ by the inductive hypothesis,}\\ &= \pi_{\varphi(\overline{\alpha}')\beta} \circ \pi^{\varphi\restriction\overline{\alpha}'}_{\overline{\alpha}'\varphi(\overline{\alpha}')}(x'),\end{aligned}$$

so that the definition of $\pi^{\varphi}_{\alpha\beta}(y)$ is independent from the choice of $\overline{\alpha} < \alpha$ and $x \in M_\alpha$ with $y = \pi_{\overline{\alpha}\alpha}(x)$. It is easy to verify the inductive hypothesis. □

We now aim to make a measurable cardinal κ singular in a generic extension without collapsing cardinals. The following definition is reminiscent of the definition of MATHIAS forcing, cf. p. 176.

Definition 10.5 Let M be a transitive model of ZFC, and let $\kappa, U \in M$ be such that $M \models$ "U is a normal $< \kappa$-complete uniform ultrafilter on κ." We let $\mathbb{P}_U = \mathbb{P} = (\mathbb{P}; \leq)$ denote the following poset, called PRIKRY *forcing*. We let $p \in \mathbb{P}$ iff $p = (a, X)$ where $a \in [\kappa]^{<\omega}$ and $X \in U, \min(X) > \max(a)$ (if $a \neq \emptyset$). We let $(b, Y) \leq (a, X)$ iff $b \supset a$, $Y \subset X$, and $b \setminus a \subset X$ (in particular, b is an end-extension of a).

If $p = (a, X) \in \mathbb{P}$ then a is called the *stem* of p. Notice that any two conditions with the same stem are compatible, so that $\mathbb{P}$ has the κ^+-c.c. in M. Hence no M-cardinal strictly above κ will be collapsed by κ.

Assume G to be $\mathbb{P}$-generic over M. Let

$$A = \bigcup\{a : \exists X\, (a, X) \in G\}.$$

It is clear that A hast order-type $\leq \omega$; in fact, an easy density argument shows that $\text{otp}(A) = \omega$ and A is cofinal in κ. In particular, κ will have cofinality ω in $M[G]$.

We shall now prove that no M-cardinal $\leq \kappa$ will be collapsed in $M[G]$. As κ is a limit cardinal, it suffices to prove that no M-cardinal $\lambda < \kappa$ will be collapsed in $M[G]$. For this in turn it is (more than) enough to prove the following.

Lemma 10.6 *Let M be a transitive model of* ZFC, *let $\kappa, U \in M$ be such that U witnesses that κ is measurable in M, let $\mathbb{P} = \mathbb{P}_U$, and let G be $\mathbb{P}$-generic over M. Then*

$$(V_\kappa)^{M[G]} = (V_\kappa)^M.$$

Proof Let us first verify the following.

Claim 10.7 (PRIKRY-Lemma) *For all $p \in \mathbb{P}$, for all formulae φ, and for all names $\tau_1, \ldots, \tau_k \in M^{\mathbb{P}_U}$ there is some $q \leq p$ with the same stem as p deciding $\varphi(\tau_1, \ldots, \tau_k)$.*

Proof This is an application of ROWBOTTOM's Theorem 4.59. Fix $p = (a, X)$ and φ. (We shall supress the parameters $\tau_1, \ldots, \tau_k$.) Let us define $F: [X]^{<\omega} \to 3$ as follows. Let $b \in [X]^{<\omega}$. We set $F(b) = 0$ iff there is no $\overline{X}$ such that $(a \cup b, \overline{X}) \leq (a, X)$ and $(a \cup b, \overline{X})$ decides φ; otherwise we set $F(b) = 1$ (resp., 2) iff there is some $\overline{X}$ such that $(a \cup b, \overline{X}) \leq (a, X)$ and $(a \cup b, \overline{X})$ forces that φ holds (resp., does not hold).

As $f \in M$, let $Y \in U$ be given by ROWBOTTOM's Theorem 4.59, i.e., for each $n < \omega$, F is constant on $[Y]^{<\omega}$. We claim that (a, Y) decides φ.

Well, if not, then there are (b_1, Y_1) and (b_2, Y_2) such that $(b_1, Y_1) \leq (a, Y)$, $(b_2, Y_2) \leq (a, Y)$ and $(b_1, Y_1) \Vdash \varphi$ and $(b_2, Y_2) \Vdash \neg\varphi$. By extending one of these two conditions if necessary we may assume that

$$\text{Card}(b_1) = \text{Card}(b_2) = \text{Card}(a) + n$$

for some $n < \omega$. But then

$$F(b_1 \backslash a) = 1 \neq 2 = F(b_2 \backslash a),$$

although

$$b_1 \backslash a, b_2 \backslash a \in [Y]^n.$$

Contradiction! □

In order to prove Lemma 10.6, it now suffices to show that if $\lambda < \kappa$, then ${}^\lambda 2 \cap M[G] \subset M$. Let $f \in {}^\lambda 2 \cap M[G]$, $f = \tau^G$. Let $p \Vdash \tau: \check{\lambda} \to \check{2}$. It suffices to find some $q \leq p$ and some $g \in M$ with $q \Vdash \tau = \check{g}$.

Let a be the stem of p. In virtue of the PRIKRY Lemma we may let for each $\xi < \lambda$ be $q_\xi \leq p$ and $h_\xi \in 2$ such that a is the stem of q_ξ and $q_\xi \Vdash \tau(\check{\xi}) = \check{(h_\xi)}$. If $q_\xi = (a, X_\xi)$ for $\xi < \lambda$ then $q = (a, \bigcap_{\xi<\lambda} X_\xi) \leq p$ and $q \Vdash \tau = \check{g}$, where $g \in {}^\lambda 2$ and $g(\xi) = h_\xi$ for all $\xi < \lambda$. □

There is a version of $\mathbb{P}_U$, called *tree* PRIKRY *forcing*, where we don't need to assume that U is normal (in M) in order to verify the PRIKRY Lemma; cf. Problem 10.24, which produces a generalization of tree PRIKRY forcing.

Definition 10.8 Let M be a transitive model of ZFC, and let $\kappa, U \in M$ be such that U witnesses that κ is measurable in M. A strictly increasing sequence $(\kappa_n : n < \omega)$ which is cofinal in κ is called a PRIKRY *sequence over M* (*with respect to U*) iff for all $X \in \mathscr{P}(\kappa) \cap M$.

$$X \in U \iff \{\kappa_n : n < \omega\} \backslash X \text{ is finite.}$$

By an easy density argument, if G is $\mathbb{P}$-generic over M and if $(\kappa_n : n < \omega)$ is the sequence given by the first coordinates of elements of G, then $(\kappa_n : n < \omega)$ is a PRIKRY sequence. Cf. also Problem 9.9 (b).

PRIKRY sequences are also generated by iterated ultrapowers, cf. also Problem 10.2 (c):

Lemma 10.9 *Let κ be a measurable cardinal, let U be a measure on κ, and let*

$$(M_\alpha, \pi_{\alpha\beta} : \alpha \leq \beta \leq \omega)$$

be an iteration of $V = M_0$ of length $\omega + 1$ given by U. Then $(\pi_{0n}(\kappa) : n < \omega)$ is a PRIKRY sequence over M_ω with respect to $\pi_{0\omega}(U)$.

Proof This is an immediate consequence of the Shift Lemma 10.4. Let $X \subset \pi_{0\omega}(\kappa)$, $X \in M_\omega$. Say $X \in \operatorname{ran}(\pi_{n_0\omega})$, where $n_0 < \omega$. Let $m \geq n > n_0$. Let $\varphi : \omega \to \omega$ be defined by

$$\varphi(k) = \begin{cases} k & \text{if } k < n \\ k + (m - n) & \text{if } k \geq n. \end{cases}$$

By Lemma 10.4, cf. (10.3), $\pi^{\varphi}_{\omega\omega} \upharpoonright \operatorname{ran}(\pi_{n_0\omega}) = \mathrm{id}$, so that in particular $\pi^{\varphi}_{\omega\omega}(X) = X$. But then

$$\pi_{0n}(\kappa) \in X \iff \pi_{0m}(\kappa) = \pi^{\varphi}_{\omega\omega}(\pi_{0n}(\kappa)) \in X.$$

We have shown that either $\{\pi_{0n}(\kappa) : n_0 < n < \omega\} \subset X$ or else $\{\pi_{0n}(\kappa) : n_0 < n < \omega\} \cap X = \emptyset$. □

We shall prove the converse to the fact that if $(\kappa_n : n < \omega)$ is the sequence given by the first coordinates of elements of G then $(\kappa_n : n < \omega)$ is a PRIKRY sequence. By virtue of Lemma 10.9, this will mean that iterations produce PRIKRY generics.

Definition 10.10 Let $(\kappa_n : n < \omega)$ be a PRIKRY sequence (with respect to U). Define $G_{(\kappa_n : n < \omega)}$ to be the set of all $(\{\kappa_n : n < n_0\}, X)$ where $n_0 < \omega$, $X \in U$, $\min(X) > \kappa_{n_0 - 1}$ (if $n_0 > 0$), and $\{\kappa_n : n \geq n_0\} \subset X$.

Theorem 10.11 (A. Mathias) *Let M be a transitive model of* ZFC *such that $M \models$ "U is a normal measure on κ." Let $(\kappa_n : n < \omega)$ be a PRIKRY sequence over M with respect to U. Then $G_{(\kappa_n : n < \omega)}$ is $\mathbb{P}$-generic over M.*

The proof will be given in Chap. 12, cf. p. 274 ff.

Theorem 10.12 *Let M be a transitive model of* ZFC, *and suppose that there is some $\kappa \in M$ such that $M \models$ "κ is a measurable cardinal, as being witnessed by the normal measure U, and $2^\kappa \geq \kappa^{++}$." Let G be $\mathbb{P}_U$-generic over M. Then in $M[G]$,* SCH, *the Singular Cardinal Hypothesis, fails.*

Proof M and $M[G]$ have the same cardinals, and $(V_\kappa)^M = (V_\kappa)^{M[G]}$. As κ is measurable in M, this implies that κ is certainly a strong limit cardinal in $M[G]$. Therefore, by Lemma 4.16

$$M[G] \models \kappa^{\text{cf}(\kappa)} = 2^\kappa. \tag{10.4}$$

As κ has countable cofinality in $M[G]$,

$$M[G] \models 2^{\text{cf}(\kappa)} = 2^{\aleph_0} < \kappa. \tag{10.5}$$

On the other hand, by hypothesis in M there is a surjection from $\mathscr{P}(\kappa)$ onto κ^{++}, so that with (10.4)

$$M[G] \models \kappa^{\text{cf}(\kappa)} \geq \kappa^{++}. \tag{10.6}$$

(10.5) and (10.6) yield a failure of SCH in $M[G]$. □

A model M which satisfies the hypothesis of Theorem 10.12 may be produced with the help of *iterated forcing* starting from a model with a measurable cardinal κ whose MITCHELL order (cf. Problem 4.27) has rank $(2^\kappa)^+$.

10.2 The Story of $0^\sharp$, Revisited

$0^\#$ is a countable structure which "transcends" GÖDEL's Constructible Universe L in a precise way, cf. Theorem 11.56. We first need to explain what we mean by "$0^\#$ exists." For later purposes (cf. Theorem 12.27) we shall in fact introduce $x^\#$ for arbitray reals x.

We begin by introducing Σ_1-Skolem functions for J-structures.

In what follows, we shall only consider models of $\mathscr{L}_{\in,\dot{E}}$, but what we shall say easily generalizes to $\mathscr{L}_{\in,\dot{A}_1,\ldots,\dot{A}_m}$. Let us fix an enumeration $(\varphi_n : n < \omega)$ of all Σ_1-formulae of the language $\mathscr{L}_{\in,\dot{E}}$. We shall denote by $\ulcorner\varphi\urcorner$ the GÖDEL number of φ, i.e., $\ulcorner\varphi\urcorner = n$ iff $\varphi = \varphi_n$. We may and shall assume that $(\ulcorner\varphi_n\urcorner : n < \omega)$ is recursive, and if $\bar{\varphi}$ is a proper subformula of φ, then $\ulcorner\bar{\varphi}\urcorner < \ulcorner\varphi\urcorner$. We shall write $v(n)$ for the set of free variables of φ_n.

Let M be a model of $\mathscr{L}_{\in,\dot{E}}$. We shall express by

$$M \models \varphi_n[\mathsf{a}]$$

the fact that $\mathsf{a} : v(n) \to M$, i.e., a assigns elements of M to the free variables of φ_n, and φ_n holds true in M under this assignment. We shall also write $\models_M^{\Sigma_1}$ for the set of all (n, a) such that $M \models \varphi_n[\mathsf{a}]$, and we shall write $\models_M^{\Sigma_0}$ for the set of all $(n, \mathsf{a}) \in \models_M^{\Sigma_1}$ such that φ_n is a Σ_0-formula.

Lemma 10.13 *Let $M = J_\alpha[E]$ be a J-structure. Let $N \in M$ be transitive. For each $m < \omega$, there is a unique $f = f_m^N \in M$ such that* $\text{dom}(f) = m$ *and for all $n < m$, if φ_n is not a Σ_0 formula, then $f(n) = \emptyset$, and if φ_n is a Σ_0 formula, then*

$$f(n) = \{\mathsf{a} \in {}^{v(n)}N : (N; \in, E \cap N) \models \varphi_n[\mathsf{a}]\}.$$

Proof As uniqueness is clear, let us verify inductively that $f_m^N \in M$. Well, $f_0^N = \emptyset \in M$. Now suppose that $f_m^N \in M$. If φ_m is not Σ_0, then $f_{m+1}^N = f_m^N \cup \{(m, \emptyset)\} \in M$. Now let φ_m be Σ_0. We have that ${}^{v(m)}N \in M$ (cf. Corollary 5.18), and if

$$T = \{\mathsf{a} \in {}^{v(m)}N \colon (N; \in, E \cap N) \models \varphi_m[\mathsf{a}]\}$$

then $T \in \mathscr{P}({}^{v(m)}N) \cap (\underset{\sim}{\Sigma}_0)^M$, and thus $T \in M$ by Lemma 5.23. Therefore, $f_{m+1}^N = f_m^N \cup \{(m, T)\} \in M$. □

Now let $\Theta(f, N, m)$ denote the following formula.

$$\begin{aligned}
&N \text{ is transitive} \wedge m < \omega \wedge f \text{ is a function with domain } m \wedge \forall n < m \\
&((n = \ulcorner v_{i_0} \in v_{i_1} \urcorner, \text{ some } i_0, i_1 \longrightarrow f(n) = \{\mathsf{a} \in {}^{v(n)}N \colon \mathsf{a}(v_{i_0}) \in \mathsf{a}(v_{i_1})\}) \wedge \\
&\quad (n = \ulcorner v_{i_0} \in \dot{E} \urcorner, \text{ some } i_0 \longrightarrow f(n) = \{\mathsf{a} \in {}^{v(n)}N \colon \mathsf{a}(v_{i_0}) \in E\}) \wedge \\
&\quad (n = \ulcorner \psi_0 \wedge \psi_1 \urcorner, \text{ some } \psi_0, \psi_1 \longrightarrow \\
&\qquad f(n) = \{\mathsf{a} \in {}^{v(n)}N \colon \mathsf{a} \restriction v(\ulcorner \psi_0 \urcorner) \in f(\ulcorner \psi_0 \urcorner) \wedge \mathsf{a} \restriction v(\ulcorner \psi_1 \urcorner) \in f(\ulcorner \psi_1 \urcorner)\}) \wedge \\
&\quad (n = \ulcorner \exists v_{i_0} \in v_{i_1} \psi \urcorner, \text{ some } i_0, i_1, \psi, \text{ where } \psi \text{ is } \Sigma_0 \longrightarrow \\
&\qquad f(n) = \{\mathsf{a} \in {}^{v(n)}N \colon \exists x \in \mathsf{a}(v_{i_1}) \\
&\qquad\quad (\mathsf{a} \cup \{(v_{i_0}, x)\}) \restriction v(\ulcorner \psi \urcorner) \in f(\ulcorner \psi \urcorner)\}) \wedge \\
&\quad (n = \ulcorner \varphi \urcorner, \text{ some } \varphi, \text{ where } \varphi \text{ is not } \Sigma_0 \rightarrow f(n) = \emptyset)).
\end{aligned}$$

It is straightforward to check that $\Theta(f, N, m)$ holds (in M) if and only if $f = f_m^N$. Now Lemma 10.13 and the fact that every element of M is contained in a transitive element of M (cf. Lemma 5.25) immediately gives the following.

Lemma 10.14 *Let $M = J_\alpha[E]$ be a J-structure. Let φ_n be Σ_0, and let $\mathsf{a} \colon v(n) \to M$. Then $M \models \varphi_n[\mathsf{a}]$ holds true if and only if*

$$M \models \exists f \, \exists N \, (\operatorname{ran}(\mathsf{a}) \subset N \wedge \Theta(f, N, n+1) \wedge \mathsf{a} \in f(n)),$$

which in turn holds true if and only if

$$M \models \forall f \, \forall N \, ((\operatorname{ran}(\mathsf{a}) \subset N \wedge \Theta(f, N, n+1)) \rightarrow \mathsf{a} \in f(n)).$$

In particular, the relation $\models_M^{\Sigma_0}$ is Δ_1^M.

Theorem 10.15 *The Σ_1-satisfaction relation $\models_M^{\Sigma_1}$ is uniformly Σ_1^M over J- structures M, i.e., there is a Σ_1-formula Φ such that whenever $M = J_\alpha[E]$ is a J-structure, Φ defines $\models_M^{\Sigma_1}$ in that*

$$(n, \mathsf{a}) \in \models_M^{\Sigma_1} \Longleftrightarrow M \models \Phi(n, \mathsf{a}).$$

Proof We have that $M \models \varphi_n[\mathsf{a}]$ iff

$$\exists \mathsf{b} \in M\ \exists(v_{i_0},\ldots,v_{i_k},j),\ \text{some } v_{i_0},\ldots,v_{i_k},j\ [n = \ulcorner \exists v_{i_0}\ldots\exists v_{i_k}\varphi_j\urcorner \wedge \varphi_j \text{ is } \Sigma_0 \wedge \mathsf{a},\mathsf{b} \text{ are functions } \wedge \operatorname{dom}(\mathsf{a}) = v(n) \wedge \operatorname{dom}(\mathsf{b}) = v(j) \wedge \mathsf{a} = \mathsf{b} \restriction v(n) \wedge (j,\mathsf{b}) \in \models_M^{\Sigma_0}].$$

Here, $\models_M^{\Sigma_0}$ is uniformly Δ_1^M by Lemma 10.14. The rest is easy. □

If M is a J-structure, then h is a Σ_1-SKOLEM *function* for M if

$$h\colon (\bigcup_{n<\omega}\{n\} \times {}^{v(n)}M) \to M,$$

where h may be partial, such that whenever $\varphi_n = \exists v_{i_0}\varphi_j$ and $\mathsf{a}\colon v(n) \to M$, then

$$\exists y \in M \quad M \models \varphi_j[\mathsf{a} \cup \{(v_{i_0}, y)\} \restriction v(j)] \implies M \models \varphi_j[\mathsf{a} \cup \{(v_{i_0}, h(n,\mathsf{a}))\} \restriction v(j)].$$

Theorem 10.16 *There is a Σ_1 SKOLEM function h_M which is uniformly Σ_1^M over J-structures M, i.e., there is a Σ_1-formula Ψ such that whenever $M = J_\alpha[E]$ is a J-structure, then Ψ defines h_M over M in that*

$$y = h_M(n,\mathsf{a}) \iff M \models \Psi(n,\mathsf{a},y).$$

Proof The idea here is to let $y = h_M(n,\mathsf{a})$ be the "first component" of a minimal witness to the Σ_1 statement in question (rather than letting y be minimal itself). We may let $y = h_M(n,\mathsf{a})$ iff

$$\exists N\ \exists \beta\ \exists R\ \exists \mathsf{b},\ \text{all in } M,\ \exists(v_{i_0},\ldots,v_{i_k},j),\ \text{some } i_0,\ldots,i_k,j\ (N = S_\beta[E] \wedge R = <_\beta^E \wedge n = \ulcorner \exists v_{i_0}\ldots\exists v_{i_k}\varphi_j\urcorner \wedge \varphi_j \text{ is } \Sigma_0 \wedge \mathsf{a},\mathsf{b} \text{ are functions } \wedge \operatorname{dom}(\mathsf{a}) = v(n) \wedge \operatorname{dom}(\mathsf{b}) = v(j) \wedge \mathsf{a} = \mathsf{b} \restriction v(n) \wedge \operatorname{ran}(\mathsf{b}) \subset N \wedge (j,\mathsf{b}) \in \models_M^{\Sigma_0} \wedge \forall \bar{\mathsf{b}} \in N((\bar{\mathsf{b}} \text{ is a function } \wedge \operatorname{dom}(\bar{\mathsf{b}}) = v(j) \wedge \mathsf{a} = \bar{\mathsf{b}} \restriction v(n) \wedge \operatorname{ran}(\bar{\mathsf{b}}) \subset N \wedge \bar{\mathsf{b}}\ R\ \mathsf{b}) \to \neg(j,\mathsf{b}) \in \models_M^{\Sigma_0}) \wedge y = \mathsf{b}(v_{i_0})\,).$$

Here, "$N = S_\beta[E]$" and "$R = <_\beta^E$" are uniformly Σ_1^M by Lemmas 5.25 (2) and 5.26 (2), respectively, and $\models_M^{\Sigma_0}$ is uniformly Δ_1^M by Lemma 10.14. The rest is straightforward. □

If we were to define a Σ_2 Skolem function for M in the same manner then we would end up with a Σ_3 definition. JENSEN solved this problem by showing that under favourable circumstances Σ_n over M can be viewed as Σ_1 over a "reduct" of M, cf. Definition 11.7.

We also want to write $h_M(X)$ for the closure of X under h_M, more precisely: Let $M = J_\alpha[E]$ be a J-structure, and let $X \subset M$. We shall write

$$h_M(X) \text{ for } h_M``(\bigcup_{n<\omega}(\{n\} \times {}^{v(n)}X)).$$

Using Theorem 10.16, it is easy to verify that $h_M(X) \prec_{\Sigma_1} M$. There will be no danger of confusing the two usages of "h_M."

Lemma 10.17 *Let $M = J_\alpha[E]$ be a J-structure. There is then some (partial) surjective $f: [\alpha]^{<\omega} \to M$ which is Σ_1^M, and there is a (partial) surjection $g: \alpha \to M$ which is $\underset{\sim}{\Sigma}{}_1^M$.*

Proof We have that $h_M(\alpha) \prec_{\Sigma_1} M$, and hence $h_M(\alpha) = M$. But it is straightforward to construct a surjective $g': [\alpha]^{<\omega} \to \bigcup_{n<\omega}(\{n\} \times {}^{v(n)}\alpha)$ which is Σ_1^M. We may then set $f = h_M \circ g'$.

The existence of g now follows from Problem 5.18. □

We now introduce mice.

Definition 10.18 Let $x \subset \omega$. We say that the J-structure $\mathscr{M} = (J_\alpha[x]; \in, U)$ is an *x-premouse* (*x-pm*, for short) iff the following hold true.

(a) $(J_\alpha[x]; \in) \models \mathsf{ZFC}^-$ (i.e., ZFC without the power set axiom) + there is a largest cardinal, and
(b) if κ is the largest cardinal of $(J_\alpha[x]; \in)$, then $\mathscr{M} \models U$ is a non-trivial normal $< \kappa$-closed ultrafilter on κ.

Notice that in (a) of Definition 10.18, we consider the reduct of $\mathscr{M}$ with U being removed. Also recall (cf. Definition 5.24) that a J-structure has to be amenable, so that in the situation of Definition 10.18, for all $z \in J_\alpha[x]$, $z \cap U \in J_\alpha[x]$. J-structures are models of Σ_0-comprehension and more, cf. Corollary 5.18.

Lemma 10.19 *Let $x \subset \omega$. Let $\pi: L[x] \to L[x]$ be s.t. $\pi \neq \text{id}$. Let $\kappa = \text{crit}(\pi)$. Set $\alpha = \kappa^{+L[x]}$, and let $X \in U$ iff $X \in \mathscr{P}(\kappa) \cap L[x] \wedge \kappa \in \pi(X)$. Then $(J_\alpha[x]; \in, U)$ is an x-pm.*

Proof We show that $(J_\alpha[x]; \in, U)$ is amenable, using what is sometimes referred to as the "ancient KUNEN argument." Let $z \in J_\alpha[x]$. Pick $f \in J_\alpha[x]$, $f: \kappa \to z$ onto. Then $y \in z \cap U$ iff there is some $\xi < \kappa$ such that $y = f(\xi) \in z$ and $\kappa \in \pi(y) = \pi(f(\xi)) = \pi(f)(\xi)$. But $\pi(f) \in L[x]$. Hence $z \cap U$ can be computed inside $L[x]$ from f and $\pi(f)$. But of course, $z \cap U \subset z \in J_\alpha[x]$, so that in fact $z \cap U \in J_\alpha[x]$ by Theorem 5.31. □

It is easy to verify that if x, π, κ are as in Lemma 10.19 then $(J_\kappa[x]; \in) \models \mathsf{ZFC}$ (cf. also Lemma 10.21 (h)).

We now aim to iterate premice by taking ultrapowers in much the same way as in the proof of Theorem 4.55 and in Definition 10.1.

Definition 10.20 Let $x \subset \omega$, and let $\mathscr{M} = (J_\alpha[x]; \in, U)$ be an x-pm. We define the *(Σ_0-)ultrapower* $\text{ult}_0(\mathscr{M})$ of $\mathscr{M}$ as follows. Let $\kappa =$ the largest cardinal of $J_\alpha[x]$. For $f, g \in {}^\kappa J_\alpha[x] \cap J_\alpha[x]$ we write $f \sim g$ iff $\{\xi < \kappa: f(\xi) = g(\xi)\} \in U$, and we write

$f \tilde{\in} g$ iff $\{\xi < \kappa : f(\xi) \in g(\xi)\} \in U$. We let $[f]$ denote the $\sim$-equivalence class of f. We write $[f] \tilde{\in} [g]$ iff $f \tilde{\in} g$, and we also write $[f] \in \tilde{U}$ iff $\{\xi < \kappa : f(\xi) \in U\} \in U$. We let

$$\text{ult}_0(\mathcal{M}) = \{\{[f] : f \in {}^\kappa J_\alpha[x] \cap J_\alpha[x]\}; \tilde{\in}, \tilde{U}\}.$$

We may define a natural map $\pi_U^{\mathcal{M}} : \mathcal{M} \to \text{ult}_0(\mathcal{M})$ by setting $\pi_U^{\mathcal{M}}(x) = [c_x]$, where $c_x(\xi) = x$ for all $\xi < \kappa$. $\pi_U^{\mathcal{M}}$ is called the (Σ_0-)*ultrapower map*.

If $\text{ult}_0(\mathcal{M})$ is well-founded, then we identify it with its transitive collapse, we identify $[f]$ with the image under the transitive collapse, and we idenitify π with the composition of π with the transitive collapse.

In what follows, $\dot{x}$ and $\dot{U}$ are predicates which are supposed to be interepreted by x and U, respectively (or, more generally, by z and U', respectively, over $(J_\gamma[z]; \in, U')$).

Lemma 10.21 *Let $x \subset \omega$, and let $\mathcal{M} = (J_\alpha[x]; \in, U)$ be an x-pm with largest cardinal κ, and suppose that* $\text{ult}_0(\mathcal{M})$ *be well-founded. Let $\pi = \pi_U^{\mathcal{M}} : \mathcal{M} \to$* $\text{ult}_0(\mathcal{M})$ *be the ultrapower map. Then:*

(a) *π is Σ_0 elementary with respect to $\dot{\in}$, $\dot{x}$, $\dot{U}$,*
(b) *π is cofinal, i.e. for all $x \in$* $\text{ult}_0(\mathcal{M})$ *there is some (transitive) $y \in \mathcal{M}$ with $x \in \pi(y)$,*
(c) *π is Σ_1 elementary with respect to $\dot{\in}$, $\dot{x}$, $\dot{U}$,*
(d) *π is fully elementary with respect to $\dot{\in}$, $\dot{x}$,*
(e) *$J_\kappa[x] \prec_{\Sigma_\omega} J_{\pi(\kappa)}[x]$,*
(f) *$\mathcal{P}(\kappa) \cap \mathcal{M} = \mathcal{P}(\kappa) \cap$* $\text{ult}_0(\mathcal{M})$,
(g) $\text{ult}_0(\mathcal{M}) \models \alpha = \kappa^+$, *and*
(h) *κ is inaccessible (in fact Mahlo) in both $\mathcal{M}$ and* $\text{ult}_0(\mathcal{M})$.

Proof (a) We have the following version of Łoś' Theorem 4.56.

Claim 10.22 (Łoś Theorem) *Let $\varphi(v_1, ..., v_k)$ be a Σ_0-formula in the language of $\mathcal{M}$, and let $f_1, \ldots, f_k \in {}^\kappa J_\alpha[x] \cap J_\alpha[x]$. Then*

$$\text{ult}_0(\mathcal{M}) \models \varphi([f_1], \cdots, [f_k]) \Longleftrightarrow$$
$$\{\xi < \kappa : \mathcal{M} \models \varphi(f_1(\xi), \cdots, f_k(\xi))\} \in U.$$

Notice that if φ is Σ_0, then $\{\xi < \kappa : \mathcal{M} \models \varphi(f_1(\xi), \cdots, f_k(\xi))\} \in J_\alpha[x]$ by Σ_0-comprehension, cf. Corollary 5.18, so that this makes sense. The proof of Claim 10.22 is analoguous to the proof of Claim 4.56, cf. Problem 10.6.

(b) Let $x \in \text{ult}_0(\mathcal{M})$, say $x = [f]$, where $f \in \mathcal{M}$. We have that, setting $y = \text{TC}(\{\text{ran}(f)\})$, $y \in \mathcal{M}$ and $\{\xi < \kappa : f(x) \in y\} = \kappa \in U$, so that by Łoś's Theorem 10.22 $x = [f] \in [c_y] = \pi(y)$. By (a), $\pi(y)$ will be transitive.

(c) Suppose that $\text{ult}_0(\mathcal{M}) \models \exists x \varphi(x, \pi(a_1), \ldots, \pi(a_k))$, where φ is Σ_0 and a_1, $\ldots$, $a_k \in \mathcal{M}$. Let $x_0 \in \text{ult}_0(\mathcal{M})$ be a witness. By (b) there is some $y \in \mathcal{M}$ with $x_0 \in \pi(y)$, so that in fact $\text{ult}_0(\mathcal{M}) \models \exists x \in \pi(y) \varphi(x, \pi(a_1), \ldots, \pi(a_k))$. By (a), $\mathcal{M} \models \exists x \in y \varphi(x, a_1, \ldots, a_k)$, i.e., $\mathcal{M} \models \exists x \varphi(x, a_1, \ldots, a_k)$.

(d) This uses that $J_\alpha \models \mathsf{ZFC}^-$, cf. Problem 10.6.
(e) This is immediate by (d).
(f) If $X \in \mathscr{P}(\kappa) \cap \mathscr{M}$, then $X = \pi(X) \cap \kappa \in \mathrm{ult}_0(\mathscr{M})$. On the other hand, let $X = [f] \in \mathscr{P}(\kappa) \cap \mathrm{ult}_0(\mathscr{M})$. Setting $X_\alpha = \{\xi < \kappa : \alpha \in f(\xi)\}$ for $\alpha < \kappa$, $(X_\alpha : \alpha < \kappa) \in \mathscr{M}$. But then $X = \{\alpha < \kappa : X_\alpha \in U\} \in \mathscr{M}$ by Σ_0-comprehension, cf. Corollary 5.18.
(g) As κ is the largest cardinal of $J_\alpha[x]$ and $J_\alpha[x] \subset \mathrm{ult}_0(\mathscr{M})$, we must have that $\alpha \leq \kappa^+$ in $\mathrm{ult}_0(\mathscr{M})$. Suppose that $\overline{\overline{\alpha}} \leq \kappa$ in $\mathrm{ult}_0(\mathscr{M})$, and let $f : \kappa \to J_\alpha[x]$ be surjective, $f \in \mathrm{ult}_0(\mathscr{M})$. Then $X_0 = \{\xi < \kappa : \xi \notin f(\xi)\} \in \mathrm{ult}_0(\mathscr{M})$ and hence $X_0 \in J_\alpha[x]$ by (f). But then if $\xi_0 < \kappa$ is such that $X_0 = f(\xi_0)$, $\xi_0 \in X_0$ iff $\xi_0 \notin f(\xi_0) = X_0$. Contradiction!
(h) By the proof of Lemma 4.52. □

Corollary 10.23 *Let $x \subset \omega$, let $\mathscr{M}$ be an x-pm, and suppose that* $\mathrm{ult}_0(\mathscr{M})$ *is well-founded. Then* $\mathrm{ult}_0(\mathscr{M})$ *is an x-pm again.*

Proof We will prove a more general statement later, cf. Corollary 11.15. Let $\mathscr{M} = (J_\alpha[x]; \in, U)$, where κ is the largest cardinal of J_α, and write

$$\pi = \pi_U^{\mathscr{M}} : \mathscr{M} \to \mathrm{ult}_0(\mathscr{M}) = (P; \in U'),$$

where P is transitive.

We first aim to see that $P = J_\beta[x]$ for some β, and for this in turn it suffices to show that $P \models$ "$V = L[\dot{x}]$," cf. p. 77 and Lemma 5.28 But by Lemma 10.21 (d), π is fully elementary with respect to $\dot{\in}$ and $\dot{x}$, so that this follows from $J_\alpha[x] \models$ "$V = L[\dot{x}]$" by elementarity.

Let us show that $(J_\beta[x]; U')$ is amenable. Let $z \in J_\beta[x]$. By Lemma 10.21 (b) there is some transitive $y \in J_\alpha[x]$ such that $z \in \pi(y)$. We have that $\mathscr{M} \models \exists u\ u = y \cap U$. If u_0 is a witness to this fact, then $\mathrm{ult}_0(\mathscr{M}) \models \pi(u_0) = \pi(y) \cap U'$. But then $z \cap U' = z \cap (\pi(y) \cap U') = z \cap \pi(u_0) \in J_\beta[x]$.

It is now straightforward to verify that $\pi(\kappa)$ is the largest cardinal of $J_\beta[x]$, and that $(J_\beta[x]; \in, U') \models$ "U' is a non-trivial normal $< \pi(\kappa)$-closed ultrafilter on $\pi(\kappa)$." □

The following is in the spirit of Definition 10.1.

Definition 10.24 Let $x \subset \omega$, let $\mathscr{M}$ be an x-pm, and let $\alpha \in \mathrm{OR} \cup \{\infty\}$. We call $\mathscr{T} = (\mathscr{M}_i; \pi_{ij} : i \leq j < \alpha)$ a (the) *putative iteration of $\mathscr{M}$ of length α* iff the following hold true.

(a) $\mathscr{M}_0 = \mathscr{M}$,
(b) $\mathscr{M}_{i+1} = \mathrm{ult}_0(\mathscr{M}_i)$, π_{ii+1} is the ultrapower map, whenever $i + 1 < \alpha$,
(c) the maps π_{ij}, $i \leq j < \alpha$, commute,
(d) if $\lambda < \alpha$ is a limit ordinal then $(\mathscr{M}_i, \pi_{ij} : i \leq j \leq \lambda)$ is the direct limit of $(\mathscr{M}_i, \pi_{ij} : i \leq j < \lambda)$, and
(e) for all $i + 1 < \alpha$, $\mathscr{M}_i$ is transitive.

It is easy to verify that if $(\mathcal{M}_i, \pi_{ij} : i \leq j < \alpha)$ is the putative iteration of the x-pm $\mathcal{M}$ of length α, then every $\mathcal{M}_i$ (for $i + 1 < \alpha$) has to be an x-pm (for successor stages this uses Corollary 10.23; it is easy for limit stages).

Definition 10.25 Let $x \subset \omega$, let $\mathcal{M}$ be an x-pm, and let $\alpha \in \mathrm{OR} \cup \{\infty\}$. A putative iteration $(\mathcal{M}_i, \pi_{ij} : i \leq j < \alpha)$ is called an *iteration* iff either α is a limit ordinal or $\alpha = \infty$ or α is a successor ordinal and the last model $\mathcal{M}_{\alpha-1}$ is transitive.[3]

Definition 10.26 Let $x \subset \omega$, and let $\mathcal{M}$ be an x-pm. $\mathcal{M}$ is called (Σ_0-)iterable iff every putative iteration of $\mathcal{M}$ is an iteration.

Lemma 10.27 *Let $x \subset \omega$, and let $\mathcal{M}$ be an x-pm. Suppose that for all $\alpha < \omega_1$, if $\mathcal{T}$ is a putative iteration of $\mathcal{M}$ of length $\alpha + 1$, then $\mathcal{T}$ is an iteration. Then $\mathcal{M}$ is iterable.*

For the proof of Lemma 10.27, cf. Problem 10.1.

Definition 10.28 Let $\mathcal{M} = (J_\alpha[x]; \in, U)$ be an x-pm. U is called ω-complete iff for all $\{X_n : n < \omega\} \subset U$, $\bigcap_{n<\omega} X_n \neq \emptyset$.

The point is that not necessarily $(X_n : n < \omega) \in J_\alpha[x]$.

Lemma 10.29 *Let $x \subset \omega$, and let $\mathcal{M} = (J_\alpha[x]; \in, U)$ be an x-pm such that U is ω-complete. Then $\mathcal{M}$ is iterable.*

*Proof*This proof is similar to the proof of Theorem 10.3.) Let $\mathcal{T}$ be a putative iteration of $\mathcal{M}$ of length $\beta + 1$. Let $\pi : \bar{V} \to V_\theta$, where θ is large enough, $\bar{V}$ is countable and transitive, and $\mathcal{M}, \mathcal{T} \in \mathrm{ran}(\pi)$. Set $\bar{\mathcal{M}}, \bar{\mathcal{T}} = \pi^{-1}(\mathcal{M}, \mathcal{T})$. Then $\bar{\mathcal{T}}$ is a putative iteration of $\bar{\mathcal{M}}$ of length $\bar{\beta} + 1 < \omega_1$. Assuming that $\mathcal{T}$ is not an iteration, $\bar{\mathcal{T}}$ is not an iteration either.

Let $\bar{\mathcal{T}} = (\bar{\mathcal{M}}_i, \bar{\pi}_{ij} : i \leq j \leq \bar{\beta})$. We shall now recursively construct maps $\sigma_i : \bar{\mathcal{M}}_i \to_{\Sigma_0} \mathcal{M}$ for $i \leq \alpha$ s.t. $\sigma_i = \sigma_j \circ \bar{\pi}_{ij}$ whenever $i \leq j \leq \alpha$. We let $\sigma_0 = \pi \restriction \bar{\mathcal{M}}$. The construction of σ_λ for limit $\lambda \leq \alpha$ is straightforward, cf. the proof of Theorem 10.3.

Now suppose σ_i has been constructed, $i < \alpha$. Let $\bar{\kappa}$ be the largest cardinal of $\bar{\mathcal{M}}_i$, and let κ be the largest cardinal of $\mathcal{M}_i$. Let $\bar{\mathcal{M}}_i = (J_\gamma; \in, \bar{U})$. As U is ω-complete, $\bigcap \sigma_i{}''\bar{U} \neq \emptyset$. Let $\xi \in \bigcap \sigma_i{}''\bar{U}$. We may then define $\sigma_{i+1} : \bar{\mathcal{M}}_{i+1} \to \mathcal{M}$ by setting $\sigma_{i+1}([f]) = \sigma_i(f)(\xi)$. Notice that we have

$$\begin{gathered}
\bar{\mathcal{M}}_{i+1} \models \varphi([f_0], \ldots, [f_{k-1}]) \iff \\
\{\bar{\xi} < \bar{\kappa} : \bar{\mathcal{M}}_i \models \varphi(f_0(\bar{\xi}), \ldots, f_{k-1}(\bar{\xi}))\} \in \bar{U} \iff \\
\{\bar{\xi} < \kappa : \mathcal{M} \models \varphi(\sigma_i(f_0)(\bar{\xi}), \ldots, f_{k-1}(\bar{\xi}))\} \in \sigma_i{}''\bar{U} \iff \\
\mathcal{M} \models \varphi(\sigma_i(f_0)(\xi), \ldots, \sigma_i(f_{k-1})(\xi))
\end{gathered}$$

whenever φ is Σ_0.

This finishes the construction. Now notice that $\sigma_{\bar{\beta}} : \bar{\mathcal{M}}_{\bar{\beta}} \to \mathcal{M}$ witnesses that $\bar{\mathcal{M}}_{\bar{\beta}}$ is well-founded. This gives a contradiction. □

---

[3] By our convention, cf. Definition 10.20, this is tantamount to saying that $\mathcal{M}_{\alpha-1}$ is well-founded.

Definition 10.30 Let $x \subset \omega$, and let $\mathscr{M}$ be an x-pm. Then $\mathscr{M}$ is called an *x-mouse* iff $\mathscr{M}$ is iterable.

We should remark that the current-day inner model theory studies mice which are much more complicated than those objects which Definition 10.30 dubs "x-mice." But as such more complicated objects won't play a role in this book, Definition 10.30 fits our purposes here.

Lemma 10.31 *Suppose that there is a measurable cardinal. Then for every $x \subset \omega$ there is an x-mouse.*

Proof Let κ be a measurable cardinal, and let U be a normal measure on κ. Let $\pi: V \to M$ be as in Theorem 4.55 (3). Fix $x \subset \omega$. We may derive an x-pm $(J_\alpha[x]; \in, \bar{U})$ from $\pi \restriction L[x] \to L[x]$ as in the proof of Lemma 10.19. In particular, $X \in \bar{U}$ iff $X \in \mathscr{P}(\kappa) \cap L[x] \wedge \kappa \in \pi(X)$. As $\kappa \in \pi(X)$ is equivalent to $X \in U$, so that $\bar{U} \subset U$, and as U is $< \kappa$-closed, $\bar{U}$ is clearly ω-complete in the sense of Definition 10.28. Then $(J_\alpha; \in, \bar{U})$ is iterable by Lemma 10.29. □

The following Lemma is shown in exacty the same manner as is Lemma 10.4.

Lemma 10.32 (Shift Lemma) *Let $x \subset \omega$, and let $\mathscr{M} = (J_\alpha[x]; \in, U)$ be an x-mouse. Let*

$$(M_\alpha, \pi_{\alpha\beta}: \alpha \leq \beta < \infty)$$

be the iteration of $\mathscr{M} = M_0$ of length ∞. For $\alpha \in \mathrm{OR}$, set $U_\alpha = \pi_{0\alpha}(U)$ and $\kappa_\alpha = crit(U_\alpha) = \pi_{0\alpha}(\kappa)$. Let $\alpha \leq \beta$, and let $\varphi: \alpha \to \beta$ be order preserving. There is then a natural elementary embedding

$$\pi^{\varphi}_{\alpha\beta}: M_\alpha \to M_\beta,$$

called the shift map *given by φ such that $\pi^{\varphi}_{\alpha\beta}(\kappa_\alpha) = \kappa_\beta$, and for all $\overline{\alpha} < \alpha$, $\pi^{\varphi}_{\alpha\beta}(\kappa_{\overline{\alpha}}) = \kappa_{\varphi(\overline{\alpha})}$ and in fact*

$$\pi^{\varphi}_{\alpha\beta} \circ \pi_{\overline{\alpha}\alpha} = \pi_{\overline{\beta}\beta} \circ \pi^{\varphi \restriction \overline{\alpha}}_{\overline{\alpha}\overline{\beta}} \tag{10.7}$$

for all $\overline{\beta}$ with $\mathrm{ran}(\varphi \restriction \overline{\alpha}) \subset \overline{\beta} < \beta$.

Lemma 10.33 *Let $x \subset \omega$, and let $\mathscr{M} = (J_\alpha[x]; \in, U)$ be an x-mouse. Let*

$$\bar{\mathscr{M}} = (J_\beta[x]; \in, \bar{U}) \stackrel{\sigma}{\cong} h_{\mathscr{M}}(\omega) \prec_{\Sigma_1} \mathscr{M}.$$

Then $\bar{\mathscr{M}} \notin J_\alpha[x]$.

Proof To begin with, with the help of Lemma 5.28 it is straightforward to verify that $\bar{M}$ is again an x-premouse.

Claim 10.34 *$\bar{\mathscr{M}}$ is an x-mouse.*

Proof Suppose that $(\bar{\mathcal{M}}_i, \bar{\pi}_{ij}: i \le j \le \gamma)$ is the putative iteration of $\bar{\mathcal{M}}_0 = \bar{\mathcal{M}}$ of length $\gamma + 1$. We aim to see that $\bar{\mathcal{M}}_\gamma$ is well-founded.

Let $(\mathcal{M}, \pi_{ij}: i \le j \le \gamma)$ be the iteration of $\mathcal{M}_0 = \mathcal{M}$ of length $\gamma + 1$. Let us write $\bar{\mathcal{M}}_i = (J_{\beta_i}[x]; \in, \bar{U}_i)$, $\mathcal{M}_i = (J_{\alpha_i}[x]; \in, U_i)$, and let us write $\bar{\kappa}_i$ for the largest cardinal of $\bar{\mathcal{M}}_i$ and κ_i for the largest cardinal of $\mathcal{M}_i$. We aim to construct Σ_0-elementary embeddings $\sigma_i: \bar{\mathcal{M}}_i \to \mathcal{M}_i$, $i \le \gamma$, such that for all $i \le j \le \gamma$,

$$\pi_{ij} \circ \sigma_i = \sigma_j \circ \bar{\pi}_{ij}. \tag{10.8}$$

We set $\sigma_0 = \sigma$. The construction at limit stages is straightforward in the light of (10.8), so let us assume that σ_i is already constructued, $i < \gamma$. In much the same way as in (4.6), cf. p. 54, we have that

$$\bar{\mathcal{M}}_{i+1} = \{\bar{\pi}_{ii+1}(f)(\bar{\kappa}_i): f \in {}^{\bar{\kappa}_i}\bar{\mathcal{M}}_i \cap \bar{\mathcal{M}}_i\}\text{and}$$
$$\mathcal{M}_{i+1} = \{\pi_{ii+1}(f)(\kappa_i): f \in {}^{\kappa_i}\mathcal{M}_i \cap \mathcal{M}_i\}.$$

We then set

$$\sigma_{i+1}(\bar{\pi}_{ii+1}(f)(\bar{\kappa}_i)) = \pi_{ii+1}(\sigma_i(f))(\kappa_i) \tag{10.9}$$

for $f \in {}^{\bar{\kappa}_i}\bar{\mathcal{M}}_i \cap \bar{\mathcal{M}}_i$. We have that if φ is Σ_0 and $f_0, \ldots, f_{k-1} \in {}^{\bar{\kappa}_i}\bar{\mathcal{M}}_i \cap \bar{\mathcal{M}}_i$,

$$\begin{gathered}\bar{\mathcal{M}}_{i+1} \models \varphi(\bar{\pi}_{ii+1}(f_0)(\bar{\kappa}_i), \ldots, \bar{\pi}_{ii+1}(f_{k-1})(\bar{\kappa}_i)) \iff \\ \{\xi < \bar{\kappa}_i: \bar{\mathcal{M}}_i \models \varphi(f_0(\xi), \ldots, f_{k-1}(\xi))\} \in \bar{U}_i \iff \\ \{\xi < \kappa_i: \mathcal{M}_i \models \varphi(\sigma_i(f_0)(\xi), \ldots, \sigma_i(f_{k-1})(\xi))\} \in U_i \iff \\ \mathcal{M}_{i+1} \models \varphi(\pi_{ii+1}(\sigma_i(f_0))(\kappa_i), \ldots, \pi_{ii+1}(\sigma(f_{k-1}))(\kappa_i)),\end{gathered}$$

so that σ_{i+1} is well-defined and Σ_0-elementary. By (10.9), $\pi_{ii+1} \circ \sigma_i = \sigma_{i+1} \circ \bar{\pi}_{ii+1}$. As $\sigma_\gamma: \bar{\mathcal{M}}_\gamma \to \mathcal{M}_\gamma$, $\bar{\mathcal{M}}_\gamma$ inherits the well-foundedness from $\mathcal{M}_\gamma$. □

We thus have an iteration $(\bar{\mathcal{M}}_i, \pi_{ij}: i \le j < \infty)$ of $\bar{\mathcal{M}}$ of length OR, and still using the notation from the proof of Claim 10.34 we have that $\{\bar{\kappa}_i: i \in \text{OR}\}$ is a class which is club in OR, and by Lemma 1.31 (g) and (e),

$$J_{\beta_{i+1}}[x] \models \beta_i = \bar{\kappa}_i^+ \text{ and } J_{\bar{\kappa}_i}[x] \prec_{\Sigma_\omega} J_{\bar{\kappa}_{i+1}}[x]$$

for every $i \in \text{OR}$. This implies that

$$L[x] \models \beta_i = \bar{\kappa}_i^+ \text{ and } J_{\bar{\kappa}_i}[x] \prec_{\Sigma_\omega} L[x] \tag{10.10}$$

for all $i \in \text{OR}$.

By Lemma 10.29 (h), $\bar{\kappa}_i$ is inaccessible in $J_{\beta_{i+1}}[x]$, i.e., also in $J_{\bar{\kappa}_{i+1}}[x]$ and thus in $L[x]$ by (10.10). On the other hand, $\bar{\mathcal{M}}$ is countable (in every transitive model of

ZFC which contains it). Thus $\bar{\mathcal{M}} \notin L[x]$. But this trivially implies that $\bar{\mathcal{M}} \notin J_\alpha[x]$, as desired. □

Lemma 10.35 *Let $x \subset \omega$, and let $\mathcal{M} = (J_\alpha[x]; \in, U)$ and $\mathcal{N} = (J_\beta[x]; \in, U')$ be two x-mice. Let $(\mathcal{M}_i, \pi_{ij}: i \leq j < \infty)$ and $(\mathcal{N}_i, \sigma_{ij}: i \leq j < \infty)$ be the iterations of $\mathcal{M}_0 = \mathcal{M}$ and $\mathcal{N}_0 = \mathcal{N}$, respectively, of length ∞. There is then some $i \in$ OR such that*

$$\mathcal{N} = \mathcal{M}_i \text{ or } \mathcal{M} = \mathcal{N}_i.$$

Proof Let us write $\mathcal{M}_i = (J_{\alpha_i}[x]; \in, U_i)$ and $\mathcal{N}_i = (J_{\beta_i}[x]; \in, U_i')$, and let us also write κ_i for the largest cardinal of $J_{\alpha_i}[x]$ and λ_i for the largest cardinal of $J_{\beta_i}[x]$.

Let

$$(J_{\bar{\alpha}}[x]; \in, \bar{U}) \overset{\chi}{\cong} h_{\mathcal{M}}(\kappa_0) \prec_{\Sigma_1} \mathcal{M}. \tag{10.11}$$

It is easy to see that χ must be the identity, so that κ_0 is also the largest cardinal of $J_{\bar{\alpha}}[x]$ and $\kappa_0 < \bar{\alpha} \leq \alpha$. As in the proof of Lemma 10.33, we get that $\bar{\alpha} = (\kappa_0)^{+L[x]}$. But certainly $\alpha \leq (\kappa_0)^{+L[x]}$, as κ_0 is the largest cardinal of $J_\alpha[x]$. Therefore $\bar{\alpha} = \alpha = (\kappa_0)^{+L[x]}$, so that (10.11) in fact gives that

$$\mathcal{M} = h_{\mathcal{M}}(\kappa_0).$$

The same argument shows that in fact for every $i \in$ OR,

$$\mathcal{M}_i = h_{\mathcal{M}_i}(\kappa_i),\ \alpha_i = (\kappa_i)^{+L[x]},\ \mathcal{N}_i = h_{\mathcal{N}_i}(\lambda_i), \text{ and } \beta_i = (\lambda_i)^{+L[x]}. \tag{10.12}$$

Moreover, the maps π_{ij} and σ_{ij} are Σ_1-elementary by Lemma 10.21 (c), so that they preserve the Σ_1-SKOLEM function. Therefore, if $i \leq j \in$ OR, then

$$\mathcal{M}_i \overset{\pi_{ij}}{\cong} h_{\mathcal{M}_j}(\kappa_i) \prec_{\Sigma_1} \mathcal{M}_j \text{ and } \mathcal{N}_i \overset{\sigma_{ij}}{\cong} h_{\mathcal{N}_j}(\lambda_i) \prec_{\Sigma_1} \mathcal{N}_j. \tag{10.13}$$

In particular, π_{ij} is the same as the inverse of the collapsing map of $h_{\mathcal{M}_j}(\kappa_i)$ and σ_{ij} is the same as the inverse of the collapsing map of $h_{\mathcal{N}_j}(\lambda_i)$. (10.13) implies that if $i < j \in$ OR, then

$$\kappa_i \notin h_{\mathcal{M}_j}(\kappa_i) \text{ and } \lambda_i \notin h_{\mathcal{N}_j}(\lambda_i), \tag{10.14}$$

as κ_i is the critical point of π_{ij} and λ_i is the critical point of σ_{ij}.

By the proof of (4.6), cf. p. 54, we have that every element of $\mathcal{M}_{i+1}$ is of the form $\pi_{ii+1}(f)(\kappa_i)$ where $f \in {}^{\kappa_i}J_{\alpha_i} \cap J_{\alpha_i}$, so that

$$\mathcal{M}_{i+1} = h_{\mathcal{M}_{i+1}}(\text{ran}(\pi_{ii+1}) \cup \{\kappa_i\}). \tag{10.15}$$

By (10.13)

$$\text{ran}(\pi_{ii+1}) = h_{\mathcal{M}_{i+1}}(\kappa_i),$$

which together with (10.15) yields that for all $j > i$,

$$\mathcal{M}_{i+1} = h_{\mathcal{M}_{i+1}}(\kappa_i \cup \{\kappa_i\}) \overset{\pi_{i+1j}}{\cong} h_{\mathcal{M}_j}(\kappa_i \cup \{\kappa_i\})$$

and therefore

$$\kappa_{i+1} \subset h_{\mathcal{M}_j}(\kappa_i \cup \{\kappa_i\}). \tag{10.16}$$

Combining (10.14) and (10.16), we get that for every j,

$$\{\kappa_i : i < j\} = \{\kappa < \kappa_j : \kappa \geq \kappa_0 \wedge \kappa \notin h_{\mathcal{M}_j}(\kappa)\}, \tag{10.17}$$

and by the same reasoning

$$\{\lambda_i : i < j\} = \{\lambda < \lambda_j : \lambda \geq \lambda_0 \wedge \lambda \notin h_{\mathcal{N}_j}(\lambda)\}. \tag{10.18}$$

Now let us pick sequences $(i_k : k < \omega \cdot \omega)$ and $(j_k : k \leq \omega \cdot \omega)$ such that for all $k < \omega \cdot \omega$, $\kappa_{i_k} < \lambda_{j_k} < \kappa_{i_{k+1}}$ and if $\nu \leq \omega \cdot \omega$ is a limit ordinal, then $i_\nu = \sup_{k<\nu} i_k = \sup_{k<\nu} j_k = j_\nu$. Notice that $\kappa_{i_\nu} = \lambda_{i_\nu}$ for every limit ordinal $\nu \leq \omega \cdot \omega$. Write $i^* = i_{\omega \cdot \omega} = j_{\omega \cdot \omega}$ and $\mu = \kappa_{i^*} = \lambda_{i^*}$. We have $\alpha_{i^*} = \mu^{+L[x]} = \beta_{i^*}$ by (10.12). Set

$$A = \{\kappa_{i_\nu} : \nu < \omega \cdot \omega \text{ is a limit ordinal}\}.$$

By the proof of Lemma 10.9, if $X \in \mathcal{P}(\mu) \cap J_{\alpha_{i^*}}[x]$, then

$$X \in U_{i^*} \iff A \setminus X \text{ is finite} \iff X \in U'_{i^*}.$$

This shows that

$$\mathcal{M}_{i^*} = (J_{\alpha_{i^*}}[x]; \in, U_{i^*}) = (J_{\beta_{i^*}}[x]; \in, U'_{i^*}) = \mathcal{N}_{i^*}. \tag{10.19}$$

Let us assume that $\kappa_0 \leq \lambda_0$. By (10.17) and (10.18) applied to $j = i^*$ there is then some i such that $\lambda_0 = \kappa_i$. But then by (10.13)

$$\mathcal{N} = \mathcal{N}_0 \overset{\sigma_{0i^*}}{\cong} h_{\mathcal{N}_{i^*}}(\lambda_0) = h_{\mathcal{M}_{i^*}}(\kappa_i) \overset{\pi_{ii^*}}{\cong} \mathcal{M}_i,$$

so that $\mathcal{N} = \mathcal{M}_i$. Symmetrically, if $\lambda_0 \leq \kappa_0$, then there is some i with $\mathcal{M} = \mathcal{N}_i$. □

Corollary 10.36 *Let $x \subset \omega$, and assume that there is an x-mouse. There is then exactly one x-mouse $\mathcal{M}$ such that*

$$\mathcal{M} = h_{\mathcal{M}}(\omega), \tag{10.20}$$

and if $\mathscr{N}$ is any x-mouse, then there is some $\gamma + 1$ such that if $(\mathscr{M}_i, \pi_{ij}: i \leq j \leq \gamma)$ is the iteration of $\mathscr{M}_0 = \mathscr{M}$ of length $\gamma + 1$, then $\mathscr{M}_\gamma = \mathscr{N}$.

Proof Let $\mathscr{M}'$ be any x-mouse, and let

$$\mathscr{M} \overset{\sigma}{\cong} h_{\mathscr{M}'}(\omega),$$

where $\mathscr{M}$ is transitive. As σ is Σ_1-elementary and thus respects the Σ_1-SKOLEM function, we must have that (10.20) holds true.

Let $\mathscr{N}$ be any x-mouse, and let λ be the largest cardinal of $\mathscr{N}$. Suppose that $\delta > 0$ would be such that if $(\mathscr{N}_i, \sigma_{ij}: i \leq j \leq \delta)$ is the iteration of $\mathscr{N}$ of length $\delta + 1$, then $\mathscr{M} = \mathscr{N}_\delta$. Then as in (10.14) in the proof of Lemma 10.35, $\lambda \notin h_{\mathscr{N}_\delta}(\lambda)$ and thus $\lambda \notin h_{\mathscr{M}}(\omega) = \mathscr{M}$, which is nonsense. By Lemma 10.35 we must therefore have some γ such that if $(\mathscr{M}_i, \pi_{ij}: i \leq j \leq \gamma)$ is the iteration of $\mathscr{M}_0 = \mathscr{M}$ of length $\gamma + 1$, then $M_\gamma = \mathscr{N}$. Moreover, let κ be the largest cardinal of $\mathscr{M}$. If $\gamma > 0$, then as in (10.14) in the proof of Lemma 10.35, $\kappa \notin h_{\mathscr{M}_\gamma}(\kappa)$, and thus $\kappa \notin h_{\mathscr{N}}(\omega)$. This means that if we assume in addition that $\mathscr{N} = h_{\mathscr{N}}(\omega)$, then $\gamma = 0$, so that $\mathscr{N} = \mathscr{M}_0 = \mathscr{M}$. □

Definition 10.37 Let $x \subset \omega$. By $x^\#$ ("*x-sharp*") we denote the unique x-mouse $\mathscr{M}$ with $\mathscr{M} = h_{\mathscr{M}}(\omega)$, if it exists. We also write $0^\#$ ("*zero-sharp*") for $\emptyset^\#$.

In the light of Lemma 10.35, it is easy to verify that $x^\#$ is that x-mouse whose largest cardinal is smallest possible among all x-mice.

Via GÖDELization and as $x^\# = h_{x^\#}(\omega)$, the Σ_1-theory of $x^\#$, call it $\mathrm{Th}_{\Sigma_1}(x^\#)$, may be construed as a set of natural numbers. By Lemma 10.27 and with the help of Lemma 7.17, it is not hard to verify that $\{\mathrm{Th}_{\Sigma_1}(x^\#)\}$ is $\Pi^1_2(x)$. In this sense we may construe $x^\#$ itself as a $\Pi^1_2(x)$-singleton, cf. Problem 10.7.

Definition 10.38 Let W be an inner model. W is called *rigid* iff there is no non-trivial elementary embedding $\pi: W \to W$.

By Theorem 4.53, V is rigid.

Theorem 10.39 (K. Kunen) *Let $x \subset \omega$. Suppose that $L[x]$ is not rigid. Then $x^\#$ exists.*

Proof We shall make use of Problem 10.10. Let us fix an elementary embedding $\pi: L[x] \to L[x]$, $\pi \neq \mathrm{id}$. Let κ be the critical point of π, and let $U = \{X \in \mathscr{P}(\kappa) \cap L[x] : \kappa \in \pi(X)\}$ be the $L[x]$-ultrafilter derived from π as in Lemma 10.19. Setting $\alpha = \kappa^{+L[x]}$, we know that $(J_\alpha[x]; \in, U)$ is an x-pm (cf. Lemma 10.19). In order to prove that $x^\#$ exists, it suffices to verify that $(J_\alpha[x]; \in, U)$ is iterable.

In much the same way as on p. 55, we may factor $\pi: L[x] \to L[x]$ as $\pi = k \circ \pi_U$, where $\pi_U: L[x] \to \mathrm{ult}(L[x]; U)$ is the ultrapower map (cf. Problem 10.10) and $k([f]_U) = \pi(f)(\kappa)$ for every $f \in {}^\kappa L[x] \cap L[x]$. We thus have $\mathrm{ult}(L[x]; U) \cong L[x]$, and we may and shall as well assume that $k = \mathrm{id}$ and $\pi_U = \pi$.

Let Γ be the class of all strong limit cardinals of cofinality $\geq (2^\kappa)^+$. By Problem 10.10 (d), we shall have that $\pi(\lambda) = \lambda$ for all $\lambda \in \Gamma$ (this is why we opted for $\pi = \pi_U$).

Let μ be any ordinal, and let $\sigma\colon L[x] \to L[x]$ be any elementary embedding with $\mu = \text{crit}(\sigma)$. Then we write

$$U(\sigma) = \{X \in \mathscr{P}(\mu) \cap L[x] \colon \mu \in \sigma(X)\}$$

for the $L[x]$-ultrafilter derived from σ. (So $U = U(\pi)$.)

An example for how to obtain such a situation is when

$$L[x] \overset{\sigma}{\cong} h_{L[x]}(\mu \cup \Gamma) \prec L[x] \tag{10.21}$$

and $\mu \notin h_{L[x]}(\mu \cup \Gamma)$. For the purpose of this proof, let us call an x-pm $\mathscr{M} = (J_\beta[x]; \in, U')$ *certified* iff the following holds true. If μ denotes the critical point of U', and if σ is then as in (10.21), then $\mu \notin h_{L[x]}(\mu \cup \Gamma)$ and $U' = U(\sigma)$.

If $\sigma\colon L[x] \to L[x]$ is an elementary embedding with critical point μ, then

$$\pi^{L[x]}_{U(\sigma)}\colon L[x] \to_{U(\sigma)} L[x]$$

is the associated ultrapower map (cf. also Problem 10.10); we also write $\pi_{U(\sigma)}$ rather than $\pi^{L[x]}_{U(\sigma)}$, and we write $k(\sigma)\colon L[x] \to L[x]$ for the canonical factor map obtained as on p. 55 such that

$$\sigma = k(\sigma) \circ \pi_{U(\sigma)}, \tag{10.22}$$

i.e., $k(\sigma)([f]_{U(\sigma)}) = \sigma(f)(\mu)$ for $f \in {}^\mu L[x] \cap L[x]$. Notice that

$$k(\sigma) \restriction (\mu + 1) = \text{id}, \tag{10.23}$$

because $k(\sigma)(\xi) = k(\sigma)(\pi_{U(\sigma)}(\xi)) = \sigma(\xi) = \xi$ for all $\xi < \mu$ and $k(\sigma)(\mu) = k(\sigma)([\text{id}]_{U(\sigma)}) = \sigma(\text{id})(\mu) = \mu$.

Claim 10.40 *Let σ be as in (10.21). For every $X \in \mathscr{P}(\mu) \cap L[x]$ there is some $i < \omega$ and some $\mathsf{a} \in [\mu \cup \Gamma]^{<\omega}$ such that $X = h_{L[x]}(i, \mathsf{a}) \cap \mu$.*

Proof If $X \in \mathscr{P}(\mu) \cap L[x]$, then there is some $i < \omega$ and some $\mathsf{a} \in [\mu \cup \Gamma]^{<\omega}$ such that $\sigma(X) = h_{L[x]}(i, \mathsf{a})$. But $X = \sigma(X) \cap \mu$, so that we must have that $X = h_{L[x]}(i, \mathsf{a})$
$\cap\, \mu$. □

In (c) of the following Claim, $\text{ult}_0(L[x]; U')$ is as defined in Problem 10.10.

Claim 10.41 *Let $\mathscr{M} = (J_\beta[x]; \in, U')$ be an x-pm, and let μ is the critical point of U', where $\mu < (2^{\text{Card}(\kappa)})^+$ in V. The following are equivalent.*

(i) $\mathcal{M}$ *is certified.*
(ii) *For all* $i < \omega$ *and* $\mathrm{a} \in [\mu \cup \Gamma]^{<\omega}$, $h_{L[x]}(i, \mathrm{a}) \cap \mu \in U' \iff \mu \in h_{L[x]}(i, \mathrm{a})$.
(iii) $\beta = \mu^{+L[x]}$ *and* $\mathrm{ult}_0(L[x]; U')$ *is well-founded (equivalently, transitive and thus equal to* $L[x]$*).*

Proof Let σ be as in (10.21).

(i) $\Longrightarrow$ (ii) Using Claim 10.40, let $i < \omega$ and $\mathrm{a} \in [\mu \cup \Gamma]^{<\omega}$. Let $X = h_{L[x]}(i, \mathrm{a}) \cap \mu$. We have that $X \in U' = U(\sigma)$ iff $\mu \in \sigma(X) \cap (\mu + 1)$, which by (10.23) is equivalent to $\mu \in \pi_{U'}(X) \cap (\mu + 1)$. But $\pi_{U'}(X) = h_{L[x]}(i, \mathrm{a}) \cap \pi_{U'}(\mu)$, as the elements of a are fixed points under $\pi_{U'}$ (cf. Problem 10.10 (d)), so that $X \in U'$ is equivalent to $\mu \in h_{L[x]}(i, \mathrm{a})$, as desired.

(ii) $\Longrightarrow$ (i) It is easy to see that (ii) implies that $\mu = \mathrm{crit}(\sigma)$: If $\mu = h_{L[x]}(i, \mathrm{a})$, where $i < \omega$ and $\mathrm{a} \in [\mu \cup \Gamma]^{<\omega}$, then $\mu = h_{L[x]}(i, \mathrm{a}) \cap \mu \in U'$ would give $\mu \in h_{L[x]}(i, \mathrm{a}) = \mu$, a contradiction!

Now let $X \in \mathscr{P}(\mu) \cap L[x]$. By Claim 10.40 there is some $i < \omega$ and some $\mathrm{a} \in [\mu \cup \Gamma]^{<\omega}$ such that $X = h_{L[x]}(i, \mathrm{a}) \cap \mu$. We get that $X \in U'$ iff $\mu \in h_{L[x]}(i, \mathrm{a}) \cap (\mu + 1)$, which is equal to $\pi_{U(\sigma)}(h_{L[x]}(i, \mathrm{a}) \cap \mu) \cap (\mu + 1)$, as the elements of a are fixed points under $\pi_{U(\sigma)}$; but by (10.23), $\pi_{U(\sigma)}(h_{L[x]}(i, \mathrm{a}) \cap \mu) \cap (\mu + 1) = \pi_{U(\sigma)}(X) \cap (\mu + 1)$ is equal to $\sigma(X) \cap (\mu + 1)$, so that $X \in U'$ is equivalent to $X \in U(\sigma)$, as desired.

In particular, $U(\sigma) \subset J_\beta[x]$ (as $\mathcal{M}$ is a J-structure), which implies that $\beta \geq \mu^{+L[x]}$ and thus $\beta = \mu^{+L[x]}$.

(i) $\Longrightarrow$ (iii) If $\mathcal{M}$ is certified, then $U' = U(\sigma)$, which immediately gives that $\mathrm{ult}_0(L[x]; U')$ must be well-founded.

(iii) $\Longrightarrow$ (ii) Let $\pi_{U'}: L[x] \to_{U'} L[x]$ be the ultrapower map, and let $i < \omega$ and $\mathrm{a} \in [\mu \cup \Gamma]^{<\omega}$. As the elements of a are fixed points under $\pi_{U'}$, we get that $h_{L[x]}(i, \mathrm{a}) \cap \mu \in U'$ iff $\mu \in \pi_{U'}(h_{L[x]}(i, \mathrm{a}) \cap \mu) = h_{L[x]}(i, \mathrm{a}) \cap \pi_{U'}(\mu)$ iff $\mu \in h_{L[x]}(i, \mathrm{a})$, as desired. □

Let us now return to the x-pm $(J_\alpha[x]; \in, U)$ isolated from $\pi = \pi_U$ above. Let γ be a countable ordinal, and let us suppose $(\mathcal{M}_i, \pi_{ij}: i \leq j \leq \gamma)$ to be the putative iteration of $\mathcal{M}_0 = (J_\alpha[x]; \in, U)$ of length $\gamma + 1$. In the light of Lemma 10.27, it suffices to show that $\mathcal{M}_\gamma$ is transitive. We show:

Claim 10.42 *Every* $\mathcal{M}_i$, $i \leq \gamma$, *is transitive and in fact certified.*

Proof Of course, if $i < \gamma$, then $\mathcal{M}_i$ is trivially transitive by the definition of "putative iteration."

That $\mathcal{M}_0 = (J_\alpha[x]; \in, U)$ be certified follows immediately from (iii) $\Longrightarrow$ (i) of Claim 10.41, applied to $\pi = \pi_U$.

Let us now suppose that $i < \gamma$ and that $\mathcal{M}_i$ is certified. We aim to verify that $\mathcal{M}_{i+1}$ is transitive and certified.

Let us write $\mathcal{M}_i = (J_{\alpha_i}[x]; \in, U_i)$, and let κ_i be the critical point of U_i. By (i) $\Longrightarrow$ (iii) of Claim 10.41, $\mathrm{ult}_0(L[x]; U_i)$ is equal to $L[x]$, so that the ultrapower map $\pi_{U_i}^{L[x]} = \pi_{U_i}$ (cf. Problem 10.10) maps $L[x]$ to $L[x]$. It is easy to verify that the universe of $\mathcal{M}_{i+1} = \mathrm{ult}_0(\mathcal{M}_i)$ is isomorphic to $\pi_{U_i}(J_{\alpha_i}[x])$. This is because if $[f]_{U_i} \in \pi_{U_i}(J_{\alpha_i}[x])$, where $f \in {}^{\kappa_i}L[x] \cap L[x]$, then $[f]_{U_i} = [f']_{U_i}$ for some

$f' \in {}^{\kappa_i}J_{\alpha_i}[x] \cap J_{\alpha_i}$, as $\alpha_i = \kappa_i^{+L[x]}$ is regular in $L[x]$. (cf. Problem 10.10 (b).) In particular, $\mathcal{M}_{i+1}$ is transitive.

Let us now write $\mathcal{M}_{i+1} = (J_{\alpha_{i+1}}[x]; \in, U_{i+1})$ and let κ_{i+1} denote the critical point of U_{i+1}. We aim to use (ii) $\Longrightarrow$ (i) of Claim 10.41 to verify that $\mathcal{M}_{i+1}$ is certified.

Let us fix $j < \omega$ and $\mathsf{a} \in [\kappa_{i+1} \cup \Gamma]^{<\omega}$. Let $\mathsf{a}_0 \in [\kappa_{i+1}]^{<\omega}$ and $\mathsf{a}_1 \in [\Gamma]^{<\omega}$ be such that $\mathsf{a} = \mathsf{a}_0 \cup \mathsf{a}_1$. Let $\mathsf{a}_0 = [f]_{U_i} = \pi_{U_i}(f)(\kappa_i)$, where $f: \kappa_i \to [\kappa_i]^{\mathrm{Card}(\mathsf{a}_0)}$, $f \in L[x]$.
Let $g: \kappa_i \to \mathscr{P}(\kappa_i)$, $g \in L[x]$, be defined by

$$g(\eta) = h_{L[x]}(j, f(\eta) \cup \mathsf{a}_1) \cap \kappa_i$$

for $\eta < \kappa_i$. Using the Łoś Theorem and the fact that the elements of a_1 are fixed points under π_{U_i}, it is straightforward to verify that $[g]_{U_i} = h_{L[x]}(j, \mathsf{a}) \cap \kappa_{i+1}$.

We now get that $h_{L[x]}(i, \mathsf{a}) \cap \kappa_{i+1} \in U_{i+1}$ iff $\{\eta < \kappa_i : g(\eta) \in U_i\} \in U_i$. Applying (i) $\Longrightarrow$ (ii) of Claim 10.41 yields that this is equivalent to

$$\{\eta < \kappa_i : \kappa_i \in h_{L[x]}(j, f(\eta) \cup \mathsf{a}_1)\} \in U_i,$$

which by $X \in U_i$ iff $\kappa_i \in \pi_{U_i}(X)$ for all $X \in \mathscr{P}(\kappa_i) \cap L[x]$ and the fact that the elements of a_1 are fixed points under π_{U_i} is in turn equivalent to

$$\kappa_{i+1} \in h_{L[x]}(j, \pi_{U_i}(f)(\kappa_i) \cup \mathsf{a}_1) = h_{L[x]}(j, \mathsf{a}), \tag{10.24}$$

as desired.

Now let $\lambda \leq \gamma$ be a limit ordinal, and suppose that every $\mathcal{M}_i$, $i < \lambda$, is certified. We first need to see that $\mathcal{M}_\lambda$ is well-founded.

For $i < \lambda$, let us write $\mathcal{M}_i = (J_{\alpha_i}[x]; \in, U_i)$, and let $\kappa_i = \mathrm{crit}(U_i)$. Because every $\mathcal{M}_i$, $i < \lambda$, is certified, by (i) $\Longrightarrow$ (iii) of Claim 10.41 we know that the iteration map $\pi_{ij}: \mathcal{M}_i \to \mathcal{M}_j$ extends to an elementary embedding from $L[x]$ to $L[x]$, which we shall denote by $\tilde{\pi}_{ij}$, for all $i \leq j < \lambda$. (cf. Problem 10.10.)

Let us also write $\kappa_\lambda = \sup_{i<\lambda} \kappa_i$. For $i \leq \lambda$, let let

$$L[x] \overset{\sigma_i}{\cong} h_{L[x]}(\kappa_i \cup \Gamma) \prec L[x]. \tag{10.25}$$

If $i < \lambda$, then $\mathcal{M}_i$ is certified, so that $\kappa_i = \mathrm{crit}(\sigma_i)$ and $U_i = U(\sigma_i)$.

For $i \leq j \leq \lambda$ we may define $\pi^*_{ij}: L[x] \to L[x]$ by

$$\pi^*_{ij}(x) = \sigma_j^{-1} \circ \sigma_i(x).$$

This is well-defined and elementary: notice that if $i \leq j \leq \lambda$, then $\mathrm{ran}(\sigma_i) = h_{L[x]}(\kappa_i \cup \Gamma) \subset h_{L[x]}(\kappa_j \cup \Gamma) = \mathrm{ran}(\sigma_j)$.

Claim 10.43 *If $\bar{\lambda} \leq \lambda$ is a limit ordinal, then $(L[x], (\pi^*_{i\bar{\lambda}}: i < \bar{\lambda}))$ is the direct limit of $((L[x]: i < \bar{\lambda}), (\pi^*_{ij}: i \leq j < \bar{\lambda}))$. Moreover, if $i \leq j < \lambda$, then $\pi^*_{ij} = \tilde{\pi}_{ij}$.*

Proof The first part trivially follows from the fact that for a limit ordinal $\bar{\lambda} \leq \lambda$, $\mathrm{ran}(\sigma_{\bar{\lambda}}) = h_{L[x]}(\kappa_{\bar{\lambda}} \cup \Gamma) = \bigcup_{i<\bar{\lambda}} h_{L[x]}(\kappa_i \cup \Gamma) = \bigcup_{i<\bar{\lambda}} \mathrm{ran}(\sigma_i)$.

As for the second part, it therefore suffices to prove this for $j = i + 1 < \lambda$.

$$
\begin{array}{ccccccccc}
 & & & h_{L[x]}(\kappa_i \cup \{\kappa_i\} \cup \Gamma) & & & & & \\
 & & & \| & & & & & \\
\cdots \prec & h_{L[x]}(\kappa_i \cup \Gamma) & \prec & h_{L[x]}(\kappa_{i+1} \cup \Gamma) & \prec & \cdots & \prec & h_{L[x]}(\kappa_\lambda \cup \Gamma) & \prec L[x] \\
 & \sigma_i \; \wr\| & & \wr\| \; \sigma_{i+1} & & & & \wr\| \; \sigma_\lambda & \\
\cdots \longrightarrow & L[x] & \xrightarrow[\tilde{\pi}_{i,i+1}]{} & L[x] & & \xrightarrow[\pi^*_{i+1,\lambda}]{} & & L[x] & \\
 & & \| & & & & & & \\
 & & \pi^*_{i,i+1} & & & & & &
\end{array}
$$

We first verify that

$$h_{L[x]}(\kappa_{i+1} \cup \Gamma) = h_{L[x]}(\kappa_i + 1 \cup \Gamma). \tag{10.26}$$

To show this, let $\xi < \kappa_{i+1}$. There is some $f \colon \kappa_i \to \kappa_i$, $f \in L[x]$, such that $\xi = [f]_{U_i} = \tilde{\pi}_{ii+1}(f)(\kappa_i)$. As $\mathscr{M}_i$ is certified, (the proof of) Claim 10.40 yields some $j < \omega$ and some $\mathsf{a} \in [\kappa_i \cup \Gamma]^{<\omega}$ such that $f = h_{L[x]}(j, \mathsf{a}) \restriction \kappa_i$. As the elements of a are not moved by $\tilde{\pi}_{ii+1}$, $\tilde{\pi}_{ii+1}(f) = h_{L[x]}(j, \mathsf{a}) \restriction \kappa_{i+1}$. But then $\xi = \tilde{\pi}_{ii+1}(f)(\kappa_i) = h_{L[x]}(j, \mathsf{a})(\kappa_i) \in h_{L[x]}(\kappa_i + 1 \cup \Gamma)$. This shows (10.26).

Let us now define

$$\tilde{\pi}_{ii+1}(f)(\kappa_i) \overset{\chi}{\mapsto} \pi^*_{ii+1}(f)(\kappa_i), \tag{10.27}$$

where $f \colon \kappa_i \to L[x]$, $f \in L[x]$. Let E be either $=$ or $\in$, and let $f \colon \kappa_i \to L[x]$, $g \colon \kappa_i \to L[x]$, $f, g \in L[x]$. We then get that

$$
\begin{aligned}
\pi^*_{ii+1}(f)(\kappa_i) E \pi^*_{ii+1}(g)(\kappa_i) &\Longleftrightarrow \kappa_i \in \pi^*_{ii+1}(\{\xi < \kappa_i \colon f(\xi) E g(\xi)\}) \\
&\Longleftrightarrow \kappa_i \in \sigma^{-1}_{i+1} \circ \sigma_i(\{\xi < \kappa_i \colon f(\xi) E g(\xi)\}) \\
&\Longleftrightarrow \kappa_i \in \sigma_i(\{\xi < \kappa_i \colon f(\xi) E g(\xi)\}) \\
&\overset{(*)}{\Longleftrightarrow} \kappa_i \in \tilde{\pi}_{ii+1}(\{\xi < \kappa_i \colon f(\xi) E g(\xi)\}) \\
&\Longleftrightarrow \tilde{\pi}_{ii+1}(f)(\kappa_i) E \tilde{\pi}_{ii+1}(g)(\kappa_i)
\end{aligned}
$$

The equivalence marked by (*) holds true as σ_i witnesses that $\mathscr{M}_i$ is certified.

Every element of $L[x]$ is of the form $\tilde{\pi}_{ii+1}(f)(\kappa_i)$ where $f \colon \kappa_i \to L[x]$, $f \in L[x]$. By (10.26), $\operatorname{ran}(\sigma_{i+1}) = h_{L[x]}(\operatorname{ran}(\sigma_i) \cup \{\kappa_i\})$, so that every element of $L[x]$ is also of the form $\pi^*_{ii+1}(f)(\kappa_i)$, where again $f \colon \kappa_i \to L[x]$, $f \in L[x]$. By the above computation, χ, as given by (10.27), is thus a well-defined $\in$-isomorphism of $L[x]$ with $L[x]$. But this now implies that $\tilde{\pi}_{ii+1} = \pi^*_{ii+1}$. □

We now easily get that $\mathscr{M}_\lambda$ is transitive, as its universe is equal to $\tilde{\pi}_{0\lambda}(J_\alpha[x])$.

In order to show that $\mathcal{M}_\lambda$ is certified, we use (ii) $\Longrightarrow$ (i) of Claim 10.41.

Let us fix $j < \omega$ and $\mathsf{a} \in [\kappa_\lambda \cup \Gamma]^{<\omega}$. Let $\mathsf{a}_0 \in [\kappa_\lambda]^{<\omega}$ and $\mathsf{a}_1 \in [\Gamma]^{<\omega}$ be such that $\mathsf{a} = \mathsf{a}_0 \cup \mathsf{a}_1$. Let $i < \lambda$ be sufficiently big such that $\mathsf{a}_0 \in [\kappa_i]^{<\omega}$. The elements of a are then fixed points under $\tilde{\pi}_{ii+1}$. We then get that $h_{L[x]}(j, \mathsf{a}) \cap \kappa_\lambda \in U_\lambda$ iff $h_{L[x]}(j, \mathsf{a}) \cap \kappa_i \in U_i$ by applying $\pi_{i\lambda} \subset \tilde{\pi}_{i\lambda}$, iff $\kappa_i \in h_{L[x]}(j, \mathsf{a})$, by (i) $\Longrightarrow$ (ii) applied to $\mathcal{M}_i$, iff $\kappa_\lambda \in h_{L[x]}(j, \mathsf{a})$ by applying $\tilde{\pi}_{i\lambda}$, as desired. □

It is not hard to see now that Lemmas 10.32 and 10.35 and the proof of Theorem 10.39 yields the following.

Corollary 10.44 (J. Silver) *Let $x \subset \omega$, and suppose that $x^\#$ exists. Let $(\mathcal{M}_i, \pi_{ij}: i \leq j \in \mathrm{OR})$ denote the iteration of $x^\#$ of length ∞, and let κ_i be the largest cardinal of $\mathcal{M}_i$, $i \in \mathrm{OR}$. The following hold true.*

(1) $L[x] = h_{L[x]}(\{\kappa_i : i \in \mathrm{OR}\})$.

(2) *Let φ be a formula, let $k < \omega$, and let $i_1 < \ldots < i_k$ and $j_1 < \ldots < j_k$. Write $i^* = \min(\{i_1, j_1\})$, and let $z \in J_{\kappa_{i^*}}[x]$. Then*

$$L[x] \models \varphi(z, \kappa_{i_1}, \ldots, \kappa_{i_k}) \Longleftrightarrow L[x] \models \varphi(z, \kappa_{j_1}, \ldots, \kappa_{j_k}).$$

(3) *Let $e: \mathrm{OR} \to \mathrm{OR}$ be order-preserving. Then $\pi_e: L[x] \to L[x]$ is an elementary embedding, where π_e is defined by*

$$\pi_e(h_{L[x]}(i, (\kappa_{i_1}, \ldots, \kappa_{i_k}))) = h_{L[x]}(i, (\kappa_{e(i_1)}, \ldots, \kappa_{e(i_k)})), \tag{10.28}$$

for $i, k < \omega$.

(4) $\kappa_i \notin h_{L[x]}(\kappa_i \cup \{\kappa_j : j > i\})$ *for every $i \in \mathrm{OR}$.*

(5) *Let $\pi: L[x] \to L[x]$ be an elementary embedding. Then there is some order-preserving $e: \mathrm{OR} \to \mathrm{OR}$ such that $\pi = \pi_e$, where π_e is defined as in (10.28).*

Proof (1) Suppose that $L[x] \neq h_{L[x]}(\{\kappa_i : i \in \mathrm{OR}\})$. Let κ be the least ordinal such that $\kappa \notin h_{L[x]}(\{\kappa_i : i \in \mathrm{OR}\})$, so that κ is the critical point of

$$\pi: L[x] \cong h_{L[x]}(\{\kappa_i : i \in \mathrm{OR}\}) \prec L[x].$$

If $\mathcal{M} = (J_\alpha[x]; \in, U)$ is derived from π as in the proof of Lemma 10.19, then the proof of Theorem 10.39 shows that $\mathcal{M}$ is an x-mouse. By Lemma 10.35 and the definition of $x^\#$, cf. Definition 10.37, we must have $\kappa = \kappa_i$ for some $i \in \mathrm{OR}$. But then trivially $\kappa \in h_{L[x]}(\{\kappa_i : i \in \mathrm{OR}\})$. Contradiction!

(2) For $i \leq j$, $\pi_{ij} \restriction \kappa_i = \mathrm{id}$ and $\pi_{ij}(\kappa_i) = \kappa_j$, which gives that $J_{\kappa_i}[x] \prec \pi_{ij}(J_{\kappa_i}[x]) = J_{\kappa_j}[x]$. This immediately yields

$$J_{\kappa_i}[x] \prec L[x] \tag{10.29}$$

for every $i \in \mathrm{OR}$.

Now let $\ell = \max\{i_k, j_k\}$, and write $\lambda = \ell \cdot (k+1)$. It is easy to see that there is an order preserving map $\varphi: i_k + 1 \to \lambda$ with $\varphi \restriction i_1 = \mathrm{id}$ and $\varphi(i_p) = \ell \cdot p$ for

$p = 1, \ldots, k$, e.g. $\varphi \restriction i_1 = \text{id}$ and $\varphi(i_p + \xi) = (\ell \cdot p) + \xi$ for $p = 1, \ldots, k-1$ and $i_p + \xi < i_{p+1}$, or $p = k$ and $\xi = 0$. Analoguously, there is an order preserving map $\varphi^* \colon j_k + 1 \to \lambda$ with $\varphi^* \restriction j_1 = \text{id}$ and $\varphi^*(j_p) = \ell \cdot p$ for $p = 1, \ldots, k$. With the help of (10.29) and the Shift Lemma 10.32 we then get that

$$\begin{aligned} L[x] \models \varphi(z, \kappa_{i_1}, \ldots, \kappa_{i_k}) &\Longleftrightarrow J_{\kappa_{i_k+1}}[x] \models \varphi(z, \kappa_{i_1}, \ldots, \kappa_{i_k}) \\ &\Longleftrightarrow J_{\kappa_\lambda}[x] \models \varphi(z, \kappa_{\ell \cdot 1}, \ldots, \kappa_{\ell \cdot k}) \\ &\Longleftrightarrow J_{\kappa_{j_k+1}}[x] \models \varphi(z, \kappa_{j_1}, \ldots, \kappa_{j_k}) \\ &\Longleftrightarrow L[x] \models \varphi(z, \kappa_{j_1}, \ldots, \kappa_{j_k}), \end{aligned}$$

as desired.

(3) This is immediate by (2).

(4) Fix $i \in \text{OR}$. Suppose that $\kappa_i \in h_{L[x]}(\{\kappa_j \colon j \neq i\})$, say

$$\kappa_i = h_{L[x]}(i^*, (\kappa_{i_1}, \ldots, \kappa_{i_p}, \kappa_{i_{p+1}}, \ldots \kappa_{i_k})),$$

where $i^* < \omega$ and $i_1 < \cdots < i_p < i < i_{p+1} < \cdots < i_k$. Using (2), we may then derive both

$$\kappa_i = h_{L[x]}(i^*, (\kappa_{i_1}, \ldots, \kappa_{i_p}, \kappa_{i_{p+1}+1}, \ldots \kappa_{i_k+1}))$$

and

$$\kappa_{i+1} = h_{L[x]}(i^*, (\kappa_{i_1}, \ldots, \kappa_{i_p}, \kappa_{i_{p+1}+1}, \ldots \kappa_{i_k+1})),$$

which is nonsense. Therefore, $\kappa_i \notin h_{L[x]}(\{\kappa_j \colon j \neq i\})$.

But then if κ is the least ordinal with $\kappa \notin h_{L[x]}(\{\kappa_j \colon j \neq i\})$, then $\kappa = \kappa_j$ for some $j \in \text{OR}$ as in the proof of (1), and hence $j = i$, i.e., $\kappa = \kappa_i$. We have shown that

$$\kappa_i \notin h_{L[x]}(\{\kappa_j \colon j \neq i\}) = h_{L[x]}(\kappa_i \cup \{\kappa_j \colon j > i\}).$$

(5) By (1), it suffices to prove that for every $i \in \text{OR}$ there is some $j \in \text{OR}$ with $\pi(\kappa_i) = \kappa_j$. Let us fix $i \in \text{OR}$, and let us write $\kappa = \pi(\kappa_i)$. Let

$$L[x] \overset{\sigma}{\cong} h_{L[x]}((\kappa+1) \cup \text{ran}(\pi)) = h_{L[x]}((\kappa+1) \cup \{\pi(\kappa_j) \colon j \in \text{OR}\}) \prec L[x].$$

Let λ be a limit ordinal with $\kappa_\lambda > \kappa$. Let $i^* < \omega$, $\eta \in [\kappa_i + 1]^{<\omega}$, and $i < \ell_1 < \ldots < \ell_n < \lambda \leq \ell_{n+1} < \ldots < \ell_k$. If $h_{L[x]}(i^*, (\eta, \kappa_{\ell_1}, \ldots, \kappa_{\ell_k})) < \kappa_\lambda$, then by (2) we must have that

$$h_{L[x]}(i^*, (\eta, \kappa_{\ell_1}, \ldots, \kappa_{\ell_k})) < \kappa_{\ell_n+1},$$

and therefore

$$h_{L[x]}((\kappa_i + 1) \cup \{\kappa_{\ell_1}, \ldots, \kappa_{\ell_k}\}) \cap \kappa_\lambda \subset \kappa_{\ell_n+1},$$

so that by the elementarity of π,

$$h_{L[x]}((\kappa+1)\cup\{\pi(\kappa_{\ell_1}),\dots,\pi(\kappa_{\ell_k})\})\cap\pi(\kappa_\lambda)\subset\pi(\kappa_{\ell_n+1}).$$

This implies that if $\pi''\kappa_\lambda\subset\kappa_\lambda$, then

$$h_{L[x]}((\kappa+1)\cup\{\pi(\kappa_\ell)\colon \ell\in \mathrm{OR}\})\cap\pi(\kappa_\lambda)\subset\kappa_\lambda. \tag{10.30}$$

If in addition κ_λ is a cardinal (in V), so that $\overline{\overline{\{\kappa_j\colon j<\lambda\}}}=\kappa_\lambda$, then

$$\mathrm{otp}(h_{L[x]}((\kappa+1)\cup\{\pi(\kappa_\ell)\colon \ell\in \mathrm{OR}\})\cap\pi(\kappa_\lambda))=\kappa_\lambda. \tag{10.31}$$

(10.30) and (10.31) together imply that $\sigma^{-1}\circ\pi(\kappa_\lambda)=\kappa_\lambda$.

Let $\chi=\sigma^{-1}\circ\pi\colon L[x]\to L[x]$, so that $\chi(\kappa_i)=\kappa$. If $\kappa\neq\kappa_j$ for all $j\in\mathrm{OR}$, then by (1),

$$\kappa=h_{L[x]}(i^*,(\eta,\kappa_{\ell_1},\dots,\kappa_{\ell_k})) \tag{10.32}$$

for some i^*, $k<\omega$ and $\eta\in[\kappa]^{<\omega}$. By (2), we may assume here that every κ_{ℓ_p}, $p=1,\dots,k$, is a cardinal (in V) above κ with $\pi''\kappa_{\ell_p}\subset\kappa_{\ell_p}$, so that $\chi(\kappa_{\ell_p})=\kappa_{\ell_p}$. As $\chi(\kappa_i)=\kappa$, (10.32) and the elementarity of χ then yield that

$$\exists\eta'<\kappa_i\ \kappa_i=h_{L[x]}(i^*,(\eta',\kappa_{\ell_1},\dots,\kappa_{\ell_k})).$$

This contradicts (4)! □

In the light of Corollary 10.44, the club class $\{\kappa_i\colon i\in\mathrm{OR}\}$ is called the class of SILVER *indicernibles for* $L[x]$. Theorem 10.39 and Corollary 10.44 give that the existence of $x^\#$ is *equivalent to* the non-rigidity of $L[x]$. JENSEN's Covering Lemma, cf. Theorem 11.56, will produce a much deeper equivalence.

10.3 Extenders

We need to introduce "extenders" which generalize measures on measurable cardinals, cf. Definition 4.54, and which allow ultrapower constructions which generalize the one from the proof of Theorem 4.55, (1) $\Longrightarrow$ (3), cf. Theorem 10.48. They will be used in the proof of JENSEN's Covering Lemma, cf. Theorem 11.56, as well as in the proof of projective determinacy, cf. Theorem 13.7.

For the sake of this section, by a "transitive model M of a sufficiently large fragment of ZFC" we mean a transitive model M of the statements listed in Corollary 5.18, i.e., (Ext), (Fund), (Inf), (Pair), (Union), the statement that every set is an element of a transitive set, Σ_0-comprehension, and "$\forall x\forall y\ x\times y$ exists," together with (AC) in the form that every set can be well-ordered. We allow M to be a model of the form $(M;\in,A_1,\dots,A_n)$, where $A_i\subset M^{<\omega}$, $1\le i\le n$, are predicates, in which case we understand Σ_0-comprehension to be Σ_0-comprehension in the language of $(M; \in, A_1, \ldots, A_n)$ (which implies that $(M; \in, A_1, \ldots, A_n)$ be amenable, cf. p. 71).

In practice, the results of this section will be applied to models M which are either acceptable J-structures (cf. Definitions 5.24 and 11.1) or to inner models (cf. Definition 4.51).

Definition 10.45 Let M be a transitive model of a sufficiently large fragment of ZFC. Then $E = (E_a : a \in [\nu]^{<\omega})$ is called a (κ, ν)-*extender over* M *with critical points* $(\mu_a : a \in [\nu]^{<\omega})$ provided the following hold true.

(1) (Ultrafilter property) For $a \in [\nu]^{<\omega}$ we have that E_a is an ultrafilter on the set $\mathcal{P}([\mu_a]^{\mathrm{Card}(a)}) \cap M$ which is $< \kappa$-closed with respect to sequences in M, i.e., if $\alpha < \kappa$ and $(X_i : i < \alpha) \in M$ is such that $X_i \in E_a$ for every $i < \alpha$, then $\bigcap_{i<\alpha} X_i \in E_a$. Moreover, μ_a is the least μ such that $[\mu]^{\mathrm{Card}(a)} \in E_a$.
(2) (Coherency) For $a, b \in [\nu]^{<\omega}$ with $a \subset b$ and for $X \in \mathcal{P}([\mu_a]^{\mathrm{Card}(a)}) \cap M$ we have that
$$X \in E_a \Longleftrightarrow X^{ab} \in E_b.$$
(3) (Uniformity) $\mu_{\{\kappa\}} = \kappa$.
(4) (Normality) Let $a \in [\nu]^{<\omega}$ and $f : [\mu_a]^{\mathrm{Card}(a)} \to \mu_a$ with $f \in M$. If
$$\{u \in [\mu_a]^{\mathrm{Card}(a)} : f(u) < \max(u)\} \in E_a$$
then there is some $\beta < \max(a)$ such that
$$\{u \in [\mu_a]^{\mathrm{Card}(a \cup \{\beta\})} : f^{a, a\cup\{\beta\}}(u) = u_\beta^{a\cup\{\beta\}}\} \in E_{a\cup\{\beta\}}.$$

We write $\sigma(E) = \sup\{\mu_a + 1 : a \in [\nu]^{<\omega}\}$. The extender E is called *short* if $\sigma(E) = \kappa + 1$; otherwise E is called *long*.

This definition as well as the discussion to follow makes use of the following notational conventions. Let $b = \{\beta_1 < \ldots < \beta_n\}$, and let $a = \{\beta_{j_1} < \ldots < \beta_{j_m}\} \subset b$. If $u = \{\xi_1 < \ldots < \xi_n\}$ then we write u_a^b for $\{\xi_{j_1} < \ldots < \xi_{j_m}\}$; we also write $u_{\beta_i}^b$ for ξ_i. If $X \in \mathcal{P}([\mu_a]^{\mathrm{Card}(a)})$ then we write X^{ab} for $\{u \in [\mu_b]^{\mathrm{Card}(b)} : u_a^b \in X\}$. Finally, if f has domain $[\mu_a]^{\mathrm{Card}(a)}$ then we write $f^{a,b}$ for that g with domain $[\mu_b]^{\mathrm{Card}(b)}$ such that $g(u) = f(u_a^b)$ if $u_a^b \in [\mu_a]^{\mathrm{Card}(a)}$ and $g(u) = \emptyset$ otherwise. Finally, we write pr for the function which maps $\{\beta\}$ to β (i.e., $\mathrm{pr} = \bigcup$).

Notice that if E is a (κ, ν)-extender over M with critical points μ_a, $a \in [\nu]^{<\omega}$, and if N is another transitive model of a sufficiently large fragment of ZFC such that $\mathcal{P}(\mu_a) \cap N = \mathcal{P}(\mu_a) \cap M$ for all $a \in [\nu]^{<\omega}$, then E is also an extender over N.

Lemma 10.46 *Let M and N be transitive models of a sufficiently large fragment of* ZFC, *and let $\pi : M \to_{\Sigma_0} N$ be cofinal with critical point κ. Let $\nu \leq N \cap OR$. For each $a \in [\nu]^{<\omega}$ let μ_a be the least $\mu \leq M \cap OR$ such that $a \subset \pi(\mu)$, and set*
$$E_a = \{X \in \mathcal{P}([\mu_a]^{\mathrm{Card}(a)}) \cap M : a \in \pi(X)\}.$$

Then $E = (E_a : a \in [\nu]^{<\omega})$ *is a* (κ, ν)*-extender over* M.

Proof We have to verify conditions (1) through (4) of Definition 10.45.

(1) Fix $a \in [\nu]^{<\omega}$. As $a \subset \pi(\mu_a)$, $a \in \pi([\mu_a]^{\text{Card}(a)}) \cap M) = [\pi(\mu_a)]^{\text{Card}(a)}) \cap N$. If $\mu < \mu_a$, then $a \setminus \pi(\mu) \neq \emptyset$, so that $a \notin \pi([\mu]^{\text{Card}(a)}) \cap M) = [\pi(\mu)]^{\text{Card}(a)}) \cap N$, i.e., $[\mu]^{\text{Card}(a)} \notin E_a$.

It is easy to see that $X \in E_a$ and $Y \supset X$, $Y \in \mathscr{P}([\mu_a]^{\text{Card}(a)}) \cap M$ imply $Y \in E_a$, and that $\emptyset \notin E_a$. If $X \in \mathscr{P}([\mu_a]^{\text{Card}(a)}) \cap M$, then $\pi(X) \cup \pi(([\mu_a]^{\text{Card}(a)} \cap M) \setminus X) = \pi([\mu_a]^{\text{Card}(a)}) \cap M)$, so that $X \in E_a$ or $([\mu_a]^{\text{Card}(a)} \cap M) \setminus X \in E_a$.

Let $(X_i : i < \alpha) \in M$, where $\alpha < \kappa$ and $X_i \in E_a$ for each $i < \alpha$. Then $a \in \bigcap_{i<\alpha} \pi(X_i) = \pi(\bigcap_{i<\alpha} X_i)$, i.e., $\bigcap_{i<\alpha} X_i \in E_a$.

(2) Let $a \subset b \in [\nu]^{<\omega}$ and $X \in \mathscr{P}([\mu_a]^{\text{Card}(a)}) \cap M$. Then $X \in E_a$ iff $a \in \pi(X)$ iff $b \in \pi(X^{ab})$ iff $X^{ab} \in E_b$.

(3) Whereas $\{\kappa\} \subset \pi(\kappa)$, $\{\kappa\}$ is not a subset of $\pi(\mu) = \mu$ whenever $\mu < \kappa$. Thus $\mu_{\{\kappa\}} = \kappa$.

(4) Let $a \in [\nu]^{<\omega}$ and $f : [\mu_a]^{\text{Card}(a)} \to \mu_a$ with $f \in M$. Suppose that

$$a \in \pi(\{u \in [\mu_a]^{\text{Card}(a)} : f(u) < \max(u)\}),$$

which means that

$$\pi(f)(a) < \max(a).$$

Set $\beta = \pi(f)(a)$. Then

$$a \cup \{\beta\} \in \pi(\{u \in [\mu_a]^{\text{Card}(a \cup \{\beta\})} : f^{a, a\cup\{\beta\}}(u) = u_\beta^{a\cup\{\beta\}}\}),$$

as desired. □

Definition 10.47 If $\pi : M \to N$, E, κ, and ν are as in the statement of Lemma 10.46, then E is called the (κ, ν)*-extender derived from* π. If $\nu = N \cap \text{OR}$, then we shall denote this extender by E_π.

Theorem 10.48 *Let* M *be a transitive model of a sufficiently large fragment of* ZFC, *and let* $E = (E_a : a \in [\nu]^{<\omega})$ *be a* (κ, ν)*-extender over* M. *There are then* N *and* π *such that the following hold true.*

(a) $\pi : M \to_{\Sigma_0} N$ *is cofinal and has critical point* κ,
(b) *the well-founded part* $\text{wfp}(N)$ *of* N *is transitive and* $\nu \subset \text{wfp}(N)$,
(c) $N = \{\pi(f)(a) : a \in [\nu]^{<\omega} \wedge f : [\mu_a]^{\text{Card}(a)} \to M \wedge f \in M\}$, *and*
(d) *for* $a \in [\nu]^{<\omega}$ *we have that* $X \in E_a$ *if and only if* $X \in \mathscr{P}([\mu_a]^{\text{Card}(a)}) \cap M$ *and* $a \in \pi(X)$.

Moreover, N *and* π *are unique up to isomorphism.*

Proof We do not construe (c) in the stament of this Theorem to presuppose that N be well-founded; in fact, this statement makes perfect sense even if N is *not* well-founded.

Let us first argue that N and π are unique up to isomorphism. Suppose that N, π and N', π' are both as in the statement of the Theorem. We claim that

$$\pi(f)(a) \mapsto \pi'(f)(a), \tag{10.33}$$

where $a \in [\nu]^{<\omega}$ and $f : [\mu_a]^{\text{Card}(a)} \to M$, $f \in M$, defines an $\in$-isomorphism from N onto N'. Notice that for $a, b \in [\nu]^{<\omega}$ and $f : [\mu_a]^{\text{Card}(a)} \to M$, $g : [\mu_b]^{\text{Card}(a)} \to M$, $f, g \in M$, we have that $\pi(f)(a) \in \pi(g)(b)$ in N if and only if, setting $c = a \cup b$,

$$c \in \pi(\{u \in [\mu_c]^{\text{Card}(c)} : f^{a,c}(u) \in g^{b,c}(u)\}),$$

which by (d) yields that

$$\{u \in [\mu_c]^{\text{Card}(c)} : f^{a,c}(u) \in g^{b,c}(u)\} \in E_c,$$

and hence by (d) once more that

$$c \in \pi'(\{u \in [\mu_c]^{\text{Card}(c)} : f^{a,c}(u) \in g^{b,c}(u)\}),$$

i.e., $\pi'(f)(a) \in \pi'(g)(b)$ in N'. The same reasoning applies with "=" instead of "$\in$," so that we indeed get that (10.33) produces an $\in$-isomorphism from N onto N'.

The existence of N and π is shown by an ultrapower construction, similar to the proof of Theorem 4.55, (1) $\Longrightarrow$ (3).

Let us assume that M is of the form $(M; \in, A)$. Let us set

$$D = \{(a, f) : a \in [\nu]^{<\omega} \wedge f : [\mu_a]^{\text{Card}(a)} \to M \wedge f \in M\}.$$

For $(a, f), (b, g) \in D$ let us write

$$(a, f) \sim (b, g) \iff \{u \in [\mu_c]^{\text{Card}(c)} : f^{a,c}(u) = g^{b,c}(u)\} \in E_c, \text{ for } c = a \cup b.$$

We may easily use (1) and (2) of Definition 10.45 to see that $\sim$ is an equivalence relation on D. If $(a, f) \in D$ then let us write $[a, f] = [a, f]^M_E$ for the equivalence class $\{(b, g) \in D : (a, f) \sim (b, g)\}$, and let us set

$$\tilde{D} = \{[a, f] : (a, f) \in D\}.$$

Let us also define, for $[a, f], [b, g] \in \tilde{D}$,

$$[a, f] \mathrel{\tilde{\in}} [b, g] \iff \{u \in [\mu_c]^{\text{Card}(c)} : f^{a,c}(u) \in g^{b,c}(u)\} \in E_c \text{ for } c = a \cup b \text{ and}$$
$$\tilde{A}([a, f]) \iff \{u \in [\mu_a]^{\text{Card}(a)} : f(u) \in A\} \in E_a$$

Notice that the relevant sets are members of M, as M is a model of a sufficiently large fragment of ZFC. Moreover, by (1) and (2) of Definition 10.45, $\tilde{\in}$ and $\tilde{A}$, are well-defined. Let us set

$$N = (\tilde{D}; \tilde{\in}, \tilde{A}).$$

Claim 10.49 (Łoś Theorem) *Let $\varphi(v_1, ..., v_k)$ be a Σ_0 formula (in the language of M), and let $(a_1, f_1), ..., (a_k, f_k) \in D$. Then*

$$N \models \varphi([a_1, f_1], ..., [a_k, f_k]) \iff$$
$$\{u \in [\mu_c]^{\mathrm{Card}(c)};\ M \models \varphi(f_1^{a_1,c}(u), ..., f_k^{a_k,c}(u))\} \in E_c \ \text{for } c = a_1 \cup \ldots \cup a_k.$$

Proof Notice again that the relevant sets are members of M. Claim 10.49 is shown by induction on the complexity of φ, by exploiting (1) and (2) of Definition 10.45. Let us illustrate this by verifying the direction from right to left in the case that, say, $\varphi \equiv \exists v_0 \in v_1\, \psi$ for some Σ_0 formula ψ.

We assume that, setting $c = a_1 \cup \ldots \cup a_k$,

$$\{u \in [\mu_c]^{\mathrm{Card}(c)}: M \models \exists v_0 \in f_1^{a_1,c}(u)\ \psi(v_0, f_1^{a_1,c}(u), \ldots, f_k^{a_k,c}(u))\} \in E_c.$$

Let us define $f_0 \colon [\mu_c]^{\mathrm{Card}(c)} \to \mathrm{ran}(f_1) \cup \{\emptyset\}$ as follows, where $<_{\mathrm{ran}(f_1)} \in M$ is a well-ordering of $\mathrm{ran}(f_1)$.

$$f_0(u) = \begin{cases} \text{the } <_{\mathrm{ran}(f_1)}\text{-smallest } x \in \mathrm{ran}(f_1) \text{ with} & \\ \quad M \models \psi(x, f_1^{a_1,c}, \ldots, f_k^{a_k,c}(u)) & \text{if some such x exists,} \\ \emptyset & \text{otherwise.} \end{cases}$$

The point is that $f_0 \in M$. But we then have that

$$\{u \in [\mu_c]^{\mathrm{Card}(c)}: M \models f_0(u) \in f_1^{a_1,c}(u) \wedge \psi(f_0(u), f_1^{a_1,c}(u), \ldots, f_k^{a_k,c}(u))\} \in E_c,$$

which inductively implies that

$$N \models [c, f_0] \in [a_1, f_1] \wedge \psi([c, f_0], [a_1, f_1], \ldots, [a_k, f_k]),$$

and hence that

$$N \models \exists v_0 \in [a_1, f_1]\ \psi(v_0, [a_1, f_1], ..., [a_k, f_k]),$$

as desired. □

Given Claim 10.49, we may and shall from now on identify, via the MOSTOWSKI collapse, the well-founded part wfp(N) of N with a transitive structure. In particular, if $[a, f] \in \mathrm{wfp}(N)$ then we identify the equivalence class $[a, f]$ with its image under the Mostowski collapse.

Let us now define $\pi \colon M \to N$ by

$$\pi(x) = [0, c_x], \text{ where } c_x \colon \{\emptyset\} = [\mu_0]^0 \to \{x\}.$$

We aim to verify that N, π satisfy (a), (b), (c), and (d) from the statement of Theorem 10.49.

Claim 10.50 *If* $\alpha < \nu$ *and* $[a, f] \tilde{\in} [\{\alpha\}, \mathrm{pr}]$ *then* $[a, f] = [\{\beta\}, \mathrm{pr}]$ *for some* $\beta < \alpha$.

Proof Let $[a, f] \tilde{\in} [\{\alpha\}, \mathrm{pr}]$. Set $b = a \cup \{\alpha\}$. By the Łoś Theorem 10.49,

$$\{u \in [\mu_b]^{\mathrm{Card}(b)} \colon f^{a,b}(u) \in \mathrm{pr}^{\{\alpha\},b}(u)\} \in E_b.$$

By (4) of Definition 10.45, there is some $\beta < \alpha$ such that, setting $c = b \cup \{\beta\}$,

$$\{u \in [\mu_c]^{\mathrm{Card}(c)} \colon f^{a,c}(u) = \mathrm{pr}^{\{\beta\},c}(u)\} \in E_c,$$

and hence, by the Łoś Theorem again, $[a, f] = [\{\beta\}, \mathrm{pr}]$. □

Claim 10.50 implies, via a straightforward induction, that

$$[\{\alpha\}, \mathrm{pr}] = \alpha \text{ for } \alpha < \nu. \tag{10.34}$$

In particular, (b) from the statement of Theorem 10.48 holds true.

Claim 10.51 *If* $a \in [\nu]^{<\omega}$ *then* $[a, \mathrm{id}] = a$.

Proof If $[b, f] \tilde{\in} [a, \mathrm{id}]$ then by the Łoś Theorem, setting $c = a \cup b$,

$$\{u \in [\mu_c]^{\mathrm{Card}(c)} \colon f^{b,c}(u) \in u^c_a\} \in E_c.$$

However, as E_c is an ultrafilter, there must then be some $\alpha \in a$ such that

$$\{u \in [\mu_c]^{\mathrm{Card}(c)} \colon f^{b,c}(u) = u^c_\alpha\} \in E_c,$$

and hence by the Łoś Theorem and (10.34)

$$[b, f] = [\{\alpha\}, \mathrm{pr}] = \alpha.$$

On the other hand, if $\alpha \in a$ then it is easy to see that $\alpha \in [a, \mathrm{id}]$. □

Claim 10.52 $[a, f] = \pi(f)(a)$.

Proof Notice that this statement makes sense even if $[a, f] \notin \mathrm{wfp}(N)$.

Let $b = a \cup \{0\}$. We have that

$$\{u \in [\mu_b]^{\mathrm{Card}(b)} \colon f^{a,b}(u) = ((c_f)^{\emptyset,b}(u))(\mathrm{id}^{a,b}(u))\} = [\mu_b]^{\mathrm{Card}(b)} \in E_b,$$

by (1) of Definition 10.45, and therefore by the Łoś Theorem and Claim 10.51,

$$[a, f] = [0, c_f]([a, \mathrm{id}]) = \pi(f)(a).$$

□

Claim 10.52 readily implies (c) from the statement of Theorem 10.48.

Claim 10.53 $\kappa = \text{crit}(\pi)$.

Proof Let us first show that $\pi \restriction \kappa = \text{id}$. We prove that $\pi(\xi) = \xi$ for all $\xi < \kappa$ by induction on ξ.

Let $\xi < \kappa$. Suppose that $[a, f] \tilde{\in} \pi(\xi) = [0, c_\xi]$. Set $b = a \cup \{\xi\}$. Then

$$\{u \in [\mu_b]^{\text{Card}(b)} \colon f^{a,b}(u) < \xi\} \in E_b.$$

As E_b is $< \kappa$-closed with respect to sequences in M (cf. (1) of Definition 10.45), there is hence some $\bar{\xi} < \xi$ such that

$$\{u \in [\mu_b]^{\text{Card}(b)} \colon f^{a,b}(u) = \bar{\xi}\} \in E_b,$$

and therefore $[a, f] = \pi(\bar{\xi})$ which is $\bar{\xi}$ by the inductive hypothesis. Hence $\pi(\xi) \subset \xi$. It is clear that $\xi \subset \pi(\xi)$.

We now prove that $\pi(\kappa) > \kappa$ (if $\pi(\kappa) \notin \text{wfp}(N)$ we mean that $\kappa \tilde{\in} \pi(\kappa)$) which will establish Claim 10.53. Well, $\mu_{\{\kappa\}} = \kappa$, and

$$\{u \in [\kappa]^1 \colon pr(u) < \kappa\} = [\kappa]^1 \in E_{\{\kappa\}},$$

from which it follows, using the Łoś Theorem, that $\kappa = [\{\kappa\}, \text{pr}] < [0, c_\kappa] = \pi(\kappa)$.□

The following, together with the previous Claims, will establish (a) from the statement of Theorem 10.48.

Claim 10.54 *For all* $[a, f] \in N$ *there is some* $y \in M$ *with* $[a, f] \tilde{\in} \pi(y)$.

Proof It is easy to see that we can just take $y = \text{ran}(f)$. □

It remains to prove (d) from the statement of Theorem 10.48. Let $X \in E_a$. By (1) of Definition 10.45,

$$X = \{u \in [\mu_a]^{\text{Card}(a)} \colon u \in X\} \in E_a,$$

which, by the Łoś Theorem and Claim 3, gives that $a = [a, \text{id}] \tilde{\in} [0, c_X] = \pi(X)$.

On the other hand, suppose that $X \in \mathscr{P}([\mu_a]^{\text{Card}(a)}) \cap M$ and $a \in \pi(X)$. Then by Claim 3, $[a, \text{id}] = a \in \pi(X) = [0, c_X]$, and thus by the Łoś Theorem

$$X = \{u \in [\mu_a]^{\text{Card}(a)} \colon u \in X\} \in E_a.$$

We have shown Theorem 10.48. □

Definition 10.55 Let M, E, N, and π be as in the statement of Theorem 10.48. We shall denote N by $\text{ult}_0(M; E)$ and call it the (Σ_0-) ultrapower of M by E, and we call $\pi: M \to N$ the (Σ_0-) *ultrapower map* (*given by* E). We shall also write π_E^M or π_E for π.

We now turn towards criteria for $\text{Ult}_0(M; E)$ being well-founded. (This will also be a big issue in the proof of Theorem 11.56.) The easiest such criterion is given by when E is a derived extender.

Lemma 10.56 *Let* $\pi: M \to N$, κ, *and* ν *be as in the statement of Theorem 10.46, and let* E *be the* (κ, ν)*-extender derived from* π *(cf. Definition 10.47). Then* $\text{ult}_0(M; E)$ *is well-founded, and in fact there is an embedding* $k: \text{ult}_0(M; E) \to N$ *such that* $\pi = k \circ \pi_E$ *and* $k \restriction \nu = \text{id}$.

Proof We define $k: \text{ult}_0(M; E) \to N$ by setting $k([a, f]) = \pi(f)(a)$ for $a \in [\nu]^{<\omega}$ and $f: [\mu_a]^{\text{Card}(a)} \to M$, $f \in M$. We have that k is a well-defined Σ_0-elementary embedding. To see this let φ be Σ_0, and $a_j \in [\nu]^{<\omega}$ and $f_j: [\mu_{a_j}]^{\text{Card}(a_j)} \to M$, $f_j \in M$, for $j \in \{1, \ldots, k\}$. Set $a = \bigcup_{j \in \{1,\ldots,k\}} a_j$. We then get that

$$\begin{aligned} \text{ult}_0(M; E) \models \varphi([a_1, f_1], \ldots, [a_k, f_k]) &\iff \\ \{u \in [\mu_a]^{\text{Card}(a)}: M \models \varphi(f_1^{a_1,a}(u), \ldots, f_k^{a_k,a}(u))\} \in E_a &\iff \\ a \in \pi(\{u \in [\mu_a]^{\text{Card}(a)}: M \models \varphi(f_1^{a_1,a}(u), \ldots, f_k^{a_k,a}(u))\}) &\iff \\ a \in \{u \in [\pi(\mu_a)]^{\text{Card}(a)}: N \models \varphi(\pi(f_1)^{a_1,a}(u), \ldots, \pi(f_k)^{a_k,a}(u))\}) &\iff \\ N \models \varphi(\pi(f_1)(a_1), \ldots, \pi(f_k)(a_k)). & \end{aligned}$$

We have that $k(\pi_E(x)) = k([\emptyset, c_x]) = \pi(c_x)(\emptyset) = c_{\pi(x)}(\emptyset) = \pi(x)$ for all $x \in M$, so that $\pi = k \circ \pi_E$. As $k(\beta) = k([\text{pr}, \{\beta\}]) = \pi(\text{pr})(\{\beta\}) = \beta$ for every $\beta < \nu$, we have that $k \restriction \nu = \text{id}$. □

Let us consider extenders over V.

Definition 10.57 Let F be a (κ, ν)-extender over V, and suppose that $\text{ult}_0(V; F)$ is well-founded. Say $\text{ult}_0(V; F) \cong M$, where M is transitive. The *strength of* F is then the largest ordinal α such that $V_\alpha \subset M$.

In the situation of Definition 10.57, the strength of F is always at least $\kappa + 1$. If F is derived from $\pi: V \to M$ (in the sense of Definition 10.47), where $\kappa = \text{crit}(\pi)$ and $V_\alpha \subset M$ for some $\alpha > \kappa + 1$, i.e., π witnesses that κ is α-strong (cf. Definition 4.60), then the strength of F may be α; more precisely:

Lemma 10.58 *Let* $\pi: V \to M$ *be an non-trivial elementary embedding, where* M *is transitive. Let* $\kappa = \text{crit}(\pi) < \alpha \leq \pi(\kappa)$, *and suppose that* $V_\alpha \subset M$. *Let* $\nu \leq \pi(\kappa)$ *be least such that* $\nu \geq \alpha$ *and* ν *is inaccessible in* M, *and let* F *be the short* (κ, ν)*-extender over* V *derived from* π. *Then* $\text{ult}_0(V; F)$ *is well-founded, and the strength of* F *is at least* α.

Proof We may pick some $E \subset \kappa \times \kappa$ such that for every inaccessible cardinal $\gamma \leq \kappa$,

$$(\gamma; E \cap (\gamma \times \gamma)) \cong (V_\gamma; \in).$$

Identifying N with $\text{ult}_0(V; F)$, we let

$$i_F : V \to_F N$$

be the ultrapower map, and we let

$$k : N \to M$$

be the factor map which is defined as in the proof of Lemma 10.56 by $k(i_F(f)(a)) = \pi(f)(a)$ for $a \in [\nu]^{<\omega}$ and $f : [\kappa]^{\text{Card}(a)} \to V$. We have that $k \circ i_F = \pi$ and $k \upharpoonright \nu = \text{id}$.

By the elementarity of π,

$$(\gamma; \pi(E) \cap (\gamma \times \gamma)) \cong ((V_\gamma)^M; \in)$$

for every $\gamma \leq \pi(\kappa)$ which is inaccessible in M. In particular,

$$(\nu; \pi(E) \cap (\nu \times \nu)) \cong ((V_\nu)^M; \in). \tag{10.35}$$

By $k \upharpoonright \nu = \text{id}$, $i_F(E) \cap (\nu \times \nu) = \pi(E) \cap (\nu \times \nu)$. Hence $\pi(E) \cap (\nu \times \nu) \in N$. As $\alpha \leq \nu$, this gives $V_\alpha = (V_\alpha)^M \subset (V_\nu)^M \in N$ by (10.35). □

Lemma 10.58 says that short extenders may be used to witness that a given cardinal is strong. On the other hand, Lemma 10.62 below will tell us that long extenders may be used to witness the supercompactness (cf. Definition 4.62) of a given cardinal.

Definition 10.59 Let F be a short (κ, ν)-extender over V, and let $\lambda \leq \kappa$. Then F is called *λ-closed* iff for all $\{a_i : i < \lambda\} \subset [\nu]^{<\omega}$ there are $b \in [\nu]^{<\omega}$ and $g : [\kappa]^{\text{Card}(b)} \to V$ such that for every $i < \lambda$,

$$\{u \in [\kappa]^{\text{Card}(b \cup a_i)} : g^{b \cup a_i, b}(u)(i) = u^{b \cup a_i}_{a_i}\} \in E_{b \cup a_i}. \tag{10.36}$$

The following Lemma may be construed as strengthening of Lemma 4.63

Lemma 10.60 *Let F be a short (κ, ν)-extender over V, let $\lambda \leq \kappa$, and suppose that F is λ-closed and $N = \text{ult}_0(V; F)$ is transitive. Then N is λ-closed, i.e.,*

$${}^\lambda N \subset N.$$

Proof Let $\{x_i : i < \lambda\} \subset N$, say $x_i = \pi_E(f_i)(a_i)$ for $i < \lambda$. We aim to prove that $(x_i : i < \lambda) \in N$. Let b and g be as in Definition 10.59. We may assume that $\lambda \in b$.

Let $H\colon [\kappa]^{\mathrm{Card}(b)} \to V$ be such that for every $u \in [\kappa]^{\mathrm{Card}(b)}$, $H(u)$ is a function with domain u^b_λ and for every $i < \lambda$,

$$H(u)(i) = f_i(g(u)(i)).$$

Then $\pi_E(H)(b)$ is a function with domain λ, and by (10.36) and the Łoś Theorem, for every $i < \lambda$,

$$\pi_E(H)(b)(i) = \pi_E(f_i)(\pi_E(g)(b)(i)) = \pi_E(f_i)(a_i) = x_i.$$

This shows that $(x_i : i < \lambda) \in N$. □

Lemma 10.61 *Let $\pi\colon V \to M$ be an non-trivial elementary embedding, where M is transitive. Let $\kappa = \mathrm{crit}(\pi) < \nu \leq \pi(\kappa)$, and suppose that ν is an inaccessible cardinal in V and $V_\nu \subset M$. Let F be the short (κ, ν)-extender over V derived from π. Then $\mathrm{ult}_0(V; F)$ is well-founded and F is κ-closed.*

Proof This is an immediate consequence of Lemma 10.58, and we may in fact just continue the proof of Lemma 10.58, where now $\alpha = \nu$ and ν is inaccessible in V (not only in M). We have that $V_\alpha \in N = \mathrm{ult}_0(V; F)$. Let $\{a_i : i < \kappa\} \subset [\nu]^{<\omega}$. Then $(a_i : i < \kappa) \in N$, and hence there are $b \in [\nu]^{<\omega}$ and $f\colon [\kappa]^{\mathrm{Card}(b)} \to V$ such that $(a_i : i < \kappa) = \pi_E(g)(b)(i)$ for every $i < \kappa$. The Łoś Theorem 10.49 then yields that b and f are as in (10.36). □

Lemma 10.62 *Let κ be λ-supercompact, where $\lambda \geq \kappa$. There is then a long extender E witnessing that κ is λ-supercompact, i.e., $\mathrm{ult}_0(V; E)$ is transitive and if $\pi_E\colon V \to N = \mathrm{ult}_0(V; E)$ is the ultrapower embedding, then $\pi_E(\kappa) > \lambda$ and ${}^\lambda N \subset N$.*

Proof Let us fix an elementary embedding

$$\pi\colon V \to M,$$

where M is an inner model, $\kappa = \mathrm{crit}(\pi)$, $\pi(\kappa) > \lambda$, and ${}^\lambda M \subset M$. We aim to derive a long extender E from π in such a way that the ultrapower of V by E is also closed under λ-sequences.

Set $\gamma = 2^\lambda$, so that $\gamma^\lambda = \gamma$. Set $\nu = \pi(\gamma)$, and let E be the (κ, ν)-extender over V derived from π. Let

$$\pi_E\colon V \to \mathrm{ult}_0(V; E)$$

be the ultrapower embedding, and let $k\colon \mathrm{ult}_0(V; E) \to M$ be the factor map which is defined as in the proof of Lemma 10.56 by $k([a, f]) = \pi(f)(a)$ for $a \in [\nu]^{<\omega}$ and $f\colon [\mu_a]^{\mathrm{Card}(a)} \to V$. We may identify $\mathrm{ult}_0(V; E)$ with its transitive collapse, and we will denote it by N. We have that $\pi = k \circ \pi_E$, and $k \restriction \nu = \mathrm{id}$.

Let $e\colon \gamma \to [\gamma]^{\leq\lambda}$ be a bijection. Then $\pi(e)\colon \nu \to [\nu]^{\leq\pi(\lambda)} \cap M$ is bijective. As ${}^\lambda M \subset M$, we have that $[\nu]^\lambda \cap V \subset M$, so that $[\nu]^\lambda \cap V \subset \mathrm{ran}(\pi(e))$. But

$k \restriction \nu = \mathrm{id}$ and $\pi = k \circ \pi_E$, which implies that $\pi_E \restriction \gamma = \pi \restriction \gamma$, so in particular $\pi_E(\kappa) = \pi(\kappa) > \lambda$. Also $\pi_E(e) = \pi(e)$. Therefore,

$$[\nu]^\lambda \cap V \subset \mathrm{ran}(\pi_E(e)). \tag{10.37}$$

By (10.37), $\pi_E"\lambda = \pi"\lambda \in N = \mathrm{ult}_0(V; E)$, which gives that $\pi_E \restriction \lambda = \pi \restriction \lambda \in N$, and we may pick some $a \in [\nu]^{<\omega}$ and $f : [\mu_a]^{\mathrm{Card}(a)} \to V$ with

$$\pi_E \restriction \lambda = \pi \restriction \lambda = \pi_E(f)(a). \tag{10.38}$$

Now let $(x_i : i < \lambda) \subset N$. We aim to show that $(x_i : i < \lambda) \in N$. Let $x_i = [a_i, f_i]$ for $i < \lambda$, where $a_i \in [\nu]^{<\omega}$ and $f_i : [\mu_{a_i}]^{\mathrm{Card}(a_i)} \to V$. Let us write G for the function with domain λ and $G(i) = a_i$ for $i < \lambda$, i.e., $G = (a_i : i < \lambda)$. By (10.37), $(a_i : i < \lambda) \in N$, so that we may pick some $b \in [\nu]^{<\omega}$ and $g : [\mu_b]^{\mathrm{Card}(b)} \to V$ with

$$G = (a_i : i < \lambda) = \pi_E(g)(b). \tag{10.39}$$

Set $c = a \cup b$, and let us define $H : [\mu_c]^{\mathrm{Card}(c)} \to V$ as follows. For each $u \in [\mu_c]^{\mathrm{Card}(c)}$, we let $H(u)$ be a function with domain λ such that for $i < \lambda$,

$$H(u)(i) = f_i(g^{b,c}(u)((f^{a,c}(u))^{-1}(i))). \tag{10.40}$$

Here, we understand that if $f^{a,c}(u)$ is an injective function with i in its domain, then $(f^{a,c}(u))^{-1}(i)$ is the preimage of i under that function, and $(f^{a,c}(u))^{-1}(i) = \emptyset$ otherwise.

We get that $\pi_E(H)(c) : \pi_E(\lambda) \to N$, and for each $i < \lambda$,

$$\begin{aligned}
\pi_E(H)(c)(\pi_E(i)) &= \pi_E(f_i)(\pi_E(g^{b,c})(c)((\pi_E(f^{a,c})(c))^{-1})(\pi_E(i))) \\
&= \pi_E(f_i)(\pi_E(g)(b)((\pi_E(f)(a))^{-1})(\pi_E(i))) \\
&= \pi_E(f_i)(\pi_E(g)(b)(\pi_E \restriction \lambda)^{-1})(\pi_E(i))) \\
&= \pi_E(f_i)(G(i)) \\
&= \pi_E(f_i)(a_i) = x_i.
\end{aligned}$$

Using $\pi_E \restriction \lambda \in N$ once more, we then get that the function with domain λ which maps

$$i \mapsto \pi_E(H)(c)((\pi_E \restriction \lambda)(i)) = x_i$$

also exists inside N. We have shown that $(x_i : i < \lambda) \in N$. □

The following concepts and techniques will be refined in the next section.

Definition 10.63 Let M be a transitive model of a sufficiently large fragment of ZFC, and let $E = (E_a : a \in [\nu]^{<\omega})$ be a (κ, ν)-extender over M. Let $\lambda < \mathrm{Card}(\kappa)$ be an infinite cardinal (in V). Then E is called λ-*complete* provided the following holds true. Suppose that $((a_i, X_i): i < \lambda)$ is such that $X_i \in E_{a_i}$ for all $i < \lambda$. Then there is some order-preserving map $\tau: \bigcup_{i<\lambda} a_i \to \sigma(E)$ such that $\tau''a_i \in X_i$ for every $i < \lambda$. E is called countably complete iff E is $\aleph_0$-complete, and E is called continuum-complete iff E is $2^{\aleph_0}$-complete.

Lemma 10.64 *Let M be a transitive model of a sufficiently large fragment of* ZFC*, and let $E = (E_a: a \in [\nu]^{<\omega})$ be a (κ, ν)-extender over M. Let $\lambda < \mathrm{Card}(\kappa)$ be an infinite cardinal. Then E is λ-complete if and only if for every $U \prec_{\Sigma_0} \mathrm{Ult}_0(M; E)$ of size λ there is some $\varphi: U \to_{\Sigma_0} M$ such that $\varphi \circ \pi_E(x) = x$ whenever $\pi_E(x) \in U$.*

Proof "$\Longrightarrow$": Let $U \prec_{\Sigma_0} \mathrm{Ult}_0(M; E)$ be of size λ. Write $U = \{[a, f]: (a, f) \in \bar{U}\}$ for some $\bar{U}$ of size λ. Let $((a_i, X_i); i < \lambda)$ be an enumeration of all pairs (c, X) such that there is a Σ_0 formula ψ and there are $(a^1, f_1), \ldots, (a^k, f_k) \in \bar{U}$ with $c = a^1 \cup \ldots \cup a^k$ and

$$X = \{u \in [\mu_c]^{\mathrm{Card}(c)}: M \models \psi(f_1^{a^1,c}(u), \ldots, f_k^{a^k,c}(u))\} \in E_c.$$

Let $\tau: \bigcup_{i<\lambda} a_i \to \sigma(E)$ be order-preserving such that $\tau''a_i \in X_i$ for every $i < \lambda$. Let us define $\varphi: U \to M$ by setting $\varphi([a, f]) = f(\tau''(a))$ for $(a, f) \in \bar{U}$.

We get that φ is well-defined and Σ_0-elementary by the following reasoning. Let $\psi(v_1, \ldots, v_k)$ be Σ_0, and let $[a^j, f_j] \in U$, $1 \leq j \leq k$. Set $c = a^1 \cup \ldots \cup a^k$. We then get that

$$\begin{aligned}
U \models &\psi([a^1, f_1], \ldots, [a^k, f_k]) &&\Longleftrightarrow\\
\mathrm{Ult}_0(M; E) \models &\psi([a^1, f_1], \ldots, [a^k, f_k]) &&\Longleftrightarrow\\
\{u \in [\mu_c]^{\mathrm{Card}(c)}: M \models &\psi(f_1^{a^1,c}(u), \ldots, f_k^{a^k,c}(u))\} \in E_c &&\Longleftrightarrow\\
\tau''c \in \{u \in [\mu_c]^{\mathrm{Card}(c)}: M \models &\psi(f_1^{a^1,c}(u), \ldots, f_k^{a^k,c}(u))\} &&\Longleftrightarrow\\
M \models &\psi(f_1(\tau''a^1), \ldots, f_k(\tau''a^k)).
\end{aligned}$$

We also get that $\varphi \circ \pi_E(x) = \varphi([\emptyset, c_x]) = c_x(\emptyset) = x$.

"$\Longleftarrow$": Let $((a_i, X_i): i < \lambda)$ be such that $X_i \in E_{a_i}$ for all $i < \lambda$. Pick $U \prec_{\Sigma_0} \mathrm{Ult}_0(M; E)$ with $\{(a_i, X_i): i < \lambda\} \subseteq U$, $\mathrm{Card}(U) = \lambda$, and let $\varphi: U \to_{\Sigma_0} M$ be such that $\varphi \circ \pi_E(x) = x$ whenever $\pi_E(x) \in U$. Set $\tau = \varphi \restriction \bigcup_{i<\lambda} a_i$. Then $\tau''a_i = \varphi(a_i) \in \varphi \circ \pi_E(X_i) = X_i$ for all $i < \lambda$. Clearly, $\mathrm{ran}(\tau) \subseteq \sigma(E)$. □

Corollary 10.65 *Let M be a transitive model of a sufficiently large fragment of* ZFC*, and let E be a countably complete (κ, ν)-extender over M. Then* $\mathrm{Ult}_0(M; E)$ *is well-founded.*

Lemma 10.66 *Let λ be an infinite cardinal, and let θ be regular. Let $\pi: \bar{H} \to H_\theta$, where $\bar{H}$ is transitive and ${}^\lambda\bar{H} \subseteq \bar{H}$. Suppose that $\pi \neq \mathrm{id}$, and set $\kappa = \mathrm{crit}(\pi)$. Let M be a transitive model of a sufficiently large fragment of* ZFC*, let ρ be regular in M, and suppose that $H_\rho^M \subseteq \bar{H}$. Set $\nu = \sup\pi''\rho$, and let E be the (κ, ν)-extender over M derived from $\pi \restriction H_\rho^M$. Then E is λ-complete.*

Proof Let $((a_i, X_i): i < \lambda)$ be such that $X_i \in E_{a_i}$, and hence $a_i \in \pi(X_i)$, for all $i < \lambda$. As ${}^{\lambda}\bar{H} \subseteq \bar{H}$, $(X_i: i < \lambda) \in \bar{H}$. Let $\sigma: \bigcup_{i<\lambda} a_i \cong \gamma = \text{otp}(\bigcup_{i<\lambda} a_i)$ be the transitive collapse; notice that $\gamma < \lambda^+ < \kappa$. For each $i < \lambda$ let $\bar{a}_i = \sigma'' a_i$. We have that $(\bar{a}_i: i < \lambda) \in \bar{H}$. But now

$$H_\theta \models \exists \text{ order-preserving } \tau: \gamma \to \text{OR } \forall i < \lambda\ \tau'' \bar{a}_i \in \pi((X_j; j < \lambda))(i),$$

as witnessed by σ^{-1}. Therefore,

$$\bar{H} \models \exists \text{ order-preserving } \tau: \gamma \to \text{OR } \forall i < \lambda\ \tau'' \bar{a}_i \in X_i. \tag{10.41}$$

Hence, if $\tau \in \bar{H}$ is a witness to (10.41), then $\tau \circ \sigma: \bigcup_{i<\lambda} a_i \to$ OR is such that $\tau \circ \sigma'' a_i \in X_i$ for every $i < \lambda$. □

10.4 Iteration Trees

Iteration trees are needed for the proof of projective determinacy, cf. Theorem 13.6. In order to prove a relevant technical tool, Theorem 10.74, we need a strengthening of the concept of countable completeness. All the extenders in this section will be short, though.

Definition 10.67 Let F be a short (κ, ν)-extender over V, and let U be any set. We say that F is complete with respect to U iff there is a map τ such that $\nu \cap U \subset \text{dom}(\tau)$, $\tau \upharpoonright (\nu \cap U): \nu \cap U \to \kappa$ is order preserving, $\tau \upharpoonright (\kappa \cap U) = \text{id}$, and for all $a \in [\nu \cap U]^{<\omega}$ and for every $X \in \mathscr{P}([\kappa]^{\text{Card}(a)}) \cap U$ which is measured by F_a,[4] we have that

$$X \in F_a \Longleftrightarrow \tau'' a \in X.$$

Hence if μ is an infinite cardinal, then F is μ-complete iff whenever U has size μ, F is complete with respect to U.

We shall be interested in a strengthening of "continuum-completeness," cf. Definition 10.63.

Definition 10.68 A formula $\psi(\alpha, \mathbf{x})$ is said to be Σ_{1+} iff $\psi(\alpha, \mathbf{x})$ is of the form

$$\exists M (M \text{ is transitive} \wedge {}^{(2^{\aleph_0})}M \subset M \wedge V_\alpha \subset M \wedge \varphi(M, \alpha, \mathbf{x})), \tag{10.42}$$

where φ is Σ_1. An ordinal β is called a reflection point iff $V_\beta \prec_{\Sigma_{1+}} V$.

It is not hard to verify that every Σ_{1+} formula is Σ_2.

Lemma 10.69 *Let λ be an inaccessible cardinal. Then λ is a reflection point.*

---

[4] i.e., $X \in F_a$ or $([\kappa]^{\text{Card}(a)}) \setminus X \in F_a$.

Proof Let ψ be Σ_{1+}, and let $\alpha, \mathbf{x} \in V_\lambda$. Let us pick some

$$\pi: P \to_{\Sigma_2} V,$$

where P is transitive, $\{V_\alpha, \mathrm{TC}(\{\mathbf{x}\})\} \in P$, ${}^{(2^{\aleph_0})}P \subset P$, and $\mathrm{Card}(P) < \lambda$. There is some such P, as λ is inaccessible.

Because every Σ_{1+}-formula can be written as a Σ_2-formula, so that if $V \models \psi(\alpha, \mathbf{x})$, then $P \models \psi(\alpha, \mathbf{x})$. But if $M \in P$ is as in (10.42) to witness $P \models \psi(\alpha, \mathbf{x})$, then M also witnesses $V_\lambda \models \psi(\alpha, \mathbf{x})$, as ${}^{(2^{\aleph_0})}P \subset P$. □

Definition 10.70 Let F be a short (κ, ν)-extender over V. Then F is called certified iff ν is also the strength[5] of F, ν is inaccessible, and for every $U \prec_{\Sigma_{1+}} V_\nu$ of size $2^{\aleph_0}$ there is some $\tau: U \to_{\Sigma_{1+}} V_\kappa$ witnessing that F is $2^{\aleph_0}$-complete with respect to U.

Lemma 10.71 *Let F be a short (κ, ν)-extender over V, and suppose that ν is also the strength of F and ν is inaccessible. Then F is certified.*

Proof By Lemma 10.69, ν is a reflection point. Let us fix $U \prec_{\Sigma_{1+}} V_\nu$ of size $2^{\aleph_0}$. We need to find some $\tau: U \to_{\Sigma_{1+}} V_\kappa$ such that $\tau \restriction (\kappa \cap U) = \mathrm{id}$ and for all $a \in [\nu \cap U]^{<\omega}$ and for all $X \in \mathscr{P}([\kappa]^{\mathrm{Card}(a)}) \cap U$, $X \in F_a$ iff $\tau" a \in X$.

Let $\pi: V \to_F M = \mathrm{ult}(V, F)$, where M is transitive. Notice that $V_\nu \subset M$ (i.e., $V_\nu \in M$), and $U \prec_{\Sigma_{1+}} V_\nu = V_\nu^M \prec_{\Sigma_{1+}} V_{\pi(\kappa)}^M \prec_{\Sigma_{1+}} M$ by Lemma 10.69 applied inside M. Let $\bar{\sigma}: \bar{U} \cong U$, where $\bar{U}$ is transitive. Then, using Lemma 10.69,

$$\bar{\sigma}: \bar{U} \to_{\Sigma_{1+}} V_\nu^M = V_\nu \prec_{\Sigma_{1+}} V_{\pi(\kappa)}^M \prec_{\Sigma_{1+}} M,$$

and $\bar{U}$ and $\bar{\sigma}$ are both elements of M by Lemmas 10.60 and 10.61.

Let $(X_i : i < 2^{\aleph_0})$ be an enumeration of $\bigcup\{\mathscr{P}([\kappa]^n) : n < \omega\} \cap U$, and let $(a_i : i < 2^{\aleph_0})$ be an enumeration of $[\nu \cap U]^{<\omega}$. Let Γ be the set of all $(i, j) \in (2^{\aleph_0}) \times (2^{\aleph_0})$ such that $X_j \in F_{a_i}$. Let $\beta < \kappa$ be such that $U \cap V_\kappa = U \cap V_\beta$. Of course, $U \cap V_\beta \in M$ and $\bar{\sigma}^{-1}"(U \cap V_\beta) \in M$.

Let $\bar{a}_i = \bar{\sigma}^{-1}(a_i)$ for $i < 2^{\aleph_0}$. Notice that $(\bar{a}_i : i < 2^{\aleph_0}) \in M$.

Now $\bar{\sigma} \in M$ witnesses that in M, the following holds true.

$$\exists k (k: \bar{U} \to_{\Sigma_{1+}} V_{\pi(\kappa)}^M \wedge k \restriction \bar{\sigma}^{-1}"(U \cap V_\beta) = \bar{\sigma} \restriction \bar{\sigma}^{-1}"(U \cap V_\beta)$$
$$\wedge \forall i, j < 2^{\aleph_0} (k(\bar{a}_i) \in \pi(X_j) \longleftrightarrow (i, j) \in \Gamma)).$$

Therefore, in V we have that

$$\exists k (k: \bar{U} \to_{\Sigma_{1+}} V_\kappa \wedge k \restriction \bar{\sigma}^{-1}"(U \cap V_\beta) = \bar{\sigma} \restriction \bar{\sigma}^{-1}"(U \cap V_\beta)$$
$$\wedge \forall i, j < 2^{\aleph_0} (k(\bar{a}_i) \in X_j \longleftrightarrow (i, j) \in \Gamma)).$$

Let $\sigma^*: \bar{U} \to_{\Sigma_{1+}} V_\kappa$ be a witness, and set $\tau = \sigma^* \circ \bar{\sigma}^{-1}$. Obviously, $\tau: U \to_{\Sigma_{1+}} V_\kappa$, and $\tau \restriction (U \cap V_\kappa) = \tau \restriction (U \cap V_\beta) = \mathrm{id}$. Moreover, if $a \in [\nu \cap U]^{<\omega}$ and

[5] Recall that the strength of an extender F is the largest ordinal α such that $V_\alpha \subset \mathrm{Ult}(V; F)$.

$X \in \mathscr{P}([\kappa]^{\text{Card}(a)}) \cap U$, say $a = a_i$ and $X = X_j$, then $X_j \in F_{a_i}$ iff $(i, j) \in \Gamma$ iff $\sigma^*(\bar{a}_i) \in X_j$ iff $\tau(a_i) \in X_j$. □

Definition 10.72 Let $0 < N \leq \omega$. A system $\mathscr{T} = ((M_i, \pi_{ij} : i \leq_T j < N), (E_i : i+1 < N), \leq_T)$ is called a putative iteration tree on V of length N iff the following holds true.

(1) $\leq_T$ is a reflexive and transitive order on N such that if $i \leq_T j$, then $i \leq j$ in the natural order, and if $i < N$, then $0 \leq_T i$.
(2) $M_0 = V$, and if $i + 1 < N$, then M_i is a (transitive) inner model.
(3) If $i \leq_T j \leq_T k < N$, then $\pi_{ij} : M_i \to M_j$ is an elementary embedding, and $\pi_{ik} = \pi_{jk} \circ \pi_{ij}$.
(4) If $i+1 < N$, then $M_i \models$ "E_i is a short extender," and if $\kappa = crit(E_i)$ and $j \leq i$ is maximal such that $j \leq_T i + 1$, then $V_{\kappa+1}^{M_j} = V_{\kappa+1}^{M_i}$, $M_{i+1} = \text{ult}(M_j; E_i)$, and π_{ji+1} is the canonical ultrapower embedding.

If $N < \omega$, then we say that $\mathscr{T}$ is *well-behaved* iff M_{N-1} is well-founded (i.e., transitive).

If $N = \omega$ and if $b \subset \omega$ is cofinal, then we say that b is an infinite branch through $\leq_T$ iff for all $i, j \in b$, if $i \leq j$, then $i \leq_T j$, and if $i \in b$ and $k \leq_T i$, then $k \in b$.

Lemma 10.73 *Let $n < \omega$, and let*

$$\mathscr{T} = ((M_i, \pi_{ij} : i \leq_T j \leq n), (E_i : i < n))$$

be a putative iteration tree on V such that for all $i < n$,

$$M_i \models \text{“}E_i \text{is certified”}.$$

Then $\mathscr{T}$ is well-behaved.

Proof For each $i < n$, ${}^{\omega}M_i \subset M_i$, and E_i is countably complete (from the point of view of V) by Lemmas 10.60 and 10.61. This implies that M_n is well-founded (i.e., transitive) by Corollary 10.65. □

The following is a key result on the "iterability" of V (cf. also Theorem 10.3) which will be used in the proof of Theorem 13.6.

Theorem 10.74 *Let*

$$\mathscr{T} = ((M_i, \pi_{ij} : i \leq_T j < \omega), (E_i : i < \omega), \leq_T)$$

be an iteration tree on V such that for all $i < \omega$,

$$M_i \models \text{“}E_i \text{ is certified.”}$$

Then there is some cofinal $b \subset \omega$, an infinite branch through $\leq_T$ such that the direct limit

$$\textit{dir lim } (M_i, \pi_{ij}: i \leq_T j \in b)$$

is well-founded.

Proof Suppose not. For each cofinal $b \subset \omega$, which is an infinite branch through $\leq_T$,[6] we may pick a sequence $(\alpha_n^b : n < \omega)$ witnessing that

$$\text{dir lim } (M_i, \pi_{ij}: i \leq_T j \in b)$$

is ill-founded, say

$$\pi_{i(n)i(m)}(\alpha_n^b) > \alpha_m^b \tag{10.43}$$

for all $n < m < \omega$, for some monotone $i: \omega \to \omega$ (which depends on b). Let

$$\sigma: \bar{V} \to_{\Sigma_{1002}} V$$

be such that $\bar{V}$ is transitive, $\text{Card}(\bar{V}) = 2^{\aleph_0}$, and $\{E_i : i < \omega\} \cup \{\alpha_n^b : n < \omega, b$ an infinite branch through $\leq_T\} \subset \text{ran}(\sigma)$. We let $U = (U, \leq_U)$ be the tree of attempts to find an infinite branch $b \subset \omega$ through $\leq_T$ together with a proof of the well-foundedness of the direct limit along b.

More precisely, U is defined as follows. Let us set $\bar{E}_i = \sigma^{-1}(E_i)$ for $i < \omega$. Obviously, there is a unique

$$\bar{\mathcal{T}} = ((\bar{M}_i, \bar{\pi}_{ij}: i \leq_T j < \omega), (\bar{E}_i : i < \omega), \leq_T) \tag{10.44}$$

such that

$$\bar{V} \models \text{“}\bar{\mathcal{T}} \text{ is a putative iteration tree on } V \text{ of length } \omega\text{”}.$$

Of course, $\bar{M}_0 = \bar{V}$. We now let $(\varphi, i) \in U$ iff $i < \omega$ and $\varphi: \bar{M}_i \to_{\Sigma_{1000}} V$ is such that $\varphi \circ \bar{\pi}_{0i} = \sigma$, and if $(\varphi, i), (\varphi', j) \in U$ then we write $(\varphi', j) \leq_U (\varphi, i)$ iff $i \leq_T j$ and $\varphi' \circ \bar{\pi}_{ij} = \varphi$.

Suppose $U = (U; \leq_U)$ to be ill-founded. It is straightforward to see that each witness $((\varphi_n, i_n): n < \omega)$ to the ill-foundedness of U gives rise to an infinite branch $b \subset \omega$ through $\leq_T$ together with an embedding

$$\tau: \text{dir lim } (\bar{M}_i, \bar{\pi}_{ij}: i \leq_T j < b) \to_{\Sigma_{1000}} V. \tag{10.45}$$

If $(\bar{M}_b, (\bar{\pi}_{ib}: i \in b))$ is the direct limit of $(\bar{M}_i, \bar{\pi}_{ij}: i \leq_T j < b)$, then τ is defined by

$$\tau(x) = \varphi_n((\bar{\pi}_{i_n b})^{-1}(x))$$

---

[6] If there is any. The current proof does not presuppose that there be some such branch. Rather, it will show the existence of some such b such that the direct limit along b is well-founded.

for some (all) large enough $n < \omega$. But as $\{\alpha_n^b : n < \omega\} \subset \text{ran}(\sigma)$, and hence if $(M_b, (\pi_{ib} : i \in b))$ is the direct limit of $(M_i, \pi_{ij} : i \leq_T j < b)$, then (10.43) yields that

$$\begin{aligned}
\bar{\pi}_{i(n)b}(\sigma^{-1}(\alpha_n^b)) &= \bar{\pi}_{i(m)b}(\bar{\pi}_{i(n)i(m)}(\sigma^{-1}(\alpha_n^b))) \\
&= \bar{\pi}_{i(m)b}(\pi_{i(n)i(m)}(\alpha_n^b)) \\
&> \pi_{i(m)b}(\alpha_m^b) \\
&= \bar{\pi}_{i(m)b}(\sigma^{-1}(\alpha_m^b))
\end{aligned}$$

for all $n < m < \omega$, so that $(\bar{\pi}_{i(n)b}(\sigma^{-1}(\alpha_n)) : n < \omega)$ witnesses that dir lim$(\bar{M}_i, \bar{\pi}_{ij} : i \leq_T j \in b)$ must be ill-founded. This contradicts the existence of τ, cf. (10.45).

We therefore must have that $U = (U; \leq_U)$ is well-founded. In order to finish the proof of the theorem, it now suffices to derive a contradiction.

In order to work towards a contradiction, we need generalized versions of U as well as "realizations" and "enlargements." Let R be a transitive model of ZC plus replacement for Σ_{1000}-formulae. We then call the triple (τ, Q, R) a *realization of* $\bar{M}_i$, where $i < \omega$, iff $\tau : \bar{M}_i \to_{\Sigma_{1000}} Q$, Q is a (not necessarily proper) rank initial segment of R, $Q \models$ "ZC plus replacement for Σ_{1000}-formulae," and ${}^{2^{\aleph_0}}Q \subset Q$ and ${}^{2^{\aleph_0}}R \subset R$. If (τ, Q, R) is a realization of $\bar{M}_i$, then we may define a tree

$$U(\tau \circ \bar{\pi}_{0i}, Q) = (U(\tau \circ \bar{\pi}_{0i}, Q), \leq_{U(\tau \circ \bar{\pi}_{0i}, Q)})$$

in the same fashion as U was defined above: we set $(\varphi, j) \in U(\tau \circ \bar{\pi}_{0i}, Q)$ iff $j < \omega$ and $\varphi : \bar{M}_j \to_{\Sigma_{1000}} Q$ is such that $\varphi \circ \bar{\pi}_{0j} = \tau \circ \bar{\pi}_{0i}$, and if $(\varphi, j), (\varphi', k) \in U(\tau \circ \bar{\pi}_{0i}, Q)$ then we write $(\varphi', k) \leq_{U(\tau \circ \bar{\pi}_{0i}, Q)} (\varphi, j)$ iff $j \leq_T k$ and $\varphi' \circ \bar{\pi}_{jk} = \varphi$. Hence $U = U(\sigma, V)$.

Let $X = \{\lambda : V_\lambda \prec_{\Sigma_{1002}} V \wedge \text{ran}(\sigma) \in V_\lambda \wedge {}^{2^{\aleph_0}}V_\lambda \subset V_\lambda\}$. Let $\alpha = \min(X)$. We have $\sigma : \bar{V} \to_{\Sigma_{1002}} V_\alpha \prec_{\Sigma_{1002}} V$ and $U(\sigma, V_\alpha)$ inherits the well-foundedness from $U = U(\sigma, V)$. We may thus write $\xi = ||(\sigma, 0)||_{U(\sigma, V_\alpha)}$, and we may let λ_0 be the ξth element of X. So there are (in order type) $\xi = ||(\sigma, 0)||_{U(\sigma, V_\alpha)}$ many $\beta < \lambda_0$ such that $(\sigma, V_\alpha, V_\beta)$ is a realization of $\bar{M}_0 = \bar{V}$ just by the choice of λ_0.

We shall now construct an "enlargement" sequence

$$(\sigma_i, Q_i, R_i : i < \omega)$$

such that for each $i < \omega$ the following holds true.

(a) (σ_i, Q_i, R_i) is a realization of $\bar{M}_i$,

(b) if $i \leq j$ and ν_i is the length of $\bar{E}_i$, then $V^{Q_i}_{\sigma_i(\nu_i)} = V^{Q_j}_{\sigma_j(\nu_i)}$ and $\sigma_i \restriction V^{\bar{M}_i}_{\nu_i} = \sigma_j \restriction V^{\bar{M}_j}_{\nu_i}$,

(c) if $U^* = U(\sigma_i \circ \bar{\pi}_{0i}, Q_i)$, then U^* is well-founded and there are (in order type) at least $||(\sigma_i, i)||_{U^*}$ many $\beta < R_i \cap OR$ such that $(\sigma_i, Q_i, V_\beta^{R_i})$ is a realization of $\bar{M}_i$, and
(d) if $i > 0$, then $R_{i+1} \in R_i$.

The last condition will give the desired contradiction.

To commence, we let $(\sigma_0, Q_0, R_0) = (\sigma, V_\alpha, V_{\lambda_0})$, where λ_0 and α are as above.

Now suppose $(\sigma_j, Q_j, R_j : j \leq i)$ to be constructed. Let $\Phi(\tau)$ abbreviate the following statement: "There is a realization (σ', Q, R) of $\bar{M}_{i+1}$ such that $V_{\tau(\nu_i)} = V^Q_{\tau(\nu_i)}$, $\sigma' \upharpoonright V^{\bar{M}_i}_{\nu_i} = \tau$, and if $U^* = U(\sigma' \circ \bar{\pi}_{0i+1}, Q)$, then U^* is well-founded and there are (in order type) at least $||(\sigma', i+1)||_{U^*}$ many $\beta < R \cap OR$ such that (σ', Q, V^R_β) is a realization of $\bar{M}_{i+1}$." We aim to verify that

$$R_i \models \Phi(\sigma_i \upharpoonright V^{\bar{M}_i}_{\nu_i}). \tag{10.46}$$

An inspection shows that $\Phi(\tau)$ is a Σ_{1+}-statement in the parameter $\bar{\mathcal{T}}$, cf. (10.44). Because

$$R_i \models \text{"}\sigma_i(\bar{E}_i) \text{ is certified"},$$

we may pick, working inside R_i and setting $\kappa = \text{crit}(\sigma_i(\bar{E}_i))$, some

$$U^* \prec_{\Sigma_{1+}} V^{R_i}_{\sigma_i(\nu_i)} \prec_{\Sigma_{1+}} R_i$$

of size $2^{\aleph_0}$ such that $\text{ran}(\sigma_i \upharpoonright V^{\bar{M}_i}_{\nu_i}) \in U^*$ and some $\tau : U^* \to_{\Sigma_{1+}} V^{R_i}_\kappa$ witnessing that $\sigma_i(\bar{E}_i)$ is $2^{\aleph_0}$-complete with respect to U^*. Notice that ${}^{2^{\aleph_0}}Q_i \subset Q_i \subset R_i$, and hence $\text{ran}(\sigma_i \upharpoonright V^{\bar{M}_i}_{\nu_i}) \in R_i$.

In order to verify (10.46) it then remains to verify that

$$V^{R_i}_\kappa \models \Phi(\tau \circ \sigma_i \upharpoonright V^{\bar{M}_i}_{\nu_i}).$$

Let $j \leq i$ be largest such that $j <_T i+1$. As τ witnesses that $\sigma_i(\bar{E}_i)$ is $2^{\aleph_0}$-complete with respect to U^*, by the proof of Lemma 10.64 we may define $\sigma' : \bar{M}_{i+1} \to_{\Sigma_{1000}} Q_j$ by setting

$$\bar{\pi}_{ji+1}(f)(a) \overset{\sigma'}{\mapsto} \sigma_j(f)(\tau(\sigma_i(a))), \tag{10.47}$$

where a is a finite subset of the length of $\bar{E}_i$ and writing $\bar{\kappa} = \text{crit}(\bar{E}_i)$, $f : [\bar{\kappa}]^{\text{Card}(a)} \to \bar{M}_j$, $f \in \bar{M}_j$. σ' is well-defined and Σ_{1000}-elementary by the following reasoning. Let φ be Σ_{1000}, and let a_ℓ, $\ell = 1, \ldots, k$, be finite subsets of the length of $\bar{E}_i$, and let f_ℓ, $\ell = 1, \ldots, k$, be such that $f_\ell : [\bar{\kappa}]^{\text{Card}(a_\ell)} \to \bar{M}_j$, $f_\ell \in \bar{M}_j$. Write $c = \bigcup_\ell a_\ell$. Then

$$\begin{aligned}
&\bar{M}_{i+1} \models \varphi(\bar{\pi}_{ji+1}(f_1)(a_1), \ldots, \bar{\pi}_{ji+1}(f_k)(a_k)) \\
&\iff \{u : \bar{M}_j \models \varphi(f_1^{a_1,c}(u), \ldots, f_k^{a_k,c}(u)\} \in (\bar{E}_i)_c \\
&\iff \sigma_i(\{u : \bar{M}_j \models \varphi(f_1^{a_1,c}(u), \ldots, f_k^{a_k,c}(u)\}) \in \sigma_i(\bar{E}_i)_{\sigma_i(c)} \\
&\overset{(*)}{\iff} \sigma_j(\{u : \bar{M}_j \models \varphi(f_1^{a_1,c}(u), \ldots, f_k^{a_k,c}(u)\}) \in \sigma_i(\bar{E}_i)_{\sigma_i(c)} \\
&\iff \{u : Q_j \models \varphi(\sigma_j(f_1^{a_1,c})(u), \ldots, \sigma_j(f_k^{a_k,c})(u)\} \in \sigma_i(\bar{E}_i)_{\sigma_i(c)} \\
&\overset{(**)}{\iff} Q_j \models \varphi(\sigma_j(f_1^{a_1,c})(\tau(\sigma_i(c))), \ldots, \sigma_j(f_k^{a_k,c})(\tau(\sigma_i(c))) \\
&\iff Q_j \models \varphi(\sigma_j(f_1)(\tau(\sigma_i(a_1))), \ldots, \sigma_j(f_k)(\tau(\sigma_i(a_k))).
\end{aligned}$$

Here, $(*)$ holds true as $\sigma_i \restriction V_{\nu_i}^{\bar{M}_i} = \sigma_j \restriction V_{\nu_i}^{\bar{M}_j}$, and $(**)$ holds true as τ witnesses that $\sigma_i(\bar{E}_i)$ is $2^{\aleph_0}$-complete with respect to U^*. (10.47) immediately yields that

$$\sigma' \circ \bar{\pi}_{0i+1} = \sigma' \circ \bar{\pi}_{ji+1} \circ \bar{\pi}_{0j} = \sigma_j \circ \bar{\pi}_{0j},$$

and hence $(\sigma', i+1) \in U(\sigma_j \circ \bar{\pi}_{0j}, Q_j)$, moreover, clearly,

$$\varepsilon = ||(\sigma', i+1)||_{U(\sigma_j \circ \bar{\pi}_{0j}, Q_j)} < ||(\sigma_j, j)||_{U(\sigma_j \circ \bar{\pi}_{0j}, Q_j)},$$

so that we may let Θ be the εth $\beta < R_j \cap OR$ such that $(\sigma_j, Q_j, V_\Theta^{R_j})$ is a realization of $\bar{M}_j$. As ${}^{2^{\aleph_0}}R_j \subset R_j$, the triple $(\sigma', Q_j, V_\Theta^{R_j}) \in R_j$ witnesses that

$$R_j \models \Phi(\tau \circ \sigma_i \restriction V_{\nu_i}^{\bar{M}_i}).$$

But this implies that

$$V_\kappa^{R_i} = V_\kappa^{R_j} \models \Phi(\tau \circ \sigma_i),$$

because $V_\kappa^{R_i} \prec_{\Sigma_{1+}} R_j$. This finishes the proof of Theorem 10.74. □

We end this section by defining the concept of Woodin cardinals. The following definition strengthens Definition 4.60.

Definition 10.75 Let κ be a cardinal, let $\delta > \kappa$ be a limit ordinal, and let $A \subset V_\delta$. We say that κ is *strong up to δ with respect to A* iff for all $\alpha < \delta$ there is some elementary embedding $\pi : V \to M$ such that M is transitive, $\text{crit}(\pi) = \kappa$, $V_\alpha \subset M$, and $\pi(A) \cap V_\alpha = A \cap V_\alpha$.

Definition 10.76 A cardinal δ is called a Woodin *cardinal* iff for every $A \subset V_\delta$ there is some $\kappa < \delta$ such that κ is strong up to δ with respect to A.

Lemma 10.77 *Let δ be a* Woodin *cardinal. Then δ is a Mahlo cardinal. In fact, for every $A \subset V_\delta$ there is a stationary subset S of δ such that for all $\kappa \in S$, for all*

$\alpha < \delta$ there is a certified extender F with critical point κ and $lh(F) \geq \alpha$ such that if $\pi_F: V \to \mathrm{Ult}(V; F)$ is the ultrapower map, then $\pi_F(A) \cap V_\alpha = A \cap V_\alpha$.[7]

Proof Fix $A \subset V_\delta$. Let $C \subset \delta$ be club in δ. Let $f: \delta \to C$ be the monotone enumeration of C, and let $g: \delta \to A$ be a surjection such that

$$g"\mu = A \cap V_\mu \text{ for every inaccessible cardinal } \mu \leq \delta. \tag{10.48}$$

Define $h: \delta \to V_\delta$ by setting

$$h(\xi) = \begin{cases} f(\lambda + n) & \text{if } \xi = \lambda + 2 \cdot n \text{ for some limit ordinal } \lambda \text{ and } n < \omega, \text{ and} \\ g(\lambda + n) & \text{if } \xi = \lambda + 2 \cdot n + 1 \text{ for some limit ordinal } \lambda \text{ and } n < \omega. \end{cases}$$

Let $\kappa < \delta$ be strong up to δ with respect to h.

Claim 10.78 *$h(\kappa) = f(\kappa) = \kappa$. In particular, $\kappa \in C$.*

Proof Suppose that $h(\kappa) > \kappa$, and let $\xi < \kappa$ be least such that ξ is even and $h(\xi) \geq \kappa$. Pick an elementary embedding $\pi: V \to M$, where M is an inner model, $\mathrm{crit}(\pi) = \kappa$, and

$$V_{h(\xi)+1} \subset M \text{ and } \pi(h) \cap V_{h(\xi)+1} = h \cap V_{h(\xi)+1}. \tag{10.49}$$

By elementarity, $\pi(h)(\xi) = \pi(h(\xi)) \geq \pi(\kappa) > \kappa$, whereas on the other hand $\pi(h)(\xi) = h(\xi)$ by (10.49). Contradiction! □

As C is arbitrary, Claim 10.78 proves that $\{\mu < \delta: \mu \text{is strong up to } \delta\}$ is stationary in δ, so that in particular δ is a Mahlo cardinal. Again as C is arbitrary, in order to finish off the proof of Lemma 10.77 it suffices to verify the following.

Claim 10.79 *For all $\alpha < \delta$ there is a certified extender F with $\mathrm{crit}(F) = \kappa$ and $\mathrm{lh}(F) \geq \alpha$ such that if $\pi_F: V \to \mathrm{ult}(V; F)$ is the ultrapower map, then $\pi_F(A) \cap V_\alpha = A \cap V_\alpha$.*

Proof Fix $\alpha < \delta$. As δ is a Mahlo cardinal, we may pick some inaccessible cardinal ν with $\max(\kappa, \alpha) < \nu < \delta$. As δ is a Woodin cardinal, there is some elementary embedding $\pi: V \to M$, where M is an inner model, $\mathrm{crit}(\pi) = \kappa$, $V_\nu \subset M$, and $h \cap V_\nu = \pi(h) \cap V_\nu$. Let F be the (κ, ν)-extender over V derived from π. By Lemmas 10.56, 10.58, and 10.71, F is certified, $(V_\nu)^{\mathrm{ult}(V;F)} = (V_\nu)^M = V_\nu$, and there is an elementary embedding $k: \mathrm{ult}(V; F) \to M$ with $\mathrm{crit}(k) \geq \nu$ and hence $k \restriction V_\nu = \mathrm{id}$. In particular,

$$\pi_F(h) \cap V_\nu = h \cap V_\nu. \tag{10.50}$$

Now we have that for $x \in V_\nu$, $x \in \pi(A)$ iff $x \in \pi_F(h)"\{\lambda + 2n + 1: \lambda < \nu \text{ a limit and } n < \omega\}$, by (10.48) and the elementarity of π_F together with the

[7] The fact that F is certified implies that $V_\alpha \subset \mathrm{Ult}(V; F)$.

choice of h, iff $x \in h''\{\lambda+2n+1 : \lambda < \nu$ a limit and $n < \omega\}$, by (10.50), iff $x \in A$, by the choice of h.

We have shown that $\pi_F(A) \cap V_\nu = A \cap V_\nu$. □

Definition 10.80 Let M be an inner model, let $\alpha < \beta$, and let $\{x_0, \cdots, x_{k-1}\} \subset (V_\beta)^M$. Then we write

$$\mathrm{type}^M(V_\beta; \in, V_\alpha, \{x_0, \cdots, x_{k-1}\})$$

for the type of $\{x_0, \cdots, x_{k-1}\}$ in $(V_\beta)^M$ with respect to the first order language of set theory with parameters in $(V_\alpha)^M$, i.e., for

$$\{\varphi(\mathbf{x}, v_0, \cdots, v_{k-1}) : \mathbf{x} \in (V_\alpha)^M \wedge (V_\beta)^M \models \varphi(\mathbf{x}, x_0, \cdots, x_{k-1})\}.$$

The following is an immediate consequence of Lemma 10.77.

Lemma 10.81 *Let δ be a* WOODIN *cardinal. Then for all $\beta > \delta$ and for all $\{x_0, \ldots, x_{k-1}\} \subset V_\beta$ there is a stationary subset S of δ such that for all $\kappa \in S$, κ is strong up to δ with respect to*

$$\mathrm{type}^V(V_\beta; \in, V_\delta, \{x_0, \cdots, x_{k-1}\}),$$

in fact for all $\alpha < \delta$ there is a certified extender F with critical point κ and $lh(F) \geq \alpha$ such that if $\pi : V \to \mathrm{Ult}(V; F)$, then

$$\begin{aligned}&\mathrm{type}^{\mathrm{Ult}(V;F)}(V_{\pi(\beta)}; \in, V_\alpha, \{\pi(x_0), \cdots, \pi(x_{k-1})\})\\ &= \mathrm{type}^V(V_\beta; \in, V_\alpha, \{x_0, \cdots, x_{k-1}\}).\end{aligned}$$

10.5 Problems

10.1. Prove Lemma 10.27, using the method from the proof of Lemma 10.29.

10.2. Let κ be a measurable cardinal, let U be a measure on κ, and let $(\mathcal{M}_i, \pi_{i,j} : i \leq j < \infty)$ be the iteration of $V = \mathcal{M}_0$ of length OR given by U. Let $\kappa_i = \pi_{0i}(\kappa)$ for $i < \infty$.
(a) Show by induction on $i \in$ OR that for all $x \in \mathcal{M}_i$ there are $k < \omega$, $i_1, \ldots, i_k < i$, and a function $f : [\kappa_0]^k \to \mathcal{M}_0$, $f \in \mathcal{M}_0$, such that $x = \pi_{0,i}(f)(\kappa_{i_1}, \ldots, \kappa_{i_k})$.
(b) Show that $\{\kappa_i : i \in \mathrm{OR}\}$ is club in OR.
(c) Let $i_0 > 0$, and let $X \in \mathscr{P}(\kappa_{i_0}) \cap \mathcal{M}_{i_0}$. Show that $X \in \pi_{0,i_0}(U)$ iff there is some $k < i_0$ such that $\{\kappa_i : k \leq i < i_0\} \subset X$. (Cf. Lemma 10.9.)
(d) Conclude that if $\lambda > 2^\kappa$ is a regular cardinal, then $\pi_{0,\mu}(U) = F_\lambda \cap \mathcal{M}_\mu$ (where F_λ is the club filter on λ, cf. Lemma 4.25).

10.3. Let κ be a measurable cardinal, let U be a measure on κ. (a) Show that, setting $\bar{U} = U \cap L[U]$, $L[U] = L[\bar{U}]$ and $L[U] \models$ "$\bar{U}$ is a measure on κ."
Let $(\mathcal{M}_i, \pi_{i,j}: i \leq j < \infty)$ be the iteration of $L[U]$ of length OR given by $\bar{U}$. (b) Show that if $\lambda > 2^\kappa$ is a regular cardinal, then $\mathcal{M}_\mu = L[F_\lambda]$, where F_λ is the club filter on μ. [Hint. Problem 10.2 (d).]

10.4. Let us call a J-structure $J_\alpha[U]$ an L^μ-*premouse* iff there is some $\kappa < \alpha$ such that $U \in J_\alpha[U]$ and $J_\alpha[U] \models$ "ZFC$^-$ + U is a measure on κ." If $M = J_\alpha[U]$ is an L^μ-premouse, then we may define (putative) iterations of M in much the same way as putative iterations of V, cf. Definition 10.1. In the spirit of Definition 10.26, an L^μ-premouse is called an L^μ-*mouse* iff every putative iteration of M is an iteration.
(a) Let $M = J_\alpha[U]$ be an L^μ-mouse and let $(\mathcal{M}_i, \pi_{i,j}: i \leq j < \infty)$ be the iteration of M of length OR. Show that there is some λ and some β such that $M_\lambda = J_\beta[F_\lambda]$. [Hint. Problem 10.3 (b).] Conclude that any two L^μ-mice M, N may be "coiterated," i.e., there are iterates M_λ of M and N_λ of N, respectively, such that $M_\lambda = J_\beta[F]$ and $N_\lambda = J_{\beta'}[F]$ for some β, β', F. Show also that if $M \models$ "U is a measure on κ," then for all i, $M = \mathcal{M}_0$ and $\mathcal{M}_i$ have the same subsets of κ.
(b) Let $\sigma: J_{\bar{\alpha}}[\bar{U}] \to J_\alpha[U]$ be an elementary embedding, where $J_{\bar{\alpha}}[\bar{U}]$ is an L^μ-premouse and $J_\alpha[U]$ is an L^μ-mouse. Show that $J_{\bar{\alpha}}[\bar{U}]$ is an L^μ-mouse also. [Hint. Cf. the proof of Lemma 10.33.]

10.5. Show in ZF + there is a measurable cardinal that there is an inner model M such that $M \models$ "GCH + there is a measurable cardinal." [Hint. Let U witness that κ is measurable, and set $M = L[U]$. To show that M satisfies GCH, verify that in M, for each infinite cardinal ν and every $X \subset \nu$ there are at most ν many $Y \subset \nu$ with $Y <_M X$ (with $<_M$ being as on p. 77), as follows. Fix X. Pick $\sigma: J_{\bar{\alpha}}[\bar{U}] \to J_\alpha[U]$ such that Card$(J_{\bar{\alpha}}[\bar{U}]) = \nu$, $\sigma \upharpoonright (\nu + 1) = \text{id}$, $X \in \text{ran}(\pi)$ and $J_{\bar{\alpha}}[\bar{U}]$ is an L^μ-premouse. We claim that if $Y \subset \nu$ with $Y <_W X$, then $Y \in J_{\bar{\alpha}}[\bar{U}]$. For this, use Theorem 10.3 and Problem 10.4.]

10.6. Prove Łoś's Theorem 10.22. Show also Lemma 10.21 (d).

10.7. Show that $\{\text{Th}_{\Sigma_1}(x^\#)\}$ is $\Pi^1_2(x)$, where $\text{Th}_{\Sigma_1}(x^\#)$ is the set of all GÖDEL numbers of Σ_1-sentences which hold true in $x^\#$.

10.8. Let $A \subset {}^\omega\omega$ be $\Sigma^1_2(x)$, $A \neq \emptyset$. Show that $A \cap x^\# \neq \emptyset$. [Hint. Corollary 7.21.]

10.9. Let α be any ordinal. A cardinal κ is called α-ERDÖS iff for every $F: [\kappa]^{<\omega} \to 2$ there is some $X \subset \kappa$ with otp$(X) = \alpha$ such that for every $n < \omega$, $F \upharpoonright [X]^n$ is constant.
Show that if κ is ω_1-Erdös, then $x^\#$ exists for every $x \subset \omega$.
Show also that if $\alpha < \omega_1^L$ and if κ is α-Erdös, then $L \models$ "κ is α-Erdös."
Let $x \subset \omega$, and let $\mathcal{M} = (J_\alpha[x]; \in, U)$ be an x-pm. Let κ be the critical point of U, and let us assume that $\alpha = \kappa^{+L[x]}$. We define the (Σ_0-)*ultrapower*, written ult$_0(L[x]; U)$ or just ult$(L[x]; U)$, of $L[x]$ as follows. For $f, g \in$

${}^{\kappa}J_\alpha[x] \cap L[x]$, set $f \sim g$ iff $\{\xi < \kappa : f(\xi) = g(\xi)\} \in U$, and write $f \tilde{\in} g$ iff $\{\xi < \kappa : f(\xi) \in g(\xi)\} \in U$. We let $[f]$ denote the $\sim$-equivalence class of f. We write $[f] \tilde{\in} [g]$ iff $f \tilde{\in} g$, and we also write $[f] \in \tilde{U}$ iff $\{\xi < \kappa : f(\xi) \in U\} \in U$. We let

$$\text{ult}(L[x]; U) = \{\{[f] : f \in {}^{\kappa}J_\alpha[x] \cap J_\alpha[x]\}; \tilde{\in}, \tilde{U}\}.$$

We may define a natural map $\pi_U^{L[x]} : L[x] \to \text{ult}(L[x]; U)$ by setting $\pi_U^{L[x]}(z) = [c_z]$, where $c_z(\xi) = x$ for all $\xi < \kappa$. $\pi_U^{L[x]}$ is called the (Σ_0-)*ultrapower map*. If $\text{ult}(L[x]; U)$ is well-founded, then we identify it with its transitive collapse, which will be equal to $L[x]$, and we identify $[f]$ with the image under the transitive collapse, and we idenitify π with the composition of π with the transitive collapse.

For $\gamma \in \text{OR}$, we may now define the putative iteration

$$((W_i : i \leq \gamma), (\tilde{\pi}_{ij} : i \leq j \leq \gamma)) \tag{10.51}$$

of $W_0 = L[x]$ by U and its images in much the same way as in Definition 10.24, with $L[x]$ and U playing the role of $\mathcal{M}$, cf. also Definition 10.1. If $i < \gamma$, then $W_i = L[x]$, and if W_γ is transitive, then also $W_\gamma = L[x]$.

10.10. Let x, $\mathcal{M} = (J_\alpha[x]; \in, U)$, and κ be as above. In particular, $\alpha = \kappa^{+L[x]}$. Let $\pi_U^{\mathcal{M}}$ be as in Definition 10.20.
(a) Show that $\pi_U^{\mathcal{M}} = \pi_U^{L[x]} \upharpoonright J_\alpha[x]$.
(b) Suppose that $\text{ult}(L[x]; U)$ is transitive. Show that $\text{ult}_0(\mathcal{M})$ is then also transitive, and if $\text{ult}_0(\mathcal{M}) = (J_{\alpha'}[x]; \in, U')$, then $J_{\alpha'}[x] = \pi_U^{L[x]}(J_\alpha[x])$.
Let $\gamma \in \text{OR}$, and let W_i and $\tilde{\pi}_{ij}$ be as in (10.51). Let $\mathcal{M}_i$ and π_{ij} be as in Definition 10.24 for a putative iteration of length $\gamma + 1$, say $\mathcal{M}_i = (J_{\alpha_i}[x]; \in, U_i)$ for $i < \gamma$.
(c) Show that if $i \leq j < \gamma$, then $J_{\alpha_i}[x] = \tilde{\pi}_{0i}(J_\alpha[x])$ and $\pi_{ij} = \tilde{\pi}_{ij} \upharpoonright J_{\alpha_i}[x]$.
(d) Show that if $\text{ult}(L[x]; U)$ is transitive and ξ is any limit ordinal with $\text{cf}(\xi) \neq \text{cf}(\kappa)$, then $\pi_U^{L[x]}$ is continuous at ξ, i.e., $\pi_U^{L[x]}(\xi) = \sup_{\eta<\xi} \pi_U^{L[x]}$"$\eta$ (cf. Lemma 4.52 (c)). Conclude that if ξ is a strong limit cardinal with $\text{cf}(\xi) \geq (2^{\text{Card}(\kappa)})^+$ and if $\gamma \leq (2^{\text{Card}(\kappa)})^+$, then $\tilde{\pi}_{0i}(\xi) = \xi$ for every $i < \gamma$.

10.11. Let $x \subset \omega$, and suppose that $x^\#$ exists. Let $A \in \mathcal{P}(\omega_1^V) \cap L[x]$. Show that either A or $\omega_1^V \setminus A$ contains a club of $L[x]$-inaccessibles. [Hint. Consider the countable SILVER indiscernibles, and exploit the arguments for Lemmas 10.9 and 10.35.]

10.12. Assume that $0^\#$ exists. Show that for every $\beta < \omega_1^L$ there is some premouse $(J_\alpha; \in, U) \in L$ such that there is a putative iteration of $(J_\alpha; \in, U)$ of length $\alpha + 1$ of which the αth model is ill-founded.

10.13. Let $x \subset \omega$, and suppose that $x^\#$ exists. Show that there is some $G \in V$ (!) such that G is $\text{Col}(\omega, < \omega_1^V)$-generic over $L[x]$. [Hint. Recursively construct initial segments of G along a club of $L[x]$-inaccessibles from Problem 10.11, exploiting the Product Lemma 6.65. As limit stages, use Lemma 6.44.]

A cardinal is called *remarkable* iff for every $\alpha > \kappa$ there are $\mu < \beta < \kappa$ such that if G is $\mathrm{Col}(\omega, <\kappa)$-generic over V, then in $V[G]$ there is an elementary embedding $\sigma\colon V_\beta \to V_\alpha$ such that $\mathrm{crit}(\sigma) = \mu$ and $\sigma(\mu) = \kappa$. (Here, V_α and V_β refer to the respective rank initial segments of V rather than $V[G]$. Compare Problems 4.29 and 10.21.)

10.14. Show that if κ is remarkable and if G is $\mathrm{Col}(\omega, <\kappa)$-generic over V, then in $V[G]$ for every $\alpha > \kappa$ the set

$$\{X \in [V_\alpha]^{\aleph_0} : X \prec V_\alpha, X \cap \kappa \in \kappa \text{ , and } \exists\beta\exists\sigma\ V_\beta \overset{\sigma}{\cong} X\}$$

is stationary in $[V_\alpha]^{\aleph_0}$.

Show that if $0^\#$ exists, then every SILVER indiscernible is remarkable in L. Show also that if κ is remarkable in V, then κ is remarkable in L. [Hint. Problem 7.4.]

10.15. Show that if κ is ω-ERDŐS, then there are $\alpha < \beta < \kappa$ such that $V_\beta \models$ "ZFC $+\, \alpha$ is remarkable." Show also that every remarkable cardinal is ineffable.

10.16. **(Martin-Solovay)** We say that V is closed under sharps iff for all α, $\Vdash_V^{\mathrm{Col}(\omega,\alpha)}$ "$x^\#$ exists for all $x \subset \omega$." Let α be any ordinal, and let G be $\mathrm{Col}(\omega, \alpha)$-generic over V. Let $z \in {}^\omega\omega \cap V$, and let $A \in V[G]$ be such that $V[G] \models$ "$A \subset {}^\omega\omega$ is $\Sigma^1_3(z)$, $A \neq \emptyset$." Show that $A \cap V \neq \emptyset$. (Compare Corollary 7.21.) [Hint. For any $X \in V$, we may make sense of $X^\#$. Let $A = \{x : \exists y\, \varphi(x, y, z)\}$, where φ is Π^1_2. Let T be a tree of attempts to find x, y, σ, $\bar{H}$, and g such that $\sigma\colon \bar{H}^\# \to (H_\theta)^\#$ is an elementary embedding, $\bar{H}$ is countable, $z \in \bar{H}$, g is $\mathbb{Q}$-generic over $\bar{H}$ for some $\mathbb{Q} \in \bar{H}$, and $\bar{H}^\#[g] \models \varphi(x, y, z)$.]

10.17. Let $E = (E_a : a \in [\nu]^{<\omega})$ be a (κ, ν)-extender over V. Show that $\mathrm{ult}(V; E)$ is well-founded iff E is ω-complete.

10.18. Let E be a *short* (κ, ν)-extender over V.
(1) If α is a limit ordinal with $\mathrm{cf}(\alpha) \neq \kappa$, then π_E is continuous at α. (Compare Lemma 4.52 (c).)
(2) If $\lambda > \nu$ is a cardinal such that $\mathrm{cf}(\lambda) \neq \kappa$ and $\mu^\kappa < \lambda$ for every $\mu < \lambda$, then λ is a fixed point of π_E, i.e., $\pi_E(\lambda) = \lambda$. (Compare Problem 4.28.)

10.19. Let $E = (E_a : a \in [\nu]^{<\omega})$ be a (κ, ν) extender over V. Let $\mathbb{P} \in V_\kappa$ be a poset, and let G be $\mathbb{P}$-generic over V. Set

$$E^*_a = \{Y \subset [\kappa]^{\mathrm{Card}(a)} : \exists X \in E_a\ Y \supset X\},$$

as defined in $V[G]$. Show that $(E^*_a : a \in [\nu]^{<\omega})$ is a (κ, ν) extender over $V[G]$. Conclude that "κ is a strong cardinal" and "κ is supercompact" are preserved by small forcing in the sense of Problem 6.18.

10.20. Show that "κ is a WOODIN cardinal" is preserved by small forcing in the sense of Problem 6.18.

10.21. **(Magidor)** Show that the conclusion of problem 4.29 yields that κ is supercompact, i.e., if for every $\alpha > \kappa$ there are $\mu < \beta < \kappa$ together with an elementary embedding $\sigma: V_\beta \to V_\alpha$ such that $\text{crit}(\sigma) = \mu$ and $\sigma(\mu) = \kappa$, then κ is supercompact. [Hint: Let $\lambda \geq \kappa$ be least such that there is no (κ, ν)-extender over V witnessing that κ is λ-supercompact. Pick $\mu < \beta < \kappa \leq \lambda < \alpha$ together with some elementary embedding $\sigma: V_\beta \to V_\alpha$ such that $\text{crit}(\sigma) = \mu$ and $\sigma(\mu) = \kappa$. Then $\lambda \in \text{ran}(\sigma)$ and one can derive from σ a (μ, ν)-extender $F \in V_\beta$ over V witnessing that μ is $\sigma^{-1}(\lambda)$-supercompact in V_β. Lift this statement up via σ.]

10.22. Show that the conclusion of problem 4.30 yields that κ is supercompact. [Hint. Design an ultrapower construction à la Theorem 10.48.]

10.23. Show that if κ is subcompact, then κ is a WOODIN cardinal.

10.24. (Supercompact tree PRIKRY forcing) Let κ be supercompact, and let $\lambda > \kappa$. Let M be an inner model with ${}^\lambda M \subset M$, and let $\pi: V \to M$ be an elementary embedding with critical point κ such that $\pi(\kappa) > \lambda$. Let U be derived from π as in Problem 4.30. Let $\mathbb{P}$ be the set of all trees T on $\mathscr{P}_\kappa(\lambda)$ (in the sense of the definition given on p. 123) such that there is some (stem) $s \in T$ such that $t \subset s \vee s \subset t$ for all $t \in T$ and for all $t \supset s, t \in T$,

$$\{x \in \mathscr{P}_\kappa(\lambda): t^\frown x \in T\} \in U.$$

$\mathbb{P}$, ordered by $U \leq_{\mathbb{P}} T$ iff $U \subset T$, is called *supercompact tree* PRIKRY *forcing*. Let G be $\mathbb{P}$-generic over V. Show that $\text{cf}^{V[G]}(\delta) = \omega$ for every $\delta \in [\kappa, \lambda]$ with $\text{cf}^V(\delta) \geq \kappa$. Show also that the PRIKRY-Lemma 10.7 holds true for the supercompact tree PRIKRY forcing $\mathbb{P}$ and conclude that V and $V[G]$ have the same V_κ.

10.25. Let E, E' be certified extenders on κ. We define $E <_M E'$ iff $E \in \text{ult}(V; E')$. Show that $<_M$ is well-founded. (Compare Problem 4.27.) [Hint. Use Theorem 10.74.] Again, $<_{\mathrm{M}}$ is called the MITCHELL *order* (this time on certified extenders).

Chapter 11
$0^\#$ and Jensen's Covering Lemma

11.1 Fine Structure Theory

Definition 11.1 Let $M = J_\alpha[E]$ be a J-structure. Then M is called *acceptable* iff for all limit ordinals $\beta < \alpha$ and for all $\delta \leq \beta$, if

$$(\mathscr{P}(\delta) \cap J_{\beta+\omega}[E]) \setminus J_\beta[E] \neq \emptyset,$$

then there is some $f \in J_{\beta+\omega}[E]$ such that $f\colon \delta \to \beta$ is surjective.

Acceptability is a strong "local" form of GCH. If $M = J_\alpha[E]$ is a J-structure and if $\kappa \in M$, then we write

$$\kappa^{+M} = \sup\{\alpha + \omega\colon \alpha \in M \wedge \exists f \in M\ (f\colon \kappa \to \alpha \text{ is surjective})\}.$$

Lemma 11.2 *Let $M = J_\alpha[E]$ be an acceptable J-structure. Let $\omega \leq \kappa \in M$, and set $\tau = \kappa^{+M}$. Then $\mathscr{P}(\kappa) \cap M \subset J_\tau[E]$. Moreover, τ is in fact the least γ with $\mathscr{P}(\kappa) \cap M \subset J_\gamma[E]$.*

Proof That $\mathscr{P}(\kappa) \cap M \subset J_\tau[E]$ follows immediately from Definition 11.1. Now suppose that there were some $\gamma < \tau$ with $\mathscr{P}(\kappa) \cap M \subset J_\gamma[E]$. As $\mathrm{Card}(\gamma) \leq \kappa$ in M, Lemma 10.17 produces some surjective $f\colon \kappa \to J_\gamma[E]$, $f \in M$. Then $A = \{\xi < \kappa\colon \xi \notin f(\xi)\} \in M$, but $A \notin J_\gamma[E]$ as in the proof of Theorem 1.3. Contradiction! □

If $M = J_\alpha[E]$ is a J-structure and ρ is a cardinal of M (or $\rho = \alpha$), then by $(H_\rho)^M$ we mean the set of all sets which are hereditarily smaller than ρ in M (or $(H_\rho)^M = M$ in case $\rho = \alpha$). Recall that for all $x \in M$, $\mathrm{TC}(\{x\}) \in M$ (cf. Corollary 5.18), so that this makes sense.

R. Schindler, *Set Theory*, Universitext, DOI: 10.1007/978-3-319-06725-4_11,

Lemma 11.3 *Let $M = J_\alpha[E]$ be an acceptable J-structure. If ρ is an infinite cardinal of M (or $\rho = \alpha$), then*

$$(H_\rho)^M = J_\rho[E].$$

Proof It suffices to prove that if $\omega \leq \kappa \in M$ and $\tau = \kappa^{+M}$, then $(H_\tau)^M = J_\tau[E]$. That $J_\tau[E] \subset (H_\tau)^M$ follows from Lemma 5.16. Let us prove $(H_\tau)^M \subset J_\tau[E]$.

Suppose not, and let x be $\in$-minimal in $(H_\tau)^M \setminus J_\tau[E]$. Then $x \subset J_\tau[E]$, and there is some surjection $g: \kappa \to x$, $g \in M$. For $\xi < \kappa$, let $\beta_\xi < \tau$ be least such that $g(\xi) \in J_{\beta_\xi}[E]$. By (the proof of) Lemma 4.15, $\beta = \sup(\{\beta_\xi : \xi < \kappa\}) < \tau$.

By Lemma 11.2, there is some $\gamma < \tau$, $\beta \leq \gamma$ such that

$$(\mathscr{P}(\kappa) \cap J_{\gamma+\omega}[E]) \setminus J_\gamma[E] \neq \emptyset,$$

which by acceptability (and Lemma 10.17) yields some surjective $f: \kappa \to J_\gamma[E]$, $f \in J_{\gamma+\omega}[E] \subset J_\tau[E]$. But now $f^{-1''}x \in \mathscr{P}(\kappa) \cap J_\tau[E]$ by Lemma 11.2, and hence $x = f''(f^{-1''}x) \in J_\tau[E]$. Contradiction! □

The following definition introduces a key concept of the fine structure theory.

Definition 11.4 The Σ_1-*projectum* (or, *first projectum*) $\rho_1(M)$ of an acceptable J-structure $M = J_\alpha[E]$ is defined by

$$\rho_1(M) = \text{ the least } \rho \in \mathrm{OR} \text{ such that } \mathscr{P}(\rho) \cap \underset{\sim}{\Sigma}{}_1^M \not\subset M.$$

Lemma 11.5 *Let $M = J_\alpha[E]$ be an acceptable J-structure. If $\rho_1(M) \in M$, then $\rho_1(M)$ is a cardinal in M. In fact, $\rho_1(M)$ is a Σ_1-cardinal in M, i.e., there is no $\underset{\sim}{\Sigma}{}_1^M$ partial map from some $\gamma < \rho_1(M)$ onto $\rho_1(M)$.*

Proof Write $\rho = \rho_1(M)$. Let us first show that ρ is a cardinal in M. Suppose not, and let $f \in M$ be such that $f: \gamma \to \rho$ is surjective for some $\gamma < \rho$. Let $A \in \mathscr{P}(\rho) \cap \underset{\sim}{\Sigma}{}_1^M$ be such that $A \notin M$. Let $\bar{A} = f^{-1''}A$. Then $\bar{A} \notin M$, since otherwise $A = f''\bar{A} \in M$. On the other hand, $\bar{A} \in M$ by the definition of ρ, since $\bar{A} \subset \gamma$ and $\bar{A} \in \underset{\sim}{\Sigma}{}_1^M$. Contradiction!

Let us now show that ρ is in fact a Σ_1-cardinal in M. Suppose not, and let $f: \gamma \to \rho$ be a possibly partial function from γ onto ρ, $f \in \underset{\sim}{\Sigma}{}_1^M$. We know that there is a $\underset{\sim}{\Sigma}{}_1^{J_\rho[E]}$ map from ρ onto $J_\rho[E]$ (cf. Lemma 10.17). Hence there is a $\underset{\sim}{\Sigma}{}_1^M$ map $g: \gamma \to J_\rho[E]$ which is surjective. Set

$$A = \{\xi \in \gamma : \xi \notin g(\xi)\}.$$

Then A is clearly in $\mathscr{P}(\gamma) \cap \underset{\sim}{\Sigma}{}_1^M$, and $A \notin J_\rho[E]$ by the proof of Theorem 1.3. We get that $A \notin M$ by Lemma 11.2. But $\gamma < \rho$, so that we must have that $A \in M$ by the definition of ρ. Contradiction! □

The following is an immediate consequence of Lemmas 11.5 and 11.3.

Corollary 11.6 *Let $M = J_\alpha[E]$ be an acceptable J-structure, and let $\rho = \rho_1(M)$.*

(a) $(H_\rho)^M = J_\rho[E]$.
(b) *If $A \subset J_\rho[E]$ is $\underset{\sim}{\Sigma}{}_1^M$, then $(J_\rho[E], A)$ is amenable.*

Recall our enumeration $(\varphi_n\colon n < \omega)$ of all Σ_1 formulae from p. 185. In what follows it will often be convenient to pretend that a given φ_n has fewer free variables than it actually has. E.g., we may always contract free variables into one as follows: if $\varphi_n \equiv \varphi_n(v_{i_1}, \ldots, v_{i_\ell})$ with all free variables shown, then we may identify, for the purposes to follow, φ_n with

$$\exists v_{i_1} \ldots \exists v_{i_\ell}(u = (v_{i_1}, \ldots, v_{i_\ell}) \wedge \varphi_n(v_{i_1}, \ldots, v_{i_\ell})).$$

If $\mathsf{a}\colon v(n) \to M$ assigns values to the free variable(s) $v_{i_1}, \ldots, v_{i_\ell}$ of φ_n then, setting $x_1 = \mathsf{a}(v_{i_1})$, ..., $x_\ell = \mathsf{a}(v_{i_\ell})$, we shall in what follows use the more suggestive $M \models \varphi_i(x_1, \ldots, x_\ell)$ rather than the notation $M \models \varphi_i[\mathsf{a}]$ from p. 185. We shall also write $h_M(i, \mathbf{x})$ instead of $h_M(i, \mathsf{a})$, where $\mathbf{x} = (x_1, \ldots, x_\ell)$.

Definition 11.7 Let $M = (J_\alpha[E], B)$ be an acceptable J-structure, write $\rho = \rho_1(M)$, and let $p \in M$. We define

$$A_M^p = \{(n, x) \in \omega \times (H_\rho)^M\colon M \models \varphi_n(x, p)\}.$$

A_M^p is called the *standard code* determined by p. The structure

$$M^p = (J_\rho[E], A_M^p)$$

is called the *reduct* determined by p.

We shall often write $A_M^p(n, x)$ instead of $(n, x) \in A_M^p$, and we shall write A^p rather than A_M^p if there is no danger of confusion.

Definition 11.8 Let M be an acceptable structure, and write $\rho = \rho_1(M)$.

$$P_M = \text{the set of all } p \in [\rho, \mathrm{OR} \cap M)^{<\omega} \text{ for which}$$
$$\text{there is a } B \in \Sigma_1^M(\{p\}) \text{ such that } B \cap \rho \notin M.$$

The elements of P_M are called *good parameters*.

Lemma 11.9 *Let M be an acceptable J-structure, $p \in [\rho_1(M), \mathrm{OR} \cap M)^{<\omega}$, and $A = A_M^p$. Then*

$$p \in P_M \iff A \cap (\omega \times \rho_1(M)) \notin M.$$

Proof "$\Longrightarrow$": Pick some B which witnesses that $p \in P_M$. Suppose B is defined by φ_n, i.e., $\xi \in B \iff (n, \xi) \in A$. As $B \cap \rho_1(M)$ is not in M, $A \cap (\omega \times \rho_1(M))$ can then not be in M either.

"$\Longleftarrow$": Suppose $A \cap (\omega \times \rho_1(M)) \notin M$. Let $f\colon \omega \times \rho_1(M) \longrightarrow \rho_1(M)$ be defined by $f(n, \lambda + i) = \lambda + 2^n \cdot 3^i$, where $n, i < \omega$ and $\lambda < \alpha$ is a limit ordinal. Clearly f is Σ_1^M, and if $\rho_1(M) \in M$ then $f \in M$. Let $B = f''A$. Then B is $\Sigma_1^M(\{p\})$, $A \cap (\omega \times \rho_1(M)) = f^{-1''}(B \cap \rho_1(M))$, and it is easy to see that $B \cap \rho_1(M) \notin M$. □

Definition 11.10 Let M be an acceptable J-structure, and write $\rho = \rho_1(M)$. We set

$$R_M = \text{ the set of all } r \in [\rho, \mathrm{OR} \cap M)^{<\omega} \text{ such that } h_M(\rho \cup \{r\}) = M.$$

The elements of R_M are called *very good parameters*.

Lemma 11.11 *Let M be an acceptable J-structure. $R_M \subset P_M \neq \emptyset$.*

Proof That $P_M \neq \emptyset$ easily follows from $h_M(\mathrm{OR} \cap M) = M$, cf. the proof of Lemma 10.17.

As to $R_M \subset P_M$, let $p \in R_M$, and define $A \subset \omega \times \mathrm{OR} \cap M$ by

$$(n, \xi) \in A \iff (n, \xi) \notin h_M(n, (\xi, p)).$$

We have that A is $\Sigma_1^M(\{p\})$, and $A \cap \omega \times \rho_1(M) \notin M$ by a diagonal argument. Using the map f from the proof of Lemma 11.9 it is easy to turn the set A into some $B \subset \mathrm{OR} \cap M$ such that B is $\Sigma_1^M(\{p\})$ and $B \cap \rho_1(M) \notin M$. □

It is not hard to see that there is a computable map $e\colon \omega \to \omega$ such that for all $n < \omega$, for all acceptable J-structures M, for all $p \in M$, and for all $m_1, \ldots, m_k < \omega$ and $x_1, \ldots, x_k \in M^p$,

$$\begin{aligned} M \models \varphi_n(h_M(m_1, (x_1, p)), \ldots, h_M(m_k, (x_k, p))) &\iff \\ M \models \varphi_{e(n)}(((m_1, x_1), \ldots, (m_k, x_k)), p) &\iff \\ (e(n), ((m_1, x_1), \ldots, (m_k, x_k))) \in A_M^p. \end{aligned} \tag{11.1}$$

If M is an acceptable J-structure and $p \in R_M$, then we may express in a uniform Π_1 fashion over M^p that A_M^p codes à la (11.1) the Σ_1-theory of some acceptable J-structure N which is given by applying the Σ_1-SKOLEM function h_N to elements of M^p. This will play a crucial role in the proof of the Upward Extension of Embeddings Lemma 11.20.

Definition 11.12 Let φ be a formula in a first order language. We say that φ is a *Q-formula* iff φ is (equivalent to a formula) of the form

$$\forall v_i \, \exists v_j \supset v_i \, \psi(v_j), \tag{11.2}$$

where ψ is Σ_1 and does not contain v_i. We also write Qv_j instead of $\forall v_i \, \exists v_j \supset v_i$ and read (11.2) as "for cofinally many v_j, $\psi(v_j)$". A map $\pi\colon M \to N$ which preserves

Q-formulae is called *Q-preserving*, in which case we write

$$\pi\colon M \to_Q N.$$

A map $\pi\colon M \to N$ is called *cofinal* iff for all $y \in N$ there is some $x \in M$ such that $y \subset \pi(x)$.

Lemma 11.13 *Let $\pi\colon U \to_{\Sigma_0} U'$, where U and U' are transitive structures.*

(a) *If π is cofinal, then π is Σ_1-elementary.*
(b) *Let π be Σ_1-elementary. Let φ be a Π_2-formula, and let $\mathbf{x} \in U$. If $U' \models \varphi(\pi(\mathbf{x}))$, then $U \models \varphi(\mathbf{x})$.*
(c) *Let π is cofinal. Let φ be a Q-formula, and let $\mathbf{x} \in U$. If $U \models \varphi(\mathbf{x})$, then $U' \models \varphi(\pi(\mathbf{x}))$.*

Proof Problem 11.3. □

We formulate the following lemma just for models of $\mathscr{L}_{\dot{\in},\dot{E},\dot{A}}$, but of course it also holds for models of different types.

Lemma 11.14 *There is a Q-sentence Ψ (of $\mathscr{L}_{\dot{\in},\dot{E},\dot{A}}$) such that for every transitive model $M = (M; \in, E, A)$ (of $\mathscr{L}_{\dot{\in},\dot{E},\dot{A}}$) which is closed under pairing, M is an acceptable J-structure iff $M \models \Psi$.*

Proof The statement "$V = L[\dot{E}]$" (cf. p. 77) may be written as

$$Qy\, \exists \beta\; y = S_\beta[E].$$

Here, "$y = S_\beta[E]$" is the Σ_1-formula from Lemma 5.25 (2). The fact that (M, A) is amenable can be expressed by

$$Qy\, \exists z\; z = A \cap y,$$

as $A \cap x \in M$ iff there is some $y \supset x$ with $A \cap y \in M$.

It remains to be checked that being acceptable can be written in a Q-fashion. Let φ be the sentence

$$\begin{aligned} &\forall \omega\xi\; \exists n < \omega\; \forall m < \omega\; \forall \tau < \omega\xi \\ &((\mathscr{P}(\tau) \cap S_{\omega\xi+m}[E]) \setminus J_{\omega\xi}[E] \neq \emptyset) \Longrightarrow \\ &\quad \exists f \in S_{\omega\xi+n}[E]\; f : \tau \to \omega\xi \wedge f \text{ surjective}). \end{aligned} \tag{11.3}$$

Clearly, if $M \models \varphi$, then M is acceptable. To show the converse, let M be acceptable, and let $\omega\xi < M \cap \mathrm{OR}$. Let τ_0 be the least τ such that

$$(\mathscr{P}(\tau) \cap J_{\omega\xi+\omega}[E]) \setminus J_{\omega\xi}[E] \neq \emptyset. \tag{11.4}$$

By acceptability, there is some $n_0 < \omega$ and some surjective $f\colon \tau_0 \to \omega\xi$, $f \in S_{\omega\xi+n_0}[E]$. But then there is some $n \geq n_0$ such that for *every* τ with (11.4) there is some surjective $f\colon \tau \to \omega\xi$, $f \in S_{\omega\xi+n}[E]$. Therefore, $M \models \varphi$.

We may now express that M be acceptable by saying that for cofinally many y, $y = S_{\omega\xi+m}[E]$ (for some $\xi, m < \omega$) and if

$$(\mathscr{P}(\tau) \cap S_{\omega\xi+m}[E]) \setminus J_{\omega\xi}[E] \neq \emptyset,$$

then there is some surjective $f\colon \tau \to \omega\xi$, $f \in S_{\omega\xi+m}[E]$. □

Lemmas 11.13 and 11.14 now immediately give:

Corollary 11.15 *Let $\bar{M}$, M be transitive structures.*

(a) *If $\pi\colon \bar{M} \to_{\Sigma_1} M$ and M is an acceptable J-structure, then so is $\bar{M}$.*
(b) *If $\pi\colon \bar{M} \to_{Q} M$ (e.g., if π is a Σ_0 preserving cofinal map) and $\bar{M}$ is an acceptable J-structure, then so is M.*

We may now turn to the *downward extension of embeddings lemma.*

Lemma 11.16 (Downward Extension of Embeddings Lemma, Part 1) *Let $\bar{M}$, M be acceptable J-structures. Let $\bar{p} \in R_{\bar{M}}$ and $p \in M$. Let $\pi\colon \bar{M}^{\bar{p}} \to_{\Sigma_0} M^p$. Then there is a unique $\tilde{\pi}\colon \bar{M} \to_{\Sigma_0} M$ such that $\tilde{\pi} \supset \pi$ and $\tilde{\pi}(\bar{p}) = p$. Moreover, $\tilde{\pi}$ is in fact Σ_1 elementary.*

Proof By Lemma 10.16, the Σ_1-Skolem function h_N is *uniformly* definable over J-structures N, i.e., there is a Σ_1-formula Ψ such that $x = h_N(n, \mathbf{y})$ iff $N \models \Psi(n, \mathbf{y}, x)$ for every J-structure N. Say $\Psi(v_1, v_2, v_3) \equiv \exists w_1 \ldots \exists w_k \bar{\Psi}(w_1, \ldots, w_k, v_1, v_2, v_3)$.

Let us first show the uniqueness of $\tilde{\pi}$. Suppose that $\tilde{\pi}$ has the above properties. Let $x \in \bar{M}$. Then $x = h_{\bar{M}}(n, (\mathbf{z}, \bar{p}))$ for some $n \in \omega$ and $\mathbf{z} \in [\rho_1(\bar{M})]^{<\omega}$. Pick $z_1, \ldots, z_k \in \bar{M}$ such that $\bar{\Psi}(z_1, \ldots, z_k, n, (\mathbf{z}, \bar{p}), x)$. Since $\tilde{\pi}$ is Σ_0 preserving, this implies $\bar{\Psi}(\tilde{\pi}(z_1), \ldots, \tilde{\pi}(z_k), n, (\tilde{\pi}(\mathbf{z}), p), \tilde{\pi}(x))$, so that we must have

$$\tilde{\pi}(x) = h_M(n, (\tilde{\pi}(\mathbf{z}), p)) = h_M(n, (\pi(\xi), p)).$$

Hence, there can be at most one such $\tilde{\pi}$.

Let us now show the existence of $\tilde{\pi}$.

Claim 11.17 *Suppose that $\varphi(v_1, \ldots, v_\ell)$ is a Σ_1-formula. For $0 < i \leq \ell$ let $\bar{x}_i = h_{\bar{M}}(n_i, (\bar{\mathbf{z}}_i, \bar{p}))$ where $n_i < \omega$ and $\bar{\mathbf{z}}_i \in [\rho_1(M)]^{<\omega}$, and let $x_i = h_M(n_i, (\mathbf{z}_i, p))$ where $\mathbf{z}_i = \pi(\bar{\mathbf{z}}_i)$. Then*

$$\bar{M} \models \varphi(\bar{x}_1, \ldots, \bar{x}_\ell) \quad \textit{iff} \quad M \models \varphi(x_1, \ldots, x_\ell).$$

Proof We shall use the map e from p. 230. Let $\varphi \equiv \varphi_n$, $n < \omega$. Then $\bar{M} \models \varphi(\bar{x}_1, \ldots, \bar{x}_\ell)$ is equivalent to

$$\bar{M} \models \varphi_n(h_{\bar{M}}(n_1, (\bar{\mathbf{z}}_1, \bar{p})), \ldots, h_{\bar{M}}(n_\ell, (\bar{\mathbf{z}}_\ell, \bar{p}))), \tag{11.5}$$

which may be written as

$$\bar{M} \models \varphi_{e(n)}(\bar{\mathbf{z}}_1, \ldots, \bar{\mathbf{z}}_\ell, \bar{p}). \tag{11.6}$$

This also works over M, i.e., $M \models \varphi(x_1, \ldots, x_\ell)$ is equivalent to

$$M \models \varphi_{e(n)}(\mathbf{z}_1, \ldots, \mathbf{z}_\ell, p). \tag{11.7}$$

Now (11.6) is equivalent to

$$A^{\bar{p}}_{\bar{M}}(e(n), (\bar{\mathbf{z}}_1, \ldots, \bar{\mathbf{z}}_\ell)), \tag{11.8}$$

and (11.7) is equivalent to

$$A^{p}_{M}(e(n), (\mathbf{z}_1, \ldots, \mathbf{z}_\ell)). \tag{11.9}$$

Since π is Σ_0 preserving, (11.8) and (11.9) are equivalent. □

Now let us define $\tilde{\pi}$ by

$$\tilde{\pi}(h_{\bar{M}}(n, (\mathbf{z}, \bar{p}))) \simeq h_M(n, (\pi(\mathbf{z}), p)) \tag{11.10}$$

for $n \in \omega$ and $\mathbf{z} \in [\rho_1(\bar{M})]^{<\omega}$. Here, "$\simeq$" is understood as saying that the left hand side is defined iff the right hand side is. Notice that $\tilde{\pi}$ is indeed well defined by (11.10); this is because if $h_{\bar{M}}(n_1, (\bar{\mathbf{z}}_1, \bar{p})) = h_{\bar{M}}(n_2, (\bar{\mathbf{z}}_2, \bar{p}))$ where $n_1, n_2 < \omega$ and $\bar{\mathbf{z}}_1, \bar{\mathbf{z}}_2 \in [\rho_1(\bar{M})]^{<\omega}$, then by Claim 11.17, $h_M(n_1, (\pi(\bar{\mathbf{z}}_1), p)) = h_M(n_2, (\pi(\bar{\mathbf{z}}_2), p))$. Claim 11.17 then also yields that $\tilde{\pi}$ is Σ_1 preserving.

To see that $\tilde{\pi} \supset \pi$ and $\tilde{\pi}(\bar{p}) = p$, pick $k_1, k_2 < \omega$ such that

$$x = h_N(k_1, (x, q)) \text{ and } q = h_N(k_2, (x, q)), \tag{11.11}$$

uniformly over all J-structures N. Then for all $\mathbf{z} \in [\rho_1(\bar{M})]^{\geq\omega}$, $\mathbf{z} = h_{\bar{M}}(k_1, (\mathbf{z}, \bar{p}))$, hence $\tilde{\pi}(\mathbf{z}) = h_M(k_1, (\pi(\mathbf{z}), p)) = \pi(\mathbf{z})$. This gives $\tilde{\pi} \supset \pi$. Also, $\bar{p} = h_{\bar{M}}(k_2, (0, \bar{p}))$, hence $\tilde{\pi}(\bar{p}) = h_M(k_2, (0, p)) = p$. □

If in addition the hypothesis of Lemma 11.16 we assume that $p \in R_M$ and $\pi : \bar{M}^{\bar{p}} \rightarrow_{\Sigma_n} M^p$, then we may show that $\tilde{\pi} : \bar{M} \rightarrow_{\Sigma_{n+1}} M$ (cf. Problem 11.4).

Lemma 11.18 (Downward Extension of Embeddings Lemma, Part 2) *Let M be an acceptable J-structure, and let $p \in M$. Suppose that N is a J-structure and $\pi \colon N \rightarrow_{\Sigma_0} M^p$. Then there are unique $\bar{M}$ and $\bar{p}$ such that $\bar{p} \in R_{\bar{M}}$ and $N = \bar{M}^{\bar{p}}$.*

Proof The uniqueness of $\bar{M}$ and $\bar{p}$ is easy to verify, arguing as in the proof of Lemma 11.16. Let us show the existence.

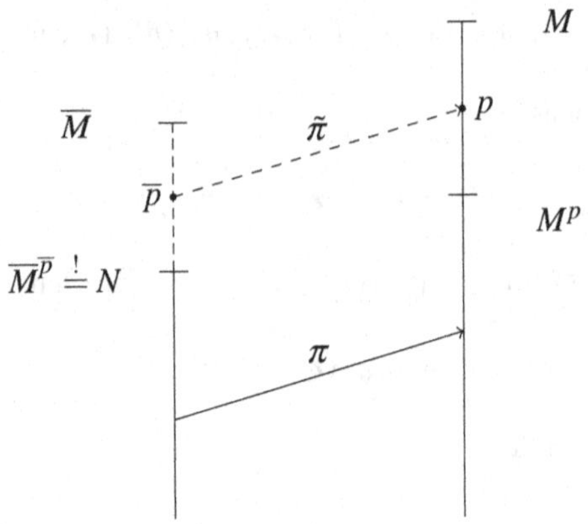

Let $M = (J_\alpha[E], B)$, $M^p = (J_\rho[E], A)$, where $\rho = \rho_1(M)$ and $A = A^p_M$, and let $N = (J_{\bar\rho}[\bar E'], \bar A)$. Let

$$\bar M \stackrel{\tilde\pi}{\cong} h_M(\text{ran}(\pi) \cup \{p\}) \prec_{\Sigma_1} M, \tag{11.12}$$

where $\bar M$ is transitive. By Corollary 11.15 (a), $\bar M$ is an acceptable J-structure, say $\bar M = (J_{\bar\alpha}[\bar E], \bar B)$.

Let us first show that

$$\text{ran}(\tilde\pi) \cap \left(\bigcup \text{ran}(\pi)\right) = \text{ran}(\pi). \tag{11.13}$$

It is easy to see "$\supset$" of (11.13). To show "$\subset$" of (11.13), suppose that $y = h_M(n, (z, p)) \in x$, where $n < \omega$ and $z, x \in \text{ran}(\pi)$. Since $y, z \in M^p$, the Σ_1 statement "$y = h_M(n, (z, p))$" can be equivalently expressed in the form $A(k, (y, z))$ for some $k \in \omega$. As also $x \in M^p$, we thus have that

$$\exists v \in x\ A(k, (v, z)),$$

a Σ_0-statement which is true in M^p, so that

$$\exists v \in \bar x\ \bar A(k, (v, \bar z)),$$

holds true in N, where $\bar x = \pi^{-1}(x)$ and $\bar z = \pi^{-1}(z)$. Let $\bar y \in \bar x$ be such that $\bar A(k, (\bar y, \bar z))$. Then $A(k, (\pi(\bar y), z))$, so that in fact $\pi(\bar y) = h_M(n, (z, p)) = y$, i.e., $y \in \text{ran}(\pi)$. We have shown (11.13).

Equation (11.13) now immediately implies that

$$\tilde\pi \supset \pi \text{ and } J_{\bar\rho}[\bar E'] = J_{\bar\rho}[\bar E]. \tag{11.14}$$

Let us now set

$$\bar{p} = \tilde{\pi}^{-1}(p).$$

We aim to verify that $\bar{M}$ and $\bar{p}$ are as desired.

We claim that

$$\rho_1(\bar{M}) = \bar{\rho}. \tag{11.15}$$

Well, by (11.12) and (11.14) and as there is a $\underset{\sim}{\Sigma}{}_1^{\bar{M}}$ map of $\bar{\rho}$ onto $J_{\bar{\rho}}[\bar{E}]$, there is a $\underset{\sim}{\Sigma}{}_1^{\bar{M}}$ map of $\bar{\rho}$ onto $\bar{M}$. This gives that $\rho_1(\bar{M}) \leq \bar{\rho}$, cf. the proof of Lemma 11.11.

To show that $\bar{\rho} \leq \rho_1(\bar{M})$, let P be $\underset{\sim}{\Sigma}{}_1^{\bar{M}}(\{\bar{q}\})$ for some $\bar{q} \in \bar{M}$, and let $\gamma < \bar{\rho}$. We aim to see that $P \cap \gamma \in \bar{M}$. By (11.12) we can find an $n < \omega$ and some $x \in N = J_{\bar{\rho}}[\bar{E}]$ such that

$$z \in P \Longleftrightarrow \bar{M} \models \varphi_n((z, x), \bar{p})$$

for all $z \in \bar{M}$. But for all $k < \omega$ and $y \in N$,

$$\bar{A}(k, y) \Longleftrightarrow A(k, \tilde{\pi}(y)) \Longleftrightarrow M \models \varphi_k(\tilde{\pi}(y), p) \Longleftrightarrow \bar{M} \models \varphi_k(y, \bar{p}). \tag{11.16}$$

In particular,

$$\bar{A}(n, (z, x)) \Longleftrightarrow \bar{M} \models \varphi_n((z, x), \bar{p})$$

for all $z \in N$. As N is a J-structure, $\bar{A} \cap (\{n\} \times (\gamma \times \{x\})) \in N$, and thus $P \cap \gamma$ is in N, too. This proves (11.15).

As an immediate consequence of (11.16) and (11.15) we get that

$$\bar{A} = A_{\bar{M}}^{\bar{p}}. \tag{11.17}$$

Because there is a $\Sigma_1^{\bar{M}}$ map of $\bar{\rho}$ onto $J_{\bar{\rho}}[\bar{E}]$, (11.12) implies that

$$\bar{p} \in R_{\bar{M}}.$$

The proof is complete. □

We now aim to prove a dual result, the *upward extension of embeddings lemma.*

Definition 11.19 Let $\bar{M}$ and M be acceptable J-structures. A map $\pi\colon \bar{M} \to M$ is called a *good embedding* iff

(a) $\pi\colon \bar{M} \to_{\Sigma_1} M$
(b) For all $\bar{R}$ and R such that $\bar{R} \subset \bar{M}^2$ is rudimentary over $\bar{M}$ and $R \subset M^2$ is rudimentary over M by the same definition,

if $\bar{R}$ is well-founded, then so is$R$.

Lemma 11.20 (Upward Extension of Embeddings Lemma) *Let $\bar{M}$ be an acceptable J-structure, and let $\bar{p} \in R_{\bar{M}}$. Suppose that N is an acceptable J-structure, and $\pi : \bar{M}^{\bar{p}} \to_{\Sigma_1} N$ is a good embedding. Then there are unique M, p such that $N = M^p$ and $p \in R_M$. Moreover, $\tilde{\pi}$ is good, where $\tilde{\pi} \supset \pi$ and $\tilde{\pi} : \bar{M} \to_{\Sigma_1} M$ with $\tilde{\pi}(\bar{p}) = p$ is given by Lemma 11.16.*

Notice that Lemma 11.16 in fact applies to the situation of Lemma 11.20.

Proof of Lemma 11.20. We shall make frequent use of the map e from p. 230.

Let us first show that M and p are unique. Suppose $\tilde{\pi}_1 : \bar{M} \to M_1$ and $\tilde{\pi}_2 : \bar{M} \to M_2$ are two extensions of π satisfying the conclusion of Lemma 11.20, and that p_1, p_2 are the corresponding parameters. Then $A_{M_1}^{p_1} = A_{M_2}^{p_2}$, call it A, and if $k \in \{1, 2\}$ and $x \in M_k$, then x is of the form $h_{M_k}(n, (\mathbf{z}, p_k))$ for some $n < \omega$ and $\mathbf{z} \in [N \cap \mathrm{OR}]^{<\omega}$. Let $\sigma \colon M_1 \to M_2$ be the map sending $h_{M_1}(n, (\mathbf{z}, p_1))$ to $h_{M_2}(n, (\mathbf{z}, p_2))$, where $n < \omega$ and $\mathbf{z} \in [N \cap \mathrm{OR}]^{<\omega}$. Then σ is a well defined surjection since

$$\exists z\, z = h_{M_1}(n, (\mathbf{z}, p_1)) \Longleftrightarrow A(m, (n, \xi)) \Longleftrightarrow \exists z\, z = h_{M_2}(n, (\mathbf{z}, p_2))$$

for an appropriate $m < \omega$ (namely, $m = e(\ulcorner \exists z\, z = v \urcorner)$). Also, σ respects $\in$, since if $x = h_{M_1}(n_1, (\mathbf{z}_1, p_1))$ and $y = h_{M_1}(n_2, (\mathbf{z}_2, p_1))$, where $n_1, n_2 < \omega$ and $\mathbf{z}_1$, $\mathbf{z}_2 \in [N \cap \mathrm{OR}]^{<\omega}$ then

$$\begin{aligned} x \in y &\Longleftrightarrow h_{M_1}(n_1, (\mathbf{z}_1, p_1)) \in h_{M_1}(n_2, (\mathbf{z}_2, p_1)) \\ &\Longleftrightarrow A(m, ((n_1, \mathbf{z}_1), (n_2, \mathbf{z}_2))) \\ &\Longleftrightarrow h_{M_2}(n_1, (\mathbf{z}_1, p_2)) \in h_{M_2}(n_2, (\mathbf{z}_2, p_2)) \\ &\Longleftrightarrow \sigma(x) \in \sigma(y) \end{aligned}$$

for an appropriate $m < \omega$ (namely, $m = e(\ulcorner v_1 \in v_2 \urcorner)$). Therefore, σ is an $\in$-isomorphism, so that σ is the identity, i.e., $M_1 = M_2$, $\tilde{\pi}_1 = \tilde{\pi}_2$ and $p_1 = p_2$.

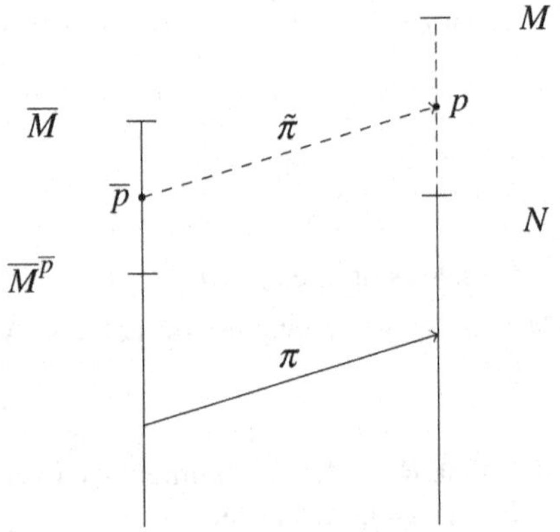

Let us now verify the existence of M, p. Suppose that $\bar{M} = (J_{\bar{\alpha}}[F], W)$. We first represent $\bar{M}$ as a term model which is definable over $\bar{M}^{\bar{p}}$, and then "unfold" the term model which is defined in the corresponding fashion over N so as to obtain M and p. Let

$$\bar{\mathscr{M}} = \{(n, z) \in \omega \times \bar{M}^{\bar{p}} : \exists y\ y = h_{\bar{M}}(n, (z, \bar{p}))\},$$

and let us define relations $\bar{I}$, $\bar{E}$, $\bar{F}$, and $\bar{W}$ over $\bar{\mathscr{M}}$ as follows, where $(n, x), (m, y) \in \bar{\mathscr{M}}$.

$$\begin{aligned}
(n, x)\bar{I}(m, y) &\iff h_{\bar{M}}(n, (x, \bar{p})) = h_{\bar{M}}(m, (y, \bar{p})) \\
(n, x)\bar{E}(m, y) &\iff h_{\bar{M}}(n, (x, \bar{p})) \in h_{\bar{M}}(m, (y, \bar{p})) \\
(n, x) \in \bar{F} &\iff h_{\bar{M}}(n, (x, \bar{p})) \in F \\
(n, x) \in \bar{W} &\iff h_{\bar{M}}(n, (x, \bar{p})) \in W
\end{aligned}$$

Obviously, $\bar{I}$ is an equivalence relation, and the predicates $\bar{E}$, $\bar{F}$, and $\bar{W}$ are $\bar{I}$-invariant. Let us write, for $(n, x), (m, y) \in \bar{\mathscr{M}}$,

$$\begin{aligned}
[n, x] &= \{[m, y] \colon (n, x)\bar{I}(m, y)\} \\
\bar{\mathscr{M}}/\bar{I} &= \{[n, x] \colon (n, x) \in \bar{\mathscr{M}}\} \\
[n, x]\bar{E}/\bar{I}[m, y] &\iff (n, x)\bar{E}(m, y) \\
[n, x] \in \bar{F}/\bar{I} &\iff (n, x) \in \bar{F} \\
[n, x] \in \bar{W}/\bar{I} &\iff (n, x) \in \bar{W}.
\end{aligned}$$

We obviously have

$$\mathscr{S} = (\bar{\mathscr{M}}/\bar{I}; \bar{E}/\bar{I}, \bar{F}/\bar{I}, \bar{W}/\bar{I}) \overset{\bar{\sigma}}{\cong} (J_{\bar{\alpha}}[F]; \in, F, W),$$

where $\bar{\sigma}$ sends $[n, x]$ to $h_{\bar{M}}(n, (x, \bar{p}))$.

Notice that $\bar{\mathscr{M}}$, $\bar{I}$, $\bar{E}$, $\bar{F}$, and $\bar{W}$ are all $\Sigma_1^{\bar{M}}(\{\bar{p}\})$, and that we may thus choose n_1, n_2, n_3, n_4, and $n_5 < \omega$ such that for all $(n, x), (m, y) \in \omega \times \bar{M}^{\bar{p}}$,

$$\begin{aligned}
(n, x) \in \bar{\mathscr{M}} &\iff (n_1, (n, x)) \in A^{\bar{p}}_{\bar{M}} \\
(n, x)\bar{I}(m, y) &\iff (n_2, ((n, x), (m, y))) \in A^{\bar{p}}_{\bar{M}} \\
(n, x)\bar{E}(m, y) &\iff (n_3, ((n, x), (m, y))) \in A^{\bar{p}}_{\bar{M}} \\
(n, x) \in \bar{F} &\iff (n_4, (n, x)) \in A^{\bar{p}}_{\bar{M}} \\
(n, x) \in \bar{W} &\iff (n_5, (n, x)) \in A^{\bar{p}}_{\bar{M}}
\end{aligned}$$

Let us write $N = (J_\rho[E'], A')$, so that $\pi \colon \bar{M}^{\bar{p}} \to_{\Sigma_1} (J_\rho[E'], A')$. We define $\mathscr{M}$, I, E, F, and W as follows, where $(n, x), (m, y) \in \omega \times N$.

$$\begin{cases} (n,x) \in \mathcal{M} \Longleftrightarrow (n_1,(n,x)) \in A' \\ (n,x)I(m,y) \Longleftrightarrow (n_2,((n,x),(m,y))) \in A' \\ (n,x)E(m,y) \Longleftrightarrow (n_3,((n,x),(m,y))) \in A' \\ (n,x) \in F \Longleftrightarrow (n_4,(n,x)) \in A' \\ (n,x) \in W \Longleftrightarrow (n_5,(n,x)) \in A' \end{cases} \tag{11.18}$$

The fact that $\bar{I}$ is an equivalence relation and that $\bar{E}$, $\bar{F}$, and $\bar{W}$ are $\bar{I}$-invariant may easily be formulated in a Π_1 fashion over $\bar{M}^{\bar{p}}$. As π is assumed to be Σ_1-elementary, I is thus also an equivalence relation, and E, F, and W are I-invariant. We may therefore write, for $(n,x), (m,y) \in \mathcal{M}$,

$$\begin{aligned} [n,x]^* &= \{[m,y]\colon (n,x)I(m,y)\} \\ \mathcal{M}/I &= \{[n,x]^*\colon (n,x) \in \mathcal{M}\} \\ [n,x]^* E/I [m,y]^* &\Longleftrightarrow (n,x)E(m,y) \\ [n,x]^* \in F/I &\Longleftrightarrow (n,x) \in F \\ [n,x]^* \in W/I &\Longleftrightarrow (n,x) \in W. \end{aligned}$$

Let us consider

$$\mathcal{S} = (\mathcal{M}/I; E/I, F/I, W/I).$$

In addition to (11.18), we shall need four more facts about A' which are inherited from $A^{\bar{p}}_{\bar{M}}$ and which will eventually enable us to show that A' is a standard code. For one thing, for all Σ_1-formulae φ and ψ and for all $(m_1,x_1), \ldots, (m_k,x_k) \in \omega \times N$,

$$\begin{aligned} &(e(\ulcorner \neg\varphi \urcorner), ((m_1,x_1),\ldots,(m_k,x_k))) \in A' \Longleftrightarrow \\ &\quad (e(\ulcorner \varphi \urcorner), ((m_1,x_1),\ldots,(m_k,x_k))) \notin A' \end{aligned} \tag{11.19}$$

and

$$\begin{aligned} &(e(\ulcorner \varphi \wedge \psi \urcorner), ((m_1,x_1),\ldots,(m_k,x_k))) \in A' \\ \Longleftrightarrow &[\, (e(\ulcorner \varphi \urcorner), ((m_1,x_1),\ldots,(m_k,x_k))) \in A' \\ &\wedge (e(\ulcorner \psi \urcorner), ((m_1,x_1),\ldots,(m_k,x_k))) \in A' \,]. \end{aligned} \tag{11.20}$$

Equations (11.19) and (11.20) just follow from the corresponding facts for $A^{\bar{p}}_{\bar{M}}$ and the Π_1-elementarity of π. We also need versions of (11.19) and (11.20) for quantification. In order to arrive at these versions, we are going to use the Σ_1-SKOLEM function for $\bar{M}$ to express Π_2-truth over $\bar{M}$ in a Π_1 fashion over $\bar{M}^{\bar{p}}$. For the sake of readability, let us pretend in what follows that if φ_n is Σ_1 but not Σ_0, then φ_n has only one free variable, w, and that it is in fact of the form $\exists v\, \psi(v, \mathbf{w})$, where ψ is Σ_0. We may pick a (partial) computable map $\bar{e}\colon \omega \to \omega$ such that for all $n < \omega$ in the domain of $\bar{e}$, $\varphi_{\bar{e}(n)}$ is Σ_0 and

$$\varphi_n(\mathbf{w}) \text{ is equivalent to } \exists v\, \varphi_{\bar{e}(n)}(v, \mathbf{w})$$

over rud–closed structures.

Now let φ_n be Σ_1 but not Σ_0. We then have that

$$\bar{M}^{\bar{p}} \models \text{“}\forall m \forall x\, ((e(n), (m, x)) \in A_{\bar{M}}^{\bar{p}} \longleftrightarrow \exists m' \exists x' (e(\bar{e}(n)), ((m', x'), (m, x))) \in A_{\bar{M}}^{\bar{p}}).\text{”} \tag{11.21}$$

Equation (11.21) is not Π_1 by itself, but it may be rephrased in a Π_1 fashion over $\bar{M}^{\bar{p}}$ as follows.

$$\bar{M}^{\bar{p}} \models \text{“}\forall m \forall x\, ((e(n), (m, x)) \in A_{\bar{M}}^{\bar{p}} \longleftrightarrow (e(\bar{e}(n)), ((k(n), (m, x)), (m, x)) \in A_{\bar{M}}^{\bar{p}}).\text{”} \tag{11.22}$$

Here, $k\colon \omega \to \omega$ is a natural computable function such that

$$\bar{M} \models \exists v\, \varphi_{k(n)}(v, ((m, x), \bar{p})) \longleftrightarrow \exists v\, \varphi_{\bar{e}(n)}(v, h_{\bar{M}}(m, (x, \bar{p}))).$$

Equation (11.22) is true as

$$\begin{aligned} ((e(n), (m, x)) \in A_{\bar{M}}^{\bar{p}} &\Longleftrightarrow \bar{M} \models \varphi_n(h_{\bar{M}}(m, (x, \bar{p}))) \\ &\Longleftrightarrow \bar{M} \models \exists v\, \varphi_{\bar{e}(n)}(v, h_{\bar{M}}(m, (x, \bar{p}))) \\ &\Longleftrightarrow \bar{M} \models \varphi_{\bar{e}(n)}(h_{\bar{M}}(k(n), ((m, x), \bar{p})), h_{\bar{M}}(m, (x, \bar{p}))) \\ &\Longleftrightarrow (e(\bar{e}(n)), ((k(n), (m, x)), (m, x)) \in A_{\bar{M}}^{\bar{p}}. \end{aligned}$$

The statement (11.22) will be transported upward via π, and we thus have

$$N \models \text{“}\forall m \forall x\, (e(n), (m, x)) \in A' \longleftrightarrow (e(\bar{e}(n)), ((k(n), (m, x)), (m, x)) \in A').\text{”} \tag{11.23}$$

We have shown that for all formulae $\varphi = \varphi_n$ which are Σ_1 but not Σ_0 and for all $(m, x) \in \omega \times N$,

$$(e(n), (m, x)) \in A' \Longleftrightarrow \exists (m', x') \in \omega \times N\, (e(\bar{e}(n)), ((m', x'), (m, x))) \in A'. \tag{11.24}$$

In a similar fashion, we may prove that if $\varphi(v_1, v_2, v_3)$ is Σ_0 and if $(m, x), (m', x') \in \omega \times N$, then

$$\begin{aligned}&(e(\ulcorner\forall v_1 \in v_2\,\varphi(v_1, v_2, v_3)\urcorner), ((m, x), (m', x')) \in A' \Longleftrightarrow\\ &\forall(m'', x'') \in \omega \times N(e(\ulcorner v_1 \in v_2 \rightarrow \varphi(v_1, v_2, v_3)\urcorner),\\ &((m'', x''), (m', x'), (m, x))) \in A'.\end{aligned} \tag{11.25}$$

We know that $\bar{E}/\bar{I}$ is extensional, which may be formulated as saying that if $(n, x), (m, y) \in \bar{\mathcal{M}}$ are such that

$$(e(\ulcorner v \neq w\urcorner), ((n, x), (m, y))) \in A^{\bar{p}}_{\bar{M}},$$

then there is some $(m', x') \in \bar{\mathcal{M}}$ with

$$(e(\ulcorner u \in v\Delta w\urcorner), ((m', x'), (n, x), (m, y))) \in A^{\bar{p}}_{\bar{M}}.$$

Equation (11.24) will now give that E/I is also extensional. Because E is rudimentary over N via the same definition as the one which gives $\bar{E}$ as being rudimentary over $\bar{M}^{\bar{p}}$, the relation E is actually well-founded by the goodness of π. Therefore,

$$\mathcal{S} = (\mathcal{M}/I; E/I, F/I, W/I) \overset{\sigma}{\cong} (J_\alpha[E^*]; \in, E^*, W^*)$$

for some α, E^*, and W^*. We shall also write $M = (J_\alpha[E^*], W^*)$ instead of $(J_\alpha[E^*]; \in, E^*, W^*)$.

It is now straightforward to use (11.18–11.24) and prove the following by induction of the complexity of φ.

Claim 11.21 *For all $n < \omega$ and $(m_1, x_1), \ldots, (m_k, x_k) \in \omega \times N$,*

$$\begin{aligned}&(e(n), ((m_1, x_1), \ldots, (m_k, x_k))) \in A' \Longleftrightarrow\\ &M \models \varphi_n(\sigma([m_1, x_1]^*), \ldots, \sigma([m_k, x_k]^*)).\end{aligned}$$

We may now define

$$\tilde{\pi} \colon \bar{M} \rightarrow M$$

by

$$h_{\bar{M}}(n, (x, \bar{p})) \mapsto \sigma([n, \pi(x)]^*),$$

where $(n, x) \in \bar{\mathcal{M}}$. Notice that $\tilde{\pi}$ is well-defined, as for $(n, x), (m, y) \in \bar{\mathcal{M}}$,

$$\begin{aligned}h_{\bar{M}}(n, (x, \bar{p})) = h_{\bar{M}}(m, (y, \bar{p})) &\Longleftrightarrow (n, x)\bar{I}(m, y)\\ &\Longleftrightarrow (n_2, ((n, x), (m, y))) \in A^{\bar{p}}_{\bar{M}}\\ &\Longleftrightarrow (n_2, ((n, \pi(x)), (m, \pi(y)))) \in A'\\ &\Longleftrightarrow (n, \pi(x))I(m, \pi(y))\\ &\Longleftrightarrow [n, \pi(x)]^* = [m, \pi(y)]^*.\end{aligned}$$

Moreover, $\tilde{\pi}$ is Σ_2-elementary, by the following reasoning. Let $n < \omega$, and say that φ_n has two free variables, v and w. Then for all $z \in \bar{M}$, say $z = h_{\bar{M}}(\ell, (y, \bar{p}))$, where $\ell < \omega$ and $y \in \bar{M}^{\bar{p}}$,

$$\begin{aligned}
\bar{M} \models \forall v\, \varphi_n(v, z) &\Longleftrightarrow \forall (m, x) \in \omega \times \bar{M}^{\bar{p}}\ \bar{M} \models \varphi_n(h_{\bar{M}}(m, (x, \bar{p})), h_{\bar{M}}(\ell, (y, \bar{p}))) \\
&\Longleftrightarrow \bar{M}^{\bar{p}} \models \forall m, x\, (e(n), ((m, x), (\ell, y))) \in A^{\bar{p}}_{\bar{M}} \\
&\Longleftrightarrow N \models \forall m, x\, (e(n), ((m, x), (\ell, \pi(y)))) \in A' \\
&\overset{(*)}{\Longleftrightarrow} \forall (m, x) \in \omega \times N\ M \models \varphi([m, x]^*, [\ell, \pi(y)]^*) \\
&\Longleftrightarrow M \models \forall v\, \varphi_n(v, \tilde{\pi}(z)).
\end{aligned}$$

Here, $(*)$ holds true by Claim 11.21.

Let us write $p = \tilde{\pi}(\bar{p})$. We claim that M, $\tilde{\pi}$, and p are as desired.

Let k_1, k_2 be as in (11.11). Then

$$x = \bar{\sigma}([k_1, x]) \text{ for all } x \in \bar{M}^{\bar{p}} \text{ and } \bar{p} = \bar{\sigma}([k_2, 0]).$$

Let us first observe that $N \subset M$. If $x \in \bar{M}^{\bar{p}}$ and $\bar{\sigma}([n, y]) \in x = \bar{\sigma}([k_1, x])$, then $\bar{\sigma}([n, y]) = \bar{\sigma}([k_1, w])$ for some $w \in x$. This may be written in a Π_1 fashion over $\bar{M}^{\bar{p}}$, so that if $x \in N$ and $\sigma([n, y]^*) \in \sigma([k_1, x]^*)$, then $\sigma([n, y]^*) = \sigma([k_1, w^*])$ for some w^*. It follows by $\in$-induction that $\sigma([k_1, x]^*) = x$ for all $x \in N$. Furthermore, given any $x \in \bar{M}^{\bar{p}}$,

$$\tilde{\pi}(x) = \tilde{\pi}(\bar{\sigma}([k_1, x])) = \sigma([k_1, \pi(x)]^*) = \pi(x),$$

and hence $\tilde{\pi} \supset \pi$.

We also have that $p = \tilde{\pi}(\bar{p}) = \tilde{\pi}(h_{\bar{M}}(k_2, (0, \bar{p}))) = \sigma([k_2, 0]^*)$. We now prove that $A' = A^p_M$.

For $x \in \bar{M}^{\bar{p}}$ and $n < \omega$, we have that

$$\begin{aligned}
rl(n, x) \in A^{\bar{p}}_{\bar{M}} &\Longleftrightarrow \bar{M} \models \varphi_n(x, \bar{p}) \\
&\Longleftrightarrow \bar{M} \models \varphi_n(h_{\bar{M}}(k_1, (x, \bar{p})), h_{\bar{M}}(k_2, (0, \bar{p}))) \\
&\Longleftrightarrow (e(n), ((k_1, x), (k_2, 0))) \in A^{\bar{p}}_{\bar{M}},
\end{aligned}$$

which implies that for all $x \in N$ and $n < \omega$,

$$(n, x) \in A' \Longleftrightarrow (e(n), ((k_1, x), (k_2, 0))) \in A' \Longleftrightarrow M \models \varphi_n(x, p).$$

To see that $A' = A^p_M$, it then suffices to show that $N \cap \mathrm{OR} = \rho = \rho_1(M)$. Our computation will also yield that $p \in R_M$.

The reader will gladly verify that for all $(n, x) \in \omega \times \bar{M}^{\bar{p}}$,

$$h_{\bar{M}}(n, (x, \bar{p})) = h_{\bar{M}}(h_{\bar{M}}(k_1, (n, \bar{p})), (h_{\bar{M}}(k_1, (x, \bar{p})), h_{\bar{M}}(k_2, (0, \bar{p}))))$$

yields that

$$(\ulcorner v_1 = h(v_2, (v_3, v_4))\urcorner, ((n, x), (k_1, n), ((k_1, x), (k_2, 0)))) \in A^{\bar{p}}_{\bar{M}},$$

so that for all $(n, x) \in \omega \times N$,

$$(\ulcorner v_1 = h(v_2, (v_3, v_4))\urcorner, ((n, x), (k_1, n), ((k_1, x), (k_2, 0)))) \in A', \quad (11.26)$$

which in turn implies that

$$\sigma([n, x]^*) = h_M([k_1, n]^*, ([k_1, x]^*, [k_2, 0]^*)) = h_M(n, (x, p)).$$

As there is a Σ_1^M map from ρ onto N and $M = \{\sigma([n, x]^*) \colon (n, x) \in \mathscr{M}\}$, (11.26) implies that

$$M = h_M(\rho \cup \{p\}), \quad \text{and hence also} \quad \rho_1(M) \leq \rho \quad (11.27)$$

by the proof of Lemma 11.11. On the other hand, if $B \in \Sigma_1^M(\{a\})$, where $a \in M$, then by (11.27) there is some $\mathbf{z} \in [\rho]^{<\omega}$ and some $\tilde{B} \in \Sigma_1^M(\{p\})$ such that for all $x \in N$,

$$x \in B \Longleftrightarrow (x, \mathbf{z}) \in \tilde{B},$$

which in turn for some fixed n (namely, the GÖDEL number of the defining formula) is equivalent to $(n, (x, \mathbf{z})) \in A'$. But (N, A') is amenable, so that if $\eta < \rho$, then $\eta \times \{\mathbf{z}\} \cap \tilde{B} \in N$, and hence also $B \cap \eta \in N$. This shows that $\rho \leq \rho_1(M)$. Thus,

$$\rho = \rho_1(M) \quad \text{and} \quad p \in R_M.$$

It only remains to show that $\tilde{\pi}$ is good. Let $\bar{R}$, R be binary relations which are rudimentary over $\bar{M}$, M, respectively, by the same rudimentary definition. Define $\bar{R}^*$, R^* as follows

$$(n, x)\bar{R}^*(m, y) \Longleftrightarrow (n, x), (m, y) \in \bar{\mathscr{M}} \wedge \bar{\sigma}([n, x])\bar{R}\bar{\sigma}([m, y])$$
$$(n, x)R^*(m, y) \Longleftrightarrow (n, x), (m, y) \in \mathscr{M} \wedge \sigma([n, x]^*)R\sigma([m, y]^*).$$

Then $\bar{R}^*$ is well-founded since $\bar{R}$ is, and $\bar{R}^*$, R^* are rudimentary over $\bar{M}^{\bar{p}}$, N, respectively, by the same rudimentary definition. As π is good, R^* must then be well-founded. Hence R must be well-founded as well. □

Definition 11.22 Let M be an acceptable J-structure. For $n < \omega$ we recursively define the *n-th projectum* $\rho_n(M)$, the *n-th standard code* $A_M^{n,p}$ and the *n-th reduct* $M^{n,p}$ as follows:

(1) $\rho_0(M) = M \cap \mathrm{OR}$, $\Gamma_M^0 = \{\emptyset\}$, $A_M^{0,\emptyset} = \emptyset$, and $M^{0,\emptyset} = M$, and
(2) $\rho_{n+1}(M) = \min\{\rho_1(M^{n,p}) \colon p \in \Gamma_M^n\}$, $\Gamma_M^{n+1} = \prod_{i \leq n} [\rho_{i+1}(M), \rho_i(M))^{\geq\omega}$, and

for $p \in \Gamma_M^{n+1}$, $A_M^{n+1,p} = A^{p(n)}_{M^{n,p\restriction n}}$, and $M^{n+1,p} = (M^{n,p\restriction n})^{p(n)}$.

We also set $\rho_\omega(M) = \min\{\rho_n(M); n < \omega\}$. The ordinal $\rho_\omega(M)$ is called the *ultimate projectum* of M.

We remark that if M is not 1-sound (cf. Definition 11.28) then it need not be the case that $\rho_2(M)$ is the least ρ such that $\mathscr{P}(\rho) \cap \underset{\sim}{\Sigma}_2^M \not\subset M$.

If $\rho_n(M) \leq \cdots \leq \rho_1(M)$, then we may identify $p = (p(0), \ldots, p(n)) \in \Gamma_M^{n+1}$ with the (finite) set $\bigcup \mathrm{ran}(p)$ of ordinals; this will play a rôle in the next section.

Definition 11.23 Let M be an acceptable J-structure. We the set

$$P_M^0 = \{\emptyset\},$$
$$P_M^{n+1} = \{p \in \Gamma_M^{n+1} \colon\ p \restriction n \in P_M^n \wedge \rho_1(M^{n,p\restriction n}) = \rho_{n+1}(M) \wedge\ p(n) \in P_{M^{n,p\restriction n}}\}, \text{ and}$$
$$R_M^{n+1} = \{p \in \Gamma_M^{n+1} \colon\ p \restriction n \in R_M^n\ \wedge\ \rho_1(M^{n,p\restriction n}) = \rho_{n+1}(M)\ \wedge\ p(n) \in R_{M^{n,p\restriction n}}\}.$$

As before, we call the elements of P_M^n *good* parameters and the elements of R_M^n *very good* parameters.

Lemma 11.24 *Let M be an acceptable J-structure.*

(a) $R_M^n \subset P_M^n \neq \emptyset$
(b) *Let $p \in R_M^n$. If $q \in \Gamma_M^n$ then $A_M^{n,q}$ is* $\mathrm{rud}_{M^{n,p}}$ *in parameters from $M^{n,p}$.*
(c) *Let $p \in R_M^n$. Then $\rho_1(M^{n,p}) = \rho_{n+1}(M)$.*
(d) *If $p \in P_M^n$, then for all $i < n$, $p(i) \in P_{M^{i,p\restriction i}}$. If $p \in R_M^n$, then for all $i < n$, $p(i) \in R_{M^{i,p\restriction i}}$. Moreover, if $p \restriction (n-1) \in R_M^{n-1}$, then $p(n-1) \in P_{M^{n-1,p\restriction n-1}}$ implies that $p \in P_M^n$ and $p(n-1) \in R_{M^{n-1,p\restriction n-1}}$ implies that $p \in R_M^n$.*

Proof (a) This is easily shown inductively by using Lemma 11.11 and amalgamating parameters.
(b) By induction on $n < \omega$. The case $n = 0$ is trivial. Now let $n > 0$, and suppose (b) holds for $n-1$. Write $m = n-1$. Let $p \in R_M^n$ and $q \in \Gamma_M^n$. We have to show that $A_{M^{m,q\restriction m}}^{q(m),\rho_n(M)}$ is $\mathrm{rud}_{M^{n,p}}$ in parameters from $M^{n,p}$. Inductively, $M^{m,q\restriction m}$ is $\mathrm{rud}_{M^{m,p\restriction m}}$ in a parameter $t \in M^{n,p}$. As $p(m) \in R_{M^{m,p\restriction m}}$, there are e_0 and e_1 and $z \in M^{n,p}$ such that

$$q(m) = h_{M^{m,p\restriction m}}(e_0, (z, p(m)))$$

and

$$t = h_{M^{m,p\restriction m}}(e_1, (z, p(m))).$$

For $i < \omega$ and $x \in M^{n,p}$, we have that

$$\begin{aligned}(i, x) \in A_{M^{m,q\restriction m}}^{q(m),\rho_n(M)} &\iff M^{m,q\restriction m} \models \varphi_i(x, q(m))\\ &\iff M^{m,q\restriction m} \models \varphi_i(x, h_{M^{m,p\restriction m}}(e_0, (z, p(m))))\\ &\iff M^{m,p\restriction m} \models \varphi_j((x, z), p(m))\\ &\iff (j, (x, z)) \in A_{M^{m,p\restriction m}}^{p(m)},\end{aligned}$$

for some J which is recursively computable from i, as $M^{m,q\restriction m}$ is $\text{rud}_{M^{m,p\restriction m}}$ in the parameter $t = h_{M^{m,p\restriction m}}(e_1, (z, p(m)))$. Therefore, $A^{q(m),\rho_n(M)}_{M^{m,q\restriction m}}$ is $\text{rud}_{A^{p(m)}_{M^{m,p\restriction m}}}$ in the parameter z. (c) Let $\rho_{n+1}(M) = \rho_1(M^{n,q})$, where $q \in \Gamma^n_M$. By (b), $M^{n,q}$ is $\text{rud}_{M^{n,p}}$ in parameters from $M^{n,p}$, which implies that $\underset{\sim}{\Sigma}{}_1^{M^{n,q}} \subset \underset{\sim}{\Sigma}{}_1^{M^{n,p}}$. But then $\rho_1(M^{n,p}) \leq \rho_1(M^{n,q}) = \rho_{n+1}(M)$, and hence $\rho_1(M^{n,p}) = \rho_{n+1}(M)$. (d) This follows inductively, using (c). □

The following is given just by the definition of R^{n+1}_M. Let M be acceptable, and let $p \in R^{n+1}_M$. Then

$$M = h_M(h_{M^{1,p\restriction 1}}(\ldots h_{M^{n,p\restriction n}}(\rho_{n+1}(M) \cup \{p(n)\}) \ldots) \cup \{p(0)\}). \quad (11.28)$$

We thus can, uniformly over M, define a function $h^{n+1,p}_M$ basically as the iterated composition of the Σ_1 Skolem functions of the ith reducts of M, $0 \leq i \leq n$, given by p such that M is the $h^{n+1,p}_M$-hull of $\rho_{n+1}(M)$ whenever $p \in R^{n+1}_M$.

More precisely, let M be acceptable, and let $p \in \Gamma^{n+1}_M$. Let us inductively define $h^{i,p}_M$, for $1 \leq i \leq n+1$, as follows. For $k < \omega$, let $g(k) =$ the largest m such that 2^m divides k, and let $u(k) =$ the largest m such that 3^m divides k. Let

$$h^{1,p}_M(k, \mathbf{x}) = h_M(k, (\mathbf{x}, p(0))) \text{ for } \mathbf{x} \in M^{1,p\restriction 1}, \text{ and}$$
$$\text{for } i > 0,\ h^{i+1,p}_M(k, \mathbf{x}) = h^{i,p}_M(g(k), h_{M^{i,p\restriction i}}(u(k), (\mathbf{x}, p(i)))) \text{ for } \mathbf{x} \in M^{i+1,p\restriction i+1}. \quad (11.29)$$

If $X \subset M^{n+1,p}$, then we shall write

$$h^{n+1,p}_M{}^{\bullet}(X) \text{ for } h^{n+1,p}_{M''}(\omega \times {}^{<\omega}X).$$

If p is clear from the context, then we may write h^{n+1}_M rather than $h^{n+1,p}_M$.

The following is straightforward.

Lemma 11.25 *Let $n < \omega$, and let M be an acceptable J-structure. If $p \in \Gamma^{n+1}_M$, then $h^{n+1,p}_M$ is in $\Sigma^M_\omega(\{p\})$, and if $p \in R^{n+1}_M$, then*

$$M = h^{n+1,p}_M{}''(\rho_{n+1}(M)). \quad (11.30)$$

Lemma 11.26 *Let $0 < n < \omega$. Let M be an acceptable J-structure, and let $p \in R^n_M$. Then $\underset{\sim}{\Sigma}{}^M_\omega \cap \mathscr{P}(M^{n,p}) = \underset{\sim}{\Sigma}{}^{M^{n,p}}_\omega$.*

Proof It is easy to verify that $\underset{\sim}{\Sigma}{}^{M^{n,p}}_\omega \subset \underset{\sim}{\Sigma}{}^M_\omega \cap \mathscr{P}(M^{n,p})$. Now let $A \in \underset{\sim}{\Sigma}{}^M_\omega \cap \mathscr{P}(M^{n,p})$, say

$$x \in A \iff M \models \exists x_1 \forall x_2 \cdots \exists/\forall x_k\, \varphi(x, y, x_1, x_2, \cdots, x_k),$$

where φ is Σ_0 and $y \in M$. By Lemma 11.25, we may write

$$x \in A \iff \exists x_1' \in M^{n,p}\, \forall x_2' \in M^{n,p} \ldots \exists/\forall x_k \in M^{n,p}$$

$$\varphi(x, h_M^n(y', p), h_M^n(x_1', p), h_M^n(x_2', p), \cdots h_M^n(x_k', p)),$$

where $y' \in M^{n,p}$. But then $A \in \underset{\sim}{\Sigma}_\omega^{M^{n,p}}$, as h_M^n is definable over M by Lemma 11.25. □

A more careful look at the proofs of Lemmata 11.25 and 11.26 shows the following.

Lemma 11.27 *Let $n < \omega$. Let M be an acceptable J-structure, and let $p \in R_M^n$. Let $A \subset M^{n,p}$ be $\underset{\sim}{\Sigma}_{n+1}^M$. Then A is $\underset{\sim}{\Sigma}_1^{M^{n,p}}$.*

Definition 11.28 Let M be an acceptable J-structure. M is *n-sound* iff $R_M^n = P_M^n$. M is *sound* iff M is n-sound for all $n < \omega$.

We shall prove later (cf. Lemma 11.53) that every J_α is sound. It is in fact a crucial requirement on "L-like" models that there proper initial segments be sound.

We may now formulate generalizations of the downward and the upward extension of embeddings Lemmas 11.16, 11.18, and 11.20.

Lemma 11.29 (General Downward Extension of Embeddings Lemma, Part 1) *Let $n > 0$. Let $\bar{M}$ and M be acceptable J-structures, and let $\pi : \bar{M}^{n,\bar{p}} \to_{\Sigma_0} M^{n,p}$, where $\bar{p} \in R_{\bar{M}}^n$. Then there is a unique map $\tilde{\pi} \supset \pi$ such that* $\mathrm{dom}(\tilde{\pi}) = \bar{M}$, $\tilde{\pi}(\bar{p}) = p$ *and, setting $\tilde{\pi}_i = \tilde{\pi} \restriction \bar{M}^{i,\bar{p}\restriction i}$,*

$$\tilde{\pi}_i : \bar{M}^{i,\bar{p}\restriction i} \to_{\Sigma_0} M^{i,p\restriction i} \text{ for } i \leq n.$$

For $i < n$, the map $\tilde{\pi}_i$ is in fact Σ_1-elementary.

In particular,

$$\tilde{\pi}\left(h_{\bar{M}}^{n,\bar{p}}(k, \mathbf{x})\right) = h_M^{n,p}(k, \pi(\mathbf{x}))$$

for every $k < \omega$ and $\mathbf{x} \in \bar{M}^{n,\bar{p}}$.

Lemma 11.30 (General Downward Extensions of Embeddings Lemma, Part 2) *Let M be an acceptable J-structure, and let $p \in M$. Let N be a J-structure, and let $\pi : N \to_{\Sigma_0} M^{n,p}$. Then there are unique $\bar{M}$, $\bar{p}$ such that $\bar{p} \in R_{\bar{M}}^n$ and $N = \bar{M}^{n,\bar{p}}$.*

The general upward extension of embeddings lemma is the conjunction of the following lemma together with Lemmas 11.29 and 11.30.

Lemma 11.31 (General Upward Extensions of Embeddings Lemma) *Let $\pi : \bar{M}^{n,\bar{p}} \to_{\Sigma_1} N$ be good, where $\bar{M}$ is an acceptable J-structure and $\bar{p} \in R_{\bar{M}}^n$. Then there are unique M, p such that M is an acceptable J-structure, $p \in R_M^n$ and $M^{n,p} = N$. Moreover, if $\tilde{\pi}$ is as in Lemma 11.29, then $\tilde{\pi}$ is good.*

If π and $\tilde{\pi}$ are as in Lemma 11.29 then $\tilde{\pi}$ is often called the *n-completion* of π.

Following [30, Sect. 2] we shall call embeddings arising from applications of the Downward and Upward Extension of Embeddings Lemma "$r\Sigma_{n+1}$ elementary." Here is our official definition, which presupposes that the structures in question possess very good parameters.

Definition 11.32 Let M, N be acceptable, let $\pi: M \to N$, and let $n < \omega$. Then π is called $r\Sigma_{n+1}$ *elementary* provided that there is $p \in R^n_M$ with $\pi(p) \in R^n_N$, and for all $i \leq n$,

$$\pi \restriction M^{i,p\restriction i}: M^{i,p\restriction i} \to_{\Sigma_1} N^{i,\pi(p)\restriction i}. \tag{11.31}$$

The map π is called *weakly* $r\Sigma_{n+1}$ *elementary* provided that there is $p \in R^n_M$ with $\pi(p) \in R^n_N$, and for all $i < n$, (11.31) holds, and

$$\pi \restriction M^{n,p}: M^{n,p} \to_{\Sigma_0} N^{n,\pi(p)}.$$

If $\pi: M \to N$ is (weakly) $r\Sigma_{n+1}$ elementary then typically both M and N will be n-sound, cf. Lemma 11.38; however, neither M nor N has to be $(n+1)$-sound. It is possible to generalize this definition so as to not assume that very good parameters exist (cf. [30, Sect. 2]).

With the terminology of Definition 11.32, Lemma 11.29 says that the map π can be extended to its n-completion $\tilde{\pi}$ which is weakly $r\Sigma_{n+1}$ elementary, and if π is Σ_1 elementary to begin with, then the n-completion $\tilde{\pi}$ in fact be $r\Sigma_{n+1}$ elementary.

Moreover, if a map $\pi: M \to N$ is $r\Sigma_{n+1}$ elementary, then π respects h^{n+1} by Theorem 10.16:

Lemma 11.33 *Let $n < \omega$, and let M and N be acceptable J-structures. Let $\pi: M \to N$ be $r\Sigma_{n+1}$ elementary. Let $p \in \Gamma^{n+1}_M$ be such that $p \restriction n \in R^n_M$ and $\pi(p \restriction n) \in R^n_N$. Then for all $k < \omega$ and $\mathbf{x} \in M^{n+1,p}$,*

$$\pi\left(h^{n+1,p}_M(k, \mathbf{x})\right) = h^{n+1,\pi(p)}_N(k, \pi(\mathbf{x})).$$

Recall the well-ordering $<^*$ of finite sets of ordinals from Problem 5.19: if $u, v \in \mathrm{OR}^{<\omega}$, then $u <^* v$ iff $\max(u \Delta v) \in v$. If M is an acceptable J-model and $n < \omega$, then the well-ordering $<^*$ induces a well-ordering of Γ^n_M by confusing $p \in \Gamma^n_M$ with $\bigcup \mathrm{ran}(p)$. We shall denote this latter well-ordering also by $<^*$.

Definition 11.34 Let M be an acceptable J-structure. The $<^*$-least $p \in P^n_M$ is called the nth standard parameter of M and is denoted by $p_n(M)$. We shall write M^n for $M^{n,p_n(M)}$. M^n is called the nth *standard reduct of* M.

Lemma 11.35 *Let M be an acceptable J-structure, let $n < \omega$, and let $p \in R^n_M$. Then there is some $\tilde{p} \in P^{n+1}_M$ with $\tilde{p} \restriction n = p$. In particular, if $n > 0$ and M is n-sound, then $p_{n-1}(M) = p_n(M) \restriction (n-1)$.*

Proof This follows immediately from Lemma 11.24 (c). □

Definition 11.36 Let M be an acceptable J-structure. Suppose that for all $n < \omega$, $p_n(M) = p_{n+1}(M) \upharpoonright n$. Then we set $p(M) = \bigcup_{n<\omega} p_n(M)$. $p(M)$ is called the *standard parameter of* M.

We shall often confuse $p(M)$ with $\bigcup \operatorname{ran}(p(M))$. Lemma 11.35 readily gives the following.

Corollary 11.37 *Let M be an acceptable J-structure which is also sound. Then $p(M)$ exists.*

Lemma 11.38 *Let M be an acceptable J-structure. M is sound iff $p_n(M) \in R^n_M$ for all $n \in \omega$.*

Proof We shall prove the non-trivial direction "$\Longleftarrow$." We need to see that for each $n > 0$,

$$p_n(M) \in R^n_M \Longrightarrow R^n_M = P^n_M. \tag{11.32}$$

Suppose $n > 0$ to be least such that (11.32) fails. Hence $P^n_M \setminus R^n_M \neq \emptyset$ by Lemma 11.24 (a). Let q be the $<^*$-least element of $P^n_M \setminus R^n_M$. This means that $p <^* q$, where $p = p_n(M)$.

We have that $q \upharpoonright (n-1) \in P^{n-1}_M = R^{n-1}_M$, $\rho_1(M^{n-1,q\upharpoonright(n-1)}) = \rho_{n+1}(M)$, and $q(n) \in P_{M^{n-1,q\upharpoonright(n-1)}} \setminus R_{M^{n-1,q\upharpoonright(n-1)}}$. Let

$$\pi \colon N \cong h_{M^{n-1,q\upharpoonright(n-1)}}(\rho_n(M) \cup \{q(n-1)\}) \prec_{\Sigma_1} M^{n-1,q\upharpoonright(n-1)},$$

where N is transitive. By the Downward Extension of Embeddings Lemma 11.29 and 11.30, there are unique $\bar{M}$, $\bar{q}$, and $\tilde{\pi}$ such that

$$\begin{aligned}
&\bar{q} \in R^{n-1}_{\bar{M}},\ N = \bar{M}^{n-1,\bar{q}},\\
&\tilde{\pi} \colon \bar{M} \to M \text{ is } r\Sigma_n \text{ elementary, and}\\
&\tilde{\pi} \supset \pi \text{ and } \tilde{\pi}(\bar{q}) = q \upharpoonright (n-1).
\end{aligned} \tag{11.33}$$

Let $\bar{q}' \in \Gamma^n_{\bar{M}}$ be such that $\bar{q}' \upharpoonright (n-1) = \bar{q}$ and $\bar{q}'(n-1) = \pi^{-1}(q(n-1))$, so that $\tilde{\pi}(\bar{q}') = q \in \operatorname{ran}(\tilde{\pi})$. Because $p = p_n(M) \in R^n_M$, there are $e < \omega$ and $z \in [\rho_n(M)]^{<\omega}$ such that $q = h^{n,p}_M(e, z)$, i.e.,

$$\exists p^* <^* q\, \exists e < \omega\, \exists z \in [\rho_n(M)]^{<\omega}\, (q = h^{n,p'}_M(e, z)). \tag{11.34}$$

We aim to verify that (11.34) also holds true in $\operatorname{ran}(\tilde{\pi})$:

Claim 11.39 *There is some $p^* \in \operatorname{ran}(\tilde{\pi})$, $p^* <^* q$, for which there are $e < \omega$ and $z \in [\rho_n(M)]^{<\omega}$ such that $q = h^{n,p^*}_M(e, z)$.*

Proof Let $i \leq n-1$ be least such that $p(i) <^* q(i)$. (So $p \upharpoonright i = q \upharpoonright i$.) Let us recursively define $((p^*(k), x(k)) \colon i \leq k \leq n-1)$ as follows.

Because

$$M^{i,q\restriction i} \models \exists r <^* q(i)\, \exists x \in [\rho_{i+1}(M)]^{<\omega}\, \exists e' < \omega\; q(i) = h_{M^{i,q\restriction i}}(e', (x, r)), \quad (11.35)$$

as being witnessed by $p(i)$, and because $\tilde{\pi} \restriction \bar{M}^{i,\bar{q}\restriction i}$ is Σ_1-elementary, we may let $p^*(i)$ be the $<^*$-least $r \in \text{ran}(\tilde{\pi})$ as in (11.35), and we let $x(i)$ be some $x \in \text{ran}(\tilde{\pi})$ as in (11.35) such that $q(i) = h_{M^{i,q\restriction i}}(e', (x, p^*(i)))$ for some $e' < \omega$. For the record, $p^*(i) <^* q(i)$.

Having defined $(p^*(k-1), x(k-1))$, where $i < k \leq n-1$ and $\{p^*(k-1), x(k-1)\} \subset \text{ran}(\tilde{\pi})$, we will have that

$$M^{k,q\restriction k} \models \exists r\, \exists x \in [\rho_{k+1}(M)]^{<\omega}\, \exists e' < \omega\; (x(k-1), q(k)) = h_{M^{k,q\restriction k}}(e', (x, r)), \quad (11.36)$$

as being witnessed by $p(k)$, and because $\tilde{\pi} \restriction \bar{M}^{k,\bar{q}\restriction k}$ is Σ_1-elementary, we may let $p^*(k)$ be the $<^*$-least $r \in \text{ran}(\tilde{\pi})$ as in (11.36), and we let $x(k)$ be some $x \in \text{ran}(\tilde{\pi})$ as in (11.36) such that $(x(k-1), q(k)) = h_{M^{k,q\restriction k}}(e', (x, p^*(k)))$ for some $e' < \omega$.

We may now set $p^* = p \restriction i \cup \{(k, p^*(k)) : i \leq k \leq n-1\}$.

It is straightforward to verify that for each $k \leq n-1$,

$$h_{M^{k,p^*\restriction k}}(\rho_{k+1}(M) \cup \{p^*(k)\}) = h_{M^{k,q\restriction k}}(\rho_{k+1}(M) \cup \{q(k)\}),$$

and $A^{p^*(k)}_{M^{k,p^*\restriction k}}$ and $A^{q(k)}_{M^{k,q\restriction k}}$ are easily computable from each other. (11.35) and (11.36) then give that $p^* \in P^n_M$. Also, $p^* <^* q$. However, $p^* \in \text{ran}(\tilde{\pi})$, whereas $q \notin R^n_M$ implies that $p \notin \text{ran}(\tilde{\pi})$, and hence $p^* \notin R^*_n$. This contradicts the choice of q. □

Solidity witnesses are witnesses to the fact that a given ordinal is a member of the standard parameter. We shall make use of witnesses in the proof of Theorem 11.64.

Definition 11.40 Let M be an acceptable J-structure, let $p \in \text{OR} \cap M^{<\omega}$, and let $\nu \in p$. Let W be an acceptable J-structure with $\nu \subset W$, and let $r \in \text{OR} \cap W^{<\omega}$. We say that (W, r), or just W, is a *witness for* $\nu \in p$ *with respect to* M, p iff for every Σ_1 formula $\varphi(v_0, \ldots, v_{l+1})$ and for all $\xi_0, \ldots, \xi_l < \nu$

$$M \models \varphi(\xi_0, \ldots, \xi_l, p \setminus (\nu+1)) \Longrightarrow W \models \varphi(\xi_0, \ldots, \xi_l, r). \quad (11.37)$$

By the proof of the following Lemma, if a witness exists, then there is also one where $\Longrightarrow$ may be replaced by $\Longleftrightarrow$ in (11.37).

Lemma 11.41 *Let M be an acceptable J-structure, and let $p \in P_M$. Suppose that for each $\nu \in p$ there is a witness W for $\nu \in p$ with respect to M, p such that $W \in M$. Then $p = p_1(M)$.*

Proof Suppose not. Then $p_1(M) <^* p$, and we may let $\nu \in p \setminus p_1(M)$ be such that $p \setminus (\nu+1) = p_1(M) \setminus (\nu+1)$. Let us write $q = p \setminus (\nu+1) = p_1(M) \setminus \nu$. Let $(W, r) \in M$ be a witness for $\nu \in p$ with respect to M, p. Let $A \in \Sigma_1^M(\{p_1(M)\})$ be such that $A \cap \rho_1(M) \notin M$.

Let $k < \omega$ and $\xi_1 < \cdots < \xi_k$ be such that $p_1(M) \cap \nu = \{\xi_1, \ldots, \xi_k\}$. Let φ be a Σ_1 formula such that for every $\xi < \rho_1(M)$,

$$\xi \in A \iff M \models \varphi(\xi, \xi_1, \ldots, \xi_k, q).$$

Because $(W, r) \in M$ is a witness for $\nu \in p$ with respect to M, p, we have that

$$M \models \psi(\xi, \xi_1, \ldots, \xi_k, q) \implies W \models \psi(\xi, \xi_1, \ldots, \xi_k, r) \tag{11.38}$$

for every $\xi < \rho_1(M) \leq \nu$ and every Σ_1 formula ψ.

Say $W = J_\alpha[E]$. Let $\beta = \sup(h_W(\nu \cup \{r\}) \cap \mathrm{OR}) \leq \alpha$, and write $\bar{W} = J_\beta[E]$. Let us define $\sigma\colon h_M(\nu \cup \{q\}) \to \bar{W}$ by setting $h_M(e, (\xi, q)) \mapsto h_{\bar{W}}(e, (\xi, r))$, where $e < \omega$ and $\xi \in [\nu]^{<\omega}$. By (11.38), σ is well-defined and Σ_0-elementary. By the choice of β, σ is cofinal and hence Σ_1-elementary by Lemma 11.13 (a). Therefore,

$$M \models \psi(\xi, \xi_1, \ldots, \xi_k, q) \iff \bar{W} \models \psi(\xi, \xi_1, \ldots, \xi_k, r) \tag{11.39}$$

for every $\xi < \rho_1(M) \leq \nu$ and every Σ_1-formula ψ. In particular, 11.39 holds for $\psi \equiv \varphi$ and every $\xi < \rho_1(M) \leq \nu$. As $\bar{W} \in M$, this shows that in fact $A \cap \rho_1(M) \in M$. Contradiction! □

Definition 11.42 Let M be an acceptable J-structure, let $p \in On \cap M^{<\omega}$, and let $\nu \in p$. We denote by $W_M^{\nu,p}$ the transitive collapse of $h_M(\nu \cup (p \setminus (\nu + 1)))$. We call $W_M^{\nu,p}$ the *standard witness for* $\nu \in p$ with respect to M, p.

Lemma 11.43 *Let M be an acceptable J-structure, and let $\nu \in p \in P_M$. The following are equivalent.*

(1) $W_M^{\nu,p} \in M$.
(2) *There is a witness (W, r) for $\nu \in p$ with respect to M, p such that $W \in M$.*

Proof We have to show (2) $\implies$ (1). Let $\tau\colon W_M^{\nu,p} \to M$ be the inverse of the transitive collapse. As in the proof of Lemma 11.41, say $W = J_\alpha[E]$, set $\beta = \sup(h_W(\nu \cup \{r\}) \cap \mathrm{OR}) \leq \alpha$, and write $\bar{W} = J_\beta[E]$. We may define a Σ_1-elementary embedding $\sigma\colon W_M^{\nu,p} \to \bar{W}$ by setting

$$\tau^{-1}(h_M(e, (\xi, p \setminus (\nu + 1)))) \mapsto h_{\bar{W}}(e, (\xi, r)),$$

where $e < \omega$ and $\xi \in [\nu]^{<\omega}$.

Now if $\tau(\nu) = \nu$ then a witness to $\rho_1(M)$ is definable over $W_M^{\nu,p}$, and hence over $\bar{W}$. But as $\bar{W} \in M$, this witness to $\rho_1(M)$ would then be in M. Contradiction!

We thus have that ν must be the critical point of τ. Thus we know that $\sigma(\nu)$ is regular in M, and hence writing $M = J_\gamma[E']$, $J_{\tau(\nu)}[E'] \models \mathsf{ZFC}^-$. We may code $W_M^{\nu,p}$ by some $a \subset \nu$, definably over $W_M^{\nu,p}$. Using σ, a is definable over $\bar{W}$, so that $a \in M$. In fact, $a \in J_{\sigma(\nu)}[B]$ by acceptability. We can thus decode a in $J_{\sigma(\nu)}[B]$, which gives $W_M^{\nu,p} \in J_{\sigma(\nu)}[B] \subset M$. □

Definition 11.44 Let M be an acceptable J-structure. We say that M is 1-*solid* iff

$$W_M^{\nu, p_1(M)} \in M$$

for every $\nu \in p_1(M)$.

Lemma 11.45 *Let $\bar{M}$, M be acceptable J-structures, and let $\pi\colon \bar{M} \to_{\Sigma_1} M$. Let $\bar{\nu} \in \bar{p} \in \mathrm{OR} \cap \bar{M}^{<\omega}$, and set $\nu = \pi(\bar{\nu})$ and $p = \pi(\bar{p})$. Let $(\bar{W}, \bar{r})$ be a witness for $\bar{\nu}$ with respect to $\bar{M}$, $\bar{p}$ such that $\bar{W} \in \bar{M}$, and set $W = \pi(\bar{W})$ and $r = \pi(\bar{r})$. Then (W, r) is a witness for ν with respect to M, p.*

Proof Let φ be a Σ_1-formula. We know that

$$\bar{M} \models \forall \xi_0 < \bar{\nu} \ldots \forall \xi_l < \bar{\nu}(\varphi(\xi_0, \ldots, \xi_l, \bar{p} \setminus (\bar{\nu}+1)) \longrightarrow \bar{W} \models \varphi(\xi_0, \ldots, \xi_l, \bar{r})).$$

As π is Π_1-elementary, this yields that

$$M \models \forall \xi_0 < \nu \ldots \xi_l < \nu(\varphi(\xi_0, \ldots, \xi_l, p \setminus (\nu+1)) \longrightarrow W \models \varphi(\xi_0, \ldots, \xi_l, r)).$$

We may thus conclude that (W, r) is a witness for ν with respect to M, p. □

Corollary 11.46 *Let $\bar{M}$, M be acceptable J-structures, and let $\pi\colon \bar{M} \to_{\Sigma_1} M$. Suppose that $\bar{M}$ is 1-solid and $\pi(p_1(\bar{M})) \in P_M$. Then $p_1(M) = \pi(p_1(\bar{M}))$, and M is 1-solid.*

The following lemma is a dual result to Lemma 11.45 with virtually the same proof.

Lemma 11.47 *Let $\bar{M}$, M be acceptable J-structures, and let $\pi\colon \bar{M} \to_{\Sigma_1} M$. Let $\bar{\nu} \in \bar{p} \in \mathrm{OR} \cap \bar{M}^{<\omega}$, and set $\nu = \pi(\bar{\nu})$ and $p = \pi(\bar{p})$. Let $(\bar{W}, \bar{r}) \in \bar{M}$ be such that, setting $W = \pi(\bar{W})$ and $r = \pi(\bar{r})$, (W, r) is a witness for ν with respect to M, p. Then $(\bar{W}, \bar{r})$ is a witness for $\bar{\nu} \in \bar{p}$ with respect to $\bar{M}$, $\bar{p}$.*

Corollary 11.48 *Let $\bar{M}$, M be acceptable J-structures, and let $\pi\colon \bar{M} \to_{\Sigma_1} M$. Suppose that M is 1-solid, and that in fact $W_M^{\nu, p_1(M)} \in \mathrm{ran}(\pi)$ for every $\nu \in p_1(M)$. Then $p_1(\bar{M}) = \pi^{-1}(p_1(M))$, and $\bar{M}$ is 1-solid.*

We now generalize Definition 11.44.

Definition 11.49 Let M be an acceptable J-structure. If $0 < n < \omega$ then we say that M is n-solid if for every $k < n$, $p_1(M^k) = p_{k+1}(M)(k) = p_n(M)(k)$ and M^k is 1-solid, i.e.,

$$W_{M^k}^{\nu, p_1(M^k)} \in M^k$$

for every $\nu \in p_1(M^k)$. We call M *solid* iff M is n-solid for every $n < \omega$, $n > 0$.

Lemma 11.50 *Let $\bar{M}$ and M be acceptable J-structures, let $n > 0$, and let $\pi: \bar{M} \to M$ be $r\Sigma_n$ elementary as being witnessed by $p_{n-1}(M)$. If $\bar{M}$ is n-solid and $\pi(p_1(\bar{M}^{n-1})) \in P_{M^{n-1}}$ then $p_n(M) = \pi(p_n(\bar{M}))$ and M is n-solid.*

Lemma 11.51 *Let $\bar{M}$ and M be acceptable J-structures, let $n > 0$, and let $\pi: \bar{M} \to M$ be $r\Sigma_n$ elementary as being witnessed by $\pi^{-1}(p_{n-1}(M))$. Suppose that M is n-solid, and in fact $W^{\nu,p_1(M^k)}_{M^k} \in \mathrm{ran}(\pi)$ for every $k < n$. If $\pi^{-1}(p_{n-1}(M)) \in P^{n-1}_{\bar{M}}$ then $p_n(\bar{M}) = \pi^{-1}(p_n(M))$ and $\bar{M}$ is n-solid.*

The ultrapower maps we shall deal with in the next section shall be elementary in the sense of the following definition. (Cf. [30, Definition 2.8.4].)

Definition 11.52 Let both M and N be acceptable, let $\pi: M \to N$, and let $n < \omega$. Then π is called an *n-embedding* if the following hold true.

(1) Both M and N are n-sound,
(2) π is $r\Sigma_{n+1}$ elementary,
(3) $\pi(p_k(M)) = p_k(N)$ for every $k \leq n$, and
(4) $\pi(\rho_k(M)) = \rho_k(N)$ for every $k < n$ and $\rho_n(N) = \sup(\pi''\rho_n(M))$.

Other examples for n-embeddings are typically obtained as follows. Let M be acceptable, and let, for $n \in \omega$, $\mathfrak{C}_n(M)$ denote the transitive collapse of $h^n_M{}''(\rho_n(M) \cup \{p_n(M)\})$. $\mathfrak{C}_n(M)$ is called the nth *core of M*. The natural map from $\mathfrak{C}_{n+1}(M)$ to $\mathfrak{C}_n(M)$ will be an n-embedding under favourable circumstances.

Lemma 11.53 *For each limit ordinal α, J_α is acceptable and sound.*

Proof by induction. Suppose that for every limit ordinal $\beta < \alpha$, J_β is acceptable and sound.

Let us first verify that J_α is acceptable. By our inductive hypothesis, this is trivial if α is a limit of limit ordinals, so let us assume that $\alpha = \beta + \omega$, where β is a limit ordinal. We need to see that if $\tau < \beta$ is such that

$$(\mathscr{P}(\tau) \cap J_{\beta+\omega}) \setminus J_\beta \neq \emptyset, \tag{11.40}$$

then there is some surjection $f: \tau \to \beta$ with $f \in J_{\beta+\omega}$. Let τ be least with (11.40). We claim that

$$\tau = \rho_\omega(J_\beta). \tag{11.41}$$

To see (11.41), note first that if n is such that $\rho_n(J_\beta) = \rho_\omega(J_\beta)$, then there is a $\undertilde{\Sigma}_1^{(J_\beta)^n}$ subset of $\rho_\omega(J_\beta)$ which is not in J_β. Such a set is $\undertilde{\Sigma}_\omega^{J_\beta}$ by Lemma 11.26 and the soundness of J_β, and it is hence in $J_{\beta+\omega} \setminus J_\beta$ by Lemma 5.15. Therefore, $\tau \leq \rho_\omega(J_\beta)$. On the other hand, let $a \subset \tau$ such that $a \in J_{\beta+\omega} \setminus J_\beta$. Then $a \in \undertilde{\Sigma}_\omega^{J_\beta}$ by Lemma 5.15. As $a \subset \tau \leq \rho_\omega(J_\beta)$ and J_β is sound, Lemma 11.27 yields that a is $\undertilde{\Sigma}_1^{(J_\beta)^n}$ for some $n < \omega$. Hence $\rho_\omega(J_\beta) \leq \tau$ and (11.41) follows.

Now by the soundness of J_β again and by Lemma 11.25 there is some $f \in \underset{\sim}{\Sigma}{}_\omega^{J_\beta}$ such that $f : \tau = \rho_\omega(J_\beta) \to J_\beta$ is surjective. By Lemma 5.15, $f \in J_{\beta+\omega}$. We have verified that J_α is acceptable.

We are now going to show that J_α is sound. We shall make use of Lemma 11.38 and verify that for every $n < \omega$,

$$p_n(J_\alpha) \in R^n_{J_\alpha}. \tag{11.42}$$

Suppose that this is false, and let n be least such that $p = p_{n+1}(J_\alpha) \notin R^{n+1}_{J_\alpha}$. Let us consider

$$(J_{\bar\rho}, \bar A) \overset{\pi}{\cong} h_{(J_\alpha)^n}(\rho_{n+1}(J_\alpha) \cup \{p(n)\}) \prec_{\Sigma_1} (J_\alpha)^n. \tag{11.43}$$

By the Downward Extension of Embeddings Lemma 11.29 and 11.30 we may extend π to a map

$$\tilde\pi : J_{\bar\alpha} \to_{\Sigma_1} J_\alpha$$

such that $p \restriction n \in \operatorname{ran}(\tilde\pi)$ and writing $\bar p = \tilde\pi^{-1}(p \restriction n)$, $\bar p \in R^n_{J_{\bar\alpha}}$ and $(J_{\bar\rho}, \bar A) = (J_{\bar\alpha})^{n,\bar p}$. Let us also write $p^* = \pi^{-1}(p(n)) = \tilde\pi^{-1}(p(n))$.

Let B be $\Sigma_1^{(J_\alpha)^n}(\{p(n)\})$ such that $B \cap \rho_{n+1}(J_\alpha) \notin J_\alpha$, say

$$B = \{x \in (J_\alpha)^n : (J_\alpha)^n \models \varphi(x, p(n))\},$$

where φ is Σ_1. Let

$$\bar B = \{x \in (J_{\bar\alpha})^{n,\bar p} : (J_{\bar\alpha})^{n,\bar p} \models \varphi(x, p^*)\}.$$

As $\pi \restriction \rho_{n+1}(J_\alpha) = \mathrm{id}$,

$$\bar B \cap \rho_{n+1}(J_\alpha) = B \cap \rho_{n+1}(J_\alpha) \notin (J_\alpha)^n. \tag{11.44}$$

If $\bar\alpha < \alpha$, then $(J_{\bar\alpha})^{n,\bar p} \in J_\alpha$, so that $\bar B \cap \rho_{n+1}(J_\alpha) \in J_\alpha$, contradicting (11.44). We must therefore have that $\bar\alpha = \alpha$. Then $\bar p \in R^n_{J_\alpha}$. For every $i < n$, we certainly have that $\bar p(i) \leq^* p(i)$, as $\tilde\pi(\bar p(i)) = p(i)$. By the choice of $p = p_{n+1}(J_\alpha)$, $p(i) \leq^* \bar p(i)$. This yields that in fact $\bar p = p \restriction n$, and therefore $(J_{\bar\rho}, \bar A) = (J_\alpha)^n$.

But $\bar B \in \Sigma_1^{(J_{\bar\alpha})^{n,\bar p}}(\{p^*\})$ and (11.44) then yield that $p^* \in P_{(J_\alpha)^n}$. We must also have $p^* \leq^* p(n)$, as $\pi(p^*) = p(n)$. By the choice of $p(n)$, $p(n) \leq^* p^*$, so that $p^* = p(n)$.

But now we must have that $\pi = \mathrm{id}$, and therefore $p(n) \in R_{(J_\alpha)^n}$, i.e., $p \in R^{n+1}_{J_\alpha}$. Contradiction! □

Lemma 11.54 *For each limit ordinal α, J_α is solid.*

Proof By Lemmas 11.53 and 11.35, it suffices to prove that if $n < \omega$ and $\nu \in p_1((J_\alpha)^n) = p(J_\alpha)(n)$, then

$$W^{\nu, p_1((J_\alpha)^n)}_{(J_\alpha)^n} \in (J_\alpha)^n. \tag{11.45}$$

Let us thus fix $n < \omega$ and $\nu \in p_1((J_\alpha)^n)$. Let us write $p = p(J_\alpha)$, so that $(J_\alpha)^n = (J_\alpha)^{n,p\restriction n}$. Let us consider

$$(J_{\bar\rho}, \bar A) \overset{\pi}{\cong} h_{(J_\alpha)^n}(\nu \cup \{p(n) \setminus (\nu+1)\}) \prec_{\Sigma_1} (J_\alpha)^n, \tag{11.46}$$

so that $(J_{\bar\rho}, \bar A) = W^{\nu, p_1((J_\alpha)^n)}_{(J_\alpha)^n}$. By the Downward Extension of Embeddings Lemma 11.29 and 11.30 we may extend π to a map

$$\tilde\pi\colon J_{\bar\alpha} \to_{\Sigma_1} J_\alpha$$

such that $p \restriction n \in \operatorname{ran}(\tilde\pi)$ and writing $\bar p = \tilde\pi^{-1}(p \restriction n)$, $\bar p \in R^n_{J_{\bar\alpha}}$ and $(J_{\bar\rho}, \bar A) = (J_{\bar\alpha})^{n,\bar p}$.

In order to verify (11.45), it suffices to prove that $\bar\alpha < \alpha$. This is because if $\bar\alpha < \alpha$, then $(J_{\bar\rho}, \bar A) = (J_{\bar\alpha})^{n,\bar p} \in J_\alpha$. But it is clear from (11.46) that $\mathrm{Card}(TC(\{(J_{\bar\rho}, \bar A)\})) = \mathrm{Card}(\nu)$ inside J_α. As $\nu < \bar\rho \le \rho_n(J_\alpha)$ and $J_{\rho_n(J_\alpha)} = (H_{\rho_n(J_\alpha)})^{J_\alpha}$ by Corollary 11.6, we then get that in fact $(J_{\bar\rho}, \bar A) \in J_{\rho_n(J_\alpha)}$, where $J_{\rho_n(J_\alpha)}$ is the universe of $(J_\alpha)^n$. Hence (11.45) follows.

We are left with having to prove that $\bar\alpha < \alpha$. Suppose that $\bar\alpha = \alpha$. Then, as $\tilde\pi(\bar p) = p \restriction n$, $\bar p \le^* p \restriction n$. However, $\bar p \in R^n_{J_{\bar\alpha}} = R^n_{J_\alpha} \subset P^n_{J_\alpha}$, so that by the choice of $p \restriction n$ we must actually have that $\bar p = p \restriction n$. That is,

$$(J_{\bar\rho}, \bar A) = (J_{\bar\alpha})^{n,\bar q} = (J_\alpha)^{n,p\restriction n}.$$

Let $B \in \underset{\sim}{\Sigma}{}_1^{(J_\alpha)^n}$ be such that $B \cap \rho_1((J_\alpha)^n) \notin (J_\alpha)^n$, say

$$B = \{x \in (J_\alpha)^n \colon (J_\alpha)^n \models \varphi(x, r)\},$$

where $r \in (J_\alpha)^n$ and φ is Σ_1. As $\nu \in p_1((J_\alpha)^n) \in [\rho_1((J_\alpha)^n), (J_\alpha)^n \cap \mathrm{OR})^{<\omega}$ and $\tilde\pi \restriction \nu = \mathrm{id}$, we have that $\tilde\pi \restriction \rho_1((J_\alpha)^n) = \mathrm{id}$. Therefore, if we let $B' \in \underset{\sim}{\Sigma}{}_1^{(J_\alpha)^n}$ be defined as

$$B' = \{x \in (J_\alpha)^n \colon (J_\alpha)^n \models \varphi(x, \pi(r))\},$$

then

$$B' \cap \rho_1((J_\alpha)^n) = B \cap \rho_1((J_\alpha)^n) \notin (J_\alpha)^n. \tag{11.47}$$

As $\mathrm{ran}(\pi) = h_{(J_\alpha)^n}(\nu \cup (p(n) \setminus (\nu + 1)))$, there is $m < \omega$ and

$$s \in (\nu \cup (p(n) \setminus (\nu + 1)))^{<\omega} \tag{11.48}$$

such that $\pi(r) = h_{(J_\alpha)^n}(m, s)$, so that $B' \in \Sigma_1^{(J_\alpha)^n}(\{s\})$. But (11.48) gives that $s <^* p(n)$, so that this contradicts the choice of $p(n)$. □

11.2 Jensen's Covering Lemma

We are now going to prove JENSEN's Covering Lemma, cf. Theorem 11.56. For this, we need the concept of a "fine ultrapower."

Definition 11.55 Let M be an acceptable J-structure, and let E be a (κ, ν)-extender over M. Let $n < \omega$ be such that $\rho_n(M) \geq \sigma(E)$. Suppose that M is n-sound, and set $p = p_n(M)$. Let

$$\pi \colon M^{n,p} \to \bar{N}$$

be the Σ_0 ultrapower map given by E. Suppose that

$$\tilde{\pi} \colon M \to N$$

is as given by the proof of Lemmas 11.20 and 11.31. Then we write $\mathrm{ult}_n(M; E)$ for N and call it the $r\Sigma_{n+1}$ *ultrapower of M by E*, and we call $\tilde{\pi}$ the $r\Sigma_{n+1}$ *ultrapower map (given by E)*.

Lemmas 11.20 and 11.31 presuppose that π is good (cf. Definition 11.19). However, the construction of the term model in the proof of Lemma 11.20 does not require π to be good, nor does it even require the target model $\bar{N}$ to be well-founded. Consequently, we can make sense of $\mathrm{ult}_n(M; E)$ even if π is not good or $\bar{N}$ is not well-founded. This is why we have "the proof of Lemmata 11.20 and 11.31" in the statement of Definition 11.55, as it does not assume anything about π or $\bar{N}$ which is not explicitly stated. We shall of course primarily be interested in situations where $\mathrm{ult}_n(M; E)$ is well-founded after all. In any event, as usual, we shall identify the well-founded part of $\mathrm{ult}_n(M; E)$ with its transitive collapse.

Recall that $\mathcal{S} \subset [\theta]^\kappa$ is called stationary iff for every algebra $\mathfrak{A} = (\theta; (f_i \colon i < \bar{\kappa}))$ with at most κ many functions f_i, $i < \bar{\kappa} \leq \kappa$, there is some $X \in \mathcal{S}$ which is closed under all the f_i, $i < \bar{\kappa}$, from $\mathfrak{A}$, cf. Definition 4.39.

Theorem 11.56 *The following statements are equivalent.*

(1) JENSEN Covering *holds, i.e., for all sets X of ordinals there is some $Y \in L$ such that $Y \supset X$ and $\overline{\overline{Y}} \leq \overline{\overline{X}} + \aleph_1$.*
(2) Strong Covering *holds, i.e., if $\kappa \geq \aleph_1$ is a cardinal and $\theta \geq \kappa$, then $[\theta]^\kappa \cap L$ is stationary in $[\theta]^\kappa$.*
(3) *L is rigid, i.e., there is no elementary embedding $\pi : L \to L$ which is not the identity.*
(4) $0^\#$ *does not exist.*

The most difficult part here is (4) $\Rightarrow$ (1) which is due to RONALD JENSEN, cf. [10], and which is called "JENSEN's Covering Lemma." There is, of course, a version of Theorem 11.56 for $L[x]$, $x \subset \omega$, which we leave to the reader's discretion.

It is not possible to cross out "$+\aleph_1$" in (1) or replace "$\geq \aleph_1$" by "$\geq \aleph_0$" in (2) of Theorem 11.56, cf. Problem 11.7 (cf. Problem 11.9, though).

(4) $\Longrightarrow$ (3) of Theorem 11.56 was shown as Theorem 10.39. (2) $\Longrightarrow$ (1) is trivial. If $0^\#$ exists, then every uncountable cardinal of V is a SILVER indiscernible, so that $\{\aleph_n : n < \omega\}$ cannot be covered in L by a set of size less than $\aleph_\omega$. This shows (1) $\Longrightarrow$ (4). We are left with having to verify (3) $\Longrightarrow$ (2).

Let us first observe that it suffices to prove (3) $\Longrightarrow$ (2) of Theorem 11.56 for the case that κ be *regular*. This follows from:

Lemma 11.57 *Let W be an inner model. Let κ be a singular cardinal, and suppose that for all $\bar{\kappa} < \kappa$, $\bar{\kappa} \geq \aleph_1$, and for all $\theta \geq \bar{\kappa}$, $[\theta]^{\bar{\kappa}} \cap W$ is stationary in $[\theta]^{\bar{\kappa}}$. Then for all $\theta \geq \kappa$, $[\theta]^\kappa \cap W$ is stationary in $[\theta]^\kappa$.*

Proof Let $\mathfrak{A} = (\theta; (f_\xi : \xi < \kappa))$ be any algebra on θ. We need to see that there is some $X \in [\theta]^\kappa \cap W$ such that for all $\xi < \kappa$, if f_ξ is n-ary, $n < \omega$, then $f_\xi''[X]^n \subset X$, i.e., X is closed under f_ξ. Let $(\kappa_i : i < \text{cf}(\kappa))$ be monotone and cofinal in κ, $\kappa_0 \geq \aleph_1$.

If $i < \text{cf}(\kappa)$, $1 \leq k < \omega$, and $X_1, \ldots, X_k \in [\theta]^{\leq \kappa_i} \cap W$, then by our hypothesis there is some $X \in [\theta]^{\kappa_i}$ such that $X \supset X_1 \cup \cdots \cup X_k$ and X is closed under all the functions f_ξ with $\xi < \kappa_i$. We may thus pick, for every $i < \text{cf}(\kappa)$ and $1 \leq k < \omega$ some

$$\Phi_i^k : [[\theta]^{\leq \kappa_i} \cap W]^k \to [\theta]^{\kappa_i} \cap W \tag{11.49}$$

such that for all $X_1, \ldots, X_k \in [\theta]^{\leq \kappa_i}$, $\Phi_i^k(X_1, \ldots, X_k) \supset X_1 \cup \cdots \cup X_k$ and $\Phi_i^k(X_1, \ldots, X_k)$ is closed under all functions f_ξ, $\xi < \kappa_i$.

Let us now consider the algebra

$$\mathfrak{A}^* = ([\theta]^{<\kappa} \cap W; (\Phi_i^k : i < \text{cf}(\kappa), 1 \leq k < \omega)).$$

It is easy to see that our hypothesis yields that if A is any set in W, then $[A]^{\text{cf}(\kappa)+\aleph_1} \cap W$ is stationary in $[A]^{\text{cf}(\kappa)+\aleph_1}$. In particular, we find some $Y \in W$ of size $\text{cf}(\kappa) + \aleph_1$ such that Y is closed under all the functions from $\mathfrak{A}^*$. Set $X = \bigcup Y$. Of course, $X \in W$. Moreover $\overline{\overline{X}} \leq \kappa$, as X is the union of $\text{cf}(\kappa) + \aleph_1 < \kappa$ many sets of size $<\kappa$. We claim that X is closed under all functions f_ξ, $\xi < \kappa$.

Let $\xi < \kappa$ and let f_ξ be n-ary, $n < \omega$. We aim to see that $f_\xi''[X]^k \subset X$. Let $x_l \in X_l \in Y$, $1 \leq l \leq k$, and let $i < \text{cf}(\kappa)$ be such that $\kappa_i > \xi$ and also $\kappa_i > \text{Card}(X_l)$, $1 \leq l \leq k$. Then

$$f_\xi(x_1, \ldots, x_k) \in \Phi_i^k(X_1, \ldots, X_k) \in Y,$$

and therefore $f_\xi(x_1, \ldots, x_k) \in X$. $\square$

Proof of Theorem 11.56, (3) $\Longrightarrow$ (2). Let us fix $\kappa \leq \theta$, where $\kappa \geq \aleph_1$ is regular. Let $\mu > \theta$ be a regular cardinal. Let $\pi : \overline{H} \to H_\mu$ be elementary s.t. $\overline{H}$ is transitive. Let

$$(\kappa_i : i < \alpha) = (\kappa_i^\pi : i < \alpha^\pi)$$

enumerate the transfinite cardinals of $L^{\overline{H}} = J_{\overline{H} \cap OR}$, and set $\kappa_\alpha = \kappa_{\alpha^\pi}^\pi = \overline{H} \cap OR$. For $i \leq \alpha$ let $\beta_i = \beta_i^\pi \geq \kappa_\alpha \in OR \cup \{\infty\}$ be largest such that κ_i is a cardinal in J_{β_i}. Hence by Lemma 11.53, if $\beta_i < \infty$, then $\rho_\omega(J_{\beta_i}) < \kappa_i$, whereas $\rho_\omega(J_\beta) \geq \kappa_i$ for all $\beta \in [\kappa_\alpha, \beta_i)$.

If $i \leq j$ then $\beta_j \leq \beta_i$, so that $\{\beta_i : i \leq \alpha\}$ is finite. For each $i \leq \alpha$ with $\beta_i < \infty$ we let $n_i = n_i^\pi$ be such that

$$\rho_{n_i+1}(J_{\beta_i}) < \kappa_i \leq \rho_{n_i}(J_{\beta_i}). \tag{11.50}$$

If $i \leq \alpha$ and $\beta_i = \infty$, then we let $n_i = n_i^\pi = 0$.

In what follows we shall make frequent use of the notation introduced by Definition 10.47. E.g., for $i \leq \alpha$, $E_{\pi \restriction J_{\kappa_i}}$ is the (long) $(\text{crit}(\pi), \sup \pi''\kappa_i)$—extender derived from $\pi \restriction J_{\kappa_i}$. Notice that by (11.50), we have that

$$\text{ult}_{n_i}(J_{\beta_i}; E_{\pi \restriction J_{\kappa_i}})$$

makes sense for all $i \leq \alpha$. If $i \leq j$ and $\beta_i = \beta_j$ then $n_j \leq n_i$. Hence $\{\rho_{n_i+1}(J_{\beta_i}) : i \leq \alpha\}$ is finite. Let $I = I^\pi$ be such that

$$\{\kappa_i : i \in I = I^\pi\} = \{\rho_{n_i+1}(J_{\beta_i}) : i \leq \alpha\} \cup \{\kappa_\alpha\},$$

cf. Lemma 11.5.

The following Claim is the key point.

Claim 11.58 *Suppose that for all $i \in I$, $\text{ult}_{n_i}(J_{\beta_i}; E_{\pi \restriction J_{\kappa_i}})$ is well-founded. Then either L is not rigid or else $\text{ran}(\pi) \cap \mu \in L$.*

Proof We may that assume $\pi \neq \text{id}$, as otherwise the conclusion is trivial. Let us assume that L is rigid and show that $\text{ran}(\pi) \cap \mu \in L$.

There must be some $i \in I$ such that $\kappa_i \leq \text{crit}(\pi)$. This is because otherwise, letting $\delta < \alpha$ be such that $\kappa_\delta = (\text{crit}(\pi))^{+L^{\overline{H}}}$, $\beta_\delta = \infty$ and the ultrapower map

$$\tilde{\pi} : L \to \text{ult}_0(L; E_{\pi \restriction J_{\kappa_\delta}}) \cong L$$

would witness that L is not rigid.

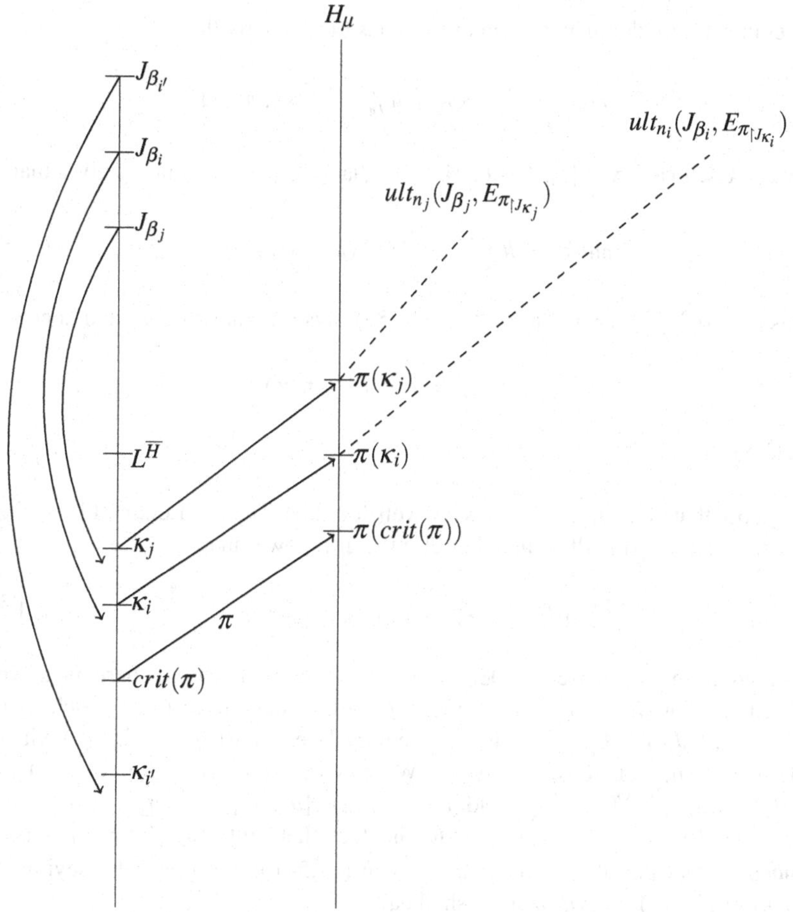

We now aim to show by induction on $i \in I$ that $\pi''\kappa_i \in L$. This is trivial for $i \in I$ such that $\kappa_i \leq \text{crit}(\pi)$, as then $\pi''\kappa_i = \kappa_i \in L$. Now suppose inductively that $\pi''\kappa_i \in L$, where $i \in I$. If $i = \alpha$, then we are done. Otherwise let J be the least element of $I \setminus (i+1)$. We may assume that $\kappa_j > \text{crit}(\pi)$, as otherwise again trivially $\pi''\kappa_j = \kappa_j \in L$. We must then have that $\beta_j < \infty$, as otherwise the ultrapower map

$$\tilde{\pi}\colon L \to \text{ult}_0(L; E_{\pi \restriction J_{\kappa_j}}) \cong L$$

would witness that L is not rigid. But then $\rho_{n_j+1}(J_{\beta_j}) \leq \kappa_i$. Let

$$\tilde{\pi}\colon J_{\beta_j} \to \text{ult}_{n_j}(J_{\beta_j}; E_{\pi \restriction J_{\kappa_j}}) \cong J_\beta \tag{11.51}$$

be the ultrapower map. Let $p = p(J_{\beta_j}) \restriction (n_j + 1)$. By Lemmas 11.53 and 11.25,

$$J_{\beta_j} = h_{J_{\beta_j}}^{n_j+1,p''}(\rho_{n_j+1}(J_{\beta_j})). \tag{11.52}$$

By Lemma 11.29, the map $\tilde{\pi}$ from (11.51) has the property that

$$\tilde{\pi}(h_{J_{\beta_j}}^{n_j+1,p}(k, \mathbf{x})) = h_{J_\beta}^{n_j+1,\tilde{\pi}(p)}(k, \pi(\mathbf{x}))$$

for every $k < \omega$ and $\mathbf{x} \in [\rho_{n_j+1}(J_{\beta_j})]^{<\omega}$, so that (11.52) immediately gives that

$$\text{ran}(\tilde{\pi}) = h_{J_\beta}^{n_j+1,\tilde{\pi}(p)''}(\pi''(\rho_{n_j+1}(J_{\beta_j}))). \tag{11.53}$$

But as $\rho_{n_j+1}(J_{\beta_j}) \leq \kappa_i$ and $\pi''\kappa_i \in L$, (11.53) says that $\text{ran}(\tilde{\pi}) \in L$, and hence

$$\pi''\kappa_j = \tilde{\pi}''\kappa_j = \tilde{\pi}(\kappa_j) \cap \text{ran}(\tilde{\pi}) \in L,$$

as desired. □

Suppose that $\text{ult}_{n_i}(J_{\beta_i}; E_{\pi \restriction J_{\kappa_i}})$ is *not* well-founded. Then by Lemma 11.31, either $\text{ult}_0((J_{\beta_i})^{n_i}; E_{\pi \restriction J_{\kappa_i}})$ is ill-founded or else the ultrapower map

$$\tilde{\pi}\colon (J_{\beta_i})^{n_i} \to_{\Sigma_1} \text{ult}_0((J_{\beta_i})^{n_i}; E_{\pi \restriction J_{\kappa_i}}) \tag{11.54}$$

is not good in the sense of Definition 11.19. In both cases, there is a well-founded relation $\bar{R} \subset ((J_{\beta_i})^n)^2$ which is rudimentary over $(J_{\beta_i})^n$ such that if $R \subset (\text{ult}_0((J_{\beta_i})^{n_i}; E_{\pi \restriction J_{\kappa_i}}))^2$ is rudimentary over $\text{ult}_0((J_{\beta_i})^{n_i}; E_{\pi \restriction J_{\kappa_i}})$ via the same definition, then R is ill-founded. We may then pick $([a_k, f_k]\colon k < \omega)$ with $[a_k, f_k] \in \text{ult}_0((J_{\beta_i})^{n_i}; E_{\pi \restriction J_{\kappa_i}})$ and $[a_{k+1}, f_{k+1}]R[a_k, f_k]$ for every $k < \omega$.

In what follows, we shall refer to the fact that "$\text{ult}_0((J_{\beta_i})^{n_i}; E_{\pi \restriction J_{\kappa_i}})$ is ill-founded or that the ultrapower map $\tilde{\pi}$ as in (11.54) is not good" by saying that "$\text{ult}_{n_i}(J_{\beta_i}; E_{\pi \restriction J_{\kappa_i}})$ *is bad*," and we shall call

$$(\bar{R}, R, ([a_k, f_k]\colon k < \omega)) \tag{11.55}$$

a "*badness witness for* $\text{ult}_{n_i}(J_{\beta_i}; E_{\pi \restriction J_{\kappa_i}})$," provided that $\bar{R}$, R, and $([a_k, f_k]\colon k < \omega)$ are as in the preceding paragraph.

If $X \prec H_\mu$ then we let π_X denote the inverse of the transfinite collapse of X. In the light of Claim 11.58, in order to finish the proof of (3) $\Longrightarrow$ (2) of Theorem 11.56 it suffices to verify the following.

Claim 11.59 *There is a stationary set $\mathcal{S} \subset [\theta]^\kappa$ such that wherever $X \in \mathcal{S}$, then for all $i \in I^{\pi_X}$,*

$$ult_{n_i}^{\pi_X}(J_{\beta_i^{\pi_X}}; E_{\pi_X \restriction J_{\kappa_i^{\pi_X}}})$$

is not bad.

Proof Let $\mathfrak{A} = (\theta; (f_i : i < \kappa))$ be given. We recursively define sequences $(Y_i : i \leq \kappa)$, $(\overline{H}_i : i \leq \kappa)$ and $(\pi_i : i \leq \kappa)$ such that the following hold true.

(1) $Y_i \prec H_\mu$, for all $i \leq \kappa$
(2) $\overline{\overline{Y_i}} < \kappa$, for all $i < \kappa$
(3) $Y_\lambda = \bigcup_{i<\lambda} Y_i$ for all limit ordinals $\lambda \leq \kappa$
(4) $Y_{i+1} \supset f_j{}''Y_i^{<\omega}$ for all $j < i < \kappa$
(5) $\pi_i : \overline{H}_i \cong Y_i$, where $\overline{H}_i$ is transitive, and
(6) Suppose that $j \in I^{\pi_i}$ and $\text{ult}_{n_j^{\pi_i}}(J_{\beta_j^{\pi_i}}; E_{\pi_i \restriction J_{\kappa_j^{\pi_i}}})$ is bad. Then for every $\bar{R}$ such that there is a badness witness $(\bar{R}, R, ([a_k, f_k] : k < \omega))$ for $\text{ult}_{n_j^{\pi_i}}(J_{\beta_j^{\pi_i}}; E_{\pi_i \restriction J_{\kappa_j^{\pi_i}}})$, there is a badness witness $(\bar{R}, R, ([a_k^{i,j}, f_k^{i,j}] : k < \omega))$ for $\text{ult}_{n_j^{\pi_i}}(J_{\beta_j^{\pi_i}}; E_{\pi_i \restriction J_{\kappa_j^{\pi_i}}})$ with the property that $\{a_k^{i,j} : k < \omega\} \subset Y_{i+1}$.

Let us write $\pi = \pi_\kappa$. We claim that for all $j \in I^\pi$, $\text{ult}_0((J_{\beta_i^\pi})^{n_j^\pi}; E_{\pi \restriction J_{\kappa_j^\pi}})$ is well-founded and the ultrapower map

$$\tilde{\pi}: (J_{\beta_i^\pi})^{n_j^\pi} \to \text{ult}_0((J_{\beta_i^\pi})^{n_j^\pi}; E_{\pi \restriction J_{\kappa_j^\pi}})$$

is good. This implies that for all $j \in I^\pi$, $\text{ult}_{n_j^\pi}(J_{\beta_i^\pi}; E_{\pi \restriction J_{\kappa_j^\pi}})$ is well-founded by Lemma 11.31. By Claim 11.58, this then finishes the proof of (3) $\Longrightarrow$ (2) of Theorem 11.56, as obviously Y_κ is closed under all functions from $\mathfrak{A}$. Let us assume π is not as claimed and work towards a contradiction.

We write $Y = Y_\kappa$ and $\overline{H} = \overline{H}_\kappa$. We also write $I = I^\pi$, $\beta_i = \beta_i^\pi$, etc. By assumption, there is some $j \in I$ such that $\text{ult}_{n_j}(J_{\beta_j}; E_{\pi \restriction J_{\kappa_j}})$ is bad. Let $(\bar{R}, R, ([a_k, f_k] : k < \omega))$ be a badness witness for $\text{ult}_{n_j}(J_{\beta_j}; E_{\pi \restriction J_{\kappa_j}})$.

Let us write $\overline{\pi}_i = \pi^{-1} \circ \pi_i : \overline{H}_i \to \overline{H}$ and $\overline{Y}_i = \pi^{-1}{}''Y_i = \text{ran}(\overline{\pi}_i)$. Hence $\overline{Y}_i \prec \overline{H}$, $\overline{Y}_i \subset \overline{Y}_l$ for $i \leq l$, $\text{Card}(\overline{Y}_i) < \kappa$ for $i < \kappa$, $\overline{Y}_\lambda = \bigcup_{i<\lambda} \overline{Y}_i$ for limit ordinals $\lambda \leq \kappa$ and $\overline{H} = \bigcup_{i<\kappa} \overline{Y}_i$. Let $\tilde{\mu} > \mu$ be some sufficiently large regular cardinal. As κ is regular, we may pick some $Z \prec H_{\tilde{\mu}}$ such that $(\overline{Y}_i : i \leq \kappa) \in Z$ and $Z \cap \kappa \in \kappa$. If $i_0 = Z \cap \kappa$, then $Z \cap \overline{H} = \overline{Y}_{i_0}$. We may thus assume that $Z \prec H_{\tilde{\mu}}$ is such that

(1)' $\overline{\overline{Z}} < \kappa$,
(2)' $\{f_k : k < \omega\} \cup \{\bar{R}, \overline{H}\} \subset Z$, and
(3)' $Z \cap \overline{H} = \overline{Y}_{i_0}$ for some $i_0 < \kappa$.

Let us write $\sigma: \tilde{H} \cong Z$, where $\tilde{H}$ is transitive. Let us also write $R' = \sigma^{-1}(\bar{R})$ and $\overline{f}_k = \sigma^{-1}(f_k)$ for $k < \omega$. Obviously, $\sigma^{-1}(\overline{H}) = \overline{H}_{i_0}$ and $\sigma \restriction \overline{H}_{i_0} = \overline{\pi}_{i_0}$. Let $k < \omega$. We then have $[a_{k+1}, f_{k+1}]R[a_k, f_k]$ in $\text{ult}_0((J_{\beta_j})^{n_j}; E_{\pi \restriction J_{\kappa_j}})$, and hence

$$
\begin{aligned}
(a_{k+1}, a_k) &\in \pi(\{(u,v) : f_{k+1}(u)\bar{R}f_k(v)\}) \\
&= \pi(\{(u,v) : \sigma(\overline{f}_{k+1})(u)\sigma(R')\sigma(\overline{f}_k)(v)\}) \\
&= \pi \circ \sigma(\{(u,v) : \overline{f}_{k+1}(u)R'\overline{f}_k(v)\}) \\
&= \pi \circ \overline{\pi}_{i_0}(\{(u,v) : \overline{f}_{k+1}(u)R'\overline{f}_k(v)\}), \text{ as } \sigma \restriction \overline{H}_{i_0} = \overline{\pi}_{i_0}, \\
&= \pi_{i_0}(\{(u,v) : \overline{f}_{k+1}(u)R'\overline{f}_k(v)\}).
\end{aligned}
$$

We may assume that $\beta_j \in \text{ran}(\sigma)$ if $\beta_j < \infty$. Let $\overline{\beta} = \sigma^{-1}(\beta_j)$ if $\beta_j < \infty$, $\overline{\beta} = \tilde{H} \cap \text{OR}$ otherwise. Write $n = n_j$, $\overline{\kappa} = \sigma^{-1}(\kappa_j)$.

We have that $\{[a_k, \overline{f}_\kappa] : k < \omega\} \subset \text{ult}_0((J_{\overline{\beta}})^n; E_{\pi_{i_0} \restriction J_{\overline{\kappa}}})$, and R' is defined over $(J_{\overline{\beta}})^n$ in the same way as $\bar{R}$ is defined over $(J_{\beta_j})^{n_j}$. Also, if $\overline{\kappa} = \kappa_l^{\pi_{i_0}}$, then $\overline{\beta} \leq \beta_l^{\pi_{i_0}}$ and $n = n_l^{\pi_{i_0}}$. Therefore,

$$
\text{ult}_{n_l^{\pi_{i_0}}}(J_{\beta_l^{\pi_{i_0}}}; E_{\pi_{i_0} \restriction J_{\kappa_l^{\pi_{i_0}}}}) \tag{11.56}
$$

is bad. Hence by (6) there is a badness witness $(R', R^*, ([a_k^{i_0,l}, f_k^{i_0,l}] : k < \omega))$ for (11.56) such that $\{a_k^{i_0,l} : k < \omega\} \subset Y_{i_0+1} \subset Y$.

Let $k < \omega$. Then $[a_{k+1}^{i_0,l}, f_{k+1}^{i_0,l}]R^*[a_k^{i_0,l}, f_k^{i_0,l}]$ gives that

$$
\begin{aligned}
\left(a_{k+1}^{i_0,l}, a_k^{i_0,l}\right) &\in \pi_{i_0}(\{(u,v) : f_{k+1}^{i_0,l}(u)R' f_k^{i_0,l}(v)\}) \\
&= \pi \circ \overline{\pi}_{i_0}(\{(u,v) : f_{k+1}^{i_0,l}(u)R' f_k^{i_0,l}(v)\}) \\
&= \pi \circ \sigma(\{(u,v) : f_{k+1}^{i_0,l}(u)R' f_k^{i_0,l}(v)\}) \\
&= \pi\left(\left\{(u,v) : \sigma\left(f_{k+1}^{i_0,l}\right)(u)\bar{R}\sigma\left(f_k^{i_0,l}\right)(v)\right\}\right).
\end{aligned}
$$

However, $a_{k+1}^{i_0,l}, a_k^{i_0,l} \in Y = \text{ran}(\pi)$, so that this gives that

$$
\left(\pi^{-1}\left(a_{k+1}^{i_0,l}\right), \pi^{-1}\left(a_k^{i_0,l}\right)\right) \in \left\{(u,v) : \sigma\left(f_{k+1}^{i_0,l}\right)(u)\bar{R}\sigma\left(f_k^{i_0,l}\right)(v)\right\},
$$

and therefore

$$
\sigma\left(f_{k+1}^{i_0,l}\right)\left(\pi^{-1}\left(a_{k+1}^{i_0,l}\right)\right)\bar{R}\sigma\left(f_k^{i_0,l}\right)\left(\pi^{-1}\left(a_k^{i_0,l}\right)\right).
$$

Because this holds for all $k < \omega$, $\bar{R}$ is ill-founded. Contradiction! □

This finishes the proof of Theorem 11.56. □

Corollary 11.60 (Weak covering for L) *Suppose that* $0^{\#}$ *does not exist. If* $\kappa \geq \aleph_2$ *is a cardinal in L, then in V,* $\text{cf}(\kappa^{+L}) \geq \text{Card}(\kappa)$. *In particular,* $\kappa^{+L} = \kappa^+$ *for every singular cardinal* κ.

Proof Assume that $0^\#$ does not exist, and let $\kappa \geq \aleph_2$ be a cardinal in L. Suppose that in V, $\text{cf}(\kappa^{+L}) < \text{Card}(\kappa)$, and let $X \subset \kappa^{+L}$ be cofinal with $\text{Card}(X) < \text{Card}(\kappa)$. By Theorem 11.56 (1), there is some $Y \in L$ such that $Y \supset X$ and $\text{Card}(Y) \leq \text{Card}(X) + \aleph_1$. As $\text{Card}(\kappa) \geq \aleph_2$, $\text{Card}(Y) < \text{Card}(\kappa)$. This implies that $\text{otp}(Y) < (\text{Card}(Y))^+ \leq \text{Card}(\kappa) < \kappa^{+L}$. Contradiction!

If κ is singular, then $\text{cf}(\kappa^{+L}) \neq \kappa$, so that $\text{cf}(\kappa^{+L}) \geq \text{Card}(\kappa) = \kappa$ implies $\kappa^{+L} = \kappa^+$. □

Corollary 11.61 *Suppose that* $0^\#$ *does not exist. Then* SCH, *the Singular Cardinal Hypothesis, holds true.*

Proof Assume that $0^\#$ does not exist. Let κ be a (singular) limit cardinal. We need to see that $\kappa^{\text{cf}(\kappa)} \leq \kappa^+ \cdot 2^{\text{cf}(\kappa)}$ (which implies that in fact $\kappa^{\text{cf}(\kappa)} = \kappa^+ \cdot 2^{\text{cf}(\kappa)}$). If $X \in [\kappa]^{\text{cf}(\kappa)}$, then by Theorem 11.56 (1) there is some $Y \in [\kappa]^{\text{cf}(\kappa)\cdot\aleph_1} \cap L$ such that $Y \supset X$. On the other hand, for any $Y \in [\kappa]^{\text{cf}(\kappa)\cdot\aleph_1}$ there are $(\text{cf}(\kappa)\cdot\aleph_1)^{\text{cf}(\kappa)} = 2^{\text{cf}(\kappa)}$ many $X \subset Y$ with $\text{Card}(X) = \text{cf}(\kappa)$. Moreover, as the GCH is true in L, $[\kappa]^{\text{cf}(\kappa)\cdot\aleph_1} \cap L$ has size at most κ^+. Therefore, $\kappa^{\text{cf}(\kappa)} \leq \kappa^+ \cdot 2^{\text{cf}(\kappa)}$. □

11.3 $\square_\kappa$ and Its Failure

We now aim to prove $\square_\kappa$ in L. This is the combinatorial principle the proof of which most heavily exploits the fine structure theory.

Definition 11.62 Let κ be an infinite cardinal, and let $R \subset \kappa^+$. We say that $\square_\kappa(R)$ holds if and only if there is a sequence $(C_\nu : \nu < \kappa^+)$ such that if ν is a limit ordinal, $\kappa < \nu < \kappa^+$, then C_ν is a club subset of ν with $\text{otp}(C_\nu) \leq \kappa$ and whenever $\bar{\nu}$ is a limit point of C_ν then $\bar{\nu} \notin R$ and $C_{\bar{\nu}} = C_\nu \cap \bar{\nu}$. We write $\square_\kappa$ for $\square_\kappa(\emptyset)$.

In order to prove $\square_\kappa$ in L, we need the "Interpolation Lemma." The proof is very similar to the proof of Lemma 10.56, and we omit it. Recall the concept of a weakly $r\Sigma_{n+1}$ elementary embedding, cf. Definition 11.32.

Lemma 11.63 *Let* $n < \omega$. *Let* $\bar{M}$, M *be an acceptable* J*-structure, and let*

$$\pi : \bar{M} \longrightarrow M$$

be $r\Sigma_{n+1}$ *elementary. Let* $\nu \leq M \cap \text{OR}$, *and let* E *be the* (κ, ν)*-extender derived from* π.

There is then a weakly $r\Sigma_{n+1}$ *elementary embedding*

$$\sigma : \text{ult}_n(\bar{M}; E) \to M$$

such that $\sigma \upharpoonright \nu = \text{id}$ *and* $\sigma \circ \pi_E = \pi$.

Theorem 11.64 (R. Jensen) *Suppose that* $V = L$. *Let* $\kappa \geq \aleph_1$ *be a cardinal. Then* $\square_\kappa$ *holds.*

Proof This proof is in need of the fact that every level of the L-hierarchy is solid, cf. Definition 11.40 and Theorem 11.54.

Let us set $C = \{\nu < \kappa^+ : J_\nu \prec_{\Sigma_\omega} J_{\kappa^+}\}$, which is a club subset of κ^+ consisting of limit ordinals above κ.

Let $\nu \in C$. Obviously, κ is the largest cardinal of J_ν. We may let $\alpha(\nu)$ be the largest $\alpha \geq \nu$ such that either $\alpha = \nu$ or ν is a cardinal in J_α. By Lemma 5.15, $\rho_\omega(J_{\alpha(\nu)}) = \kappa$. Let $n(\nu)$ be the unique $n < \omega$ such that $\kappa = \rho_{n+1}(J_{\alpha(\nu)}) < \nu \leq \rho_n(J_{\alpha(\nu)})$.

If $\nu \in C$, then we define D_ν as follows. We let D_ν consist of all $\bar{\nu} \in C \cap \nu$ such that $n(\bar{\nu}) = n(\nu)$ and there is a weakly $r\Sigma_{n(\nu)+1}$ elementary embedding

$$\sigma \colon J_{\alpha(\bar{\nu})} \longrightarrow J_{\alpha(\nu)}.$$

such that $\sigma \restriction \bar{\nu} = \text{id}$, $\sigma(p_{n(\bar{\nu})+1}(J_{\alpha(\bar{\nu})})) = p_{n(\nu)+1}(J_{\alpha(\nu)})$, and if $\bar{\nu} \in J_{\alpha(\bar{\nu})}$, then $\nu \in J_{\alpha(\nu)}$ and $\sigma(\bar{\nu}) = \nu$. It is easy to see that if $\bar{\nu} \in D_\nu$, then by Lemma 11.53 there is exactly one map σ witnessing this, namely the one which is given by

$$h_{J_{\alpha(\bar{\nu})}}^{n(\bar{\nu})+1, p_{n(\bar{\nu})+1}(J_{\alpha(\bar{\nu})})}(i, \mathbf{x}) \mapsto h_{J_{\alpha(\nu)}}^{n(\nu)+1, p_{n(\nu)+1}(J_{\alpha(\nu)})}(i, \mathbf{x}), \tag{11.57}$$

where $i < \omega$ and $x \in [\kappa]^{<\omega}$. We shall denote this map by $\sigma_{\bar{\nu},\nu}$.

Notice that if $\nu \in C$, then again by Lemma 11.53

$$J_\alpha(\nu) = h_{J_{\alpha(\nu)}}^{n(\nu)+1, p_{n(\nu)+1}(J_{\alpha(\nu)})''} \kappa,$$

so that if $\bar{\nu} \in D_\nu$, then

$$\text{ran}(\sigma_{\bar{\nu},\nu}) \subsetneq h_{J_{\alpha(\nu)}}^{n(\nu)+1, p_{n(\nu)+1}(J_{\alpha(\nu)})''} \kappa,$$

which means that there must be $i < \omega$ and $\xi < \kappa$ such that the left hand side of (11.57) is undefined, whereas the right hand side of (11.57) is defined.

Notice that the maps $\sigma_{\bar{\nu},\nu}$ trivially commute, i.e., if $\bar{\nu} \in D_\nu$ and $\nu \in D_{\nu'}$, then $\bar{\nu} \in D_{\nu'}$ and

$$\sigma_{\bar{\nu},\nu'} = \sigma_{\nu,\nu'} \circ \sigma_{\bar{\nu},\nu}.$$

Claim 11.65 *Let $\nu \in C$. The following hold true.*

(a) *D_ν is closed.*
(b) *If $\text{cf}(\nu) > \omega$, then D_ν is unbounded in ν.*
(c) *If $\bar{\nu} \in D_\nu$ then $D_\nu \cap \bar{\nu} = D_{\bar{\nu}}$.*

Proof (a) is easy. Let ν' be a limit point of D_ν. If $\mu \leq \mu' < \nu'$, $\mu, \mu' \in D_\nu$, then

$$\text{ran}(\sigma_{\mu,\nu}) \subset \text{ran}(\sigma_{\mu',\nu}).$$

We then have that the inverse of the transitive collapse of

$$\bigcup_{\mu \in D_\nu \cap \nu'} \operatorname{ran}(\sigma_{\mu,\nu})$$

proves that $\nu' \in D_\nu$.

Let us now show (b). Suppose that $\operatorname{cf}(\nu) > \omega$. Set $\alpha = \alpha(\nu)$ and $n = n(\nu)$. Let $\beta < \nu$. We aim to show that $D_\nu \setminus \beta \neq \emptyset$.

Let $\pi\colon J_{\bar\alpha} \longrightarrow_{\Sigma_{n+1}} J_\alpha$ be such that $\bar\alpha$ is countable, $\beta \in \operatorname{ran}(\pi)$, and

$$\{W^{\nu, p_1(J^k_\alpha)}_{J^k_\alpha} : \nu \in p_1(J^k_\alpha), k \leq n\} \subset \operatorname{ran}(\pi).$$

Let $\bar\nu = \pi^{-1}(\nu)$ (if $\nu = \alpha$, we mean $\bar\nu = \bar\alpha$). Let

$$\pi' = \pi_{E_\pi \restriction J_{\bar\nu}} : J_{\bar\alpha} \longrightarrow_{r\Sigma_{n+1}} \operatorname{ult}_n(J_{\bar\alpha}; E_\pi \restriction J_{\bar\nu}).$$

By Lemma 11.63, we may define a weakly $r\Sigma_{n+1}$ elementary embedding

$$k\colon \operatorname{ult}_n(J_{\bar\alpha}; E_\pi \restriction J_{\bar\nu}) \longrightarrow J_\alpha$$

with $k \circ \pi' = \pi$. In particular, $\operatorname{ult}_n(J_{\bar\alpha}; E_\pi \restriction J_{\bar\nu})$ is well-founded and we may identify it with its transitive collapse. Let us write $J_{\alpha'} = \operatorname{ult}_n(J_{\bar\alpha}; E_\pi \restriction J_{\bar\nu})$. As $\beta \in \operatorname{ran}(\pi)$, $k^{-1}(\nu) > \beta$. Moreover, $k^{-1}(\nu) = \sup \pi''\bar\nu < \nu$, as $\operatorname{cf}(\nu) > \omega$. Therefore $\beta < k^{-1}(\nu) \in D_\nu$. This shows (b).

Let us now verify (c). If $\bar\nu \in D_\nu$, then $D_{\bar\nu} \subset D_\nu$. To show (c), we thus let $\mu < \bar\nu$ be such that μ and $\bar\nu$ are both in D_ν. We need to see that $\mu \in D_{\bar\nu}$, i.e., that $\sigma_{\mu,\bar\nu}$ is well-defined. For the purpose of this proof, let us, for $\gamma \in \{\mu, \bar\nu, \nu\}$, abbreviate

$$h^{n(\gamma)+1, p_{n(\gamma)+1}(J_{\alpha(\gamma)})}_{J_{\alpha(\gamma)}}(i, \mathbf{x}) \text{ by } h_\gamma(i, \mathbf{x}),$$

where $i < \omega$ and $x \in [\kappa]^{<\omega}$. We need to see that if $i < \omega$, $\mathbf{x} \in [\kappa]^{<\omega}$, and $h_\mu(i, \mathbf{x})$ exists, then $h_{\bar\nu}(i, \mathbf{x})$ exists as well.

Let $e < \omega$ and $\mathbf{y} \in [\kappa]^{<\omega}$ be such that $\mu = h_{\bar\nu}(e, \mathbf{y})$. We obviously cannot have that $h_\mu(e, \mathbf{y})$ exists, as otherwise

$$\sigma_{\mu,\nu}(h_\mu(e, \mathbf{y})) = h_\nu(e, \mathbf{y}) = \sigma_{\bar\nu,\nu}(h_{\bar\nu}(e, \mathbf{y})) = \sigma_{\bar\nu,\nu}(\mu) = \mu,$$

but $\mu \notin \operatorname{ran}(\sigma_{\mu,\nu})$. Write $n = n(\mu) = n(\bar\nu) = n(\nu)$. Let $\beta < \rho_n(J_{\alpha(\bar\nu)})$ be least such that S_β contains a witness to "$\exists v\, v = h_{(J_{\alpha(\bar\nu)})^n}(u(e), (\mathbf{y}, p(J_{\alpha(\bar\nu)})(n)))$," cf. (11.29). We claim that

$$\operatorname{ran}(\sigma_{\mu,\nu}) \cap (J_{\alpha(\nu)})^n \subset \sigma_{\bar\nu,\nu}(S_\beta). \tag{11.58}$$

If (11.58) is false, then we may pick some $\gamma < \rho_n(J_{\alpha(\mu)})$ such that $\sigma_{\mu,\nu}(\gamma) \geq \sigma_{\bar\nu,\nu}(\beta)$. We may then write "$h_\nu(e, \mathbf{y}) = \mu$" as a statement over $J_{\alpha(\nu)}$ in the parameter

$J_{\sigma_{\mu,\nu}(\gamma)}$ in a way that it is preserved by the weak $r\Sigma_{n+1}$ elementarity of $\sigma_{\mu,\nu}$ to yield that $h_\mu(e, \mathbf{y})$ exists after all. Contradiction! Therefore, (11.58) holds true.

But now if $i < \omega$, $\mathbf{x} \in [\kappa]^{<\omega}$, and $h_\mu(i, \mathbf{x})$ exists, then "$\exists v\, v = h_\mu(i, \mathbf{x})$" may be expressed by a statement over $J_{\alpha(\nu)}$ in the parameter $J_{\bar{\nu},\nu}(\beta)$ in a way that it is preserved by the weak $r\Sigma_{n+1}$ elementarity of $\sigma_{\bar{\nu},\nu}$ to give that $h_{\bar{\nu}}(i, \mathbf{x})$ also exists.

Claim 11.65 is shown. $\square$

Now let $\nu \in C$. We aim to define C_ν. Set $\alpha = \alpha(\nu)$ and $n = n(\nu)$. Recursively, we define sequences $(\mu_i^\nu : i \leq \theta(\nu))$ and $(\xi_i^\nu : i < \theta(\nu))$ as follows. Set $\mu_0^\nu = \min(D_\nu)$. Given μ_i^ν with $\mu_i^\nu < \nu$, we let ξ_i^ν be the least $\xi < \kappa$ such that

$$h_{J_\alpha}^{n+1, p_{n+1}(J_\alpha)}(k, \mathbf{x}) \notin \operatorname{ran}(\sigma_{\mu_i^\nu,\nu})$$

for some $k < \omega$ and some $\mathbf{x} \in [\xi]^{<\omega}$. Given ξ_i^ν, we let μ_{i+1}^ν be the least $\bar{\nu} \in D_\nu$ such that

$$h_{J_\alpha}^{n+1, p_{n+1}(J_\alpha)}(k, \mathbf{x}) \in \operatorname{ran}(\sigma_{\bar{\nu},\nu})$$

for all $k < \omega$ and $\mathbf{x} \in [\xi_i^\nu]^{<\omega}$ such that $h_{J_\alpha}^{n+1, p_{n+1}(J_\alpha)}(k, \mathbf{x})$ exists. Finally, given $(\mu_i^\nu : i < \lambda)$, where λ is a limit ordinal, we set $\mu_\lambda^\nu = \sup(\{\mu_i^\nu : i < \lambda\})$. Naturally, $\theta(\nu)$ will be the least i such that $\mu_i^\nu = \nu$. We set $C_\nu = \{\mu_i^\nu : i < \theta(\nu)\}$.

Claim 11.66 *Let $\nu \in C$. The following hold true.*

(a) *$(\xi_i : i < \theta(\nu))$ is strictly increasing.*
(b) *$\operatorname{otp}(C_\nu) = \theta(\nu) \leq \kappa$.*
(c) *C_γ is closed.*
(d) *If $\bar{\nu} \in C_\nu$ then $C_\nu \cap \bar{\nu} = C_{\bar{\nu}}$.*
(e) *If D_ν is unbounded in ν then so is C_ν.*

Proof (a) is immediate, and it implies (b). (c) and (e) are trivial.

Let us show (d). Let $\bar{\nu} \in C_\nu$. We have $C_\nu \subset D_\nu$, and $D_{\bar{\nu}} = D_\nu \cap \bar{\nu}$ by Claim 1 (c). We may then show that $(\mu_i^{\bar{\nu}} : i < \theta(\bar{\nu})) = (\mu_i^\nu : i < \theta(\bar{\nu}))$ and $(\xi_i^{\bar{\nu}} : i < \theta(\bar{\nu})) = (\xi_i^\nu : i < \theta(\bar{\nu}))$ by an induction. Say $\mu_i^{\bar{\nu}} = \mu_i^\nu$, where $i + 1 \in \theta(\bar{\nu}) \cap \theta(\nu)$. Write $\mu = \mu_i^{\bar{\nu}} = \mu_i^\nu$. As $\sigma_{\mu,\nu} = \sigma_{\bar{\nu},\nu} \circ \sigma_{\mu,\bar{\nu}}$, for all $k < \omega$ and $\mathbf{x} \in [\kappa]^{<\omega}$,

$$h_{J_{\alpha(\bar{\nu})}}^{n(\bar{\nu})+1, p_{n(\bar{\nu})+1}(J_{\alpha(\bar{\nu})})}(k, \mathbf{x}) \in \operatorname{ran}(\sigma_{\mu,\bar{\nu}}) \Longrightarrow$$

$$h_{J_{\alpha(\nu)}}^{n(\nu)+1, p_{n(\nu)+1}(J_{\alpha(\nu)})}(k, \xi) \in \operatorname{ran}(\sigma_{\mu,\nu}) \neq \emptyset.$$

This gives $\mu_{i+1}^\nu \leq \mu_{i+1}^{\bar{\nu}}$ On the other hand, $\operatorname{ran}(\sigma_{\mu_{i+1}^\nu,\nu})$ contains the relevant witness so as to guarantee conversely that $\mu_{i+1}^{\bar{\nu}} \leq \mu_{i+1}^\nu$. $\square$

Now let $f : \kappa^+ \to C$ be the monotone enumeration of C. For $\nu < \kappa^+$, let us set

$$\bar{B}_\nu = f^{-1\prime\prime} C_{f(\nu)}.$$

Of course, $\text{otp}(\bar{B}_\nu) \leq \kappa$ for every $\nu < \kappa^+$, and because $C_\mu \subset D_\mu \subset C \cap \mu$ for every $\mu \in C$ we have that every $\bar{B}_\nu$ is closed, if $\text{cf}(\nu) > \omega$, then $\bar{B}_\nu$ is unbounded in ν, and if $\bar{\nu} \in \bar{B}_\nu$, then $\bar{B}_{\bar{\nu}} = \bar{B}_\nu \cap \bar{\nu}$. For every $\nu < \kappa^+$ such that $\text{cf}(\nu) = \omega$ and C_ν is not unbounded in ν, let us pick some $\bar{B}'_\nu$ of order type ω which is cofinal in ν. For $\nu < \kappa^+$, let

$$B_\nu = \begin{cases} \bar{B}'_\nu & \text{if } \bar{B}'_\nu \text{ is defined, and} \\ \bar{B}_\nu & \text{otherwise.} \end{cases}$$

It is easy to see now that $(B_\nu \colon \nu < \kappa^+)$ witnesses that $\square_\kappa$ holds true. $\square$

Corollary 11.60 and Theorem 11.64 immediately imply the following.

Corollary 11.67 *Let κ be a singular cardinal, and suppose that $\square_\kappa$ fails. Then $0^\#$ exists.*

Let $(C_\nu \colon \nu < \kappa^+)$ witness that $\square_\kappa$ holds true. By FODOR's Theorem 4.32, there must be some stationary $R \subset \kappa^+$ such that $\text{otp}(C_\nu)$ is constant on R, say $\theta = \text{otp}(C_\nu)$ for all $\nu \in R$. For any $\nu < \kappa^+$, C_ν can have at most one (limit) point β such that $C_\nu \cap \beta = C_\beta$ has order type θ. We may then define

$$C^*_\nu = \begin{cases} C_\nu \setminus (\alpha+1) & \text{if } \alpha \in C_\nu \text{ and } \text{otp}(C_\nu \cap \alpha) = \theta \\ C_\nu & \text{otherwise.} \end{cases}$$

$(C^*_\nu \colon \nu < \kappa^+)$ then witnesses $\square_\kappa(R)$, cf. also Problem 11.17.

The following result generalizes Lemma 5.36, as $\square_{\aleph_0}$ is provable in ZFC.

Lemma 11.68 (Jensen) *Let κ be an infinite cardinal. Suppose that there is some stationary set $R \subset \kappa^+$ such that both $\diamondsuit_{\kappa^+}(R)$ and $\square_\kappa(R)$ hold true. There is then a κ^+-Souslin tree.*

Proof Let $R \subset \kappa^+$ be stationary such that $\diamondsuit_{\kappa^+}(R)$ and $\square_\kappa(R)$ both hold true. Let $(S_\alpha \colon \alpha \in R)$ witness $\diamondsuit_{\kappa^+}(R)$, and let $(C_\alpha \colon \alpha < \kappa^+)$ witness $\square_\kappa(R)$. We will construct $T \upharpoonright \alpha$ by induction on $\alpha < \kappa^+$ in such a way that the underlying set of $T \upharpoonright \alpha$ will always be an ordinal below κ^+ (which we will also denote by $T \upharpoonright \alpha$).

We set $T \upharpoonright 1 = \{0\}$. Now let $\alpha < \kappa$ be such that $T \upharpoonright \alpha$ has already been constructed. If α is a successor ordinal, then we use the next $(T \upharpoonright \alpha - T \upharpoonright (\alpha-1)) \cdot 2$ ordinals above $T \upharpoonright \alpha$ to provide each top node of $T \upharpoonright \alpha$ with an immediate successor in $T \upharpoonright (\alpha+1)$.

Let us then suppose α to be a limit ordinal. Let us set $S = \{s \in T \upharpoonright \alpha \colon \exists t \in S_\alpha (t = s \vee t <_T s)\}$ if $\alpha \in R$ and S_α happens to be a maximal antichain of $T \upharpoonright \alpha$, and let us otherwise set $S = T \upharpoonright \alpha$.

Let us for a moment fix $s \in S$. We aim to define a chain c_s through $T \upharpoonright \alpha$ as follows. Let $\varepsilon \in C_\alpha$ be least such that in fact $s \in T \upharpoonright \varepsilon$. Let $(\delta_i \colon i < \gamma)$ be the monotone enumeration of $C_\alpha \setminus \varepsilon$. We then let s_0 be the least ordinal such that $s_0 \in T \upharpoonright (\delta_0+1) \setminus T \upharpoonright \delta_0$ and $s <_T s_0$, and for $i > 0$, $i < \gamma$, we let s_i be the least ordinal such that $s_i \in T \upharpoonright (\delta_i+1) \setminus T \upharpoonright \delta_i$ and $s_j <_T s_i$ for all $j < i$, if there is one (if not, then we let the construction break down). We set $c_s = \{s_i \colon i < \gamma\}$.

We now use at most the next Card(S) ordinals above $T \upharpoonright \alpha$ to construct $T \upharpoonright (\alpha+1)$ in such a way that for every $s \in S$ there is some $t \in T \upharpoonright (\alpha+1) \setminus T \upharpoonright \alpha$ such that $s' <_T t$ for all $s' \in c_s$ and for every $t \in T \upharpoonright (\alpha+1) \setminus T \upharpoonright \alpha$ there is some $s \in S$ such that $s' <_T t$ for all $s' \in c_s$.

This finishes the construction. It is not hard to verify that the construction in fact never breaks down and produces a κ^+-tree. Otherwise, for some limit ordinal $\alpha < \kappa^+$ there would be some $s \in S$ and some limit ordinal $i < \gamma$ such that there is no $t \in T \upharpoonright (\delta_i + 1) \setminus T \upharpoonright \delta_i$ with $s_j <_T t$ for all $j < i$ (for S, γ, $(\delta_i \colon i < \gamma)$ as above). However, $\delta_i \in C_\alpha$ gives that $\delta_i \notin R$, so that $C_{\delta_i} = C_\alpha \cap \delta_i$ is easily seen to yield that in the construction of $T \upharpoonright (\delta_i + 1) \setminus T \upharpoonright \delta_i$, we did indeed add some $t \in T \upharpoonright (\delta_i + 1) \setminus T \upharpoonright \delta_i$ such that $s_j <_T t$ for all $j < i$.

Finally, suppose that T would not be κ^+-Souslin, and let $A \subset T$ be an antichain of size κ^+. As R is stationary, we may then pick some limit ordinal $\alpha \in R$ such that $A \cap \alpha = S_\alpha$ is a maximal antichain in $T \upharpoonright \alpha$, so that also $S = \{s \in T \upharpoonright \alpha \colon \exists t \in S_\alpha (t = s \vee t <_T s)\}$ (for S as above). The construction of T then gives that every node in $T \setminus T \upharpoonright \alpha$ is above some node in S_α, so that in fact $A = A \cap \alpha = S_\alpha$. But because T is a κ^+-tree, this means that A has size at most κ. Contradiction! □

Lemma 11.69 (Jensen) *Let κ be subcompact. Then $\square_\kappa$ fails.*

Proof Suppose $(C_\alpha \colon \alpha < \kappa^+)$ witnesses $\square_\kappa$. As κ is subcompact, there must then be some $\lambda < \kappa$ together with a witness $(D_\alpha \colon \alpha < \lambda^+)$ to $\square_\lambda$ and an elementary embedding

$$\sigma \colon (H_{\lambda^+}; \in, (D_\alpha \colon \alpha < \lambda^+)) \to (H_{\kappa^+}; \in, (C_\alpha \colon \alpha < \kappa^+))$$

with $\text{crit}(\sigma) = \lambda$. Set $\tau = \sup(\sigma''\lambda^+) < \kappa^+$. As $\sigma''\lambda^+$ is $< \lambda$-club in τ,

$$C = C_\tau \cap \sigma''\lambda^+$$

is also $< \lambda$-club in τ.

Let $\sigma(\alpha) < \sigma(\beta) < \tau$ both be limit points of C. Then

$$C_{\sigma(\beta)} \cap \sigma(\alpha) = C_\tau \cap \sigma(\alpha) = C_{\sigma(\alpha)}. \tag{11.59}$$

But $C_{\sigma(\alpha)} = \sigma(D_\alpha)$ and $C_{\sigma(\beta)} = \sigma(D_\beta)$, so that by (11.59)

$$D_\beta \cap \alpha = D_\alpha.$$

Setting

$$D = \bigcup_{\pi(\alpha) \in C} D_\alpha,$$

we then have that D is cofinal in λ^+ and $D_\alpha = D \cap \alpha$ for every $\pi(\alpha) \in C$. Pick $\pi(\alpha) \in C$ such that $\text{otp}(D \cap \alpha) > \lambda$. Then $\text{otp}(D_\alpha) > \lambda$. Contradiction! □

Lemma 11.70 (Solovay) *Let κ be λ^+-supercompact, where $\lambda \geq \kappa$. Then $\Box_\lambda$ fails. In particular, if κ is supercompact, then $\Box_\lambda$ fails for all $\lambda \geq \kappa$.*

Proof Let κ be λ^+-supercompact, where $\lambda \geq \kappa$, and suppose that $(D_\alpha \colon \alpha < \lambda^+)$ witnesses that $\Box_\lambda$ holds. Let

$$\pi \colon V \to M$$

be such that M is an inner model, $\text{crit}(\pi) = \kappa$, $\pi(\kappa) > \lambda$ and ${}^{\lambda^+}M \subset M$. This implies that $\tau = \sup(\pi''\lambda^+) < \pi(\lambda^+)$. Let $(C_\alpha \colon \alpha < \pi(\lambda^+)) = \pi((D_\alpha \colon \alpha < \lambda^+))$. As $\sigma''\lambda^+$ is $< \kappa$-club in τ,

$$C = C_\tau \cap \sigma''\lambda^+$$

is also $< \kappa$-club in τ. The rest is virtually as in the proof of Lemma 11.69. □

A fine structure theory for inner models is developped e.g. in [30] and [47], and fine structural models with significant large cardinals are constructed e.g. in [2, 31, 40]; cf. also [45] and [46]. Generalizations of JENSEN's Covering Lemma 11.56 are shown in [29] and [28]. In the light of Theorem 11.70, an ultimate generalization of Theorem 11.64 is shown in [35], and an application in the spirit of Corollary 11.67 is given in [17].

11.4 Problems

11.1. Assume GCH to hold in V. Show that there is some $E \subset \text{OR}$ such that $V = L[E]$ and $L[E]$ is acceptable. [Hint. Use Problem 5.12.]

11.2. Let $L[U]$ be as in Problem 10.3. Show that $L[U]$ is not acceptable. Show also that $L[U]$ is *weakly acceptable* in the following sense. If $(\mathscr{P}(\rho) \cap J_{\alpha+\omega}[U]) \setminus J_\alpha[U] \neq \emptyset$, then there is some surjection $f \colon \rho \to \mathscr{P}(\rho) \cap J_\alpha[U]$, $f \in J_{\alpha+\omega}[U]$. [Hint. Problem 10.5.]

11.3. Prove Lemma 11.13!

11.4. Let $\bar{M}, M, \bar{p}, p, \pi, \tilde{\pi}$ be as in Lemma 11.16. Suppose moreover that $p \in R_M$, and that $\pi : \bar{M}^{\bar{p}} \to_{\Sigma_n} M^p$. Then $\tilde{\pi} : \bar{M} \to_{\Sigma_{n+1}} M$.

11.5. Let M be an acceptable J-structure, and let $n < \omega$. Show that if κ is a cardinal of M such that $\rho_{n+1}(M) \leq \kappa < \kappa^{+M} \leq \rho_n(M)$, then $\text{cf}(\kappa^{+M}) = \text{cf}(\rho_n(M))$.
Let κ be a regular uncountable cardinal, and let $\lambda > \kappa$ be a cardinal. By $\diamondsuit^*_{\lambda,\kappa}$ we mean the following statement. There is a family $(\mathscr{A}_x \colon x \in [\lambda]^{<\kappa})$ such that for every $x \in [\lambda]^{<\kappa}$, $\mathscr{A}_x \subset \mathscr{P}(x)$ and $\text{Card}(\mathscr{A}_x) \leq \text{Card}(x)$, and for every $A \subset \lambda$ there is some club $\mathscr{C} \subset [\lambda]^{<\kappa}$ such that for every $x \in \mathscr{C}$, $A \cap x \in \mathscr{A}_x$.

11.6. **(R. Jensen)** Assume $V = L$. Let κ be a regular uncountable cardinal, and let $\lambda > \kappa$ be a cardinal. Then $\diamondsuit^*_{\lambda,\kappa}$ holds true. [Hint. If $x = X \cap \lambda$, where

$J_\alpha \overset{\pi}{\cong} X \prec J_\lambda$, then let $\beta > \alpha$ be least such that $\rho_\omega(J_\beta) < \alpha$ and set $\mathscr{A}_x = \{\pi''y \cap x \colon y \in J_\beta \cap \mathscr{P}(\lambda)\}$. Cf. the proof of Theorem 5.39.]

11.7. Let $\mathbb{N}$ be the set of all perfect trees on ω_2. $\mathbb{N}$, ordered by $U \leq_{\mathbb{N}} T$ iff $U \subset T$, is called NAMBA *forcing*. Show that if G is $\mathbb{N}$-generic over V, then $\mathrm{cf}^{V[G]}(\omega_2^V) = \omega$ and $\omega_1^{V[G]} = \omega_1^V$. [Hint. Let G be $\mathbb{N}$-generic over V. Then $\bigcup\bigcap G$ is cofinal from ω into ω_2^V. Let $T \Vdash \tau \colon \omega \to \omega_1$. Choose $(t_s, T_s, \alpha_s \colon s \in {}^{<\omega}\omega_2)$ such that $T_\emptyset = T, t_\emptyset = \emptyset, t_s \in T_s, \{t_{s^\frown\xi} \colon \xi < \omega_2\} \subset T_s$ is a set of $\aleph_2$ extensions of t_s of the same length, $t_{s^\frown\xi} \neq t_{s^\frown\xi'}$ for $\xi \neq \xi'$, $T_{s^\frown\xi} \leq_{\mathbb{N}} T_s, T_s \subset \{t \in T \colon t_s \supset t \vee t_s \subset t\}$, and $T_s \Vdash \mathrm{ran}(\tau \restriction (\mathrm{lh}(s)))\check{} \subset \check{\alpha}_n$, $\alpha_n < \omega_1$. For $\alpha < \omega$, set

$$T^\alpha = \bigcap\left\{\bigcup\{T_s \colon \mathrm{lh}(s) = n \wedge \alpha_s < \alpha\} \colon n < \omega\right\}.$$

Let us write $||t||^{CB}_{T^\alpha}$ for the CANTOR-BENDIXSON rank of t in T^α, cf. Problem 7.5. It suffices to prove that there is some $\alpha < \omega_1$ such that $||\emptyset||^{CB}_{T^\alpha} = \infty$, as then $T^\alpha \in \mathbb{N}$, $T^\alpha \leq_{\mathbb{N}} T$, and $T^\alpha \Vdash \mathrm{ran}(\tau) \subset \check{\alpha}$. Otherwise we may construct some $x \in {}^\omega(\omega_2)$ such that for all $n < \omega$ and for all (sufficiently big) $\alpha < \omega_1$, $||t_{x\restriction(n+1)}||^{CB}_{T^\alpha} < ||t_{x\restriction n}||^{CB}_{T^\alpha}$.] Show also that if CH holds in V, then forcing with $\mathbb{N}$ does not add a new real.
Conclude that it is not possible to cross out "$+\aleph_1$" in (1) or replace "$\geq \aleph_1$" by "$\geq \aleph_0$" in (2) of Theorem 11.56.

11.8. Show that $0^{\#}$ is not generic over an inner model which does not contain $0^{\#}$, i.e., if W is an inner model with $0^{\#} \notin W$, if $\mathbb{P} \in W$ is a poset, and if $G \in V$ is $\mathbb{P}$-generic over W, then $0^{\#} \notin W[G]$.

11.9. Assume that $0^{\#}$ does not exist. Let W be any inner model such that $(\aleph_2)^W = \aleph_2$. Show that for every θ, $[\theta]^\omega \cap W$ is stationary in $[\theta]^\omega$.

11.10. Suppose $V = W[x]$, where $x \subset \omega$ is $\mathbb{P}$-generic over W for some $\mathbb{P} \in W$. Suppose that W and V have the same cardinals, $W \models \mathsf{CH}$, but $V \models \neg\mathsf{CH}$. Show that $0^{\#} \in W$. [Hint. Use Strong Covering.]

11.11. **(M. Magidor)** Assume ZF plus both $\aleph_1$ and $\aleph_2$ are singular. Show that $0^{\#}$ exists. (Compare Theorem 6.69.)

11.12. Show that if $0^{\#}$ does not exist and if κ is weakly compact, then $\kappa^{+L} = \kappa^+$. [Hint. Use Problem 4.23.]

11.13. Let $\bar\kappa \leq \kappa$ be limit ordinals, and let $\pi \colon J_{\bar\kappa} \to J_\kappa$ be Σ_0-cofinal. Assume that $\mathrm{cf}(\bar\kappa) = \mathrm{cf}(\kappa) > \omega$. Let $\beta \geq \bar\kappa$ be such that $\bar\kappa$ is a cardinal in J_β, and let $n < \omega$ be such that $\rho_n(J_\beta) \geq \bar\kappa$. Show that $\mathrm{ult}_n(J_\beta; E_\pi)$ is well-founded. [Hint. Let $\sigma \colon J_{\bar\beta} \to J_\beta$ be elementary, where $\bar\beta$ is countable, and let $\sigma' \colon J_{\bar\beta} \to \mathrm{ult}_n(J_{\bar\beta}; E_{\sigma\restriction\sigma^{-1}(J_{\bar\kappa})})$. By Lemma 11.63, this ultrapower is well-founded, say equal to $J_{\tilde\beta}$, and there is an embedding $k \colon J_{\tilde\beta} \to J_\beta$. By hypothesis, $\tilde\beta < \bar\kappa$. We may assume that $\sigma \subset \tilde\sigma \colon H \to H_\theta$, where θ

is sufficiently big and H is (countable and) transitive. We may then embed $\tilde{\sigma}^{-1}(\text{ult}_n(J_\beta; E_\pi))$ into $\pi(J_{\bar{\beta}})$.]

11.14. Assume that $0^\#$ does not exist. Let κ be a cardinal in L, and let $X \prec J_\kappa$ be such that $X \cap \omega_2^V \in \omega_2^V$ and if $\omega_2^V \leq \mu < \kappa$ is an L-cardinal, then $\text{cf}(X \cap \mu^{+L}) > \omega$. Show that we may write $X = \bigcup_{n<\omega} X_n$, where $X_n \in L$ for each $n < \omega$. [Hint. Problem 11.13.]

11.15. **(M. Foreman, M. Magidor)** Assume $V = L$, and let $X \prec J_{\omega_\omega}$ be such that $\text{cf}(X \cap \omega_{n+1}) > \omega$ for all $n < \omega$. Show that there is some $n_0 < \omega$ such that for all $n \geq n_0$, $\text{cf}(X \cap \omega_n) = \text{cf}(X \cap \omega_{n_0})$. [Hint. Problems 11.13 and 11.5.]

11.16. Assume $V = L$, and $\kappa \geq \aleph_1$ be a cardinal. Let C and $n(\nu)$ (for $\nu \in C$) be defined as in the proof of Theorem 11.64. Show that for every $n < \omega$, $\{\nu \in C : n(\nu) = n\}$ is stationary in κ^+.

11.17. Show that if $\square_\kappa$ holds, then there is some stationary $R \subset \kappa^+$ such that $\square_\kappa(R)$ holds. Also, if $\square_\kappa$ holds, then there is some stationary $S \subset \kappa^+$ such that for no $\alpha < \kappa^+$, $S \cap \alpha$ is stationary in α.

11.18. **(R. Jensen)** ("Global $\square$") Assume $V = L$. Let S be the class of all ordinals α such that $\text{cf}(\alpha) < \overline{\overline{\alpha}}$. Show that there is a sequence $(C_\alpha : \alpha \in S)$ such that for every $\alpha \in S$, C_α is club in α, $\text{otp}(C_\alpha) < \alpha$, and if β is a limit point of C_α, then $\beta \in S$ and $C_\beta = C_\alpha \cap \beta$. [Hint. Imitate the proof of Theorem 11.64.]
Let λ be a limit ordinal. A sequence $(C_\alpha : \alpha < \lambda)$ is called *coherent* iff for all $\alpha < \lambda$, C_α is a club subset of α and whenever $\bar{\alpha}$ is a limit point of C_α, then $C_{\bar{\alpha}} = C_\alpha \cap \bar{\alpha}$. An ordinal λ is called *threadable* iff every coherent sequence $(C_\alpha : \alpha < \lambda)$ admits a *thread* C, i.e., $C \subset \lambda$ is club and for every limit point α of C, $C \cap \alpha = C_\alpha$.

11.19. Let λ be a limit ordinal with $\text{cf}(\lambda) > \omega$. Show that λ is threadable iff $\text{cf}(\lambda)$ is threadable.

11.20. Show that if κ^+ is threadable, then $\square_\kappa$ fails. Also, if λ is weakly compact, then λ is threadable.

11.21. **(R. Jensen)** Assume $V = L$, and let κ be not weakly compact. Show that there is an unthreadable coherent sequence $(C_\alpha : \alpha < \kappa)$. [Hint. Let T be a κ-ARONSZAJN tree. Let

$$S = \{\alpha < \kappa : \exists \beta > \alpha \; J_\beta \models \text{"ZFC}^- \text{ and } \alpha \text{ is regular and there is a cofinal branch through the } \alpha\text{-tree } T \cap J_\alpha''\}.$$

Notice that if $\alpha \notin S$, then α is singular. Now imitate the argument from Theorem 11.64 or rather from Problem 11.18, working separately on S and on $\kappa \setminus S$.]

11.22. Let κ be an uncountable regular cardinal which is not threadable. Show that if G is $\text{Col}(\omega_1, \kappa)$-generic over V, then $\aleph_1$ is not threadable in $V[G]$. [Hint. Use Problem 11.19. Let $(C_\alpha : \alpha < \kappa) \in V$ witness that κ is not threadable in V. If $\aleph_1$ were threadable in $V[G]$, we could pick H such that G, H are mutually $\mathrm{Col}(\omega_1, \kappa)$-generic over V such that both $V[G]$ and $V[H]$ contain a thread. As $\mathrm{cf}(\kappa) = \omega_1$ in $V[G][H]$, such a thread would then also be in $V[G] \cap V[H]$ and hence in V by Problem 6.12.]

Let $\kappa \geq \aleph_2$ be regular, and let $S \subset \kappa$ be stationary. We say that S *reflects* iff there is some $\alpha < \kappa$ with $\mathrm{cf}(\alpha) > \omega$ such that $S \cap \alpha$ is stationary in α.

11.23. Let κ be regular. Show that $S = \{\alpha < \kappa^+ : \mathrm{cf}(\alpha) = \kappa\}$ does not reflect.

11.24. Show that if κ is weakly compact, then every stationary $S \subset \kappa$ reflects.

11.25. Assume $V = L$. Let $\kappa \geq \aleph_2$ be regular but not weakly compact. Suppose that $S \subset \kappa$ is stationary. Show that there is some stationary $T \subset S$ which does not reflect. [Hint. If $\kappa = \lambda^+$, then use $\Box_\lambda$, cf. Theorem 11.64. If κ is inaccessible, then exploit the argument from Problem 11.21.]

11.26. **(J. Baumgartner)** Let κ be weakly compact, and let G be $\mathrm{Col}(\omega_1, <\kappa)$-generic over V. Show that the following is true in $V[G]$. Let $S \subset \kappa = \aleph_2$ be stationary such that $\mathrm{cf}(\alpha) = \omega$ for all $\alpha \in S$; then S reflects. [Hint. In V let us pick $\sigma \colon H \to H^*$ exactly as in Problem 4.23, and let $\tilde{\sigma} \colon H[G] \to H^*[K]$ be as in Problem 6.17. We may assume that our given S is in $H[G]$. We have ${}^{\omega_1}H \cap V[G] = {}^{\omega_1}H \cap H[G] \subset H^*[G] \subset V[G]$. It suffices to prove that S is stationary in $H^*[K]$, which follows from Problem 6.15(b).]

Chapter 12
Analytic and Full Determinacy

12.1 Determinacy

E. ZERMELO observed that finite two player games (which don't allow a tie) are determined in that one of the two players has a winning strategy. Let X be any non-empty set, and let $n < \omega$. Let $A \subset {}^{2n}X$, and let players I and II alternate playing elements $x_0, x_1, x_2, \ldots, x_{2n-1}$ of X. Say that I wins iff $(x_0, x_1, x_2, \ldots, x_{2n-1}) \in A$, otherwise II wins. Then either I has a winning strategy, i.e.,

$$\exists x_0 \forall x_1 \exists x_2 \ldots \forall x_{2n-1} \ (x_0, x_1, x_2, \ldots, x_{2n-1}) \in A,$$

or else II has a winning strategy, i.e.,

$$\forall x_0 \exists x_1 \forall x_2 \ldots \exists x_{2n-1} \ (x_0, x_1, x_2, \ldots, x_{2n-1}) \in {}^{2n}X \setminus A.$$

JAN MYCIELSKI (* 1932) and HUGO STEINHAUS (1887–1972) proposed studying infinite games and their winning strategies, which led to a deep structural theory of definable sets of reals. Let X be a non-empty set, and let $A \subset {}^{\omega}X$. We associate to A a game, called $G(A)$, which we define as follows. In a run of this game two players, I and II, alternate playing elements $x_0, x_1, x_2, \ldots$ of X as follows.

I	x_0		x_2		$\ldots$	
II		x_1		x_3		$\ldots$

After ω moves they produced an element $x = (x_0, x_1, x_2, \ldots)$ of ${}^{\omega}X$. We say that I *wins* this run of $G(A)$ iff $x \in A$, otherwise II *wins*. A *strategy for I* is a function

$$\sigma \colon \bigcup_{n<\omega} {}^{2n}X \to X,$$

and a *strategy for II* is a function

R. Schindler, *Set Theory*, Universitext, DOI: 10.1007/978-3-319-06725-4_12,

$$\tau\colon \bigcup_{n<\omega} {}^{2n+1}X \to X.$$

If σ is a strategy for I and $z \in {}^{\omega}X$, then we let $\sigma * z$ be the unique $x \in {}^{\omega}X$ with

$$x(2n) = \sigma(x(0), x(1), x(2), \ldots x(2n-1)) \text{ and}$$
$$x(2n+1) = z(n)$$

for $n < \omega$, i.e., x is the element of ${}^{\omega}X$ produced by a run of $G(A)$ in which I follows σ and II plays z. We say that σ is a *winning strategy for I in* $G(A)$ iff

$$\{\sigma * z\colon z \in {}^{\omega}X\} \subset A.$$

Symmetrically, if τ is a strategy for II and $z \in {}^{\omega}\omega$, then we let $z * \tau$ be the unique $x \in {}^{\omega}X$ with

$$x(2n+1) = \sigma(x(0), x(1), x(2), \ldots x(2n)) \text{ and}$$
$$x(2n) = z(n)$$

for $n < \omega$, i.e., x is the element of ${}^{\omega}X$ produced by a run of $G(A)$ in which II follows τ and I plays z. We say that τ is a *winning strategy for II in* $G(A)$ iff

$$\{z * \tau\colon x \in {}^{\omega}X\} \subset {}^{\omega}\omega \setminus A.$$

Of course, at most one of the two players can have a winning strategy.

Definition 12.1 Let X be a non-empty set, and let $A \subset {}^{\omega}X$. We say that $G(A)$ is *determined* iff player I or player II has a winning strategy in $G(A)$. In this case, we also call A itself *determined*. The *Axiom of Determinacy*, abbreviated by AD, states that every $A \subset {}^{\omega}\omega$ is determined.

We also refer to A as the "payoff" of the game $G(A)$.

We shall mostly be interested in the case where X is countable, in fact $X = \omega$ in which A is a set of reals. It can be shown in ZFC that if $A \subset {}^{\omega}\omega$ is Borel, then A is determined, cf. [20, Chap. 20, pp. 137–148]. We here aim to show that every analytic set is determined, cf. Theorem 12.20. It turns out that this cannot be done in ZFC, though, cf. Corollary 12.27. We shall prove later (cf. Theorem 13.7) that in fact every projective set of reals is determined. The full Axiom of Choice, AC, though, is incompatible with AD.

Lemma 12.2 *Assume* ZF + AD. *Then* AC *is false, but* AC_{ω} *holds for sets of reals, i.e., if* $(A_n\colon n < \omega)$ *is a sequence of non-empty sets of reals, then there is some* $f\colon \omega \to {}^{\omega}\omega$ *with* $f(n) \in A_n$ *for every* $n < \omega$.

Proof In the presence of AC, we may enumerate all strategies for I as $(\sigma_\alpha\colon \alpha < 2^{\aleph_0})$ and all strategies for II as $(\sigma_\alpha\colon \alpha < 2^{\aleph_0})$. Using AC again, let us pick sequences

$(x_\alpha : \alpha < 2^{\aleph_0})$ and $(y_\alpha : \alpha < 2^{\aleph_0})$ such that for all $\alpha < 2^{\aleph_0}$, $x_\alpha = \sigma_\alpha * z$ for some $z \in {}^\omega\omega$ such that $\sigma_\alpha * z \notin \{y_\beta : \beta < \alpha\}$, and $y_\alpha = z * \tau_\alpha$ for some $z \in {}^\omega\omega$ such that $z * \tau_\alpha * z \notin \{x_\beta : \beta \leq \alpha\}$. It is then easy to see that

$$A = \{y_\alpha : \alpha < 2^{\aleph_0}\}$$

cannot be determined.

Now let $(A_n : n < \omega)$ be a sequence of non-empty sets of reals, and let us consider the following game (in which I just keeps passing after his first move).

I	n				$\ldots$
II		n_1	n_2	x_3 n_4	$\ldots$

I plays some $n \in \omega$, and then II plays some $x = (n_0, n_1, n_2, \ldots)$. We say that II wins iff $x \in A_n$.

Of course, I cannot have a winning strategy in this game. Hence II has a winning strategy, as our game may be construed as $G({}^\omega\omega \setminus A)$, where $A = \{x : (x(1), x(3), \ldots) \in A_{x(0)}\}$, and II's winning strategy then gives rise to a function f as desired. □

With a some extra work beyond Theorem 13.7 one can construct models of ZF + DC + AD, which are of the form of the models of Theorems 6.69 or 8.30. The moral of this is that whereas AD is false it holds for "definable" sets of reals, and the results of this section should be thought of being applied inside models of ZF + AD which contain all the reals.

We first want to show that open games on are determined. Let still X be an arbitrary non-empty set. Recall, cf. p. 123, that we may construe ${}^\omega X$ as a topological space as follows. For $s \in {}^{<\omega}X$, set $U_s = \{x \in {}^\omega X : s \subset x\}$. The sets U_s are declared to be the basic open sets, so that a set $A \subset {}^\omega X$ is called open iff there is some $Y \subset {}^{<\omega}X$ with $A = \bigcup_{s \in Y} U_s$.

If σ, τ is a strategy for player I, II, respectively, then we say that x is *according to* σ, τ iff there is some y such that $x = \sigma * y$, $x = y * \tau$, respectively.

Theorem 12.3 (Gale, Stewart) *Let $A \subset {}^\omega X$ be open. Then A is determined.*

Proof Let us suppose I not to have a winning strategy in $G(A)$. We aim to produce a winning strategy for II in $G(A)$. Let us say that I has a winning strategy in $G_s(A)$, where $s \in {}^{<\omega}X$ has even length, iff I has a winning strategy in $G(\{x \in {}^\omega X : s^\frown x \in A\})$. By our hypothesis, I doesn't have a winning strategy in $G_\emptyset(A)$.

Claim 12.4 *Let $s \in {}^{<\omega}X$ have even length, and suppose I not to have a winning strategy in $G_s(A)$. Then for all $y \in X$ there is some $z \in X$ such that I doesn't have a winning strategy in $G_{s^\frown y^\frown z}(A)$.*

Proof Otherwise there is some $y \in X$ such that for all $z \in X$, I has a winning strategy in $G_{s^\frown y^\frown z}(A)$. But then I has a winning strategy in $G_s(A)$: he first plays such a y, and subsequently, after II played z, follows his strategy in $G_{s^\frown y^\frown z}(A)$. □

Let us now define a strategy τ for *II* in $G(A)$ as follows. Let $s = (x_0, \ldots, x_n)$ be a position in $G(A)$ where it's *II*'s turn to play, i.e., n is odd. Then let $\tau(s)$ be some $z \in X$ such that *I* doesn't have a winning strategy in $G_{s^\frown z}(A)$, if some such z exists; otherwise we let $\tau(s)$ be an arbitrary element of X.

Claim 12.5 *τ is a winning strategy for II in $G(A)$.*

Proof Let

I	x_0		x_2		$\ldots$
II		x_1		x_3	$\ldots$

be a play of $G(A)$ which is according to τ. Claim 12.4 can easily be used to show inductively that for each even $n < \omega$, I does not have a winning strategy in $G_{(x_0,\ldots,x_{n-1})}(A)$.

Now suppose that II looses, i.e., $x = (x_0, x_1, x_2, \ldots) \in A$. Because A is open, there is some basic open set U_s such that $x \in U_s \subset A$. We may assume $\text{lh}(s)$ to be even. Then I has a trivial winning strategy in $G_s(A)$: he may play as he pleases, as every $s^\frown x'$, $x' \in {}^\omega X$, will be in A. But this is a contradiction! □

Lemma 12.6 *Let $A \subset {}^\omega X$. Suppose that for every $y \in X$, $\{x \in {}^\omega X: y^\frown x \in A\}$ is determined. Then ${}^\omega X \setminus A$ is determined.*

Proof Let us first suppose that there is a $y \in X$ such that II has a winning strategy τ^* in $G(\{x \in {}^\omega X: y^\frown x \in A\})$. We claim that in this case I has a winning strategy σ in $G({}^\omega X \setminus A)$. We let $\sigma(\emptyset) = y$, and we let $\sigma(y^\frown s) = \tau^*(s)$, where $\text{lh}(s)$ is odd. It is easy to see that if $x \in {}^\omega X$ is produced by a run which is according to σ, then $x \in {}^\omega X \setminus A$.

Now let us suppose that for all $y \in X$, I has a winning strategy σ_y^* in $G(\{x \in {}^\omega X: y^\frown x \in A\})$. We claim that in this case *II* has a winning strategy τ in $G({}^\omega X \setminus A)$. We let $\tau(y^\frown s) = \sigma_y^*(s)$, where $\text{lh}(s)$ is even. It is easy to see that if $x \in {}^\omega X$ is produced by a run which is according to τ, then $x \in A$. □

Corollary 12.7 *Let $A \subset {}^\omega X$ be closed. Then A is determined.*

As an application, we now give a

Proof of Theorem 10.11, which uses an argument from a paper by I. NEEMAN. Recall that Theorem 10.11 says that if M is a transitive model of ZFC such that $M \models$ "U is a normal measure on κ," and if $(\kappa_n : n < \omega)$ is a PRIKRY sequence over M with respect to U, then $G_{(\kappa_n : n<\omega)}$ is $\mathbb{P}_U$-generic over M. We shall write $\mathbb{P} = \mathbb{P}_U$.

Set $G = G_{(\kappa_n : n<\omega)}$, cf. Definition 10.10. It is easy to see that G is a filter. We shall prove that G is generic over M.

To this end, fix $D \in M$ which is open and dense in $\mathbb{P}$ for the rest of this proof. We aim to show that $D \cap G \neq \emptyset$.

To each $s \in [\kappa]^{<\omega}$ we shall associate a game G_s. Let $s = \{\lambda_0 < \cdots < \lambda_{k-1}\}$ where $k < \omega$. The game has two players, I and II, and starts with round k.

$$\begin{array}{c|cccc} \text{I} & X_k & & X_{k+1} & & \dots \\ \hline \text{II} & & \lambda_k & & \lambda_{k+1} & & \dots \end{array}$$

In round $l \geq k$, I has to play some $X_l \in U$. II has to reply with some $\lambda_l \in X_l$ such that $\lambda_l > \lambda_{l-1}$ (if $l > 0$). II wins the game iff there is some $n < \omega$ and some $X \in U$ such that

$$(\{\lambda_0, \dots, \lambda_{n-1}\}, X) \in D.$$

G_s is a closed game. Moreover, $G_s \in M$. Therefore, one of the two players has a winning strategy in M which works for plays in V.

Claim 12.8 *I does not have a winning strategy in M.*

Proof Suppose $\sigma \in M$ to be a winning strategy for I. We claim that there is some $Z \in U$ such that all $\lambda_k < \lambda_{k+1} < \cdots$ in Z are compatible with σ, by which we mean that there is a play of G_s in which I follows σ and II plays $\lambda_k, \lambda_{k+1}, \dots$. In order to get Z, define $F: [\kappa]^{<\omega} \to 2$ by $F(\{\lambda_k < \cdots < \lambda_{n-1}\}) = 1$ iff $\lambda_k < \cdots < \lambda_{n-1}$ are compatible with σ. As $F \in M$, we may let $Z \in U$ be such that F is constant on $[Z]^l$ for every $l < \omega$. It cannot be that $F''Z]^l = \{0\}$ for some $l < \omega$; this is because in round m, where $k \leq m < k + l$, if σ tells I to play X_m then II can reply with some $\lambda_m \in X_m \cap Z$.

Now let us look at $(\{\lambda_0, \dots, \lambda_{k-1}\}, Z) \in \mathbb{P}$. As $D \in M$ is dense in $\mathbb{P}$ there is some

$$(\{\lambda_0, \dots, \lambda_{n-1}\}, \overline{Z}) \leq (\{\lambda_0, \dots, \lambda_{k-1}\}, Z)$$

with

$$(\{\lambda_0, \dots, \lambda_{n-1}\}, \overline{Z}) \in D.$$

Because $\lambda_k < \cdots < \lambda_{n-1} \in Z$, $\lambda_k < \cdots < \lambda_{n-1}$ are compatible with σ; on the other hand, II wins if he plays $\lambda_k, \dots, \lambda_{n-1}$ in rounds $k, \dots, n-1$. Contradiction! □

Let us still fix $s = \{\lambda_0 < \cdots < \lambda_{k-1}\}$ for a while, and let $\tau = \tau_s \in M$ be a winning strategy for II in G_s (which also works for plays in V).

We shall now for $l \geq k$ and $t = \{\lambda_0 < \cdots < \lambda_{k-1} < \cdots < \lambda_{l-1}\} \supset s$ and $\lambda > \lambda_{l-1}$ define sets Y_s^t and $X_s^{t,\lambda}$ which are in U. The definition will be by recursion on the length of t. We shall call $t \cup \{\lambda_l\}$ "realizable" iff

$$\begin{array}{c|cccc} \text{I} & X_s^{s,\lambda_k} & & \dots & & X_s^{t,\lambda_l} \\ \hline \text{II} & & \lambda_k & & \dots & & \lambda_l \end{array}$$

is a position in the game G_s in which I obeyed the rules and II played according to τ.

Now fix t and λ_l and assume that $X_s^{t\upharpoonright m, \lambda_m}$ and $Y_s^{t\upharpoonright m}$ have been defined for all $k \leq m < l$, where $l \geq k$. We first aim to define Y_s^t.

Claim 12.9 *Suppose t to be realizable. There is then some $Y \in U$ such that for all $\lambda \in Y$ there is an X such that*

$$\begin{array}{c|ccccccc} I & X_s^{s,\lambda_k} & & \dots & X_s^{t\restriction l-1} & & X & \\ \hline II & & \lambda_k & & \dots & \lambda_{l-1} & & \lambda \end{array}$$

is a position in G_s in which I *obeyed the rules and* II *played according to τ.*

Proof Let Y denote the set of λ such that there is some X such that

$$\begin{array}{c|ccccccc} \mathrm{I} & X_s^{s,\lambda_k} & & \dots & X_s^{t\restriction l-1} & & X & \\ \hline \mathrm{II} & & \lambda_k & & \dots & \lambda_{l-1} & & \lambda \end{array}$$

is a position in G_s in which I obeyed the rules and II played according to τ. We want to see that $Y \in U$. Suppose that $Y \notin U$, i.e., $\kappa \setminus Y \in U$. Consider the following position in G_s in which λ is as dictated by τ.

$$\begin{array}{c|ccccccc} \mathrm{I} & X_s^{s,\lambda_\kappa} & & \dots & X_s^{t\restriction l-1} & & \kappa\setminus Y & \\ \hline \mathrm{II} & & \lambda_\kappa & & \dots & \lambda_{l-1} & & \lambda \end{array}$$

As II follows τ and thus obeys the rules, $\lambda \in \kappa \setminus Y$. On the other hand, $\lambda \in Y$ by definition of Y. Contradiction! □

We now let Y_s^t be as given by Claim 12.9. For each $\lambda \in Y_s^t$ we let $X_s^{t,\lambda}$ be a witness to the fact that $\lambda \in Y_s^t$, i.e., some X as in the statement of Claim 12.9 for $Y = Y_s^t$.

We finally also assign some $Z_s \in U$ to s as follows. If there is some Z with $(s, Z) \in D$ then we let Z^s be some such Z; otherwise we set $Z = \kappa$.

We now let[1]

$$X_0 = \Delta_s Z_s \cap \Delta_{s \subset t} Y_s^t.$$

By our hypothesis on $(\kappa_i : i < \omega)$, there is some $n < \omega$ such that $\{\kappa_n, \kappa_{n+1}, \dots\} \subset X_0$. Set $s = (\kappa_l : l < n)$.

Claim 12.10 *The following is a play of G_s in which I obeys the rules and II follows τ_s.*

$$\begin{array}{c|cccccc} I & X_s^{s,\kappa_n} & & \dots & X_s^{x\restriction m,\kappa_m} & & \dots \\ \hline II & & \kappa_n & \dots & & \kappa_m & \dots \end{array}$$

Proof by induction on m. Notice that $\kappa_m \in X_0$ and hence $\kappa_m \in Y_s^{x\restriction m}$. □

Because II follows τ_s in the play above, there is some $m \geq n$ and some X with $(\{\kappa_0, \dots, \kappa_{m-1}\}, X) \in D$. But then $(\{\kappa_0, \dots, \kappa_{m-1}\}, Z^{x\restriction m}) \in D$. However, for any $l \geq m$, $\kappa_l \in X_0$, and hence $\kappa_l \in Z^{x\restriction m}$. We thus have that

[1] cf. Problem 4.26 on "$\Delta_s X_s$".

$$(\{\kappa_0, \ldots, \kappa_{m-1}\}, Z^{x \restriction m}) \in G.$$

Therefore, $D \cap G \neq \emptyset$. □

We shall now prove that AD, the Axiom of Determinacy, proves the ultimate generalization of the CANTOR- BENDIXSON Theorem 1.9. If AD were not to contradict the Axiom of Choice, it could be construed as providing a solution to CANTOR's project of proving the Continuum Hypothesis, cf. p. 3. The following Theorem says that *all of* $\mathscr{P}({}^{\omega}\omega)$ has the perfect subset property (cf. p. 138).

Theorem 12.11 (M. Davis) *Assume* AD. *Every uncountable* $A \subset {}^{\omega}\omega$ *has a perfect subset.*

Proof Fix $A \subset {}^{\omega}\omega$. Let $f : {}^{\omega}\omega \to {}^{\omega}2$ be a continuous bijection (cf. Problem 7.2), and write $B = f''A$. It suffices to prove that B is either countable or else has a perfect subset.

Let us consider the following game, $G^{\mathrm{p}}(B)$.

I	s_0		s_1		$\ldots$
II		n_0		n_1	$\ldots$

In this game, I plays finite 0–1-sequences $s_i \in {}^{<\omega}2$ (with $s_i = \emptyset$ being explicitly allowed), and II plays $n_i \in 2 = \{0, 1\}$. Player I wins iff

$$s_0{}^\frown n_0{}^\frown s_1{}^\frown n_1{}^\frown \ldots \in B,$$

otherwise II wins. We may construe $G^{\mathrm{p}}(B)$ as $G(B')$ for some $B' \subset {}^{\omega}\omega$, so that $G^{\mathrm{p}}(B)$ is determined.

If I has a winning strategy in $G^{\mathrm{p}}(B)$, σ, then

$$\{s_0{}^\frown n_0{}^\frown s_1{}^\frown n_1{}^\frown \ldots : (n_0, n_1, \ldots) \in {}^{\omega}2 \wedge \forall i < \omega \; s_i = \sigma(s_0{}^\frown n_0{}^\frown \ldots {}^\frown n_{i-1})\}$$

is a perfect subset of B.

So let us suppose τ to be a winning strategy for II in $G^{\mathrm{p}}(B)$. We say that a finite sequence $p = (s_0, n_0, \ldots, s_k, n_k)$ is *according to* τ iff $n_i = \tau(s_0{}^\frown n_0{}^\frown \ldots {}^\frown s_i)$ for every $i \leq k$, and if $x \in {}^{\omega}2$, then we say that $p = (s_0, n_0, \ldots, s_k, n_k)$ is *compatible with* x iff $s_0{}^\frown n_0{}^\frown \ldots {}^\frown s_i{}^\frown n_i \subset x$. Because τ is a winning strategy for II, we have that $x \notin B$ follows from the fact that for all $p = (s_0, n_0, \ldots, s_k, n_k)$, if p is according to τ and compatible with x, then there exists some $s \in {}^{\omega}2$ such that $(s_0, n_0, \ldots, s_k, n_k, s, \tau((s_0, n_0, \ldots, s_k, n_k, s))$ is compatible with x. In other words, if $x \in B$, then there is some $p_x = (s_0, n_0, \ldots, s_k, n_k)$ which is according to τ and compatible with x such that for all $s \in {}^{\omega}2$, $(s_0, n_0, \ldots, s_k, n_k, s, \tau((s_0, n_0, \ldots, s_k, n_k, s))$ is not compatible with x.

Notice that $x \mapsto p_x$ is injective for $x \in B$. This is because if $x \in B$, $p_x = (s_0, n_0, \ldots, s_k, n_k)$, and

$$x = s_0{}^\frown n_0{}^\frown \dots {}^\frown s_k{}^\frown n_k{}^\frown m_0{}^\frown m_1{}^\frown m_2{}^\frown \dots,$$

then we must have that $m_i = 1 - \tau(s_0{}^\frown n_0{}^\frown \dots {}^\frown s_k{}^\frown n_k{}^\frown m_0{}^\frown \dots {}^\frown m_{i-1})$ for every $i < \omega$.

But as $x \mapsto p_x$, $x \in B$, is injective, B is countable. □

Lemma 12.12 *Assume* AD. *Let $A \subset {}^\omega 2$ be such that every* LEBESGUE *measurable $D \subset A$ is a null set. Then A is a null set.*

Proof Let us fix $A \subset {}^\omega 2$ such that every LEBESGUE measurable $D \subset A$ is a null set. Let $\varepsilon > 0$ be arbitrary. We aim to show that A can be covered by a countable union B of basic open sets such that $\mu(B) \leq \varepsilon$. Let us consider the following game $G^{\text{cov}}(A)$, the *covering game* for A.

I	n_0		n_1		$\dots$	
II		K_0		K_1		$\dots$

In this game, I plays $n_i \in \{0,1\}$, $i < \omega$, and II plays K_i, $i < \omega$, where each K_i is a finite union of basic open sets such that $\mu(K_i) \leq \frac{\varepsilon}{2^{2(i+1)}}$. Player I wins iff

$$(n_0, n_1, n_2, \dots) \in A \setminus \bigcup_{i<\omega} K_i.$$

Notice that for each $i < \omega$ there are only countably many candidates for K_i, so that $G^{\text{cov}}(A)$ may be simulated by a game in which both players play just natural numbers. Therefore, $G^{\text{cov}}(A)$ is determined.

We claim that I cannot have a winning strategy. Suppose not, let σ be a winning strategy for I, and let $D \subset {}^\omega 2$ be the set of all $(n_0, n_1, \dots)$ such that there is a play of $G^{\text{cov}}(A)$ in which II plays some sequence $(K_i : i < \omega)$ and $(n_i : i < \omega)$ is the sequence of moves of I obtained by following the winning strategy σ in response to $(K_i : i < \omega)$. The set D is then analytic and hence LEBESGUE measurable by Corollary 8.15. As σ is a winning strategy for I, $D \subset A$, so that D is in fact a null set by our hypothesis on A.

Let $D \subset \bigcup_{n<\omega} I_n$, where each I_n is a basic open set and $\mu\left(\bigcup_{n<\omega} I_n\right) \leq \frac{2}{3} \cdot \varepsilon$. Notice that $\frac{2}{3} = \sum_{i=0}^{\infty} \frac{1}{2^{2(i+1)}}$, so that by cutting and relabelling if necessary we may assume that there is a strictly increasing sequnce $(\ell_i : i < \omega)$ of natural numbers with $\ell_0 = 0$ such that $\mu\left(\bigcup_{k=\ell_i}^{\ell_{i+1}-1} I_k\right) \leq \frac{\varepsilon}{2^{2(i+1)}}$. But now II can defeat I's alleged winning strategy σ by playing $\bigcup_{k=\ell_i}^{\ell_{i+1}-1} I_k$ in her ith move. If I plays by following σ, he will lose. Contradiction!

By AD, player II therefore has a winning strategy τ in $G^{\text{cov}}(A)$. For $s \in {}^{<\omega} 2$ with $\text{lh}(s) = i+1 \in \omega \setminus \{0\}$, let us write K_s for the ith move K_i of II in a play in which I's first $i+1$ moves are $s(0), \dots, s(i)$ and II's first $i+1$ moves $K_0, \dots, K_i$ are obtained by following τ in response to $s(0), \dots, s(i)$. Write

$$B = \bigcup_{s \in {}^{<\omega}2 \setminus \emptyset} K_s.$$

Because τ is a winning strategy for *II*, $A \subset B$. Moreover, $\mu(K_s) \leq \frac{1}{2^{2(\mathrm{lh}(s))}}$, so that for every $i < \omega$,

$$\mu\left(\bigcup_{\mathrm{lh}(s)=i+1} K_s\right) \leq 2^{i+1} \cdot \frac{1}{2^{2(i+1)}} = \frac{1}{2^{i+1}}$$

and hence

$$\mu(B) \leq \sum_{i=0}^{\infty} \frac{\varepsilon}{2^{i+1}} = \varepsilon.$$

We covered A by a countable union B of closed intervals such that $\mu(B) \leq \varepsilon$. □

Theorem 12.13 (Mycielski-Swierczkowski) *Assume* AD. *Every set* $A \subset {}^{\omega}2$ *is* LEBESGUE *measurable.*

Proof Fix $A \subset {}^{\omega}2$. Let $B \supset A$, $B \subset {}^{\omega}2$ be LEBESGUE measurable such that for all LEBESGUE measurable $D \subset B \setminus A$, D is a null set. Then $B \setminus A$ is a null set by Lemma 12.12, so that A is LEBESGUE measurable. □

The hypothesis of Theorem 12.13 therefore also gives that every $A \subset {}^{\omega}\omega$ is LEBESGUE measurable.

Lemma 12.14 *Assume* AD. *Let* $A \subset {}^{\omega}\omega$ *be non-meager. There is then some* $s \in {}^{<\omega}\omega$ *such that* $A \cap U_s$ *is comeager in the space* U_s.

Proof Let us fix $A \subset {}^{\omega}\omega$. For $s \in {}^{<\omega}\omega$, let us consider the following game, $G_s^{\mathrm{bm}}(A)$, called the BANACH- MAZUR *game.*

I	s_0		s_2		...	
II		s_1		s_3		...

In this game, *I* and *II* alternate playing *non-empty* $s_n \in {}^{<\omega}\omega$, $n < \omega$, and *I* wins iff

$$s^\frown s_0{}^\frown s_1{}^\frown s_2{}^\frown s_3{}^\frown \ldots \in A.$$

Claim 12.15 *II has a winning strategy in* $G_s^{\mathrm{bm}}(A)$ *iff* $A \cap U_s$ *is meager.*

Proof This proof does not use AD. Suppose first that $A \cap U_s$ is meager, so that we may write $A \cap U_s = \bigcup_{n<\omega} A_n$, where each A_n is nowhere dense. Let us define a strategy σ for *II* as follows. Let $\sigma(s_0, s_1, \ldots, s_{2n})$ be some non-empty $t \in {}^{\omega}\omega$ such that

$$U_{s^\frown s_0{}^\frown s_1{}^\frown \ldots{}^\frown s_{2n}{}^\frown t} \cap A_n = \emptyset.$$

This is always well-defined, as all A_n are nowhere dense, and σ is easily seen to be a winning strategy for *II* in $G_s^{\text{bm}}(A)$.

Now let us assume that *II* has a winning strategy, σ, in $G_s^{\text{bm}}(A)$. We may define what it means for a finite sequence $p = (s_0, s_1, \dots, s_n)$ of non-empty elements of ${}^{<\omega}\omega$ to be *according to* σ in much the same way as in the proof of Theorem 12.11. Let $x \in {}^{\omega}\omega$ with $s \subset x$. Because σ is a winning strategy for *II*, if for all sequences $(s_0, s_1, \dots, s_{2n-1})$ of non-empty elements of ${}^{<\omega}\omega$ which are according to σ and such that $s^\frown p = s^\frown s_0{}^\frown s_1{}^\frown \dots {}^\frown s_{2n-1} \subset x$ there is some non-empty $t \in{}^{<\omega}\omega$ such that $s^\frown p^\frown t^\frown \sigma(p^\frown t) \subset x$, then $x \notin A$. Therefore,

$$A \cap U_s \subset \bigcup B_p,$$

where, for a sequence $p = (s_0, s_1, \dots, s_{2n-1})$ of non-empty elements of ${}^{<\omega}\omega$ which is according to σ, B_p is the set of all $x \in {}^{\omega}\omega$ such that $s^\frown p = s^\frown s_0{}^\frown s_1{}^\frown \dots {}^\frown s_{2n-1} \subset x$, but for all non-empty $t \in {}^{<\omega}\omega$, $s^\frown p^\frown t^\frown \sigma(p^\frown t)$ is *not* an initial segment of x, i.e.,

$$B_p \cap U_{s^\frown p^\frown t^\frown \sigma(p,t)} = \emptyset.$$

Every B_p is thus nowhere dense, so that $A \cap U_s$ is in fact meager. □

Let us now assume AD. Let us suppose A to be non-meager. By Claim 12.15, *II* does not have a winning strategy in $G_\emptyset^{\text{bm}}(A)$, so that *I* has a winning strategy τ in $G_\emptyset^{\text{bm}}(A)$. Setting $s = \tau(\emptyset)$, τ esily induces a winning strategy for *II* in $G_s^{\text{bm}}({}^{\omega}\omega \backslash A)$. Again by the Claim, $({}^{\omega}\omega \backslash A) \cap U_s$ is now meager, so that $A \cap U_s$ is comeager in U_s. □

Theorem 12.16 (Mazur) *Assume* AD. *Every* $A \subset {}^{\omega}\omega$ *has the property of* BAIRE.

Proof Let us fix $A \subset {}^{\omega}\omega$, and set

$$O = \bigcup\{U_s : s \in {}^{<\omega}\omega \wedge U_s \backslash A \text{ is meager}\}.$$

Trivially, $O \backslash A$ is meager. If $A \backslash O$ were non-meager, then by Lemma 12.14 there is some $s \in {}^{<\omega}\omega$ such that $(A \backslash O) \cap U_s$ is comeager in U_s. This means that $U_s \backslash A \subset U_s \backslash (A \backslash O)$ is meager in U_s, so that by the definition of O, $U_s \subset O$. So $(A \backslash O) \cap U_s = \emptyset$ is not comeager in U_s after all. Contradiction! □

For $x, y \in {}^{\omega}\omega$, we let $x \leq_T y$ denote that x is TURING *reducible to* y, and we write $x \equiv_T y$ for $x \leq_T y$ and $y \leq_T x$. A set $A \subset {}^{\omega}\omega$ is called TURING *invariant* iff for all $x \in A$ and $y \in {}^{\omega}\omega$, if $y \equiv_T x$, then $y \in A$. A set $A \subset {}^{\omega}\omega$ is called a (TURING) *cone* iff there is some $x \in {}^{\omega}\omega$ such that $A = \{y : x \leq_T y\}$, in which case x is also called a *base* of the cone A.

Set S be a set of ordinals. A set $A \subset {}^{\omega}\omega$ is called S-*invariant* iff for all $x \in A$ and $y \in {}^{\omega}\omega$, if $L[S, x] = L[S, y]$, then $y \in A$. We call $A \subset {}^{\omega}\omega$ an S-*cone* iff there is some $x \in {}^{\omega}\omega$ such that $A = \{y : x \in L[S, y]\}$, in which case x is also called a *base* of the S-cone A.

Theorem 12.17 (D. A. Martin) *Assume* AD. *Let* $A \subset {}^\omega\omega$ *be* TURING *invariant. Then either* A *or* ${}^\omega\omega \setminus A$ *contains a* (TURING) *cone. Also, if* S *is a set of ordinals and* $A \subset {}^\omega\omega$ *is* S*-invariant, then either* A *or* ${}^\omega\omega \setminus A$ *contains an* S*-cone.*

Proof We prove the first part. Let $A \subset {}^\omega\omega$ be TURING invariant. If σ is a winning strategy for I in $G(A)$, then $x \equiv_T \sigma * x \in A$ whenever $\sigma \leq_T x$, and if τ is a winning strategy for II in $G(A)$, then $x \equiv_T x * \tau \in {}^\omega\omega \setminus A$ whenever $\tau \leq_T x$. The proof of the second part is the same. □

In what follows it will be very convenient to use the $\oplus$-notation. If $x, y \in {}^\omega\omega$, then we write $x \oplus y$ for that $z \in {}^\omega\omega$ with $z(2n) = x(n)$ and $z(2n+1) = y(n)$. If $x_n \in {}^\omega\omega$, $n < \omega$, then $\oplus x_n$ is that $z \in {}^\omega\omega$ such that $z(\langle n, k\rangle) = z_n(k)$, where $\cdot, \cdot \mapsto \langle \cdot, \cdot \rangle$ is the GÖDEL pairing function (cf. p. 33).

Theorem 12.18 *Assume* AD. *The following hold true.*

(a) ω_1 *is inaccessible to the reals.*
(b) ω_1 *is a measurable cardinal.*
(c) *For every* $x \subset \omega$, $x^\#$ *exists.*
(d) *For every* $X \subset \omega_1$ *there is a real* x *such that* $X \in L[x]$.
(e) *The club filter on* ω_1 *is an ultrafilter.*
(f) ω_2 *is a measurable cardinal.*

Proof (a) Immediately follows from Theorem 12.11 via Corollary 7.29.

(b) Set $B = \{\omega_1^{L[x]} : x \in {}^\omega\omega\}$. By (a), B is a subset of ω_1 of size $\aleph_1$. Let $\pi : \omega_1 \to B$ be the monotone enumeration of B. For $X \subset \omega_1$, let $A(X)$ be the Turing invariant set $\{x \in {}^\omega\omega : \omega_1^{L[x]} \in \pi''X\}$. Let

$$U = \{X \subset \omega_1 : A(X) \text{ contains a cone}\}.$$

Notice that if x is a base for the cone A and y is a base for the cone B, then $x \oplus y$ is a cone for a base contained in $A \cap B$. It is thus straightforward to verify that U is a filter, and U is in fact an ultrafilter by Theorem 12.17. If $\{X_n : n < \omega\} \subset U$, then by AC_ω (cf. Lemma 12.2) we may pick a sequence $(x_n : n < \omega)$ of reals such that for each $n < \omega$, x_n is a base for a cone of reals contained in $A(X_n)$. But then $\oplus_{n<\omega} x_n$ is a base for a cone of reals contained in $\bigcap_{n<\omega} A(X_n) = A(\bigcap_{n<\omega} X_n)$. This shows that U is $<\aleph_1$-complete, so that U witnesses that ω_1 is a measurable cardinal.

(c) This follows from (b) and the proof of Lemma 10.31 which does not need AC. If U is a $<\aleph_1$-complete measure on ω_1, then for every $x \subset \omega$, $U \cap L[x]$ is ω-complete, so that $\mathrm{ult}(L[x]; U \cap L[x])$ is well-founded by Lemma 10.29, and we get $x^\#$ by Theorem 10.39.

(d) Fix $X \subset \omega_1$. Let us consider the following game, called the SOLOVAY *game*.

I	n_0		n_1		...	
II		m_1		m_2		...

Players I and II alternate playing natural numbers. Let us write $x = (n_0, n_1, \ldots)$ and $y_i = (m_{2^{i+1}\cdot 3^{k+1}} : k < \omega)$. We say that II wins iff

$$x \in \mathrm{WO} \Longrightarrow \{\alpha \in X \cap ||x|| + 1\} \subset \{||y_i|| : i < \omega \wedge y_i \in \mathrm{WO}\} \subset X.$$

If I had a winning strategy, σ, then $A = \{\sigma * x : x \in {}^\omega\omega\}$ would be analytic (cf. the proof of Lemma 12.12). Also $A \subset \mathrm{WO}$, and by the Boundedness Lemma 7.12 there would be some countable ξ such that $\{||\sigma * x|| : x \in^\omega \omega\} \subset \xi$. But then II can easily defeat σ by playing some y such that $y_i \in \mathrm{WO}$ for all $i < \omega$ and $\{||y_i|| : i < \omega\} = X \cap (\xi + 1)$.

Therefore, II has a winning strategy, τ. Let $G \in V$ be $\mathrm{Col}(\omega, < \omega_1^V)$-generic over $L[\tau]$ (Cf. Problem 10.13). Then $\alpha \in X$ iff $L[\tau][G] \models$ "there is some $x \in \mathrm{WO}$ such that $\alpha = ||x||$ and $\alpha \in \{||y_i|| : i < \omega \wedge y_i \in \mathrm{WO}\}$, where y is the result of having II play according to τ in a play in which I plays x." By the homogeneity of $\mathrm{Col}(\omega, < \omega_1^V)$, cf. Lemma 6.54, X is therefore in $L[\tau]$, cf. Corollary 6.62.

(e) Let $X \subset \omega_1$, and let, using (d), $x \in^\omega \omega$ be such that $X \in L[x]$. By (c), $x^\#$ exists, so that X either contains a club or is disjoint from a club. (cf. Problem 10.11.)

(f) Let $C = \{(\omega_1^V)^{+L[x]} : x \in {}^\omega\omega\}$. As $x^\#$ exists for every $x \in {}^\omega\omega$, $C \subset \omega_2$. By (d), C is cofinal in ω_2^V, so that we may let $\sigma : \omega_2 \to C$ be the monotone enumeration of C. In a fashion similar to (a), for $X \subset \omega_2$ we may let $D(X)$ be the Turing invariant set $\{x \in {}^\omega\omega : \omega_1^{+L[x]} \in \sigma''X\}$, and we may define

$$F = \{X \subset \omega_2 : D(X) \text{ contains a cone}\}.$$

Using Theorem 12.17 and AC_ω as in (a), F can be verified to be a $< \aleph_1$-complete ultrafilter on ω_2.

It remains to be shown that F is $<\aleph_2$-complete. Let us fix a sequence $(X_i : i < \omega)$ such that $X_i \in F$ for every $i < \omega_1$. Let us consider the following game.

I	n_0		n_1		$\ldots$	
II		m_1		m_2		$\ldots$

Players I and II alternate playing natural numbers. Let us write $x = (n_0, n_1, \ldots)$ and $y = (m_i : i < \omega)$. We say that II wins iff

$$x \in \mathrm{WO} \Longrightarrow \{z \in {}^\omega\omega : y \leq_T z\} \subset \bigcap_{i \leq ||x||} D(X_i).$$

The Boundedness Lemma 7.12 implies as in (b) that I cannot have a winning strategy in this game. Let τ be a winning strategy for player II. We aim to verify that

$$\{z \in {}^\omega\omega : \tau \leq_T z\} \subset D(X_i)$$

for every $i < \omega$.

Let us fix $z \in {}^\omega\omega$ such that $\tau \leq_T z$, and let us also fix $i < \omega_1$. By the existence of $z^\#$, we may pick $g \in V$ to be $\mathrm{Col}(\omega, i)$-generic over $L[z]$. Let $x \in \mathrm{WO} \cap L[z][g]$ be such that $||x|| = i$. As $x * \tau \leq_T x \oplus z$, $\omega_1^{+L[x\oplus z]} \in X_i$. However, $\omega_1^{+L[z]} \leq \omega_1^{+L[x\oplus z]} \leq \omega_1^{+L[z][g]}$, which is equal to $\omega_1^{+L[z]}$, as $\mathrm{Col}(\omega, i)$ is smaller than ω_1^V in $L[z]$. Therefore $\omega_1^{+L[z]} = \omega_1^{+L[x\oplus z]} \in X_i$, i.e., $z \in D(X_i)$ as desired. □

By OD_S-*determinacy* we mean the statement that if $A \subset {}^\omega\omega$ is OD_S, cf. Definition 5.42, then A is determined.

Theorem 12.19 (A. Kechris) *Assume* AD. *Let $S \subset OR$. For an S-cone of reals x we have*

$$L[S, x] \models \mathsf{OD}_S\text{-}determinacy.$$

In particular, $\omega_1^{L[S,x]}$ is measurable in $\mathsf{HOD}_S^{L[S,x]}$.

Proof Let us assume that there is no S-cone of reals x such that in $L[S, x]$, all OD_S-sets of reals are determined. By Theorem 12.17 there is thus an S-cone $C \subset {}^\omega\omega$ such that for every $x \in C$, in $L[S, x]$ there is an non-determined $\mathsf{OD}_S^{L[S,x]}$-set of reals. Define, for $x \in C$, $x \mapsto A_x$ by letting A_x be the least $\mathsf{OD}_S^{L[S,x]}$-set of reals which is not determined in $L[S, x]$. ("Least" in the sense of a well-ordering of the OD_S-sets, cf. the proof of Theorem 5.45) I.e., if $G(A_x)$ is the usual game with payoff A_x, as defined in $L[S, x]$, then $\mathscr{G}_{A_x}$ is not determined in $L[S, x]$. Notice that A_x only depends on the S-constructibility degree of x, i.e., if $L[S, x] = L[S, y]$, then $A_x = A_y$.

Let G be the game in which I, II alternate playing natural numbers so that if

I	n_0		n_2		...	
II		n_1		n_3		...

is a play of G, then I wins iff, setting $x = (n_{4i} : i < \omega)$, $a = (n_{4i+2} : i < \omega)$, $y = (n_{4i+1} : i < \omega)$, and $b = (n_{4i+3} : i < \omega)$ (which we shall also refer to by saying that I produces the reals x, a and II produces the reals y, b), then

$$a \oplus b \in A_{x\oplus y}.$$

Let us suppose that I has a winning strategy, τ, in G. Let $\tau \in L[S, z]$, where z is in C. Let τ^* be a strategy for I in G_{A_z} played in $L[S, z]$ so that if II produces the real b, and if τ calls for I to produce the reals a, x in a play of G in which II plays b, $z \oplus b$, then τ^* calls for I to produce the real a. Then for every $b \in L[S, z]$, if $a = \tau^* * b$, in fact if $a, x = \tau(b, z \oplus b)$, then

$$a \oplus b \in A_{x\oplus(z\oplus b)} = A_z.$$

So τ^* is a winning strategy for I in the game $G(A_z)$ played in $L[S, z]$. Contradiction! We may argue similarily if II has a winning strategy in G.

We have shown that for an S-cone of x, $L[S,x] \models \mathsf{OD}_S$-determinacy. Let $x \in {}^\omega\omega$ be such that $L[S,x] \models \mathsf{OD}_S$-determinacy. Working inside $L[S,x]$, we may then define a filter μ on $\omega_1^{L[S,x]}$ as follows.

For reals x, let $|x| = \sup\{||y||\colon y \equiv_T x \wedge y \in \mathrm{WO}\}$. Let $S = \{|x|\colon x \in \mathbb{R}\}$. Let $\pi\colon \omega_1 \to S$ be the order isomorphism. Now if $A \subset \omega_1$, then we put $A \in \mu$ iff

$$\{x\colon |x| \in \pi'' A\}$$

contains an S-cone of reals. It is easy to verify that $\mu \cap \mathsf{HOD}_S$ witnesses that ω_1 is measurable in HOD_S. □

12.2 Martin's Theorem

Theorem 12.20 (D. A. Martin) *Suppose that $x^\#$ exists for every $x \in {}^\omega\omega$. Then every analytic set $B \subset {}^\omega\omega$ is determined. In fact, if $x \in {}^\omega\omega$ and $x^\#$ exists, then every $\Sigma^1_1(x)$ set $B \subset {}^\omega\omega$ is determined.*

Proof Let us fix an analytic set B, set $A = {}^\omega\omega \setminus B$. Recall that a set $A \subset {}^\omega\omega$ is coanalytic iff there is some map $s \mapsto <_s$, where $s \in {}^{<\omega}\omega$, such that for all $s, t \in {}^{<\omega}\omega$ with $s \subset t$, $<_t$ is an order on $lh(t)$ which extends $<_s$, and for all $x \in {}^\omega\omega$,

$$x \in A \iff <_x = \bigcup_{s \subset x} <_s \text{ is a well-ordering}$$

(Cf. Lemma 7.8 and Problem 7.7).

We have to consider the game $G(B)$,

I	n_0		n_2	…
II		n_1		n_3 …

in which I and II alternate playing integers $n_0, n_1, \ldots$, and I wins iff $x = (n_0, n_1, \ldots) \in B$. We have to prove that $G(B)$ is determined.

The key idea is to first consider the following auxiliary game, $G^*(A)$.

I	n_0		n_2		n_4		…
II		n_1, α_0		n_3, α_1		n_5, α_2	…

In this game, I and II also alternate playing integers $n_0, n_1, \ldots$. In addition, II has to play countable ordinals $\alpha_0, \alpha_1, \ldots$ such that for all $k < \omega$,

$$\left(k+1, <_{(n_0,\ldots,n_k)}\right) \overset{\pi}{\cong} (\{\alpha_0, \ldots, \alpha_k\}, <),$$

where $\pi(i) = \alpha_i$ for every $i \leq k$ (and $<$ is the natural order on ordinals). The first player to disobey one of the rules loses. If the play is infinite, then *II* wins.

Notice that what *II* has to do is playing a witness to the fact that $<_x$ is a well-order, where $x = (n_0, n_1, \ldots)$.

Notice also that $G^*(A)$ is an open game in the space ${}^{\omega}\omega \times {}^{\omega}\omega_1$ [(which we identify with ${}^{\omega}(\omega \times \omega_1)$] and hence by Theorem 12.3, $G^*(A)$ is determined in every inner model which contains $s \mapsto <_s$.

Fix a real x such that the map $s \mapsto <_s$ is in $L[x]$. E.g., let x be such that B is $\Sigma^1_1(x)$.

Let us first assume that *II* has a winning strategy for $G^*(A)$ in $L[x]$, call it τ. Obviously, $\tau \in L[x]$ is then also a winning strategy for $G^*(A)$ for all plays in V (not only the ones in $L[x]$). But then *II* will win $G(B)$ in V by just following τ and hiding her "side moves" $\alpha_0, \alpha_1, \ldots$. If $x = (n_0, n_1, \ldots)$ is the real produced at the end of a play, then

$$(\omega, <_x) \overset{\pi}{\cong} (\{\alpha_0, \alpha_1, \alpha_2, \ldots\}, <),$$

where $\pi(i) = \alpha_i$ for all $i < \omega$, so that $<_x$ must be a well-order, and thus $x \in A$, i.e. $x \notin B$.

Let us now suppose that *I* has a winning strategy for $G^*(A)$ in $L[x]$, call it σ. Whenever $\alpha_0, \ldots, \alpha_k$ and $\alpha'_0, \ldots, \alpha'_k$ are countable x-indiscernibles with

$$(\{\alpha_0, \ldots, \alpha_k\}, <) \overset{\pi}{\cong} (\{\alpha'_0, \ldots, \alpha'_k\}, <),$$

where $\pi(\alpha_i) = \alpha'_i$ for every $i \leq k$, then

$$L[x] \models \varphi(\sigma, \alpha_0, \ldots, \alpha_k) \Longleftrightarrow L[x] \models \varphi(\sigma, \alpha'_0, \ldots, \alpha'_k)$$

for every $\mathscr{L}_\in$-formula φ, cf. Corollary 10.44 (2). In particular, then,

$$\sigma(n_0, n_1, \alpha_0, \ldots, n_{2k}, n_{2k+1}, \alpha_k) = \sigma(n_0, n_1, \alpha'_0, \ldots, n_{2k}, n_{2k+1}, \alpha'_k)$$

for all integers $n_0, n_1, \ldots, n_{2k+1}$. We may therefore define a strategy $\overline{\sigma}$ for *I* in $G(B)$ as follows. Let

$$\overline{\sigma}(n_0, n_1, \ldots, n_{2k}, n_{2k+1}) = \sigma(n_0, n_1, \alpha_0, \ldots, n_{2k}, n_{2k+1}, \alpha_k)$$

where $\alpha_0, \ldots, \alpha_k$ are countable x-indiscernibles with

$$\left(k+1, <_{(n_0, \ldots, n_k)}\right) \overset{\pi}{\cong} (\{\alpha_0, \ldots, \alpha_k\}, <),$$

$\pi(i) = \alpha_i$ for $i \leq k$. We claim that $\overline{\sigma}$ is a winning strategy for *I* in $G(B)$.

Let us assume that this is not the case, so that there is a play of $G(B)$ in which *I* follows $\overline{\sigma}$ and which produces $x = (n_0, n_1, \ldots) \in A$. Then $<_x$ is a well-order and there is certainly a set $\{\alpha_0, \alpha_1, \ldots\}$ of countable x-indiscernibles such that

$$(\omega, <_x) \overset{\pi}{\cong} (\{\alpha_0, \alpha_1, \ldots\}, <),$$

where $\pi(i) = \alpha_i$ for $i < \omega$. I.e.,

$$\left(k + 1, <_{(n_0,\ldots,n_k)}\right) \overset{\pi\restriction(k+1)}{\cong} (\{\alpha_0, \ldots, \alpha_k\}, <)$$

for all $k < \omega$, and this means that for every $k < \omega$,

$$n_{2k} = \sigma(n_0, n_1, \alpha_0, \ldots, n_{2k-2}, n_{2k-1}, \alpha_{k-1}),$$

that is, $n_0, n_1, \alpha_0, n_2, n_3, \alpha_1, \ldots$ is a play of $G^*(A)$ in which I follows σ.

Let us now define the tree T of attempts to find an infinite play of $G^*(A)$ in which I follows σ as follows. We set $s \in T$ iff $s = (n_0, n_1, \alpha_0, \ldots, n_{2k-2}, n_{2k-1}, \alpha_{k-1}, n_{2k})$ for some $n_0, n_1, \ldots, n_{2k} \in \omega$ and $\alpha_0, \ldots, \alpha_{k-1} \in \omega_1$ such that for all $l \leq k$,

$$n_{2l} = \sigma(n_0, n_1, \alpha_0, \ldots, n_{2l-2}, n_{2l-1}, \alpha_{l-1}).$$

If $s, t \in$T, then we let $s \leq t$ iff $s \supset t$. Notice that $(T; \leq) \in L[x]$.

Now $(T; \leq)$ is ill-founded in V by what was shown above. Hence $(T; \leq)$ is ill-founded in $L[x]$ as well by the absoluteness of well-foundedness, cf. Lemma 5.6. Therefore, in $L[x]$ there is a play of $G^*(A)$ in which I follows σ and loses. But there cannot be such a play in $L[x]$, as σ is a winning strategy for I in $G^*(A)$. Contradiction! □

12.3 Harrington's Theorem

We now aim to prove the converse to the previous theorem. We'll first need the following

Lemma 12.21 *Let $x \in {}^\omega\omega$. Suppose that there is a real y such that whenever α is a countable ordinal with $J_\alpha[x, y] \models \mathsf{ZFC}^-$, then α is a cardinal of $L[x]$. Then $x^\#$ exists.*

Proof Suppose not. Let $y \in {}^\omega\omega$ be such that whenever α is a countable ordinal with $J_\alpha[x, y] \models \mathsf{ZFC}^-$, then α is a cardinal of $L[x]$. Let κ be a singular cardinal, and let α be such that $\kappa < \alpha < \kappa^+$ and $J_\alpha[x, y] \models \mathsf{ZFC}^-$. As we assume that $x^\#$ does not exist, Weak Covering, Corollary 11.60, yields that $\alpha < \kappa^{+L[x]} = \kappa^{+V}$. Let $\pi\colon J_\beta[x, y] \to J_{\kappa^+}[x, y]$ be elementary, where β is countable and $\alpha \in \operatorname{ran}(\pi)$. Write $\bar{\alpha} = \pi^{-1}(\alpha)$. Obviously, $J_{\bar{\alpha}}[x, y] \models \mathsf{ZFC}^-$, but $\bar{\alpha}$ is not an $L[x]$-cardinal ($\bar{\alpha}$ is not even a cardinal in $J_\beta[x]$). Contradiction! □

Any real code y for $x^{\#}$ satisfies the hypothesis of Lemma 12.21, cf. Problem 12.11.

There is a proof of Lemma 12.21 which avoids the use of Corollary 11.60 and which just makes use of the argument for Lemma 10.29. Cf. problem 12.9.

Theorem 12.22 (L. Harrington) *If analytical determinacy holds then for every $x \in {}^{\omega}\omega$, $x^{\#}$ exists. In fact, if $x \in {}^{\omega}\omega$ and every $\Sigma^1_1(x)$ set $B \subset {}^{\omega}\omega$ is determined, then $x^{\#}$ exists.*

Proof Forcing with $\mathrm{Col}(\omega, \alpha)$ adds reals coding ordinals below $\alpha + 1$. There is a forcing which adds such reals more directly, namely STEEL *forcing* which we shall denote by $\mathrm{TCol}(\omega, \alpha)$.

Let α be an infinite ordinal. We let $\mathrm{TCol}(\omega, \alpha)$ consist of all (t, h) such that t is a finite tree on ω, i.e., t is a non-empty finite subset of ${}^{<\omega}\omega$ such that $s \in t$ and $n \leq \mathrm{lh}(s)$ implies $s \upharpoonright n \in t$, and h is a "ranking" of t in the following sense: $h : t \to \alpha \cup \{\infty\}$ is such that $h(\emptyset) = \infty$, and if $s \in t$, $n < \mathrm{lh}(s)$, and $h(s \upharpoonright n) \in \alpha$, then $h(s) \in \alpha$ and $h(s) < h(s \upharpoonright n)$. For $(t, h), (t', h') \in \mathrm{TCol}(\omega, \alpha)$, we let $(t', h') \leq (t, h)$ iff $t' \supset t$ and $h' \supset h$.

Let G be $\mathrm{TCol}(\omega, \alpha)$-generic over V, and set

$$\begin{cases} T = \bigcup\{t : \exists h (t, h) \in G\} \text{ and} \\ H = \bigcup\{h : \exists t (t, h) \in G\}. \end{cases} \tag{12.1}$$

By easy density arguments, T must be an infinite tree on ω, and $H : T \to \alpha \cup \{\infty\}$ is surjective. For $s \in T$, write $T \upharpoonright s = \{s' \in {}^{\omega}\omega : s ^\frown s' \in T\}$. Straightforward density arguments also yield that $T \upharpoonright s$ is a well-founded tree on ω iff $H(s) \in \alpha$, and if $H(s) \in \alpha$, then $H(s)$ is the rank of $\emptyset$ in $T \upharpoonright s$ (i.e., the rank of s in T). If $\beta < \alpha$ and $H(s) = \beta$, then $T \upharpoonright s$ "codes" β in the sense that $T \upharpoonright s$, ordered by $\supset$, is a well-founded relation of rank β.

If $(t, h) \in \mathrm{TCol}(\omega, \alpha)$ and $\xi \leq \alpha$, then we may construe (t, h) as an element of $\mathrm{TCol}(\omega, \xi)$ by identifying ordinals in $[\xi, \alpha)$ with ∞. We define $(t, h)|\xi$ as (t, h'), where, for $s \in t$, $h'(s) = h(s)$ if $h(s) \in \xi$ and $h'(s) = \infty$ if $h(s) \notin \xi$.

The following combinational fact will be crucial for later purposes.

Claim 12.23 *Let $\omega \leq \xi < \xi + \omega \leq \alpha, \alpha'$, and let $(t, h) \in \mathrm{TCol}(\omega, \alpha)$ and $(t', h') \in \mathrm{TCol}(\omega, \alpha')$ be such that*

$$(t, h)|\xi + \omega = (t', h')|\xi + \omega.$$

Let $(u, g) \leq (t, h)$ in $\mathrm{TCol}(\omega, \alpha)$. Then there is $(u', g') \leq (t', h')$ in $\mathrm{TCol}(\omega, \alpha')$ such that

$$(u, g)|\xi = (u', g')|\xi.$$

Proof The hypothesis implies that $t' = t$. Set $u' = u$. We now define g'. Set $u^* = t \cup \{s \in u \backslash t : g(s) \in \xi\}$, and let $g^* = h' \cup g \upharpoonright \{s \in u \backslash t : g(s) \in \xi\}$. We are forced to let $g' \restriction u^* = g^*$. For $s \in u \setminus u^*$, we let $g'(s) = \xi + k$, where $k < \omega$ is the rank of $\emptyset$ in the tree $u \restriction s = \{s' : s^\frown s' \in u\}$ (i.e., the rank of s in u).

Let $s \in u$ and $n < lh(s)$. We need to see that if $g'(s \restriction n) \in \alpha'$, then $g'(s) < g'(s \restriction n)$. This is clear if $s \restriction n$ and s are both in t or both in $\{s \in u \backslash t : g(s) \in \xi\}$. If $s \restriction n \in t$ and $s \in u \backslash t$ with $g(s) \in \xi$, then $g'(s) = g^*(s) = g(s)$ and $h(s \restriction n) = g(s \restriction n) > g(s)$. By $(t, h)|\xi + \omega = (t', h')|\xi + \omega$, we must then have $g'(s \restriction n) = g^*(s \restriction n) = h'(s \restriction n) > g'(s)$.

Finally let $s \restriction n \in u \backslash t$ and hence $s \in u \backslash t$. If $g(s \restriction n) \in \xi$, then clearly $g(s) \in \xi$, too, so $g'(s) = g^*(s) = g(s) < g(s \restriction n) = g^*(s \restriction n) = g'(s \restriction n)$. If $g(s \restriction n) \notin \xi$ and $g(s) \in \xi$, then $g'(s \restriction n) \in [\xi, \xi + \omega)$ and $g'(s) = g^*(s) = g(s)$, so clearly $g'(s) < g'(s \restriction n)$. If $g(s \restriction n) \notin \xi$ and $g(s) \notin \xi$, then $g'(s \restriction n) = \xi + k$ and $g'(s) = \xi + k'$, where $k > k'$, so $g'(s) < g'(s \restriction n)$. □

We shall now be interested in forcing with $\mathrm{TCol}(\omega, \alpha)$ over (initial segments of) $L[x]$, where x is a real. If G is $\mathrm{TCol}(\omega, \alpha)$-generic over $L[x]$ and T and H are defined from G as in (12.1), then truth about initial segments of $L[x][T]$ can be decided by the right "restrictions" $(t, h)|\xi$ of elements (t, h) from G. In order to formulate this precisely, we need to rank sentences expressing truths about initial segments of $L[x][T]$ as follows.

Recall (cf. p. 70) that the $\mathrm{rud}_{x,T}$ functions are *simple* in the sense that if $\varphi(v_0, \ldots, v_{k-1})$ is a Σ_0-formula (in the language for $L[x, T]$) and $f_0, \ldots, f_{k-1}$ are $\mathrm{rud}_{x,T}$ functions, then there is a Σ_0-formula φ' (again in the language for $L[x, T]$) such that

$$\varphi(f_0(\boldsymbol{x_0}), \ldots, f_{k-1}(\boldsymbol{x_{k-1}})) \longleftrightarrow \varphi'(\boldsymbol{x_0}, \ldots, \boldsymbol{x_{k-1}})$$

holds true over all transitive $\mathrm{rud}_{x,T}$-closed models which contain $\boldsymbol{x_0}, \ldots, \boldsymbol{x_{k-1}}$. In particular, we may associate to each pair f, g of $rud_{x,T}$ functions a Σ_0-formula φ' and hence a Σ_ω-formula φ^* such that for all limit ordinals α and for all $\boldsymbol{x}, \boldsymbol{y} \in J_\alpha[x, T]$,

$$\begin{aligned} f(J_\alpha[x, T], \boldsymbol{x}) \in g(J_\alpha[x, T], \boldsymbol{y}) &\Longleftrightarrow J_{\alpha+\omega}[x, T] \models \varphi'(J_\alpha[x, T], \boldsymbol{x}, \boldsymbol{y}) \\ &\Longleftrightarrow J_\alpha[x, T] \models \varphi^*(\boldsymbol{x}, \boldsymbol{y}). \end{aligned}$$

We shall write $\varphi(f, g)$ for φ^* in what follows. The choice of φ' and $\varphi^* \equiv \varphi(f, g)$ can in fact be made uniformly in x, T.

Let us now pretend that the language for $L[x, T]$ has function symbols for $\mathrm{rud}_{x,T}$ functions available; we shall in fact confuse a given $\mathrm{rud}_{x,T}$ function f with the function symbol denoting it. We then define "terms of rank α" recursively as follows. A term of rank α is an expression of the form

$$f(J_\alpha[x, T], \boldsymbol{y}),$$

where f is (the function symbol for) a $\mathrm{rud}_{x,T}$ function, $\boldsymbol{y}$ is a vector of terms of rank $< \alpha$, and $J_\alpha[x, T]$ stands for the term denoting $J_\alpha[x, T]$. Inductively, every element of $J_{\alpha+\omega}[x, T]$ is thus denoted by a term of rank α.

The following Claim will be crucial.

Claim 12.24 *Let $\alpha \geq \omega$ be a limit ordinal, let φ be a formula of complexity $n \in \omega$ (in the language for $L[x,T]$), and let $\tau_1, \ldots, \tau_k$ be terms of rank $< \alpha$. If $\beta \geq \omega + (\alpha + 2n) \cdot \omega$, then for all $(t,h) \in \mathrm{TCol}(\omega, \beta)$,*

$$(t,h) \Vdash_{L[x]}^{\mathrm{TCol}(\omega,\beta)} J_\alpha[x,T] \models \varphi(\tau_1, \ldots, \tau_k) \Longleftrightarrow$$
$$(t,h)|\omega + (\alpha+2n)\cdot\omega \Vdash_{L[x]}^{\mathrm{TCol}(\omega,\omega+(\alpha+2n)\cdot\omega)} J_\alpha[x,T] \models \varphi(\tau_1, \ldots, \tau_k).$$

Proof The proof is, of course, by induction on $\alpha + n$.

Let us first assume $n = 0$, i.e. that φ is an atomic formula. Let us assume that $\tau_1 \equiv f(J_\beta[x,T], y)$ and $\tau_2 \equiv g(J_\beta[x,T], z)$, where $\beta < \alpha$, and that $\varphi(\tau_1, \tau_2) \equiv f(J_\beta[x,T], y) \in g(J_\beta[x,T], z)$. Then for $\xi \geq \omega + \alpha \cdot \omega$ and $(t,h) \in \mathrm{TCol}(\omega, \xi)$,

$$(t,h) \Vdash_{L[x]}^{\mathrm{TCol}(\omega,\xi)} J_\alpha[x,T] \models \varphi(\tau_1,\tau_2) \Longleftrightarrow$$
$$(t,h) \Vdash_{L[x]}^{\mathrm{TCol}(\omega,\xi)} J_{\beta+\omega}[x,T] \models \varphi(\tau_1,\tau_2) \Longleftrightarrow$$
$$(t,h) \Vdash_{L[x]}^{\mathrm{TCol}(\omega,\xi)} J_\beta[x,T] \models \varphi(f,g)(y,z).$$

Therefore, the desired statement easily follows from the inductive hypothesis.

Now let $n > 0$. Let us assume that $\varphi \equiv \exists v_0 \psi$. The cases $\varphi \equiv \neg\psi$ and $\varphi \equiv \psi_1 \wedge \psi_2$ are similar and easier.

Let us assume that

$$(t,h) \Vdash_{L[x]}^{\mathrm{TCol}(\omega,\beta)} J_\alpha[x,T] \models \exists v_0 \psi(v_0, \tau_1, \ldots, \tau_k).$$

Let $(t',h') \leq (t,h)|\omega + (\alpha+2n)\cdot\omega$ in $\mathrm{TCol}(\omega, \omega + (\alpha+2n)\omega)$. By Claim 12.23, there is $(t',h'') \leq (t,h)$ in $\mathrm{TCol}(\omega,\beta)$ such that $(t',h')|\omega + (\alpha+2n-1)\cdot\omega = (t',h'')|\omega + (\alpha+2n-1)\cdot\omega$. Let $(t^*,h^*) \leq (t',h'')$ in $\mathrm{TCol}(\omega,\beta)$ be such that

$$(t^*,h^*) \Vdash_{L[x]}^{\mathrm{TCol}(\omega,\beta)} J_\alpha[x,T] \models \psi(\tau_0, \tau_1, \ldots, \tau_k)$$

for some term τ_0 of rank $< \alpha$. Let $(t^*,h^{**}) \leq (t',h')$ in $\mathrm{TCol}(\omega, \omega + (\alpha+2n)\omega)$ such that $(t^*,h^{**})|\omega + (\alpha+2n-2)\cdot\omega = (t^*,h^*)|\omega + (\alpha+2n-2)\cdot\omega$, which may again be chosen by Claim 12.23. By the induction hypothesis,

$$(t^*,h^{**}) \Vdash_{L[x]}^{\mathrm{TCol}(\omega,\omega+(\alpha+2n)\cdot\omega)} J_\alpha[x,T] \models \psi(\tau_0, \tau_1, \ldots, \tau_k).$$

We have shown that the set of $(\bar{t},\bar{h}) \leq (t,h)|\omega + (\alpha+2n)\cdot\omega$ in $\mathrm{TCol}(\omega, \omega + (\alpha + 2n)\cdot\omega)$ such that

$$(\bar{t},\bar{h}) \Vdash_{L[x]}^{\mathrm{TCol}(\omega,\omega+(\alpha+2n)\cdot\omega)} J_\alpha[x,T] \models \varphi(\tau_1, \ldots, \tau_k).$$

is dense below $(t,h)|\omega + (\alpha+2n)\cdot\omega$, so that in fact

$$(t,h)|\omega + (\alpha+2n)\cdot\omega \Vdash^{\mathrm{TCol}(\omega,\omega+(\alpha+2n)\cdot\omega)}_{L[x]} J_\alpha[x,T] \models \varphi(\tau_1,\ldots,\tau_k).$$

The converse direction is shown in exactly the same fashion. □

Let us now assume analytic determinacy.

Let us fix $x \in {}^\omega\omega$ and a natural bijection $e\colon \omega \to {}^{<\omega}\omega$. Let us consider the following game G.

I	n_0		n_2		$\ldots$	
II		n_1		n_3		$\ldots$

Let us write $z_0 = (n_0, n_2, \ldots)$ and $z_1 = (n_1, n_3, \ldots)$. We say that player *II* wins iff the following holds true: if $z_0 \in {}^\omega\omega$ codes a well-founded tree T, i.e.,

$$T = \{e(n_{2i})\colon i < \omega\}$$

is a well-founded tree on ω, then z_1 codes a model $(\omega; E)$ of $\mathsf{ZFC}^- +$ "$V = L[x]$," say $E = \{(k,l)\colon z_1(\langle k,l\rangle) = 1\}$,[2] such that $||T||$ is contained in the transitive collapse of the well-founded part of $(\omega; E)$. It is straightforward to verify that the payoff set for G is analytic, in fact $\Sigma^1_1(x)$, so that G is determined.

Claim 12.25 *I does not have a winning strategy in G.*

Proof Suppose that σ is a winning strategy for I. Let D^* be the set of all real codes for well-founded trees, and let

$$D = \{\sigma * z_1\colon z_1 \in {}^\omega\omega\}.$$

Then D is an analytic set, $D \subset D^*$. It is easy to define a continuous function $f\colon {}^\omega\omega \to {}^\omega\omega$ such that for all $z \in {}^\omega\omega$,

$$z \in \mathrm{WO} \iff f(z) \in D^*.$$

By the Boundedness Lemma 7.12 there is some $\alpha < \omega_1$ with

$$\{||z||\colon f(z) \in D\} \subset \alpha,$$

i.e.,

$$\{||T||\colon \exists z_1 \in {}^\omega\omega(\sigma * z_1 \text{ codes } T)\} \subset \alpha.$$

But then *II* can easily defeat σ by playing a code for a transitive model of $\mathsf{ZFC}^- +$ "$V = L[x]$" which contains α. □

By Claim 12.25 and $\Sigma^1_1(x)$-determinacy, we may let τ be a winning strategy for player *II* in G. By Lemma 12.21, the following will produce $x^\#$.

[2] Here and in what follows we use the notation $\langle\cdot,\cdot\rangle$ from p. 33.

Claim 12.26 *Let α be a countable ordinal such that $J_\alpha[x,\tau] \models \mathsf{ZFC}^-$. Then α is a cardinal of $L[x]$. In fact, if α is countable and $x \oplus \tau$-admissible (cf. p. 88), then α is a cardinal of $L[x]$.*

Proof Let $\kappa < \alpha$ be an infinite cardinal of $L[x]$. It suffices to verify that if $b \subset \kappa$, $b \in L[x]$, then $b \in J_\alpha[x,\tau]$ (cf. Problem 12.10). So let us fix some such b, and let $b \in J_\delta[x]$, where without loss of generality $\omega_1 > \delta \geq \alpha$. Let $\gamma > \delta$ be such that $J_\gamma[x,\tau] \models \mathsf{ZFC}^-$ and γ is countable. Let G be $\mathrm{TCol}(\omega,\delta+1)$-generic over $J_\gamma[x,\tau]$, and let T and H be given by G as in (12.1). Obviously, there is a real z_0 in $J_\alpha[x,\tau,T]$ (even in $J_{\omega+\omega}[x,\tau,T]$) which codes a well-founded tree S such that $||S|| = \delta$. E.g., let $S = T \restriction s$, where $s \in T$ with $H(s) = \delta$. Therefore, $z_0 * \tau \in J_\alpha[x,\tau,T]$ (in fact $\in J_{\omega+\omega}[x,\tau,T]$) codes a model $(\omega; E)$ of $\mathsf{ZFC}^- + V = L[x]$ such that δ is contained in the transitive collapse of the well-founded part of $(\omega; E)$. Let

$$\pi\colon (J_\beta[x];\in) \cong \mathrm{wfp}(\omega; E)$$

be the transitive collapse of the well-founded part $\mathrm{wfp}(\omega; E)$ of $(\omega; E)$, so that $\beta \geq \delta$.

By $E \in J_{\omega+\omega}[x,\tau,T]$, it is easy to verify inductively that $\pi \restriction J_{\bar\beta}[x]$ is uniformly $\Sigma_1^{J_{\bar\beta}[x,\tau,T]}(\{E\})$ and $\pi \restriction J_{\bar\beta}[x] \in J_{\bar\beta+\omega}[x,\tau,T]$ for all $\bar\beta \leq \beta$, so that in particular

$$\pi \restriction J_\kappa[x] \in J_{\kappa+\omega}[x,\tau,T]. \tag{12.2}$$

As $b \in J_\delta[x] \subset J_\beta[x]$, there is some $n_0 \in \omega$ such that for all $\xi < \kappa$,

$$\xi \in b \Longleftrightarrow (\omega; E) \models \text{“}m \in n_0\text{,” where } m = (\pi \restriction J_\kappa[x])(\xi). \tag{12.3}$$

By (12.2), there is some formula φ and terms τ_0 and τ_1 for E and $\pi \restriction J_\kappa[x]$, respectively, where τ_0 and τ_1 are of rank $\leq \kappa$ and such that

$$\forall \xi < \kappa(\xi \in b \Longleftrightarrow J_{\kappa+\omega}[x,\tau,T] \models \varphi(\xi,\tau_0,\tau_1)). \tag{12.4}$$

Let $(t,h) \in \mathrm{TCol}(\omega,\delta+1)$ force (12.4) to hold true, i.e,

$$(t,h) \Vdash^{\mathrm{TCol}(\omega,\delta+1)}_{L[x,\tau]} \quad \forall \xi < \check\kappa(\xi \in \check b \longleftrightarrow J_{\check\kappa+\omega}[x,\tau,T] \models \varphi(\xi,\tau_0,\tau_1)),$$

which may be rewritten as saying that for all $\xi < \kappa$,

$$\xi \in b \Longleftrightarrow (t,h) \Vdash^{\mathrm{TCol}(\omega,\delta+1)}_{L[x,\tau]} \; J_{\check\kappa+\omega}[x,\tau,T] \models \varphi(\xi,\tau_0,\tau_1). \tag{12.5}$$

The point is now that because τ_0 and τ_1 are of rank $\leq \kappa$, letting $\beta = (\kappa+\omega\cdot 2)\cdot\omega = \omega + (\kappa+\omega\cdot 2)\cdot\omega$, we may use Claim 12.24 to rewrite (12.5) further to say that for every $\xi < \kappa$,

$$\xi \in b \Longleftrightarrow (t,h)|\beta \Vdash^{\mathrm{TCol}(\omega,\beta)}_{L[x,\tau]} \; J_{\check\kappa+\omega}[x,\tau,T] \models \varphi(\xi,\tau_0,\tau_1) \tag{12.6}$$

But we may replace $L[x, \tau]$ by $J_\alpha[x, \tau]$ here (we could in fact replace it by $J_{\kappa+\omega\cdot 2}[x, \tau]$), so that we may therefore define b over $J_\alpha[x, \tau]$ as follows:

$$b = \{\xi < \kappa : (t, h) | \beta \Vdash^{\mathrm{TCol}(\omega,\beta)}_{J_\alpha[x,\tau]} J_{\kappa+\omega}[x, \tau, T] \models \varphi(\check{\xi}, \tau_0, \tau_1)\}.$$

Therefore $b \in J_\alpha[\tau]$ as desired. □

We may now add the following to our list of equivalences to "$x^\#$ exists," cf. Theorem 11.56.

Corollary 12.27 *The following statements are equivalent.*

(1) *Every analytic $A \subset {}^\omega\omega$ is determined.*
(2) *For every $x \in {}^\omega\omega$, $x^\#$ exists.*

The paper [8] gives information on the Axiom of Determinacy. Cf. also [25].

12.4 Problems

12.1 Assume AD. Show that if $(x_i : i < \theta)$ is a sequence of pairwise different reals, then $\theta < \omega_1$ [Hint. Let U witness that ω_1 is measurable, cf. Theorem 12.18 (b). Consider $L[U, (x_i : i < \theta)]$, cf. proof of Problem 10.3 (a)].

12.2 Show (in ZF) that there is some $A \subset {}^\omega(\omega_1)$ which is not determined [Hint. If AD holds, then ask for *II* to play some $x \in {}^\omega\omega$ with $||x|| = \alpha$ in response to *I* playing $\alpha < \omega_1$].

12.3. Assume AD. Show that for every set A, $\mathrm{OD}_{\{A\}} \cap {}^\omega\omega$ is countable. Fixing A, show that there is no $f : {}^\omega\omega \to {}^\omega\omega$ such that $f(x) \in {}^\omega\omega \setminus \mathrm{OD}_{\{x,A\}}$ for all $x \in {}^\omega\omega$ and $f \in \mathrm{OD}_{{}^\omega\omega \cup \{A\}}$. Conclude that $\mathrm{HOD}_{{}^\omega\omega \cup \{A\}} \models$ "AD and there is some $(A_x : x \in {}^\omega\omega)$ with $\emptyset \neq A_x \subset {}^\omega\omega$ for all $x \in {}^\omega\omega$ with no choice function."

12.4. Assume ZF plus "$x^\#$ exists for every real x." Show that there is some $f : {}^\omega\omega \to {}^\omega\omega$ such that $f(x) \in {}^\omega\omega \setminus L[x]$ for all $x \in {}^\omega\omega$ and $f \in \mathrm{OD}$. (In fact, we may pick f to be Σ^1_3, cf. Problem 10.7.) Show also that there is a function $f : {}^\omega\omega \to \mathrm{HC}$ such that for every $x \in {}^\omega\omega$, $f(x)$ is a $\mathbb{C}$-generic filter over $L[x]$ (Hint. Use $x^\#$ to enumerate the dense sets of $L[x]$).

12.5. For $A, B \subset {}^\omega\omega$, write $A \leq_{\mathrm{Wadge}} B$ iff there is some continuous $f : {}^\omega\omega \to {}^\omega\omega$ such that for all $x \in {}^\omega\omega$, $x \in A \iff f(x) \in B$, or for all $x \in {}^\omega\omega$, $x \in A \iff f(x) \notin B$. Assume AD.

(a) **(Wadge)** Show for all $A, B \subset {}^\omega\omega$, $A \leq_{\mathrm{Wadge}} B$ or $B \leq_{\mathrm{Wadge}} A$. [Hint. Let $G_{\mathrm{Wadge}}(A, B)$ be the game so that if *I* plays x and *II* plays y, then *I* wins iff $x \in B \iff y \in A$]. Show that $\leq_{\mathrm{Wadge}}$ is reflexive and transitive, so that $A \sim_{\mathrm{Wadge}} B$ iff $A \leq_{\mathrm{Wadge}} B \wedge B \leq_{\mathrm{Wadge}} A$ is an equivalence relation. Show that $\leq_{\mathrm{Wadge}}$ is not symmetric.

(b) **(Martin, Monk)** Show that $\leq_{\text{Wadge}}$ is well-founded [Hint. Otherwise there are $A_n \subset {}^\omega\omega$, $n < \omega$, such that for all $n < \omega$, I has winning strategies σ_n^0 and σ_n^1 for $G_{\text{Wadge}}(A_{n+1}, A_n)$ and $G_{\text{Wadge}}({}^\omega\omega \setminus A_{n+1}, A_n)$, respectively. For $z \in {}^\omega 2$, we get $(x_n^z : n < \omega)$ such that $x_n^z = \sigma_n^{z(n)} * x_{n+1}^z$ for all $n < \omega$. Let $z, z' \in {}^\omega 2$, and $n < \omega$ be such that for all m, $z(m) = z'(m)$ iff $m \neq n$. Then $x_{n+1}^z = x_{n+1}^{z'}$, and $x_n^z \in A_n \iff x_n^{z'} \notin A_n, \ldots, x_0^z \in A_0 \iff x_0^{z'} \notin A_0$. Hence $\{z \in {}^\omega 2 : x_0^z \in A_0\}$ is a flip set, cf. Problem 8.3].

(c) Show that for all $A \subset {}^\omega\omega$ there is some $J(A) \subset {}^\omega\omega$ with $A <_{\text{Wadge}} J(A)$ [Hint. For $x \in {}^\omega\omega$ write f_x for the "canonical" continuous function given by x. Let $x \in B$ iff $f_x(x) \in A$, and set $J(A) = \{x \oplus y : x \in B \text{ and } y \notin B\}$].

(d) Let $\Theta = \sup(\{\alpha : \exists \text{ surjective } f : {}^\omega\omega \to \alpha\})$. Show that $|| <_{\text{Wadge}} || = \Theta$ [Hint. To show that $|| <_{\text{Wadge}} || \geq \Theta$, let $f : {}^\omega\omega \to \alpha$ be surjective, and let $(A_\nu : \nu < \alpha)$ be such that if $\nu < \alpha$, then $A_\nu = J(\{x \oplus y : f(x) < \nu \wedge y \in A_{f(x)}\})$, where J is as in (c)]. Let $X = {}^\omega\omega$. If $A \subset {}^\omega({}^\omega\omega)$, then in a run of the game $G(A)$ players I and II alternate playing real numbers, i.e., elements of ${}^\omega\omega$. The *Axiom of Real Determinacy*, abbreviated by $\mathsf{AD}_\mathbb{R}$, states that $G(A)$ is determined for every $A \subset {}^\omega({}^\omega\omega)$.

12.6. Assume $\mathsf{AD}_\mathbb{R}$.

(a) Show that for all $(A_x : x \in {}^\omega\omega)$ such that $\emptyset \neq A_x \subset {}^\omega\omega$ for every $x \in {}^\omega\omega$, there is a choice function.

(b) Show that there is a $< \aleph_1$-closed ultrafilter U on $[{}^\omega\omega]^{\aleph_0}$ such that every member of U is uncountable, $\{a \in [{}^\omega\omega]^{\aleph_0} : x \in a\} \in U$ for every $x \in {}^\omega\omega$, and U is normal in the following sense: if $(A_x : x \in {}^\omega\omega)$ is such that $A_x \in U$ for every $x \in {}^\omega\omega$, then there is some $A \in U$ such that whenever $x \in a \in A$, then $a \in A_x$ (Compare Problem 4.30) [Hint. For $A \subset [{}^\omega\omega]^{\aleph_0}$, let $A \in U$ iff I has a winning strategy in $G(\{f \in {}^\omega({}^\omega\omega) : \text{ran}(f) \in A\})$. To show normality, argue as follows. Let $(A_x : x \in {}^\omega\omega)$ be such that $A_x \in U$ for every $x \in {}^\omega\omega$. Let σ_x be a winning strategy for I in the game corresponding to A_x, $x \in {}^\omega\omega$. Let σ be a strategy for I such that if

I	x_0		x_2	...
II		x_1		x_3 ...

is a play, then for each $n < \omega$ there is an infinite $X_n \subset \omega \setminus (n+1)$, say $X_n = \{m(n,0) < m(n,1) < \ldots\}$, such that $x_{m(n,i)}$ is according to σ_{x_n} in a play where so far I played $x_{m(n,0)}, \ldots, x_{m(n,i-1)}$, and II played the first i many reals from $x_0, x_1, \ldots$ that were not played by I].

The SOLOVAY *sequence* $(\Theta_i : i \leq \Omega)$ is defined as follows. Let Θ be as in Problem 12.5 (d). Let $\Theta_0 = \sup(\{\alpha : \exists \text{ surjective } f : {}^\omega\omega \to \alpha,\ f \in \mathsf{OD}_{{}^\omega\omega}\})$. If Θ_i has been defined, then set $\Omega = i$ provided that $\Theta_i = \Theta$; otherwise let $\Theta_{i+1} = \sup(\{\alpha : \exists \text{ surjective } f : {}^\omega\omega \to \alpha,\ f \in \mathsf{OD}_{{}^\omega\omega \cup \{A\}}\})$ for some (all) $A \subset {}^\omega\omega$ with $||A||_{<_{\text{Wadge}}} = \Theta_i$. If $\lambda > 0$ is a limit ordinal and Θ_i has been defined for all $i < \lambda$, then set $\Theta_\lambda = \sup_{i<\lambda} \Theta_i$. We call Ω the *length* of the SOLOVAY sequence.

12.7. Assume AD.

(a) Show that if $A \subset {}^\omega\omega$ is such that $||A||_{<\text{Wadge}} < \Theta_0$, then $A \in \mathsf{OD}_{^\omega\omega}$ [Hint. Writing $\alpha = ||A||_{<\text{Wadge}}$, pick an $\mathsf{OD}_\mathbb{R}$ surjection $f: {}^\omega\omega \to \alpha + 1$, and let $(A_\nu : \nu \leq \alpha)$ be as in Problem 12.5 (d). Then $||A_\alpha||_{<\text{Wadge}} \geq \alpha$ and $A_\alpha \in \mathsf{OD}_{^\omega\omega}$, which yields $A \in \mathsf{OD}_{^\omega\omega}$]. Conclude that if $\Theta = \Theta_0$, then $\mathscr{P}({}^\omega\omega) \subset \mathsf{HOD}_{^\omega\omega}$.

(b) Show also if $A \subset {}^\omega\omega$ is such that $||A||_{<\text{Wadge}} < \Theta_{i+1}$, then $A \in \mathsf{OD}_{^\omega\omega \cup \{B\}}$ for some (all) $B \subset {}^\omega\omega$ with $||B||_{<\text{Wadge}} = \Theta_i$. Conclude that if the length of the SOLOVAY sequence is a successor ordinal, then there is some $B \subset {}^\omega\omega$ with $\mathscr{P}({}^\omega\omega) \subset \mathsf{HOD}_{^\omega\omega \cup \{B\}}$.

12.8. Show that if $\mathsf{AD}_\mathbb{R}$ holds, then the length of the SOLOVAY sequence is a limit ordinal and there is no $B \subset {}^\omega\omega$ with $\mathscr{P}({}^\omega\omega) \subset \mathsf{HOD}_{^\omega\omega \cup \{B\}}$. Conclude that AD does not imply $\mathsf{AD}_\mathbb{R}$ [Hint. Use Problems 12.3 and 12.7 (b)].

12.9. Show Lemma 12.21 by using the argument for Lemma 10.29.

12.10. (a) Show that there is a transitive model M of ZFC^- with $(M \cap L \cap \mathscr{P}(\omega)) \setminus J_\alpha \neq \emptyset$, where $\alpha = M \cap \mathrm{OR}$, so that $J_\alpha = L^M$ (Hint: Let α be countable in L such that $J_\alpha \models \mathsf{ZFC}^-$ and pick $G \in L$ which is $\mathbb{C}$-generic over J_α).

(b) Show that if M is admissible with $\alpha = M \cap \mathrm{OR}$ and $\mathscr{P}(\kappa) \cap L \subset M$ for every $\kappa < \alpha$, then α is a cardinal in L [Hint: Let $\beta < \kappa^{+L}$, let $f: \kappa \to J_\beta$ be bijective, $f \in L$, and let nEm iff $f(n) \in f(m)$. Then $E \in M$, and hence $J_\beta \in M$ by Problem 5.28].

12.11. Suppose that $0^\#$ exists, and let x be a real code for $0^\#$. Let κ be an infinite L-cardinal. Show that if $\alpha > \kappa$ is x-admissible, then $\mathscr{P}(\kappa) \cap L \subset J_\alpha[x]$ (Hint. The κth iterate of $0^\#$ exists in $J_\alpha[x]$, cf. Problem 5.28). Conclude that every x-admissible is a cardinal of L. In fact, every x-admissible is a SILVER indiscernible.

12.12. (**A. Mathias**) Let M be a transitive model of ZFC such that $M \models$ "U is a selective ultrafilter on ω." Let $\mathbb{M}$ be $(\mathbb{M}_U)^M$, i.e., MATHIAS forcing for U, as being defined in M, cf. p. 176. Let $x \in [\omega]^\omega$ be such that $x \setminus X$ is finite for every $X \in U$, and let

$$G = \{(s, X) \in \mathbb{M} : \exists n < \omega\, s = x \cap n\}$$

(cf. Definition 10.10.). Show that G is $\mathbb{M}$-generic over M (This is the converse to Problem Problem 9.9.) [Hint. Use Problem 9.3 (b) and the proof of Theorem 10.11, cf. p. 274 ff].

12.13. (**A. Mathias**) Show that in the model of Theorem 8.30, every uncountable $A \subset [\omega]^\omega$ is RAMSEY, cf. Definition 8.17 (Hint. Imitate the proof of Lemma 8.17, replacing $\mathbb{C}$ by MATHIAS forcing for some selective ultrafilter on ω, cf. Problem 9.4. Then use Problem 12.12).

Chapter 13
Projective Determinacy

We shall now use large cardinals to prove stronger forms of determinacy. We shall always reduce the determinacy of a complicated game in ${}^\omega\omega$ to the determinacy of a simple (open or closed) game in a more complicated space as in the proof of MARTIN's Theorem 12.20.

13.1 Embedding Normal Forms

Definition 13.1 Let $A \subset {}^\omega\omega$. We say that A has an *embedding normal form*,

$$(M_s, \pi_{s,t} : s \subset t \in {}^{<\omega}\omega),$$

iff $M_\emptyset = V$, each M_s is an inner model, each $\pi_{s,t}\colon M_s \to M_t$ is an elementary embedding, $\pi_{t,r} \circ \pi_{s,t} = \pi_{s,r}$ whenever $s \subset t \subset r$, and for each $x \in {}^\omega\omega$,

$$x \in A \iff \text{dir lim}(M_s, \pi_{s,t}\colon s \subset t \subset x) \text{ is well-founded.}$$

Such an embedding normal form is *κ-closed*, where κ is an infinite cardinal, iff ${}^\kappa M_s \subset M_s$ for each $s \in {}^{<\omega}\omega$.

Even though the following result had already been implicit in MARTIN's proof of Theorem 12.20 (cf. also [26]), it was first explicitly isolated and verified in [43].

Theorem 13.2 (K. Windßus) *Let $A \subset {}^\omega\omega$. If A has a $2^{\aleph_0}$-closed embedding normal form, then A is determined.*

Proof Let $(M_s, \pi_{s,t}\colon s \subset t \in {}^{<\omega}\omega)$ be a $2^{\aleph_0}$-closed embedding normal form for A. For $x \in {}^\omega\omega$, we shall write

$$(M_x, (\pi_{s,x} : s \subset x)) = \text{dir lim}(M_s, \pi_{s,t}\colon s \subset t \subset x).$$

R. Schindler, *Set Theory*, Universitext, DOI: 10.1007/978-3-319-06725-4_13,

We first construct a natural tree T, which we call "the" WINDßUS *tree* for A, such that $A = p[T]$ as follows. We first define a sequence $(\alpha_s : s \in {}^{<\omega}\omega)$. If $x \notin A$, so that M_x is ill–founded, we pick a sequence $(\alpha_x^n : n < \omega)$ of ordinals witnessing that M_x is ill–founded in the sense that

$$\pi_{x\restriction n, x\restriction(n+1)}(\alpha_x^n) > \alpha_x^{n+1} \tag{13.1}$$

for all $n < \omega$. If $x \in A$, then we let $\alpha_x^n = 0$ for all $n < \omega$. We then let, for $s \in {}^{<\omega}\omega$, $\alpha_s : {}^{\omega}\omega \to \mathrm{OR}$ be defined by

$$\alpha_s(x) = \begin{cases} \alpha_x^{\mathrm{lh}(s)} & \text{if } x \notin A \wedge s = x \restriction \mathrm{lh}(s) \\ 0 & \text{otherwise.} \end{cases}$$

Let β be an ordinal which is bigger than all α_x^n.

We define T by setting $(s, f) \in T$ iff $s \in {}^{<\omega}\omega$, $f = (f_i : i < \mathrm{lh}(s))$, where each f_i is a function from ${}^{\omega}\omega$ to β, and for all $i + 1 < \mathrm{lh}(s)$ and for all $x \in {}^{\omega}\omega$, if $x \notin A \wedge s \restriction i+1 = x \restriction i+1$, then $f_{i+1}(x) < f_i(x)$, and if $x \in A \vee s \restriction i+1 \neq x \restriction i+1$, then $f_{i+1}(x) = 0$. The order on T is reverse inclusion.

Now if $(f_i : i < \omega)$ witnesses that $x \in p[T]$, then $x \in A$, because otherwise $f_{i+1}(x) < f_i(x)$ for each $i < \omega$. Hence $p[T] \subset A$. On the other hand, if $x \in A$, then we may define a witness $(f_i : i < \omega)$ to the fact that $x \in p[T]$ as follows: Let $x' \in {}^{\omega}\omega$. If $x' \notin A$, then let $k < \omega$ be maximal with $x' \restriction k = x \restriction k$ and define $f_i(x') = k + 1 - i$ for $i \leq k$ and $f_i(x') = 0$ for $i > k$. If $x' \in A$, then define $f_i(x') = 0$ for all $i < \omega$. This shows $A \subset p[T]$, and hence $A = p[T]$.

Let us now consider the following game, called $G^*(A)$:

I	n_0, f_0		n_2, f_1		$\ldots$
II		n_1		n_3	$\ldots$

Here, each n_i is a natural number, and each f_i is a function from ${}^{\omega}\omega$ to β. I wins iff $((n_i : i < \omega), (f_i : i < \omega)) \in [T]$. The payoff set is thus a closed subset of

$${}^{\omega}\omega \times {}^{\omega}({}^{({}^{\omega}\omega)}\beta),$$

and is hence determined by Theorem 12.3.

Let us first suppose I to have a winning strategy σ^* for $G^*(A)$. Then a winning strategy σ for I in $G(A)$ is obtained by playing as according to σ^*, but hiding the "side moves" f_i. Recall that $((n_i : i < \omega), (f_i : i < \omega)) \in [T]$ proves that $(n_i : i < \omega) \in A$, so that σ is indeed a winning strategy for I.

Now let us suppose II to have a winning strategy τ^* in $G^*(A)$. We aim to produce a winning strategy τ for II for $G(A)$.

Let $k < \omega$, and let

I	n_0		n_2		$\ldots$	n_{2k}
II		n_1		n_3	$\ldots$	

be a position in $G(A)$, so that it's II's turn to play. Notice that, because each M_s is $2^{\aleph_0}$–closed, $\alpha_s \in M_s$ for every $s \in {}^{<\omega}\omega$. Moreover, setting $s = (n_0, \ldots, n_k)$,

$$((n_0, \ldots, n_k), (\pi_{\emptyset,s\restriction k}(\alpha_\emptyset), \ldots, \pi_{s\restriction k-1,s\restriction k}(\alpha_{s\restriction k-1}), \alpha_{s\restriction k})) \in \pi_{\emptyset,s\restriction k}(T) \quad (13.2)$$

by the construction of the $\alpha_{s\restriction i}$ and of T. Equation (13.2) holds true because if $x \notin A \wedge s = x \restriction \mathrm{lh}(s)$, then (13.1) yields that for all $i < \mathrm{lh}(s)$, $\alpha_{s\restriction i+1}(x) < \pi_{s\restriction i,s\restriction i+1}(x)$, and thus $\pi_{s\restriction(i+1),s}(\alpha_x^{i+1}) = \pi_{s\restriction i+1,s}(\alpha_{s\restriction(i+1)})(x) < \pi_{s\restriction i+1,s} \circ \pi_{s\restriction i,s\restriction i+1}(\alpha_{s\restriction i})(x) = \pi_{s\restriction i,s}(\alpha_{s\restriction i})(x) = \pi_{s\restriction i,s}(\alpha_x^i)$. Moreover, if $x \in A$ or $s \neq x \restriction \mathrm{lh}(s)$, then $\pi_{s\restriction(i+1)s}(\alpha_{s\restriction(i+1)})(x) = 0$.

We may therefore define τ by letting $\tau((n_0, \ldots, n_{2k}))$ be the unique $n < \omega$ such that

$$M_s \models n = \pi_{\emptyset,s}(\tau^*)((n_0, \pi_{\emptyset,s}(\alpha_\emptyset), n_1, \ldots, n_{2k}, \alpha_s)).$$

Suppose τ not to be a winning strategy for II for $G(A)$. There is then a play $(n_0, n_1, \ldots)$ of $G(A)$ in which II follows τ, but I wins, i.e., setting $x = (n_0, n_1, \ldots)$, $x \in A$. In particular, M_x is well–founded (i.e., transitive), and by the elementarity of $\pi_{\emptyset,x}$,

$$M_x \models \pi_{\emptyset,x}(\tau^*) \text{is a winning strategy for } II \text{ in } \pi_{\emptyset,x}(G^*(A)).$$

By (13.1) and the elementarity of $\pi_{\emptyset,x}$, in V there is a play of $\pi_{\emptyset,x}(G^*(A))$ in which II follows $\pi_{\emptyset,x}(\tau^*)$ and in which II loses, namely

I	$n_0, \pi_{\emptyset,x}(\alpha_\emptyset)$	$n_2, \pi_{x\restriction 1,x}(\alpha_{x\restriction 1})$	...
II	n_1		...

We may now exploit the absoluteness of well–foundedness between V and M_x, cf. Lemma 5.6, and argue exactly as in the second last paragraph of the proof of Theorem 12.20 to deduce that there is hence a play of $\pi_{\emptyset,x}(G^*(A))$ in M_x in which II follows $\pi_{\emptyset,x}(\tau^*)$ and in which II looses. But this is a contradiction! □

It is not hard to show that if there is a measurable cardinal, then every set of reals has an embedding normal form, cf. Problem 13.1. It is much harder to get embedding normal forms which are sufficiently closed. The proof of the following result is similar to the proof of Theorem 12.20.

Theorem 13.3 *Let κ be a measurable cardinal, and let $A \subset {}^\omega\omega$ be coanalytic. Then A has a κ–closed embedding normal form.*

Proof Recall again that a set $A \subset {}^\omega\omega$ is coanalytic iff there is some map $s \mapsto <_s$, where $s \in {}^{<\omega}\omega$, such that for all $s, t \in {}^{<\omega}\omega$ with $s \subset t$, $<_t$ is an order on $\mathrm{lh}(t)$ which extends $<_s$, and for all $x \in {}^\omega\omega$,

$$x \in A \iff <_x = \bigcup_{s \subset x} <_s \text{ is a well–ordering.} \quad (13.3)$$

(Cf. Lemma 7.8 and Problem 7.7.)

Let $s \subsetneq t$, where $\text{lh}(t) = \text{lh}(s) + 1$. Write $n = \text{lh}(s)$. Suppose that n is the kth element of $n + 1 = \{0, \ldots, n\}$ according to $<_t$, i.e.,

$$m_0 <_t \ldots <_t m_{k-1} <_t n <_t m_k <_t \ldots <_t m_{n-1},$$

where m_l is the lth element of $n = \{0, \ldots, n-1\}$ according to $<_s$, $l < n$. We then define $\varphi(s,t): n \to n+1$ by $\varphi(s,t)(l) = l$ for $l < k$ and $\varphi(s,t)(l) = l+1$ for $l \geq k$, $l < n$.

If $s \subsetneq t$, where $\text{lh}(t) = \text{lh}(s) + m$, then we define $\varphi(s,t): \text{lh}(s) \to \text{lh}(t)$ by

$$\varphi(s,t) = \varphi(t \restriction \text{lh}(t) - 1, t) \circ \ldots \circ \varphi(s, t \restriction \text{lh}(s) + 1).$$

The map $\varphi(s,t)$ then tells us how the $\text{lh}(s), \text{lh}(s) + 1, \ldots, \text{lh}(t) - 1$ sit inside $0, 1, \ldots, \text{lh}(t) - 1$ according to $<_t$.

Let us now define an embedding normal form $(M_s, \pi_{st}: s \subset t \in {}^{<\omega}\omega)$ for A as follows. Let U be a measure on κ, and let

$$(M_\alpha, \pi_{\alpha\beta}: \alpha \leq \beta \in \text{OR})$$

be the (linear) iteration of $V = M_0$ given by U. Let us write $U_\alpha = \pi_{0\alpha}(U)$ and $\kappa_\alpha = \pi_{0\alpha}(\kappa)$, where $\alpha \in \text{OR}$. We set $M_s = M_{\text{lh}(s)}$ and

$$\pi_{s,t} = \pi^{\varphi(s,t)}_{\text{lh}(s)\text{lh}(t)},$$

where $\pi^{\varphi(s,t)}_{\text{lh}(s)\text{lh}(t)}$ is the shift map given by $\varphi(s,t)$, cf. the Shift Lemma 10.4. For $x \in {}^\omega\omega$, let

$$(M_x, (\pi_{s,x}: s \subset x)) = \text{dir lim}(M_s, \pi_{s,t}: s \subset t \subset x).$$

Notice that $\{\pi_{sx}(\kappa_n): n < \omega, s \subset x, \text{lh}(s) = n+1\}$, ordered by the $\in$–relation of M_x, is always isomorphic to ω, ordered by $<_x$. By (13.3), this readily implies that if $x \notin A$, i.e., if $<_x$ is ill–founded, then M_x is ill–founded.

But it also implies that if $x \in A$, i.e., $<_x$ is well–founded, then M_x is well–founded as follows. Let $x \in A$ and let $\gamma = \text{otp}(<_x)$. We may define maps $\varphi(s,x): \text{lh}(s) \to \gamma$ by setting

$$\varphi(s,x)(n) = ||n||_{<_x}$$

for $n < \text{lh}(s)$, $s \subset x$. Notice that $\varphi(s,x) = \varphi(t,x) \restriction \text{lh}(s)$ whenever $s \subset t \subset x$. By Lemma 10.4, we have, for $s \subset x$,

$$\pi^{\varphi(s,x)}_{\text{lh}(s),\gamma}: M_s \to M_\gamma,$$

where for $s \subset t \subset x$,

$$\pi^{\varphi(t,x)}_{\text{lh}(t),\gamma} \circ \pi^{\varphi(s,t)}_{\text{lh}(s),\text{lh}(t)} = \pi^{\varphi(s,x)}_{\text{lh}(s),\gamma}.$$

Therefore, we may define an elementary embedding $k\colon M_x \to M_y$ by setting

$$k(\pi_{sx}(y)) = \pi^{\varphi(s,x)}_{\mathrm{lh}(s),y}(y)$$

for $y \in M_s$.

As ${}^{\kappa}M_n \subset M_n$ for all $n < \omega$ by Lemma 4.63, we have thus shown that A has a κ–closed embedding normal form. □

Theorems 13.2 and 13.3 reprove Martin's Theorem 12.20.

We shall now turn towards proving Projective Determinacy.

Definition 13.4 Projective Determinacy, PD, is the statement that all projective subsets of ${}^{\omega}\omega$ are determined.

The key new ingredients to show that Projective Determinacy holds true are iteration trees which are produced by Woodin cardinals.

Definition 13.5 Let $A \subset {}^{\omega}\omega$, and let

$$\mathscr{E} = (M_s, \pi_{st} \colon s \subset t \in {}^{<\omega}\omega)$$

be an embedding normal form for A. If α is an ordinal, then we say that the *additivity* of $\mathscr{E}$ is bigger than α iff $\pi_{st} \restriction (\alpha + 1) = \mathrm{id}$ for all $s, t \in {}^{<\omega}\omega$, $s \subsetneq t$.

13.2 The Martin–Steel Theorem

The following seminal result was produced in [26].

Theorem 13.6 (D. A. Martin, J. Steel) *Let δ be a Woodin cardinal, let $B \subset {}^{\omega}\omega$, and suppose that B has a $2^{\aleph_0}$–closed embedding normal form whose additivity is bigger than δ. Then for every $\alpha < \delta$,*

$$\{x \in {}^{\omega}\omega \colon \forall y \in {}^{\omega}\omega\ x \oplus y \notin B\}$$

has a $2^{\aleph_0}$–closed embedding normal form whose additivity is bigger than α.

Theorems 13.2 and 13.3 immediately give the following.

Corollary 13.7 *Let $n < \omega$, and suppose that there is a measurable cardinal above n Woodin cardinals. Then every $\underset{\sim}{\Pi}^1_{n+1}$ subset of ${}^{\omega}\omega$ is determined. In particular, if there are infinitely many Woodin cardinals, then Projective Determinacy holds.*

The proofs of Theorems 12.11, 12.13, and 12.16 also give the following.

Corollary 13.8 *Suppose that there are infinitely many* WOODIN *cardinals. If A is projective set of reals, then A is* LEBESGUE *measurable and has the* BAIRE *property, and if A is uncountable, then A has a perfect subset.*

In particular, the collection of all projective sets of reals has the perfect subset property (cf. p. 142), i.e., here are no "definable" counterexamples to CANTOR's program, cf. 3.

Proof of Theorem 13.6. Let us fix

$$(N_s, \sigma_{s,t} \colon s \subset t \in {}^{<\omega}\omega),$$

a $2^{\aleph_0}$–closed embedding normal form for B whose additivity is bigger than δ.

If $s, t \in {}^{<\omega}\omega$ are such that $\mathrm{lh}(s) = \mathrm{lh}(t)$, then we define $s \oplus t$ to be that $r \in {}^{<\omega}\omega$ such that $\mathrm{lh}(r) = 2 \cdot \mathrm{lh}(s)$, and for all $n < \mathrm{lh}(s)$, $r(2n) = s(n)$ and $r(2n+1) = t(n)$. Let us write

$$A = \{(x, y) \in ({}^{\omega}\omega)^2 : x \oplus y \in B\},$$

so that trivially $\{x \in {}^{\omega}\omega \colon \forall y \in {}^{\omega}\omega\ x \oplus y \notin B\} = \{x \in {}^{\omega}\omega \colon \forall y \in {}^{\omega}\omega \colon (x, y) \notin A\}$. Let us also write $N_{s,t}$ for $N_{s \oplus t}$, and $\sigma_{(s,t),(s',t')}$ for $\sigma_{s \oplus t, s' \oplus t'}$, where $s, s', t, t' \in {}^{<\omega}\omega$ with $\mathrm{lh}(s) = \mathrm{lh}(t) < \mathrm{lh}(s') = \mathrm{lh}(t')$. Then

$$(N_{s,t}, \sigma_{(s,t),(s',t')} \colon s \subset s' \in {}^{<\omega}\omega, t \subset t' \in {}^{<\omega}\omega, \mathrm{lh}(s) = \mathrm{lh}(t), \mathrm{lh}(s') = \mathrm{lh}(t')) \tag{13.4}$$

is a $2^{\aleph_0}$–closed embedding normal form for A whose additivity is bigger than δ in the sense that for all $x, y \in {}^{\omega}\omega$,

$$(x, y) \in A \iff \text{dir lim}(N_{s,t}, \sigma_{(s,t),(s',t')} \colon s \subset t \subset x, t \subset t' \subset y) \text{ is well–founded,} \tag{13.5}$$

every $N_{s,t}$ is $2^{\aleph_0}$–closed, and $\sigma_{(s,t),(s',t')} \restriction (\delta + 1) = \mathrm{id}$ for all relevant $\sigma_{(s,t),(s',t')}$.

Let us first construct "the" WINDßUS *tree* T for A in much the same way as in the proof of Theorem 13.2, as follows. We start by defining a sequence $(\alpha_{s,t} \colon s, t \in {}^{<\omega}\omega, \mathrm{lh}(s) = \mathrm{lh}(t))$. If $(x, y) \notin A$, so that

$$\text{dir lim}(N_{s,t}, \sigma_{(s,t),(s',t')} \colon s \subset s' \subset x, t \subset t' \subset y)$$

is ill–founded, we pick a sequence $(\alpha^n_{x,y} \colon n < \omega)$ of ordinals witnessing that this direct limit is ill–founded, i.e.,

$$\sigma_{(x \restriction n, y \restriction n),(x \restriction (n+1), y \restriction (n+1))}(\alpha^n_{x,y}) > \alpha^{n+1}_{x,y} \tag{13.6}$$

for all $n < \omega$. If $(x, y) \in A$, we let $\alpha^n_{x,y} = 0$ for all $n < \omega$. Let, for $s, t \in {}^{<\omega}\omega$, $\mathrm{lh}(s) = \mathrm{lh}(t)$, $\alpha_{s,t} \colon {}^{\omega}\omega \times {}^{\omega}\omega \to \mathrm{OR}$ be defined by

$$\alpha_{s,t}(x,y)=\begin{cases}\alpha^{\mathrm{lh}(s)}_{x,y} & \text{if } (x,y)\notin A\wedge s=x\restriction \mathrm{lh}(s)\wedge t=y\restriction \mathrm{lh}(s)\\ 0 & \text{otherwise}\end{cases} \tag{13.7}$$

Let β be an ordinal which is bigger than all $\alpha^n_{x,y}$. We define T by setting $(s,t,f)\in T$ iff $s,t\in {}^{<\omega}\omega$, $\mathrm{lh}(s)=\mathrm{lh}(t)$, $f=(f_i : i<\mathrm{lh}(s))$, where each f_i is a function from ${}^{\omega}\omega\times{}^{\omega}\omega$ to β, and for all $i+1<\mathrm{lh}(s)$ and for all $x,y\in{}^{\omega}\omega$, if $(x,y)\notin A\wedge s\restriction i+1=x\restriction i+1\wedge t\restriction i+1=y\restriction i+1$, then $f_{i+1}(x,y)<f_i(x,y)$, and if $(x,y)\in A\vee s\restriction i+1\neq x\restriction i+1\vee t\restriction i+1\neq y\restriction i+1$, then $f_{i+1}(x,y)=0$. The order on T is again reverse inclusion.

Now if $(f_i : i<\omega)$ witnesses that $(x,y)\in p[T]$, then we cannot have that $(x,y)\notin A$, as otherwise $f_{i+1}(x,y)<f_i(x,y)$ for each $i<\omega$. Hence $p[T]\subset A$. On the other hand, if $(x,y)\in A$, then we may define a witness $(f_i : i<\omega)$ to the fact that $(x,y)\in p[T]$ as follows: Let $x',y'\in{}^{\omega}\omega$. If $(x',y')\notin A$, then let $k<\omega$ be maximal with $x'\restriction k=x\restriction k\wedge y'\restriction k=y\restriction k$, and define $f_i(x',y')=k+1-i$ for $i\leq k+1$ and $f_i(x',y')=0$ for $i\geq k+1$. If $(x',y')\in A$, then define $f_i(x',y')=0$ for all i. This shows $A\subset p[T]$, and hence $A=p[T]$.

Notice that, as each $N_{s,t}$ is $2^{\aleph_0}$–closed, we have that $\alpha_{s,t}\in N_{s,t}$ for every s,t. In fact, for all $s,t\in{}^{<\omega}\omega$, $\mathrm{lh}(s)=\mathrm{lh}(t)$,

$$((s,t,\sigma_{(s\restriction i,t\restriction i),(s,t)}(\alpha_{s\restriction i,t\restriction i})) : i<\mathrm{lh}(s))\in\sigma_{(\emptyset,\emptyset),(s,t)}(T) \tag{13.8}$$

To see this, notice that if $x,y\notin A$, $s,t\in{}^{<\omega}\omega$, $\mathrm{lh}(s)=\mathrm{lh}(t)$, $x\restriction\mathrm{lh}(s)=s$ and $y\restriction\mathrm{lh}(s)=t$, and $i<j<\mathrm{lh}(s)$, then by (13.6)

$$\begin{aligned}\sigma_{(s\restriction i,t\restriction i),(s,t)}(\alpha_{s\restriction i,t\restriction i})(x,y)&=\sigma_{(s\restriction i,t\restriction i),(s,t)}(\alpha^i_{x,y})\\&>\sigma_{(s\restriction j,t\restriction j),(s,t)}(\alpha^j_{x,y})\\&=\sigma_{(s\restriction j,t\restriction j),(s,t)}(\alpha_{s\restriction j,t\restriction j})(x,y),\end{aligned}$$

and if $(x,y)\in A\vee x\restriction\mathrm{lh}(s)\neq s\vee y\restriction\mathrm{lh}(s)\neq t$, then $\sigma_{(s\restriction i,t\restriction i),(s,t)}(\alpha_{s\restriction i,t\restriction i})(x,y)=0$.

Let us now fix $\alpha<\delta$. We aim to construct a $2^{\aleph_0}$–closed embedding normal form for $\{x : \forall y(x,y)\notin A\}$ whose additivity is bigger than α. The embedding normal form will be produced by iteration trees on V.

Let $(s_n : n<\omega)$ be an enumeration of ${}^{<\omega}\omega$ such that if $s_n\subsetneq s_m$, then $n<m$. Let $\preceq$ be the following order on ω: we set $n\preceq m$ iff $n=0$, or n,m are both even and $n\leq m$, or n,m are both odd and

$$s_{\frac{n+1}{2}}\subset s_{\frac{m+1}{2}}.$$

We intend to have s_n correspond to the node $2n\dot{-}1$ in the tree order $\preceq$.[1]

For $s\in{}^{<\omega}\omega$, we shall produce objects

[1] We here and in what follows use $k\dot{-}l$ to denote $k-l$, unless $l>k$ in which case $k\dot{-}l=0$.

$$\begin{cases} \mathcal{T}_s & \\ \kappa_{s,k} & \text{for } k \leq 2 \cdot \text{lh}(s), \\ \beta_{s,k} & \text{for } k \leq \text{lh}(s), \text{ and} \\ \eta_{s,k} & \text{for } k \leq \text{lh}(s). \end{cases} \tag{13.9}$$

We will arrange that the following statements (PD, 1) through (PD, 4) hold true.

(PD, 1) Each $\mathcal{T}_s$ is an iteration tree on V of length $2 \cdot \text{lh}(s) + 1$,

$$\mathcal{T}_s = ((M_{s,k}, \pi_{s,k,l} \colon k \preceq l \leq 2 \cdot \text{lh}(s)), (E_{s,k} \colon k < 2 \cdot \text{lh}(s)), \preceq\restriction (2 \cdot \text{lh}(s) + 1)),$$

such that for each $k < 2 \cdot \text{lh}(s)$,

$$M_{s,k} \models \text{“}E_{s,k} \text{ is a } 2^{\aleph_0}\text{-closed and certified extender with critical point } \kappa_{s,i} > \alpha \text{ and } E_{s,k} \in V_\delta,\text{”}$$

where

$$i = \begin{cases} 2m & \text{if } k \text{ is even, say } k = 2n \text{ and } s_m = s_{n+1} \restriction (\text{lh}(s_{n+1}) - 1), \text{ and} \\ k & \text{if } k \text{ is odd.} \end{cases}$$

Moreover, $\kappa_{s,0} < \kappa_{s,1} < \ldots < \kappa_{s,2\cdot\text{lh}(s)}$, and for all $k < l \leq 2 \cdot \text{lh}(s)$,

$$(V_\lambda)^{M_{s,k}} = (V_\lambda)^{M_{s,l}},$$

where λ is the least inaccessible cardinal of $M_{s,k}$ which is bigger than $\kappa_{s,k+1}$.

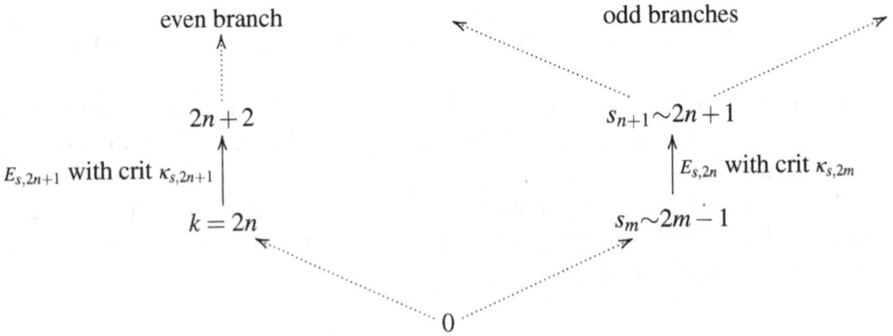

(PD, 2) If $s \subset s'$, then $\mathcal{T}_{s'}$ end–extends $\mathcal{T}_s$, i.e., $M_{s',k} = M_{s,k}$ for all $k \leq 2 \cdot \text{lh}(s)$ and $E_{s',k} = E_{s,k}$ for all $k < 2 \cdot \text{lh}(s)$ (of course, the latter implies the former); also, $\kappa_{s',k} = \kappa_{s,k}$ for $k \leq 2 \cdot \text{lh}(s)$, $\beta_{s',k} = \beta_{s,k}$ for $k \leq \text{lh}(s)$, and $\eta_{s',k} = \eta_{s,k}$ for $k \leq \text{lh}(s)$.

(PD, 3) Say $n = \mathrm{lh}(s)$. Then

$$(s \restriction \mathrm{lh}(s_n), s_n, (\pi_{s,2l\dot{-}1,2n\dot{-}1}(\eta_{s,l})\colon 2l\dot{-}1 \preceq 2n\dot{-}1)) \in \pi_{s,0,2n\dot{-}1}(T).$$

The forth condition which will guarantee that the "even branches" of $\mathscr{T}_s$, $s \subset x$, where $x \in {}^\omega\omega$, produce an embedding normal form. In order to state it, we need a notation for the models and embeddings of $\mathscr{T}_s$ from the point of view of $N_{s',t'}$, the models from (13.4). Let $s \in {}^{<\omega}\omega$, and let $s', t' \in {}^{<\omega}\omega$, $\mathrm{lh}(s') = \mathrm{lh}(t')$. Because for all $k < 2 \cdot \mathrm{lh}(s)$, $E_{s,k} \in V_\delta^{M_{s,k}}$, and because the additivity of our given embedding normal form (13.4) for A is bigger than δ, the sequence

$$(E_{s,k}\colon k < 2 \cdot \mathrm{lh}(s))$$

is easily seen to generate[2] $(\mathscr{T}_s)^{N_{s',t'}}$ such that

$$N_{s',t'} \models \text{"}(\mathscr{T}_s)^{N_{s',t'}} \text{ is an iteration tree on } N_{s',t'} \text{ of length } 2 \cdot \mathrm{lh}(s) + 1,\text{"}$$

where we may write

$$(\mathscr{T}_s)^{N_{s',t'}} = ((M_{s,k}^{s',t'}, \pi_{s,k,l}^{s',t'}\colon k \preceq l \leq 2 \cdot \mathrm{lh}(s)), (E_{s,k}\colon k < 2 \cdot \mathrm{lh}(s)),$$
$$\preceq \restriction (2 \cdot \mathrm{lh}(s) + 1)).$$

We will have that

$$\begin{aligned} M_{s,k}^{s',t'} &= \sigma_{(\emptyset,\emptyset),(s',t')}(M_{s,k}), \\ \pi_{s,k,l}^{s',t'} &= \sigma_{(\emptyset,\emptyset),(s',t')}(\pi_{s,k,l}), \text{ and} \\ V_\delta^{M_{s,k}^{s',t'}} &= V_\delta^{M_{s,k}} \end{aligned}$$

for all $k \preceq l \leq 2 \cdot \mathrm{lh}(s)$.[3] The models $M_{s,k}^{s',t'}$ are "the models $M_{s,k}$ from the point of view of $N_{s',t'}$," and the embeddings $\pi_{s,k,l}^{s',t'}$ are "the embeddings $\pi_{s,k,l}$ from the point of view of $N_{s',t'}$."

Notice that we shall have

$$\sigma_{(s',t'),(s'',t'')} \circ \pi_{s,k,l}^{s',t'} = \pi_{s,k,l}^{s'',t''} \circ (\sigma_{(s',t'),(s'',t'')} \restriction M_{s,k}^{s',t'}), \tag{13.10}$$

[2] There is a slight abuse of notation here, as $\mathscr{T}_s$ is not a set but rather a (sequence of) proper class(es). By $(\mathscr{T}_s)^{N_{s',t'}}$ we mean that object which is defined over $N_{s',t'}$ from the parameter $(E_{s,k}\colon k < 2\cdot\mathrm{lh}(s))$ by the very same formula which defines $\mathscr{T}_s$ over V from the same parameter $(E_{s,k}\colon k < 2\cdot 2\mathrm{lh}(s))$.

[3] For a proper class X, we write $\sigma_{(\emptyset,\emptyset),(s',t')}(X) = \bigcup\{\sigma_{(\emptyset,\emptyset),(s',t')}(X \cap V_\alpha)\colon \alpha \in \mathrm{OR}\}$. Cf. the footnote on p. 184. As a matter of fact, in what follows we shall only need $M_{s,k}^{s',t'}$ in case s and s' are compatible.

where this is an embedding from $M^{s',t'}_{s,k}$ to $M^{s'',t''}_{s,l}$ whenever s, s', s'', t', $t'' \in {}^{<\omega}\omega$, $k \leq l \leq 2 \cdot \text{lh}(s)$, $\text{lh}(s') = \text{lh}(t') \leq \text{lh}(s'') = \text{lh}(t'')$. Equation (13.10) holds true because for all $x \in M^{s',t'}_{s,k} \subset N_{s',t'}$,

$$\begin{aligned}\sigma_{(s',t'),(s'',t'')} \circ \pi^{s',t'}_{s,k,l}(x) &= \sigma_{(s',t'),(s'',t'')}(\sigma_{(\emptyset,\emptyset),(s',t')}(\pi_{s,k,l})(x)) \\ &= \sigma_{(s',t'),(s'',t'')}(\sigma_{(\emptyset,\emptyset),(s',t')}(\pi_{s,k,l}))(\sigma_{(s',t'),(s'',t'')}(x)) \\ &= \sigma_{(\emptyset,\emptyset),(s'',t'')}(\pi_{s,k,l})(\sigma_{(s',t'),(s'',t'')}(x)) \\ &= \pi^{s'',t''}_{s,k,l} \circ \sigma_{(s',t'),(s'',t'')}(x).\end{aligned}$$

Our forth condition now runs as follows.

(PD, 4) If $m < n < \text{lh}(s)$ and $s_m = s_{n+1} \restriction (\text{lh}(s_{n+1}) - 1)$, then

$$\beta_{s,n+1} < \pi^{s\restriction\text{lh}(s_{n+1}),s_{n+1}}_{s,2m,2n+2}(\sigma_{(s\restriction\text{lh}(s_m),s_m),(s\restriction\text{lh}(s_{n+1}),s_{n+1})}(\beta_{s,m}).$$

Suppose that we manage producing objects $\mathscr{T}_s$, $\kappa_{s,k}$, $\beta_{s,k}$, and $\eta_{s,k}$ with the properties (PD, 1) through (PD, 4). We may then verify the following

Claim 13.9

$$((M_{s,2\cdot\text{lh}(s)} : s \in {}^{<\omega}\omega), (\pi_{s',2\cdot\text{lh}(s),2\cdot\text{lh}(s')} : s \subset s', s, s' \in {}^{<\omega}\omega))$$

is a $2^{\aleph_0}$*–closed embedding normal form for* $\{x \in {}^\omega\omega : \forall y \in {}^\omega\omega\, (x, y) \notin A\}$.

Proof $2^{\aleph_0}$–closedness is clear by (PD, 1), as every extender used is $2^{\aleph_0}$–closed in the model where it is taken from. Cf. Lemmas 10.60 and 10.61.

Let us fix $x \in {}^\omega\omega$. Let

$$(M_{\text{Even}}, (\pi_{2k,\text{Even}} : k < \omega)) = \text{dir lim}((M_{x\restriction k,2k} : k < \omega), (\pi_{x\restriction l,2k,2l} : k \leq l < \omega)).$$

We need to see that

$$\exists y (x, y) \in A \Longrightarrow M_{\text{Even}} \text{ is ill–founded} \tag{13.11}$$

and

$$\forall y (x, y) \notin A \Longrightarrow M_{\text{Even}} \text{ is well–founded.} \tag{13.12}$$

Let us first show (13.11). If $s, t \in {}^{<\omega}\omega$ with $\text{lh}(s) = \text{lh}(t)$, then by the coherence condition (PD, 2), we may write $M^{s,t}_{x,k}$ for $M^{s,t}_{x\restriction k,k}$ ($= M^{s,t}_{x\restriction m,k}$ for all sufficiently large m) and $\pi^{s,t}_{x,k,l}$ for $\pi^{s,t}_{x\restriction l,k,l}$ ($= \pi^{s,t}_{x\restriction m,k,l}$ for all sufficiently large m).

Let us now pick some $y \in {}^\omega\omega$ such that $(x, y) \in A$. Let

$$(N, (\sigma_{(s,t),\infty}: s \subset x, t \subset y)) = \text{dir lim}(N_{s,t}, \sigma_{(s,t),(s',t')}: s \subset s' \subset x, t \subset t' \subset y), \tag{13.13}$$

so that N is transitive by (13.5).

For any $s \in {}^{<\omega}\omega$, the sequence

$$(E_{s,k}: k < 2 \cdot \text{lh}(s))$$

is easily seen to generate $(\mathscr{T}_s)^N$ such that

$$N \models \text{“}(\mathscr{T}_s)^N \text{ is an iteration tree on } N \text{ of length } 2 \cdot \text{lh}(s) + 1,\text{”}$$

where we may write

$$(\mathscr{T}_s)^N = ((M^{x,y}_{s,k}, \pi^{x,y}_{s,k,l}: k \preceq l \leq 2 \cdot \text{lh}(s)), (E_{s,k}: k < 2 \cdot \text{lh}(s)), \preceq \restriction (2 \cdot \text{lh}(s) + 1)).$$

By the coherence condition (PD, 2) we may write $M^{x,y}_{x,k}$ for $M^{x,y}_{x \restriction k,k}$ ($= M^{x,y}_{x \restriction m,k}$ for all sufficiently large m), $\pi^{x,y}_{x,k,l}$ for $\pi^{x,y}_{x \restriction l,k,l}$ ($= \pi^{x,y}_{x \restriction m,k,l}$ for all sufficiently large m), and

$$(M^{x,y}_{x,\infty}, (\pi^{x,y}_{x,2k,\infty} : k < \omega)) = \text{dir lim}(M^{x,y}_{x,2k}, \pi^{x,y}_{x,2k,2l} : k \leq l < \omega). \tag{13.14}$$

Virtually the same proof as the one of (13.10) shows that

$$\sigma_{(x \restriction k, y \restriction k),\infty} \circ \pi^{x \restriction k, y \restriction k}_{x,2n,\infty} = \pi^{x,y}_{x,2n,\infty} \circ (\sigma_{(x \restriction k, y \restriction k),\infty} \restriction M^{x \restriction k, y \restriction k}_{x,2n}) \tag{13.15}$$

for all $n, k < \omega$.

Again by (PD, 2) we also write $\beta_{x,k}$ for $\beta_{x \restriction k,k}$ ($= \beta_{x \restriction k',k}$ for all $k' \geq k$). By (PD, 4), if $k \leq m \leq n < \omega$, $s_m = y \restriction k$ and $s_{n+1} = y \restriction (k+1)$, then

$$\beta_{x,n+1} < \pi^{x \restriction (k+1), y \restriction (k+1)}_{x,2m,2n+2}(\sigma_{(x \restriction k, y \restriction k),(x \restriction (k+1)), y \restriction (k+1))}(\beta_{x,m})).$$

which implies that

$$\pi^{x \restriction (k+1), y \restriction (k+1)}_{x,2n+2,\infty}(\beta_{x,n+1}) < \pi^{x \restriction (k+1), y \restriction (k+1)}_{x,2m,\infty}(\sigma_{(x \restriction k, y \restriction k),(x \restriction (k+1)), y \restriction (k+1))}(\beta_{x,m})). \tag{13.16}$$

Then (13.15) and (13.16) yield the following.

$$\begin{aligned}
&\pi^{x,y}_{x,2n+2,\infty}(\sigma_{(x \restriction (k+1), y \restriction (k+1)),\infty}(\beta_{x,n+1})) \\
&= \sigma_{(x \restriction (k+1), y \restriction (k+1)),\infty}(\pi^{x \restriction (k+1), y \restriction (k+1)}_{x,2n+2,\infty}(\beta_{x,n+1})) \\
&< \sigma_{(x \restriction (k+1), y \restriction (k+1)),\infty}(\pi^{x \restriction (k+1), y \restriction (k+1)}_{x,2m,\infty}(\sigma_{(x \restriction k, y \restriction k),(x \restriction (k+1)), y \restriction (k+1))}(\beta_{x,m}))) \\
&= \pi^{x,y}_{x,2m,\infty}(\sigma_{(x \restriction k, y \restriction k),\infty}(\beta_{x,m})).
\end{aligned}$$

This shows that the sequence

$$(\pi^{x,y}_{x,2m,\infty}(\sigma_{(x\restriction \mathrm{lh}(s_m),y\restriction \mathrm{lh}(s_m)),\infty}(\beta_{x,m}))\colon m<\omega,\, s_m\subset y)$$

witnesses that $M^{x,y}_{x,\infty}$ is ill–founded. However, the direct limit (13.14) which produces $M^{x,y}_{x,\infty}$ can be formed within N, cf. (13.13), which is transitive. We may therefore use absoluteness of well–foundedness, Lemma 5.6, to see that such a sequence witnessing that $M^{x,y}_{x,\infty}$ be ill–fouded must also be an element of N. By the elementarity of $\pi_{(\emptyset,\emptyset),\infty}\colon V\to N$, which maps $(M_{x,2m},\pi_{x,2m,2k}\colon m\le k<\omega)$ to $(M^{x,y}_{x,2m},\pi^{x,y}_{x,2m,2k}\colon m\le k<\omega)$, there is hence a sequence which witnesses that

$$M_{\text{Even}}=\operatorname{dir\,lim}(M_{x,2m},\pi_{x,2m,2k}\colon m\le k<\omega)$$

is ill–founded. We have verified (13.11).

Let us now show (13.12). For this, Theorem 10.74 is the key tool. Let us thus suppose that for all $y\in{}^{\omega}\omega$, $(x,y)\notin A$. As every extender used is certified in the model where it is taken from, in order to verify that M_{Even} is well–founded, it suffices by Theorem 10.74 to show that for all $y\in{}^{\omega}\omega$, if

$$(M_y,(\pi_{k,y}\colon k<\omega))=\operatorname{dir\,lim}(M_{x\restriction k,2k\dot-1},\pi_{x\restriction l,2k\dot-1,2l\dot-1}\colon s_k\subset s_l\subset y),$$

then M_y is ill–founded.

To this end, let $y\in{}^{\omega}\omega$ be arbitrary. As $(x,y)\notin p[T]$,

$$T_{x,y}=\{f\colon (x\restriction \mathrm{lh}(f),y\restriction \mathrm{lh}(f),f)\in T\}$$

is well–founded. By (PD, 3), if $k<\omega$ and $s_n=y\restriction k$, then

$$(x\restriction k,y\restriction k,(\pi_{x\restriction n,2l\dot-1,2n\dot-1}(\eta_{x\restriction n,l})\colon 2l\dot-1\preceq 2n\dot-1))\in\pi_{x\restriction n,0,2n\dot-1}(T).$$

Let $(n_i\colon i<\omega)$ be the monotone enumeration of $\{n<\omega\colon s_n\subset y\}$, and let, for $i<\omega$,

$$\gamma_i=||(\pi_{x\restriction n_i,2l\dot-1,2n_i\dot-1}(\eta_{x\restriction n_i,l})\colon 2l\dot-1\preceq 2n_i\dot-1)||_{\pi_{x\restriction n_i,0,2n_i\dot-1}(T_{x,y})}.$$

If $i<\omega$, then the node $(\pi_{x\restriction n_{i+1},2l\dot-1,2n_{i+1}\dot-1}(\eta_{x\restriction n_{i+1},l})\colon 2l\dot-1\preceq 2n_{i+1}\dot-1)$ extends the node $(\pi_{x\restriction n_{i+1}2l\dot-1,2n_{i+1}\dot-1}(\eta_{x\restriction n_{i+1},l})\colon 2l\dot-1\preceq 2n_i\dot-1)$ in the tree $\pi_{x\restriction n_{i+1},0,2n_{i+1}\dot-1}(T_{x,y})$, so that

$$\begin{aligned}&\pi_{x\restriction n_{i+1},2n_i\dot-1,2n_{i+1}\dot-1}(\gamma_i)\\&=||(\pi_{x\restriction n_{i+1},2l\dot-1,2n_{i+1}\dot-1}(\eta_{x\restriction n_{i+1},l})\colon 2l\dot-1\preceq 2n_i\dot-1)||_{\pi_{x\restriction n_{i+1},0,2n_{i+1}\dot-1}(T_{x,y})}\\&>||(\pi_{x\restriction n_{i+1},2l\dot-1,2n_{i+1}\dot-1}(\eta_{x\restriction n_{i+1},l})\colon 2l\dot-1\preceq 2n_{i+1}\dot-1)||_{\pi_{x\restriction n_{i+1},0,2n_{i+1}\dot-1}(T_{x,y})}\\&=\gamma_{i+1}.\end{aligned}$$

This proves that M_y is ill–founded, as witnessed by $(\gamma_i\colon i<\omega)$.

We have verified (13.12). □

We are left with having to produce the objects $\mathcal{T}_s$, $\kappa_{s,k}$, $\beta_{s,k}$, and $\eta_{s,k}$ such that (PD, 1) through (PD, 4) hold true.

We first need a set of "indiscernibles." Let us fix λ, a cardinal which is "much bigger than" δ; in particular, we want that $T \in V_\lambda$ and that all $\pi_{0,k}^{s',t'}(\alpha_{s,t})$ will be in V_λ also. Let $c_0 < c_1 < \eta$ be strong limit cardinals above λ of cofinality $> \delta$ such that

$$\text{type}^V(V_\eta; \in, V_\lambda, \{\lambda, c_0\})) = \text{type}^V(V_\eta; \in, V_\lambda, \{\lambda, c_1\})). \tag{13.17}$$

An easy pigeonhole argument shows that such objects exist: e.g., let η be a strong limit cardinal of cofinality $\overline{\overline{V_{\lambda+1}}}^+$. In the construction to follow, we shall use the "descending" chain of ordinals

$$c_0 + 1 > c_0 \sim c_1 > c_0 + 1 > c_0 \sim c_1 > \dots$$

and we may and shall assume that λ, c_0, c_1, and η are fixed points of all the elementary embeddings which we will encounter. Cf. Lemma 10.56 Problem 10.18. δ will always be a fixed point anyway.

In order to keep our recursion going, we shall need a fifth condition.

(PD, 5) For each $l < \omega$, $\eta_l \in V_\lambda^{M_{2l-1}}$.

The objects $\mathcal{T}_s$, $\kappa_{s,k}$, $\beta_{s,k}$, and $\eta_{s,k}$ will be constructed by recursion on the length of s. To get started, we let $\mathcal{T}_\emptyset$ be the trivial tree of length 1 which just consists of V. We also set $\beta_{\emptyset,0} = c_0$, and we pick $\kappa_{\emptyset,0} < \delta$, $\kappa_{\emptyset,0} > \alpha$, such that in V, $\kappa_{\emptyset,0}$ is strong up to δ with respect to

$$\text{type}^V(V_{c_0+1}; \in, V_\delta, \{\delta, \lambda, T, \alpha_{\emptyset,\emptyset}\}).$$

This choice of $\kappa_{\emptyset,0}$ is certainly possible, as δ is a Woodin cardinal, cf. Lemma 10.81. We also set $\eta_{\emptyset,0} = \alpha_{\emptyset,\emptyset}$.

Now let us fix $s \in {}^{<\omega}\omega$ of positive length throughout the rest of this proof. Write $n + 1 = \text{lh}(s)$. Let us suppose that

$$\begin{array}{ll} \mathcal{T}_{s \restriction n} & \\ \kappa_{s,k} & \text{for } k \leq 2 \cdot n, \\ \beta_{s,k} & \text{for } k \leq n, \text{ and} \\ \eta_{s,k} & \text{for } k \leq n \end{array}$$

have already been constructed. We are forced to set $\kappa_{s,k} = \kappa_{s \restriction n,k}$ for $k \leq 2 \cdot n$, $E_{s,k} = E_{s \restriction n,k}$ for $k < 2 \cdot n$, $M_{s,k} = M_{s \restriction n,k}$ for $k \leq 2 \cdot n$, $\pi_{s,k,l} = \pi_{s \restriction n,k,l}$ for $k \leq l \leq 2 \cdot n$, $\beta_{s,k} = \beta_{s \restriction n,k}$ for $k \leq n$, and $\eta_{s,k} = \eta_{s \restriction n,k}$ for $k \leq n$. Let us write $t = s_{n+1}$, $k = \text{lh}(t) \leq n + 1$, and $s_m = t \restriction (k - 1)$. We now need to define $\kappa_{s,2n+1}$, $\kappa_{s,2n+2}$, $\beta_{s,n+1}$, and $\eta_{s,n+1}$, and we also need to define $M_{s,2n+1}$ as an ultrapower of

$M_{s,2m\dot{-}1}$ by an extender $E_{s,2n}$ with critical point $\kappa_{s,2m}$ and $M_{s,2n+2}$ as an ultrapower of $M_{s,2n}$ by an extender $E_{s,2n+1}$ with critical point $\kappa_{s,2n+1}$.

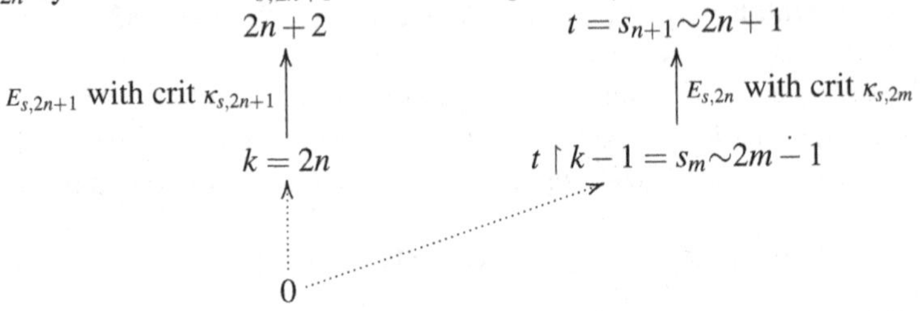

As $s \in {}^{<\omega}\omega$ will be fixed from now on, we shall mostly suppress the *sub*script s and write

$$\begin{array}{ll} M_k & \text{for } M_{s,k}, \\ \pi_{k,l} & \text{for } \pi_{s,k,l}, \\ M_k^{s',t'} & \text{for } M_{s,k}^{s',t'}, \\ \pi_{k,l}^{s',t'} & \text{for } \pi_{s,k,l}^{s',t'}, \\ \kappa_k & \text{for } \kappa_{s,k}, \\ \beta_k & \text{for } \beta_{s,k}, \text{ and} \\ \eta_k & \text{for } \eta_{s,k}. \end{array}$$

Inductively, we shall assume that the following two statements, (A) and (B), are satisfied. Here, $\alpha_{s\upharpoonright i,t\upharpoonright i}$ is as in (13.7).

(A)

$$\begin{aligned} &\text{type}^{M_{2m}^{s\upharpoonright k-1,t\upharpoonright k-1}}(V_{\beta_m+1}; \in, V_{\kappa_{2m}}, \{\delta, \lambda, \pi_{0,2m}^{s\upharpoonright k-1,t\upharpoonright k-1}(\sigma_{(\emptyset,\emptyset),(s\upharpoonright k-1,t\upharpoonright k-1)}(T)), \\ &(\pi_{0,2m}^{s\upharpoonright k-1,t\upharpoonright k-1}(\sigma_{(s\upharpoonright i,t\upharpoonright i),(s\upharpoonright k-1,t\upharpoonright k-1)}(\alpha_{s\upharpoonright i,t\upharpoonright i})) : i \leq k-1)\}) \\ &\quad = \text{type}^{M_{2m\dot{-}1}}(V_{c_0+1}; \in, V_{\kappa_{2m}}, \{\delta, \lambda, \pi_{0,2m\dot{-}1}(T), \\ &\qquad (\pi_{2i\dot{-}1,2m\dot{-}1}(\eta_i)) : 2i\dot{-}1 \preceq 2m\dot{-}1)\}). \end{aligned}$$

(B) Inside $M_{2m}^{s\upharpoonright k-1,t\upharpoonright k-1}$, κ_{2m} is strong up to δ with respect to

$$\begin{aligned} &\text{type}^{M_{2m}^{s\upharpoonright k-1,t\upharpoonright k-1}}(V_{\beta_m+1}; \in, V_{\delta}, \{\delta, \lambda, \pi_{0,2m}^{s\upharpoonright k-1,t\upharpoonright k-1}(\sigma_{(\emptyset,\emptyset),(s\upharpoonright k-1,t\upharpoonright k-1)}(T)), \\ &(\pi_{0,2m}^{s\upharpoonright k-1,t\upharpoonright k-1}(\sigma_{(s\upharpoonright i,t\upharpoonright i),(s\upharpoonright k-1,t\upharpoonright k-1)}(\alpha_{s\upharpoonright i,t\upharpoonright i})) : i \leq k-1)\}). \end{aligned}$$

Notice that this is trivially true for $n = 0$ (in which case $m = k-1 = 0$) by the choices of $\beta_0 = \beta_{\emptyset,0} = c_0$, $\kappa_0 = \kappa_{\emptyset,0}$, and $\eta_0 = \alpha_{\emptyset,\emptyset}$.

Because δ is a WOODIN cardinal inside $M_{2n}^{s\upharpoonright k,t}$, we may pick some $\kappa_{2n+1} < \delta$, $\kappa_{2n+1} > \kappa_{2n}$, such that

(C) inside $M_{2n}^{s\upharpoonright k,t}$, κ_{2n+1} is strong up to δ with respect to

$$\text{type}^{M_{2n}^{s\restriction k,t}}(V_{\pi_{2m,2n}^{s\restriction k,t}(\sigma_{(s\restriction k-1,t\restriction k-1),(s\restriction k,t)}(\beta_m))}; \in, V_\delta,$$
$$\{\delta, \lambda, \pi_{0,2n}^{s\restriction k,t}(\sigma_{(\emptyset,\emptyset),(s\restriction k,t)}(T)), (\pi_{0,2n}^{s\restriction k,t}(\sigma_{(s\restriction i,t\restriction i),(s\restriction k,t)}(\alpha_{s\restriction i,t\restriction i}) \colon i \le k)\}),$$

cf. Lemma 10.81. We may apply the map $\pi_{2m,2n}^{s\restriction k,t} \circ \sigma_{(s\restriction k-1,t\restriction k-1),(s\restriction k,t)}$ to (B), which by (13.10) and the fact that $\text{crit}(\pi_{2m,2n}) = \kappa_{2m+1} > \kappa_{2m}$ produces the assertion that

$$\text{inside } M_{2n}^{s\restriction k,t},\ \kappa_{2m} \text{ is strong up to } \delta \text{ with respect to}$$
$$\text{type}^{M_{2n}^{s\restriction k,t}}(V_{\pi_{2m,2n}^{s\restriction k,t\restriction k}(\sigma_{(s\restriction k-1,t\restriction k-1),(s\restriction k,t)}(\beta_m))+1}; \in, V_\delta, \tag{13.18}$$
$$\{\delta, \lambda, \pi_{0,2n}^{s\restriction k,t}(\pi_{(\emptyset,\emptyset),(s\restriction k,t)}(T)), (\pi_{0,2n}^{s\restriction k,t}(\sigma_{(s\restriction i,t\restriction i),(s\restriction k,t)}(\alpha_{s\restriction i,t\restriction i})) \colon i < k)\}).$$

We may thus let

$$\pi_{2m\dot{-}1,2n+1} \colon M_{2m\dot{-}1} \to_{E_{s,2n}} M_{2n+1},$$

where $E_{s,2n} \in M_{2n}$ is a $2^{\aleph_0}$–closed certified extender in M_{2n} which witnesses that inside $M_{2n}^{s\restriction k,t}$, κ_{2m} is strong up to the least inaccessible cardinal λ^* above κ_{2n+1} with respect to the type from (13.18); notice that $(V_{\kappa_{2m}+1})^{M_{2n}} = (V_{\kappa_{2m}+1})^{M_{2m\dot{-}1}}$, cf. (PD, 1), immediately gives that

$$(V_{\lambda^*})^{M_{2n+1}} = (V_{\lambda^*})^{M_{2n}}. \tag{13.19}$$

We have that

$$\text{type}^{M_{2n+1}}(V_{c_0+1}; \in, V_{\kappa_{2n+1}+1}, \{\delta, \lambda,$$
$$\pi_{0,2n+1}(T), (\pi_{2i\dot{-}1,2m\dot{-}1}(\eta_i) \colon 2i \dot{-} 1 \preceq 2m \dot{-} 1)\ ) \tag{13.20}$$
$$= \text{type}^{M_{2n}^{s\restriction k,t}}(V_{\pi_{2m,2n}^{s\restriction k,t\restriction k}(\sigma_{(s\restriction k-1,t\restriction k-1),(s\restriction k,t)}(\beta_m))+1}; \in, V_{\kappa_{2n+1}+1}, \{\delta, \lambda,$$
$$\pi_{0,2n}^{s\restriction k,t}(\sigma_{(\emptyset,\emptyset),(s\restriction k,t)}(T)), (\pi_{0,2n}^{s\restriction k,t}(\sigma_{(s\restriction i,t\restriction i),(s\restriction k,t)}(\alpha_{s\restriction i,t\restriction i})) \colon i \le k-1)\}).$$

This is because by the choice of $E_{s,2n}$, the right hand side of (13.20) is equal to

$$\pi_{2m\dot{-}1,2n+1}(\text{type}^{M_{2n}^{s\restriction k,t}}(V_{\pi_{2m,2n}^{s\restriction k,t\restriction k}(\sigma_{(s\restriction k-1,t\restriction k-1),(s\restriction k,t)}(\beta_m)+1}; \in, V_{\kappa_{2m}}, \{\delta, \lambda,$$
$$\pi_{0,2n}^{s\restriction k,t}(\sigma_{(\emptyset,\emptyset),(s\restriction k,t)}(T)), (\pi_{0,2n}^{s\restriction k,t}(\sigma_{(s\restriction i,t\restriction i),(s\restriction k,t)}(\alpha_{s\restriction i,t\restriction i})) \colon i \le k-1)\})),$$

restricted to parameters in $(V_{\kappa_{2n+1}+1})^{M_{2n+1}} = (V_{\kappa_{2n+1}+1})^{M_{2n}}$, which by (13.10) and the elementarity of the map $\sigma_{(s\restriction k-1,t\restriction k-1),(s\restriction k,t)} \circ \pi_{2m,2n}^{s\restriction k-1,t\restriction k-1}$ is equal to

$$\pi_{2m\dot{-}1,2n+1}(\text{type}^{M_{2m}^{s\restriction k-1,t\restriction k-1}}(V_{\beta_m+1};\in,V_{\kappa_{2m}},\{\delta,\lambda,$$
$$\pi_{0,2m}^{s\restriction k-1,t\restriction k-1}(\sigma_{(\emptyset,\emptyset),(s\restriction k-1,t\restriction k-1)}(T)),$$
$$(\pi_{0,2m}^{s\restriction k-1,t\restriction k-1}(\sigma_{(s\restriction i,t\restriction i),(s\restriction k-1,t\restriction k-1)}(\alpha_{s\restriction i,t\restriction i})):i\leq k-1)\})),$$

restricted to parameters in $(V_{\kappa_{2n+1}+1})^{M_{2n+1}}$, which in turn by (A) is equal to

$$\pi_{2m\dot{-}1,2n+1}(\text{type}^{M_{2m\dot{-}1}}(V_{c_0+1};\in,V_{\kappa_{2m}},\{\delta,\lambda,$$
$$\pi_{0,2m\dot{-}1}(T),\ (\pi_{2i\dot{-}1,2m\dot{-}1}(\eta_i):2i\dot{-}1\preceq 2m\dot{-}1)\})),$$

restricted to parameters in $(V_{\kappa_{2n+1}+1})^{M_{2n+1}}$, and thus to the left hand side of (13.20).

Let us write

$$\tau=\text{type}^{M_{2n}^{s\restriction k,t}}(V_{\pi_{2m,2n}^{s\restriction k,t\restriction k}(\sigma_{(s\restriction k-1,t\restriction k-1),(s\restriction k,t)}(\beta_m)};\in,V_{\kappa_{2n+1}},\{\delta,\lambda,$$
$$\pi_{0,2n}^{s\restriction k,t}(\sigma_{(\emptyset,\emptyset),(s\restriction k,t)}(T)), \tag{13.21}$$
$$(\pi_{0,2n}^{s\restriction k,t}(\sigma_{(s\restriction i,t\restriction i),(s\restriction k,t)}(\alpha_{s\restriction i,t\restriction i})):i\leq k)\}).$$

By (C), we have that $\eta=\pi_{0,2n}^{s\restriction k,t}(\alpha_{s\restriction k,t})$ witnesses that

$$M_{2n}^{s\restriction k,t}\models\exists\eta\in V_\lambda\ (\tau=\text{type}^{M_{2n}^{s\restriction k,t}}(V_{\pi_{2m,2n}^{s\restriction k,t\restriction k}(\sigma_{(s\restriction k-1,t\restriction k-1),(s\restriction k,t)}(\beta_m)};\in,V_{\kappa_{2n+1}},$$
$$\{\delta,\lambda,\pi_{0,2n}^{s\restriction k,t}(\sigma_{(\emptyset,\emptyset),(s\restriction k,t)}(T)),$$
$$(\pi_{0,2n}^{s\restriction k,t}(\sigma_{(s\restriction i,t\restriction i),(s\restriction k-1,t\restriction k-1)}(\alpha_{s\restriction i,t\restriction i})):i\leq k-1)^\frown\eta\})\wedge$$
$$\kappa_{2n+1}\text{ is strong up to }\delta\text{ with respect to} \tag{13.22}$$
$$\text{type}^{M_{2n}^{s\restriction k,t}}(V_{\pi_{2m,2n}^{s\restriction k,t\restriction k}(\sigma_{(s\restriction k-1,t\restriction k-1),(s\restriction k,t)}(\beta_m)};\in,V_\delta,$$
$$\{\delta,\lambda,\pi_{0,2n}^{s\restriction k,t}(\sigma_{(\emptyset,\emptyset),(s\restriction k,t)}(T)),$$
$$(\pi_{0,2n}^{s\restriction k,t}(\sigma_{(s\restriction i,t\restriction i),(s\restriction k-1,t\restriction k-1)}(\alpha_{s\restriction i,t\restriction i})):i\leq k-1)^\frown\eta\})\ ).$$

As

$$\tau\in(V_{\kappa_{2n+1}+1})^{M_{2n}^{s\restriction k,t}},$$

the statement "$\exists\eta\in V_\lambda\,(\ldots)$" in (13.22) can be written as an element of the type from the right hand side of (13.20), so that by (13.20),

$$\begin{aligned} M_{2n+1} \models \exists \eta \in V_\lambda(\tau = \text{type}^{M_{2n+1}}(&V_{c_0}; \in, V_{\kappa_{2n+1}}, \\ &\{\delta, \lambda, \pi_{0,2n+1}(T), (\pi_{2i \dot{-} 1,2n+1}(\eta_i) \colon 2i \dot{-} 1 \preceq 2m \dot{-} 1)^\frown \eta\}) \wedge \\ &\kappa_{2n+1} \text{ is strong up to } \delta \text{ with respect to} \\ &\text{type}^{M_{2n+1}}(V_{c_0}; \in, V_\delta, \\ &\{\delta, \lambda, \pi_{0,2n+1}(T), (\pi_{2i \dot{-} 1,2n+1}(\eta_i) \colon 2i \dot{-} 1 \preceq 2m \dot{-} 1)^\frown \eta\})\ ). \end{aligned} \tag{13.23}$$

Let $\eta_{n+1} \in V_\lambda^{M_{2n-1}}$ be a witness to (13.23), so that we shall now have that

(D)

$$\begin{aligned} \tau &= \text{type}^{M_{2n}^{s \restriction k,t}}(V_{\pi_{2m,2n}^{s \restriction k,t \restriction k}(\sigma_{(s \restriction k-1,t \restriction k-1),(s \restriction k,t)}(\beta_m)}; \in, V_{\kappa_{2n+1}}, \{\delta, \lambda, \pi_{0,2n}^{s \restriction k,t}(\sigma_{(\emptyset,\emptyset),(s \restriction k,t)}(T)), \\ &\quad (\pi_{0,2n}^{s \restriction k,t}(\sigma_{(s \restriction i,t \restriction i),(s \restriction k,t)}(\alpha_{s \restriction i,t \restriction i})) \colon i \leq k)\}) \\ &= \text{type}^{M_{2n+1}}(V_{c_0}; \in, V_{\kappa_{2n+1}}, \{\delta, \lambda, \pi_{0,2n+1}(T), \\ &\quad (\pi_{2i \dot{-} 1,2n+1}(\eta_i) \colon 2i \dot{-} 1 \preceq 2n+1)\}) \end{aligned}$$

and

(C)'

$$\begin{aligned} \text{inside } M_{2n+1}, \kappa_{2n+1} \text{ is strong up to } \delta \text{ with respect to} \\ \text{type}^{M_{2n+1}}(V_{c_0}; \in, V_\delta, \{\delta, \lambda, \pi_{0,2n+1}(T), \\ (\pi_{2i \dot{-} 1,2n+1}(\eta_i) \colon 2i \dot{-} 1 \preceq 2n+1)\}). \end{aligned}$$

Also, if we inductively assume (PD, 5) for $l \leq n$, then $\eta_{n+1} \in V_\lambda^{M_{2n+1}}$ (i.e., (PD, 5) for $l = n+1$) yields that

$$(\pi_{2i \dot{-} 1,2n+1}(\eta_i) \colon 2i \dot{-} 1 \preceq 2n+1) \in V_\lambda^{M_{2n+1}}. \tag{13.24}$$

Now again because δ is a WOODIN cardinal inside M_{2n+1}, we may pick some $\kappa_{2n+2} < \delta$, $\kappa_{2n+2} > \kappa_{2n+1}$, such that

(E) inside M_{2n+1}, κ_{2n+2} is strong up to δ with respect to

$$\text{type}^{M_{2n+1}}(V_{c_0+1}; \in, V_\delta, \{\delta, \lambda, \pi_{0,2n+1}(T), (\pi_{2i \dot{-} 1,2n+1}(\eta_i) \colon 2i \dot{-} 1 \preceq 2n+1)\}),$$

cf. Lemma 10.81.

Let

$$\pi_{2n,2n+2} \colon M_{2n} \to_{E_{s,2n+1}} M_{2n+2},$$

where $E_{s,2n+1} \in M_{2n+1}$ is a $2^{\aleph_0}$–closed certified extender which in M_{2n+1} and witnesses that inside M_{2n+1}, κ_{2n+1} is strong up to the least inaccessible cardinal λ^* above κ_{2n+2} with respect to

$$\text{type}^{M_{2n+1}}(V_{c_0}; \in, V_\delta, \{\delta, \lambda, \pi_{0,2n+1}(T), (\pi_{2i \dot{-} 1,2n+1}(\eta_i) \colon 2i \dot{-} 1 \preceq 2n+1)\}).$$

This choice is possible by (C)'; notice that $(V_{\kappa_{2n+1}+1})^{M_{2n+1}} = (V_{\kappa_{2n+1}+1})^{M_{2n}}$, cf. (13.19), which immediately gives that $(V_{\lambda^*})^{M_{2n+2}} = (V_{\lambda^*})^{M_{2n+1}}$. We shall have that

(F)

$$\begin{aligned}
&\text{type}^{M_{2n+2}^{s\restriction k,t}}(V_{\pi^{s\restriction k,t}_{2m,2n+2}(\sigma_{s\restriction k-1,t\restriction k-1),(s\restriction k,t)}(\beta_m)}; \in, V_{\kappa_{2n+2}+1},\\
&\{\delta, \lambda, \pi^{s\restriction k,t}_{0,2n+2}(\sigma_{(\emptyset,\emptyset),(s\restriction k,t)}(T)),\\
&\pi^{s\restriction k,t}_{0,2n+2}(\sigma_{(s\restriction i,t\restriction i),(s\restriction k,t)}(\alpha_{(s\restriction i,t\restriction i)}) : i \le k)\})\\
&\quad = \text{type}^{M_{2n+1}}(V_{c_1}; \in, V_{\kappa_{2n+2}+1},\\
&\qquad \{\delta, \lambda, \pi_{0,2n+1}(T),\\
&\qquad (\pi_{2i\dot{-}1,2n+1}(\eta_i) : 2i \dot{-} 1 \preceq 2n+1)\}).
\end{aligned}$$

This is because the left hand side of (F) is equal to

$$\begin{aligned}
&\pi^{s\restriction k,t}_{2n,2n+2}(\text{type}^{M_{2n}^{s\restriction k,t}}(V_{\pi^{s\restriction k,t}_{2m,2n}(\sigma_{s\restriction k-1,t\restriction k-1),(s\restriction k,t)}(\beta_m)}; \in, V_{\kappa_{2n+1}},\\
&\{\delta, \lambda, \pi^{s\restriction k,t}_{0,2n}(\sigma_{(\emptyset,\emptyset),(s\restriction k,t)}(T)),\\
&\pi^{s\restriction k,t}_{0,2n}(\sigma_{(s\restriction i,t\restriction i),(s\restriction k,t)}(\alpha_{(s\restriction i,t\restriction i)}) : i \le k)\})\\
&\text{restricted to parameters in } (V_{\kappa_{2n+2}+1})^{M_{2n+2}^{s\restriction k,t}}\\[1em]
&\quad = \pi_{E_{2n+1}}(\text{type}^{M_{2n+1}}(V_{c_0}; \in, V_{\kappa_{2n+1}}, \{\delta, \lambda, \pi_{0,2n+1}(T),\\
&\quad (\pi_{2i\dot{-}1,2n+1}(\eta_i) : 2i \dot{-} 1 \preceq 2n+1)\})),\\
&\quad \text{restricted to parameters in } (V_{\kappa_{2n+2}+1})^{M_{2n+1}}
\end{aligned}$$

by (D),

$$= \text{type}^{M_{2n+1}}(V_{c_0}; \in, V_{\kappa_{2n+2}+1}, \{\delta, \lambda, \pi_{0,2n+1}(T), (\pi_{2i\dot{-}1,2n+1}(\eta_i) : 2i \dot{-} 1 \preceq 2n+1)\}),$$

by the choice of $E_{s,2n+1}$, which is equal to the right hand side of (F) by (13.24) and the choice of c_0 and c_1, cf. (13.17). We verified (F).

Let us write

$$\sigma = \text{type}^{M_{2n+1}}(V_{c_0+1}; \in, V_{\kappa_{2n+2}}, \{\delta, \lambda, \pi_{0,2n+1}(T), (\pi_{2i\dot{-}1,2n+1}(\eta_i) : 2i \dot{-} 1 \preceq 2n+1)\}).$$

With (E),

$$\begin{aligned}
(V_{c_1})^{M_{2n+1}} \models \exists\beta(&\sigma = \text{type}^{M_{2n+1}}(V_{\beta+1}; \in, V_{\kappa_{2n+2}}, \{\delta, \lambda, \pi_{0,2n+1}(T),\\
&(\pi_{2i\dot{-}1,2n+1}(\eta_i) : 2i \dot{-} 1 \preceq 2n+1)\}) \wedge\\
&\kappa_{2n+2} \text{ is strong up to } \delta \text{ with respect to}\\
&\text{type}^{M_{2n+1}}(V_{\beta+1}; \in, V_\delta, \{\delta, \lambda, \pi_{0,2n+1}(T),\\
&(\pi_{2i\dot{-}1,2n+1}(\eta_i) : 2i \dot{-} 1 \preceq 2n+1)\}).\ ).
\end{aligned} \tag{13.25}$$

We have that

$$\sigma \in (V_{\kappa_{2n+2}+1})^{M_{2n+1}},$$

so that the statement "$\exists \beta$ (...)" in (13.25) can be written as an element of the type from the right hand side of (F), and (F) yields the following.

$$\begin{aligned}
&(V_{\pi^{s\restriction k,t}_{2m,2n+2}(\sigma_{(s\restriction k-1,t\restriction k-1),(s\restriction k,t)}(\beta_m))})^{M^{s\restriction k,t}_{2n+2}} \models \\
&\exists \beta(\sigma = \text{type}^{M^{s\restriction k,t}_{2n+2}}(V_{\beta+1}; \in, V_{\kappa_{2n+2}}, \{\delta, \lambda, \pi^{s\restriction k,t}_{0,2n+2}(\sigma_{(\emptyset,\emptyset),(s\restriction k,t)}(T)), \\
&\pi^{s\restriction k,t}_{0,2n+2}(\sigma_{(s\restriction i,t\restriction i),(s\restriction k,t)}(\alpha_{(s\restriction i,t\restriction i)}: i \leq k))\})\wedge \\
&\kappa_{s,2n+2} \text{ is strong up to } \delta \text{ with respect to} \qquad (13.26)\\
&\text{type}^{M^{s\restriction k,t}_{2n+2}}(V_{\beta+1}; \in, V_\delta, \{\delta, \lambda, \pi^{s\restriction k,t}_{0,2n+2}(\sigma_{(\emptyset,\emptyset),(s\restriction k,t)}(T)), \\
&\pi^{s\restriction k,t}_{0,2n+2}(\sigma_{(s\restriction i,t\restriction i),(s\restriction k,t)}(\alpha_{(s\restriction i,t\restriction i)}: i \leq k))\})).
\end{aligned}$$

Let β_{n+1} be a witness to this fact. In particular, with a brief show of the subscript s,

$$\beta_{s,n+1} < \pi^{s\restriction k,t}_{2m,2n+2}(\sigma_{(s\restriction k-1,t\restriction k-1),(s\restriction k,t)}(\beta_{s,m})). \qquad (13.27)$$

By (13.26) and the definition of σ, we now have the following.

(G)

$$\begin{aligned}
&\text{type}^{M^{s\restriction k,t}_{2n+2}}(V_{\beta_{n+1}+1}; \in, V_{\kappa_{2n+2}}, \{\delta, \lambda, \pi^{s\restriction k,t}_{0,2n+2}(\sigma_{(\emptyset,\emptyset),(s\restriction k,t)}(T)), \\
&(\pi^{s\restriction k,t}_{0,2n+2}(\sigma_{(s\restriction i,t\restriction i),(s\restriction k,t)}(\alpha_{s\restriction i,t\restriction i})): i \leq k)\}) \\
&\quad = \text{type}^{M_{2n+1}}(V_{c_0+1}; \in, V_{\kappa_{2n+2}}, \{\delta, \lambda, \pi_{0,2n+1}(T), \\
&\quad (\pi_{2i\dot{-}1,2n+1}(\eta_i): 2i - 1 \leq 2n + 1)\}),
\end{aligned}$$

and

(H) inside $M^{s\restriction k,t}_{2n+2}$, κ_{2n+2} is strong up to δ with respect to

$$\begin{aligned}
&\text{type}^{M^{s\restriction k,t}_{2n+2}}(V_{\beta_{n+1}+1}; \in, V_\delta, \{\delta, \lambda, \pi^{s\restriction k,t}_{0,2n+2}(\sigma_{(\emptyset,\emptyset),(s\restriction k,t)}(T)), \\
&(\pi^{s\restriction k,t}_{0,2n+2}(\pi_{(s\restriction i,t\restriction i),(s\restriction k,t)}(\alpha_{s\restriction i,t\restriction i})): i \leq k)\}).
\end{aligned}$$

We are back to where we started from, cf. (A) and (B).

It is now straightforward to verify that (PD, 1) through (PD, 4) hold true. Notice that (13.27) above gives (PD, 4). Also, (PD, 3) follows from (A) (or, (G)) by virtue of (13.8).

This finishes the proof of Theorem 13.6. □

It can be shown that the conclusion of Thorem 13.6 implies the consistency of WOODIN cardinals, cf. e.g. [22]; in fact, PD turns out to be equivalent to the existence of mice with WOODIN cardinals, for a proof cf. [36].

Results which are stronger than Theorem 13.6 but build upon its proof method are presented e.g. in [32, 33] and [41].

The reader might also want to consult [34] and [42] on recent developments concerning determinacy hypotheses and large cardinals.

13.3 Problems

13.1. Show that if there is a measurable cardinal, then every set of reals has an embedding normal form. [Hint. The embedding normal form will not be $2^{\aleph_0}$–closed.]
Let $n \leq m < \omega$, let μ be an ultrafilter on a set of functions with domain n, and let μ' be an ultrafilter on a set of functions with domain m. We say that μ, μ' *cohere* iff for all X,

$$X \in \mu \Longleftrightarrow \{f \in {}^m\kappa : f \upharpoonright n \in X\} \in \mu'.$$

We may define $\pi_\mu : V \to \text{ult}_0(V; \mu)$ and $\pi_{\mu'} : V \to \text{ult}_0(V; \mu')$, and we may also define a canonical elementary embedding $\pi_{\mu,\mu'} : \text{ult}_0(V; \mu) \to \text{ult}_0(V; \mu)$.
Let $A \subset {}^\omega\omega$, and let $\delta \geq \aleph_0$. We say that A is *δ–homogeneously* SOUSLIN iff there is some α and a tree T on $\omega \times \alpha$ such that $A = p[T]$ and there is $(\mu_s : s \in {}^{<\omega}\omega)$ such that for all $s \in {}^{<\omega}\omega$, μ_s is a $< \delta^+$–closed ultrafilter on $T_s = \{t : (s,t) \in T\}$, if $s \subset t \in {}^{<\omega}\omega$, then μ_s and μ_t cohere, and if $x \in A$, then

$$\text{dir lim}_{n<\omega}(\text{ult}_0(V; \mu_{x\upharpoonright n}), \pi_{\mu_{x\upharpoonright n},\mu_{x\upharpoonright m}} : n \leq m < \omega) \text{ is well–founded.} \quad (13.28)$$

A is called *homogeneously* SOUSLIN iff A is $\aleph_0$–homogeneously.

13.2. Show that in the situation of the preceeding paragraph, if (13.28) holds true, then $x \in A$. [Hint. If T_x is well–founded, then look at $||[\text{id}]_{\mu_{x\upharpoonright n}}||_{\pi_{\mu_{x\upharpoonright n}}(T_x)}$, $n < \omega$.]

13.3. (**K. Windßus**) Let $A \subset {}^\omega\omega$, and let $\delta \geq \aleph_0$. Show that the following are equivalent.

(a) A has a $2^{\aleph_0}$–closed embedding normal form whose additivity is bigger than δ.
(b) A is δ–homogeneously SOUSLIN.

Conclude that every homogeneously SOUSLIN set of reals is determined. [Hint. For $(a) \Longrightarrow (b)$, construct the WINDßUS tree and define μ_s, $s \in {}^{<\omega}\omega$, via (13.2).]
Let $A \subset {}^\omega\omega$, and let $\delta \geq \aleph_0$. We say that A is *δ–weakly homogeneously* SOUSLIN iff $A = \{x \in {}^\omega\omega : \exists y \in {}^\omega\omega\ x \oplus y \in B\}$, where B is δ–homogeneously

SOUSLIN, and A is *weakly homogeneously* SOUSLIN iff A is $\aleph_0$–weakly homogeneously SOUSLIN.

13.4. **(D.A. Martin, R. Solovay)** Show that if $A \subset {}^\omega\omega$ is δ–weakly homogeneously SOUSLIN, then A is $< \delta^+$–universally BAIRE. [Hint. Let $(s_n : n < \omega)$ be a reasonable enumeration of ${}^{<\omega}\omega$. For $s \in {}^{<\omega}\omega$ with $k = \mathrm{lh}(s)$, let $(s, (\alpha_0, \ldots, \alpha_{k-1})) \in S$ iff for all $i < j < k$, if $s_i \subsetneq s_j$, then

$$\alpha_j < \pi_{\mu_{s\restriction \mathrm{lh}(s_i)\oplus s_i}, \mu_{s\restriction \mathrm{lh}(s_j)\oplus s_j}}(\alpha_i),$$

where $\alpha_0, \ldots, \alpha_{k-1} < \gamma$ for some sufficiently big γ.]
Conclude that if κ is a measurable cardinal, then all $\underset{\sim}{\Sigma}^1_2$–sets of reals are $< \kappa$–universally BAIRE. [Hint: Use Problem 6.18 and the construction from Theorem 13.3.]
Conclude also that if κ is a measurable cardinal and if $\delta_1 < \cdots < \delta_n < \kappa$ are WOODIN cardinals, then all $\underset{\sim}{\Sigma}^1_{n+2}$–sets of reals are $< \delta_1$–universally BAIRE.

13.5. Suppose that κ is a measurable cardinal and $\delta_1 < \cdots < \delta_n < \kappa$ are WOODIN cardinals. Let $A \subset {}^\omega\omega$ be $\underset{\sim}{\Pi}^1_{n+2}$, so that by Problem 13.4, A is $< \delta_1$–universally BAIRE. Let $\mathbb{P} \in V_{\delta_1}$, and let g be $\mathbb{P}$–generic over V. Let A^* be the new version of A in $V[g]$ (cf. p. 150). Show that if $V[g] \models A^* \neq \emptyset$, then $V \models A \neq \emptyset$. (Compare Problem 10.16.)

13.6. **(W. H. Woodin)** Let κ be a strong cardinal, and let $A \subset {}^\omega\omega$ be κ–universally BAIRE. (Equivalently, A is universally BAIRE, cf. Problem 8.10.)
Let g be $\mathrm{Col}(\omega, 2^{(2^\kappa)})$–generic over V, and let A^* be the new version of A in $V[g]$ (cf. p. 150). Show that $\exists^{\mathbb{R}}({}^\omega\omega \cap V[g]) \setminus A^* = \{x \in {}^\omega\omega : \exists y \in {}^\omega\omega \; x \oplus y \notin A^*\}$ is universally BAIRE in $V[g]$. [Hint. In V, let T and U on $\omega \times \kappa$ witness that A is κ–universally BAIRE. In $V[g]$, we construct T^* and U^* by amalgamating set–sized trees. We get T^* by rearranging stretched versions of U. For every (short) $\aleph_0$–complete (κ, ν)–extender E let us define an approximation U^*_E to U^* as follows. Let $\pi_E : V \to M$ be the ultrapower map, where M is transitive. In $V[g]$, fix a reasonable enumeration $(\mu_n : n < \omega)$ of all $\pi_E(E_a)$, $a \in [\nu]^{<\omega}$, and write $\ell(n)$ for $\mathrm{Card}(a)$ in case $\mu_n = \pi_E(E_a)$. If μ_i and μ_j cohere [cf. Definition 10.45 (2)], then we write π_{ij} for the canonical embedding from $\mathrm{ult}(M; \mu_i)$ to $\mathrm{ult}(M; \mu_j)$. For $s \in {}^{<\omega}\omega$, say $k = \mathrm{lh}(s)$, we set $(s, (\alpha_0, \ldots, \alpha_{k-1})) \in U^*_E$ iff

$$\forall i < j < k\, (\pi^V_E(U_{s\restriction \ell(i)}) \in \mu_i \wedge \pi^V_E(U_{s\restriction \ell(j)}) \in \mu_j \wedge \mu_j \text{ projects to } \mu_i \longrightarrow \pi_{ij}(\alpha_i) > \alpha_j).$$

Show that this works.]
Conclude that if λ is the supremum of infinitely many strong cardinals and if G is $\mathrm{Col}(\omega, \lambda)$–generic over $V[G]$, then in $V[G]$ every projective set of reals is LEBESGUE measurable and has the property of BAIRE.

References

1. Abraham, U., Magidor, M.: Cardinal arithmetic. In: Foreman, M., Kanamori, A. (eds.) Handbook of Set Theory, vol. 2, pp. 1149–1228. Springer, Berlin (2010)
2. Andretta, A., Neeman, I., Steel, J.: The domestic levels of k^c are iterable. Isr. J. Math. **125**, 157–201 (2001)
3. Bartoszynski, T., Judah, H.: Set Theory. On the Sructure of the Real Line. A K Peters, Wellesley, MA (1995)
4. Becker, H., Kechris, A.: The descriptive set theory of Polish group actions. London Mathematical Society Lecture Notes Series, vol. 232 (1996)
5. Blass, A.: Combinatorial cardinal characteristics of the continuum. In: Kanamori, A., Foreman, M. (eds.) Handbook of Set Theory, vol. 1, pp. 395–489. Springer, Berlin (2010)
6. Blau, U.: Die Logik der Unbestimmtheiten und Paradoxien. Synchron-Verlag (2008)
7. Bilinsky, E., Gitik, M.: A model with a measurable which does not carry a normal measure. Archive Math. Logic **51**, 863–876 (2012)
8. Caicedo, A., Ketchersid, R.: A trichotomy theorem in natural models of AD^+. In: Babinkostova, L., Caicedo, A., Geschke, S., and Scheepers, M. (eds.) Set Theory and Its Applications, Contemporary Mathematics, vol. 533, pp. 227–258, American Mathematical Society, Providence, RI (2011)
9. Cummings, J.: Iterated forcing and elementary embeddings. In: Foreman, M., Kanamori, A. (eds.) Handbook of Set Theory, vol. 2, pp. 775–883. Springer, Berlin (2010)
10. Devlin, K., Ronald, B.: Jensen. Marginalia to a theorem of Silver. Lecture Notes in Mathematics # 499, pp. 115–142. Springer, Berlin (1975)
11. Enderton, H.B.: A Mathematical Introduction to Logic, 2nd edn. Harcourt Academic Press, Burlington (2011)
12. Gao, S.: Invariant Descriptive Set Theory. CRC Press, Boca Raton (2009)
13. Halbeisen, L.: Combinatorial Set Theory. Springer, Berlin (2012)
14. Hjorth, G., Kechris, A.: New dichotomies for Borel equivalence relations. Bull. Symbolic Logic **3**, 329–346 (1997)
15. Jech, T.: Set Theory, 3rd edn. Springer, Berlin (2002)
16. Jensen, Ronald B.: The fine structure of the constructible hierarchy. Ann. Math. Logic **4**, 229–308 (1972)
17. Jensen, R.B., Schimmerling, E., Schindler, R., Steel, J.: Stacking mice. J. Symb. Logic **74**, 315–335 (2009)
18. Kanamori, A.: The Higher Infinite, 2nd edn. Springer, Berlin (2009)
19. Kanovei, V., Sabok, M., Zapletal, J.: Canonical Ramsey theory on Polish spaces. Cambridge Tracts in Mathematics. Cambridge University Press, Cambridge (2013)

R. Schindler, *Set Theory*, Universitext, DOI: 10.1007/978-3-319-06725-4,

20. Kechris, A.: Classical Descriptive Set Theory Graduate Texts in Mathematics # 156. Springer, Berlin (1994)
21. Kechris, A., Miller, B.: Topics in orbit equivalence. Lecture Notes in Mathematics 1852. Springer, Berlin (2004).
22. Koellner, P., Woodin, W.H.: Large cardinals from determinacy. In: Foreman, M., Kanamori, A. (eds.) Handbook of Set Theory, vol. 3, pp. 1951–2119. Springer, Berlin (2010)
23. Kunen, K.: Set Theory. An Introduction to Independence Proofs. Elsevier, Amsterdam (1980)
24. Larson, P.: The Stationary Tower. In: Woodin, W.H. (ed.) Notes on a Course. AMS, New York (2004)
25. Larson, P.: AD^+. Monograph in preparation
26. Martin, D.A., Steel, J.: A proof of projective determinacy. J. Am. Math. Soc. **2**, 71–125 (1989)
27. Mochovakis, Y.: Descriptive Set Theory, 2nd edn. AMS, New York (2009)
28. Mitchell, W.J., Schimmerling, E.: Weak covering without countable closure. Math. Res. Lett. **2**, 595–609 (1995)
29. Mitchell, W.J., Schimmerling, E., Steel, J.: The covering lemma up to a woodin cardinal. Ann. Pure Appl. Logic **84**, 219–255 (1997)
30. Mitchell, W.J., Steel, J.R.: Fine structure and iteration trees. Lecture Notes in Logic, vol. 3 (1994).
31. Neeman, Itay: Inner models in the region of a woodin limit of woodin cardinals. Ann. Pure Appl. Logic **116**, 67–155 (2002)
32. Neeman, I.: The Determinacy of Long Games. de Gruyter, Berlin (2004)
33. Neeman, I.: Determinacy in $L(\mathbb{R})$. In: Foreman, M., Kanamori, A. (eds.) Handbook of Set Theory, vol. 3, pp. 1877–1950. Springer, Berlin (2010)
34. Sargsyan, G.: Descriptive inner model theory. Bull. Symbolic Logic **19**, 1–55 (2013)
35. Schimmerling, E., Zeman, M.: A characterization of $\square_\kappa$ in core models. J. Math. Logic **4**, 1–72 (2004)
36. Schindler, R., Steel, J.: The core model induction. Monograph in preparation. http://wwwmath.uni-muenster.de/logik/Personen/rds/core_model_induction.pdf
37. Shelah, S.: Proper and Improper Forcing, 2nd edn. Springer, Berlin (1998)
38. Shiryaev, A.: Probability Theory, 2nd edn. Springer, New York (1996)
39. Steel, J.: An outline of inner model theory. In: Kanamori, A., Foreman, M. (eds.) Handbook of Set Theory, vol. 3, pp. 1595–1684. Springer, Berlin (2010)
40. Steel, J.: The core model iterability problem. Lecture Notes in Logic, vol. 8 (1996).
41. Steel, J.: A stationary tower free proof of the derived model theorem. In: Gao, S., Zhang, J. (eds.) Advances in Logic. The North Texas Logic Conference 8–10 Oct 2004, pp. 1–8, AMS, New York (2007).
42. Steel, J.: Derived models associated to mice. In: Chong, C., Feng, Q., Slaman, T.A., Woodin, W.H., Yong, Y. (eds.) Computational Aspects of Infinity, Part I: Tutorials pp. 105–194, Singapore (2008).
43. Windßus, K.: Projektive Determiniertheit. Diplomarbeit, Universität Bonn (1993)
44. Woodin, W.H.: The Axiom of Determinacy, Forcing Axioms, and the Nonstationary Ideal, 2nd edn. de Gruyter, Berlin (2010)
45. Woodin, W.H.: Suitable extender models i. J. Math. Logic **10**, 101–339 (2010)
46. Woodin, W.H.: Suitable extender models ii: beyond ω-huge. J. Math. Logic **11**, 115–436 (2011)
47. Zeman, M.: Inner models and large cardinals. de Gruyter Series in Logic and Its Applications, vol. 5 (2002).

Index

R. Schindler, *Set Theory*, Universitext, DOI: 10.1007/978-3-319-06725-4,